PURE AND APPLIED MATHEMATICS

A Series of Texts and Monographs

Edited by: R. COURANT · L. BERS · J. J. STOKER

Vol. I: **Supersonic Flow and Shock Waves**
By R. Courant and K. O. Friedrichs

Vol. II: **Non-linear Vibrations in Mechanical and Electrical Systems**
By J. J. Stoker

Vol. III: **Dirichlet's Principle, Conformal Mapping, and Minimal Surfaces**
By R. Courant

Vol. IV: **Water Waves**
By J. J. Stoker

Vol. V: **Integral Equations**
By F. G. Tricomi

Vol. VI: **Differential Equations: Geometric Theory**
By Solomon Lefschetz

Vol. VII: **Linear Operators—Parts I and II**
By Nelson Dunford and Jacob T. Schwartz

Vol. VIII: **Modern Geometrical Optics**
By Max Herzberger

Vol. IX: **Orthogonal Expansions**
By G. Sansone

Vol. X: **Lectures on Differential and Integral Equations**
By K. Yosida

Vol. XI: **Representation Theory of Finite Groups and Associative Algebras**
By C. W. Curtis and I. Reiner

Vol. XII: **Electromagnetic Theory and Geometrical Optics**
By Morris Kline and Irvin W. Kay

Vol. XIII: **Combinatorial Group Theory**
By Wilhelm Magnus, Abraham Karrass, and Donald Solitar

Additional volumes in preparation

PURE AND APPLIED MATHEMATICS

A Series of Texts and Monographs

VOLUME XIII

COMBINATORIAL GROUP THEORY:
Presentations of Groups in Terms of Generators and Relations

WILHELM MAGNUS
COURANT INSTITUTE OF MATHEMATICAL SCIENCES
NEW YORK UNIVERSITY

ABRAHAM KARRASS
ADELPHI UNIVERSITY

DONALD SOLITAR
ADELPHI UNIVERSITY

INTERSCIENCE PUBLISHERS.
A division of John Wiley & Sons, Inc. New York London Sydney

LIBRARY OF CONGRESS CATALOG CARD NUMBER: 65-27657
PRINTED IN THE UNITED STATES OF AMERICA

Dedicated
To the Memory of
MAX DEHN
1878–1952

Preface

This book contains an exposition of those parts of group theory which arise from the presentation of groups in terms of generators and defining relations. Groups appear naturally in this form in certain topological problems, and the first serious contributions to this part of group theory were made by Poincaré, Dehn, Tietze, and other topologists. The name "Combinatorial Group Theory" refers to the frequent occurrence of combinatorial methods, which seem to be characteristic of this discipline.

The book is meant to be used as a textbook for beginning graduate students who are acquainted with the elements of group theory and linear algebra. The first two chapters are of a fairly elementary nature, and a particularly large number of exercises were included in these parts. The exercises are not always easy ones, but the hints given are usually broad enough to make them so. Some interesting results have been presented in the form of exercises; the text proper does not make use of these results except where specifically indicated. (It is a good idea for the reader to examine the exercises even if he does not wish to attempt them.)

There is not very much overlapping of the topics presented here with those treated in the books on group theory by A. Kurosh and by Marshall Hall, Jr. The subjects of Nielsen Transformations (Chapter 3), Free and Amalgamated Products (Chapter 4), and Commutator Calculus (Chapter 5) are treated here in a more detailed fashion than in the works of Kurosh and of Hall.

All theorems which are labeled with a number are proved in full. However, we have stated some advanced results without proof, whenever the original proofs were long and could not be amalgamated with the main body of the text. Such results are stated either as theorems labeled with the name of the author (e.g., Grushko's Theorem) or with a letter and number (e.g., Theorems N1 to N13 on Nielsen transformations, or T1 to T5 on topological aspects).

We have tried to give references to relevant papers and monographs in the later parts of the book (after the first two chapters). Usually, such references are collected at the end of each section under the heading "References and Remarks."

The sixth (and last) chapter contains a brief survey of some recent developments. It is hardly necessary to say that we could not even try to give a complete account. We are painfully aware of the many gaps. Some methods and results, as well as references, may have escaped our attention altogether.

Over the years, we have received suggestions and criticisms from many mathematicians, and we owe much to comments from our colleagues as well as from our students. We also wish to acknowledge the help given to us by the National Science Foundation which, through several grants given to New York University and Adelphi University, facilitated the cooperation of the authors.

This book is dedicated to the memory of Max Dehn. We believe this to be more than an acknowledgment of a personal indebtedness by one of the authors who was Dehn's student. The stimulating effect of Dehn's ideas on presentation theory was propagated not only through his publications, but also through talks and personal contacts; it has been much greater than can be documented by his papers. Dehn pointed out the importance of fully invariant subgroups in 1923 in a talk (which was mimeographed and widely circulated but never published). His insistence on the importance of the word problem, which he formulated more than fifty years ago, has by now been vindicated beyond all expectations.

New York University and
Adelphi University
December 1965

WILHELM MAGNUS
ABRAHAM KARRASS
DONALD SOLITAR

Contents

Technical Remarks

1. Equations are numbered (1), (2), (3), etc., in each section. If referred to in another section, their numbers are preceded by the numbers of the chapter and section in which they appear; e.g., (5.2.1) refers to Equation 1 in Section 2 of Chapter 5.
2. A list of theorems, corollaries, lemmas, and definitions may be found before the index at the end of the book. It is followed by a list of symbols and notations which are either explained immediately, or marked with the number of the page on which they are defined.
3. References are given by the name of the author and the year in which his paper appeared. If there are several papers by the same author in the same year, they are distinguished by letters a, b, c. For example, Higman, 1951c. The bibliography appears before the index.
4. The end of a proof is indicated by ◀.

Chapter 1

Basic Concepts

1.1. Introduction

In beginning a discussion of special topics in group theory, it does not seem inappropriate to review the definition of a group. Recall that a group $(G, \cdot)$ is a non-empty set G of elements $a, b, c, \ldots$ together with a binary operation $\cdot$ defined in G for which the following four postulates are satisfied:

I. If a, b is an ordered pair of elements of G ($a \neq b$ or $a = b$), then there is a uniquely determined element c of G such that

$$a \cdot b = c. \tag{1}$$

(c is called the *product* of a and b; we often omit the dot and write simply $ab = c$.)

II. The operation defined by $\cdot$ in G is *associative*, i.e., for any elements a, b, c in G we have

$$(ab)c = a(bc). \tag{2}$$

III. There exists an element of G, denoted by 1, for which

$$a \cdot 1 = 1 \cdot a = a, \tag{3}$$

where a is any element in G. (1 is called the *unit* or *identity element* in G.)

IV. If a is any element in G, then there exists an element in G, denoted a^{-1}, for which

$$a \cdot a^{-1} = a^{-1} \cdot a = 1. \tag{4}$$

(a^{-1} is called the *inverse* of a.)

It should be noted that we do not assume that $\cdot$ is a *commutative* operation, i.e., that

$$a \cdot b = b \cdot a \tag{5}$$

for all a and b in G. If (5) holds for each a, b in G, $(G, \cdot)$ is called a *commutative* or *Abelian* group.

Postulates I and II allow us to define the product of a sequence of n elements $a_1, a_2, \ldots, a_n$, which is independent of the partitioning of the factors, e.g.,

$$(a_1a_2)(a_3a_4) = ((a_1a_2)a_3)a_4.$$

In particular, if $a_1 = a_2 = \ldots = a_n = a$, then we denote the product $a_1a_2 \ldots a_n$ by a^n. Using postulates III and IV, we can extend the definition of a^n to the case where n is zero or a negative integer; indeed, we define $a^0 = 1$ and $a^{-n} = (a^{-1})^n$, for positive integers n. It easily follows that $a^m \cdot a^n = a^{m+n}$, and $(a^m)^n = a^{mn}$ for all integers m, n.

A group $(G, \cdot)$ is called *cyclic* with *generator* a if every element g of G is a *power* of a, i.e., $g = a^n$ for some integer n. The group of integers under addition (here the integer zero is the identity element and a^n is usually denoted na) is an example of a cyclic group; indeed, the integer 1 (or -1) can be used as a generator. The group of complex numbers $1, -1, i, -i$ under multiplication is another example of a cyclic group; here i (or $-i$) can be used as a generator because $1 = i^0$, $i = i^1$, $-1 = i^2$, $-i = i^{-1}$. Certainly, not every group G is cyclic. For if $g_1 = a^m$ and $g_2 = a^n$ then

$$g_1g_2 = a^ma^n = a^{m+n} = a^{n+m} = a^na^m = g_2g_1,$$

i.e., any two elements of G commute; but there are non-Abelian groups, e.g., the group of permutations on three objects under resultant composition. Moreover, not every Abelian group is cyclic, e.g., the Klein four-group V of functions x, $-x$, $1/x$ and $-1/x$ under resultant composition is Abelian but non-cyclic.

A cyclic group G has a particularly simple structure; indeed, G can be characterized (up to isomorphism) by the *order* of its generator a, i.e., the smallest positive integer n such that $a^n = 1$ (if $a^n = 1$ implies $n = 0$, we say a has *infinite order*). The elements of G can be represented by powers of a and multiplication in G is determined by the rule $a^m \cdot a^n = a^{m+n}$ (which holds in any group) and the relation $a^r = 1$ where r is the order of a. For example, if b is a generator of order twelve of a cyclic group B then

$$b^5 \cdot b^9 = b^{14} = b^2 \cdot b^{12} = b^2 \cdot 1 = b^2.$$

Moreover, from the relation $b^{12} = 1$, we can see that the elements of B can be represented as $1, b, b^2, b^3, b^4, b^5, b^6, b^7, b^8, b^9, b^{10}, b^{11}$. For, since B is cyclic with generator b, every element can be represented by b^n for some integer n. Using the division algorithm, $n = 12q + s$ where $0 \leq s < 12$. Hence,

$$b^n = b^{12q+s} = b^{12q} \cdot b^s = (b^{12})^q \cdot b^s = 1^q \cdot b^s = 1 \cdot b^s = b^s.$$

Specifying that B is a cyclic group whose generator has order twelve is more economical than defining B by means of its multiplication table; indeed, the multiplication table for B requires 144 entries.

The Klein four-group V of functions x, $-x$, $1/x$, $-1/x$ under resultant composition does not have a single generator. However, it can be generated by the functions $-x$ and $1/x$, i.e., every element of V is some product (i.e., composite) of these two functions. If a represents $-x$ and b represents $1/x$, then $a^2 = 1$, $b^2 = 1$, and $ab = ba$; here 1, the identity element of V, is, of course, the function x. Are there any other algebraic relations (involving only resultant composition) satisfied by a and b? Clearly yes. For $a^3 = a$, $(ab)^2 = 1$ are algebraic relations satisfied by a and b; however, these can be derived from the three given ones. For example, $abab = a(ba)b = a(ab)b$ follows from $ab = ba$; $a(ab)b = a^2b^2 = 1 \cdot 1 = 1$ follows from $a^2 = 1$ and $b^2 = 1$. Indeed, we shall show that every algebraic relation between a and b can be derived from the relations $a^2 = 1$, $b^2 = 1$, and $ab = ba$ (and the general properties of a group, such as $aa^{-1} = 1$, $a \cdot 1 = a$).

Let

$$a^{\alpha_1}b^{\beta_1} \ldots a^{\alpha_r}b^{\beta_r} = a^{\gamma_1}b^{\delta_1} \ldots a^{\gamma_s}b^{\delta_s} \tag{6}$$

be an algebraic relation satisfied in V when a is replaced by $-x$ and b is replaced by $1/x$. The relation (6) is equivalent to

$$a^{\alpha_1}b^{\beta_1} \ldots a^{\alpha_r}b^{\beta_r-\delta_s}a^{-\gamma_s} \ldots b^{-\delta_1}a^{-\gamma_1} = 1. \tag{7}$$

Using $a^2 = 1$ and $b^2 = 1$, one can derive $a = (a^2)a^{-1} = 1a^{-1} = a^{-1}$ and similarly $b = b^{-1}$. Thus (7) may be changed to an equivalent relation in which all exponents are non-negative; moreover, using $a^2 = 1$, $b^2 = 1$ all exponents in the relation may be reduced to 0 or 1. Since in any group $a^0 = b^0 = 1$, any a or b occurring with a zero exponent can be deleted. Moreover, using the relation $ab = ba$, as well as $a^2 = 1$ and $b^2 = 1$, the relation (7) may be further reduced to the form

$$a^{\epsilon_1}b^{\epsilon_2} = 1 \tag{8}$$

where ϵ_1 and ϵ_2 can be either 0 or 1. But if ϵ_1 or ϵ_2 is 1, then the left side of (8) becomes a, b, or ab which represent the functions $-x$, $1/x$, $-1/x$, respectively, in V and not the identity function x. Thus $\epsilon_1 = \epsilon_2 = 0$ and (8) reduces to $1 = 1$. By reversing our steps then we may derive the relation (6) from $a^2 = 1$, $b^2 = 1$ and $ab = ba$.

The Klein four-group may thus be described as generated by two elements a and b satisfying the relations $a^2 = 1$, $b^2 = 1$, $ab = ba$, and such that all other relations between a and b may be derived from these three. Moreover, any group which has two generators c, d satisfying the relations $c^2 = 1$, $d^2 = 1$ and $cd = dc$, and such that all other relations between c and d may be derived from these three, is isomorphic to the Klein four-group (see Theorem 1.1).

In a similar manner, the group Σ_3, of permutations of the three objects x, y, z is generated by the three-cycle (xyz) and the two-cycle (xy). If a denotes the three-cycle and b the two-cycle, then Σ_3 is generated by a and b which satisfy the relations $a^3 = 1$, $b^2 = 1$, $ab = ba^2$. Moreover, one can show that all relations between a and b can be derived from these three relations (see Problem 6a). Conversely, any group generated by two generators c and d satisfying the relations $c^3 = 1$, $d^2 = 1$, and $cd = dc^2$, and such that all other relations between c and d can be derived from these three relations, is isomorphic to Σ_3 (see Theorem 1.1).

Describing a group in terms of a set of generators and a set of defining relations for these generators, i.e., a set of relations from which all other relations can be derived, is called presenting a group. Thus the cyclic group of order twelve can be presented on a single generator b with the relation $b^{12} = 1$ as a set of defining relations; the Klein four-group can be presented on two generators a and b with the relations $a^2 = 1$, $b^2 = 1$, $ab = ba$ as a set of defining relations; the group of permutations on three objects can be presented on two generators a and b with the relations $a^3 = 1$, $b^2 = 1$, $ab = ba^2$ as a set of defining relations.

Before proceeding further, we shall generalize the preceding examples, and at the same time give more precision to the ideas introduced, in particular to those of "algebraic relation" and "a relation can be derived from given relations."

Let $a, b, c, \ldots$ be distinct symbols and form the new symbols $a^{-1}, b^{-1}, c^{-1}, \ldots$. A *word* W *in the symbols* $a, b, c, \ldots$ is a finite sequence

$$f_1, f_2, \ldots, f_{n-1}, f_n \tag{9}$$

where each of the f_ν is one of the symbols

$$a, b, c, \ldots, a^{-1}, b^{-1}, c^{-1}, \ldots;$$

the *length* $L(W)$ of W is the integer n. For convenience we introduce the *empty word* of *length zero* and denote it by 1. If we wish to exhibit the symbols involved in W we write $W(a, b, c, \ldots)$.

It is customary to write the sequence (9) without the commas as

$$f_1 f_2 \cdots f_{n-1} f_n. \tag{10}$$

It is also customary to abbreviate a block of n consecutive symbols a by a^n, and to abbreviate a block of n consecutive symbols a^{-1} by a^{-n}; e.g., the word $a^3b^2b^{-1}a^{-2}c^{-1}$ is the same as the word $aaabbb^{-1}a^{-1}a^{-1}c^{-1}$, but is different from the word $a^3ba^{-2}c^{-1}$.

Thus aa^{-1} is a word in a of length two; $a^2ba^{-1}b^{-2}b$ is a word in a and b of length seven; b^2c^{-2} is a word in a, b, and c of length four, and is also a word in b and c of length four; 1 is a word in a, b, and c of length zero.

The *inverse* W^{-1} of a word W given by (10) is the word

(11) $$f_n^{-1}f_{n-1}^{-1}\cdots f_2^{-1}f_1^{-1}$$

where if f_ν is a or a^{-1} then f_ν^{-1} is a^{-1} or a, respectively, and similarly if f_ν is one of the symbols b or b^{-1}, c or c^{-1}, . . . ; the *inverse* of the empty word is itself.

For example, $(aa^{-1})^{-1} = aa^{-1}$,

$$(aaba^{-1}b^{-1}b^{-1}b)^{-1} = b^{-1}bbab^{-1}a^{-1}a^{-1}, \qquad 1^{-1} = 1.$$

Clearly, $L(W) = L(W^{-1})$, and $(W^{-1})^{-1} = W$.

If W is the word $f_1f_2\cdots f_n$ and U is the word $f_1'f_2'\cdots f_r'$ then we define their *juxtaposed product* WU as the word $f_1f_2\cdots f_nf_1'f_2'\cdots f_r'$.

Clearly, $(WU)^{-1} = U^{-1}W^{-1}$ and $L(WU) = L(W) + L(U)$.

A word is used to represent a formal product of elements in a group in the same way that a polynomial with integer coefficients in several variables is used to represent a formal algebraic combination of elements in an integral domain. Just as we can assign values to the variables in a given polynomial from the elements in any integral domain, and compute the value of the polynomial in this domain, so we can assign values to the symbols in a given word from the elements in any group, and compute the value of the word in this group.

Given a mapping α of the symbols $a, b, c, \ldots$ into a group G with $\alpha(a) = g$, $\alpha(b) = h$, $\alpha(c) = k, \ldots$, then we say that (*under* α) a *defines* g, b *defines* h, c *defines* k, . . . , a^{-1} *defines* g^{-1}, b^{-1} *defines* h^{-1}, c^{-1} *defines* k^{-1}, . . . ; moreover, if W is given by (10) then W *defines* the element, denoted $W(g, h, k, \ldots)$, in G given by

(12) $$g_1g_2\cdots g_{n-1}g_n,$$

where f_ν defines g_ν; the empty word 1 *defines* the identity element 1 of G.

Clearly, if the words U and V define the elements p and q of G, then U^{-1} defines p^{-1} and UV defines pq.

If every element of G is defined by some word in $a, b, c, \ldots$, then $a, b, c, \ldots$ are called *generating symbols* for G (*under* α) and $g, h, k, \ldots$ are called *generating elements* for G; if the context makes it clear, both generating symbols and generating elements may be referred to as *generators* for G.

For example, if V is the Klein four-group of functions x, $-x$, $1/x$, $-1/x$, then under the mapping $a \to -x$, $b \to 1/x$, the word a^2 defines the function x and the word ab defines the function $-1/x$; therefore a and b are generating symbols for V (under this mapping). Under the mapping $a \to -x$, $b \to -1/x$, a^2 defines the function x and ab defines the function $1/x$; therefore a and b are generating symbols for V (under this mapping). Under the mapping $a \to x$, $b \to -x$, any word in a and b will only define a power of the function $-x$, i.e., x or $-x$; therefore a and b are not generating symbols for V (under this mapping).

We shall assume in the remainder of this section that $a, b, c, \ldots$ are generating symbols (under a mapping α) for the group G.

A word $R(a, b, c, \ldots)$ which defines the identity element 1 in G is called a *relator*. (Intuitively, a relator is simply the left-hand side of a relation whose right-hand side is 1.) The equation

$$R(a, b, c, \ldots) = S(a, b, c, \ldots)$$

is called a *relation* if the word RS^{-1} is a relator (or equivalently, if R and S define the same element in G).

In any group, the empty word and the words aa^{-1}, $a^{-1}a$, bb^{-1}, $b^{-1}b$, cc^{-1}, $c^{-1}c, \ldots$ are always relators; they are called the *trivial relators*.

Suppose $P, Q, R, \ldots$ are any relators of G. We say that the word W *is derivable from* $P, Q, R, \ldots$, if the following operations, applied a finite number of times, change W into the empty word:

(i) *Insertion of one of the words* $P, P^{-1}, Q, Q^{-1}, R, R^{-1}, \ldots$ *or one of the trivial relators between any two consecutive symbols of* W, *or before* W, *or after* W.

(ii) *Deletion of one of the words* $P, P^{-1}, Q, Q^{-1}, R, R^{-1}, \ldots$ *or one of the trivial relators, if it forms a block of consecutive symbols in* W.

This definition of derivability for relators allows us to capture the intuitive notion of derivability for relations by first "transposing the right-hand side" and thereby reducing the problem to the simpler notion of derivability among relators. Formally, we say the equation $W = V$ is *derivable from the relations* $P_1 = P_2$, $Q_1 = Q_2$, $R_1 = R_2, \ldots$ if and only if the word WV^{-1} is derivable from the relators $P_1P_2^{-1}$, $Q_1Q_2^{-1}$, $R_1R_2^{-1}, \ldots$.

Again, consider the Klein four-group V and let α be the mapping $a \to -x$, $b \to 1/x$. Then a^2, b^2, and $aba^{-1}b^{-1}$ (this last word is the "transposed form" of the relation $ab = ba$) are relators since they define the function x, the identity of V. The word $abab$ is also a relator, and in fact, it can be derived from the preceding three relators. To show this, consider the following sequence of words, each obtained from the preceding by (i) or (ii):

$$abab,\ ab \cdot (aa)^{-1} \cdot ab = aba^{-1}a^{-1}ab = aba^{-1}(a^{-1}a)b,\ aba^{-1}b,$$
$$aba^{-1}(bb)^{-1}b = aba^{-1}b^{-1}b^{-1}b = aba^{-1}b^{-1}(b^{-1}b),\ aba^{-1}b^{-1},\ 1.$$

Moreover, we can show that every relator can be derived from the relators a^2, b^2, and $aba^{-1}b^{-1}$.

It is clear that if the word W is derivable from the relators P, Q, $R, \ldots$, then W is itself a relator; for, the operations (i) and (ii) applied to a word do not change the element of the group defined by the word, and since the empty word is reached, W must define the identity element of G.

If every relator is derivable from the relators $P, Q, R, \ldots$, then we call $P, Q, R, \ldots$ a *set of defining relators* or a *complete set of relators* for the group G on the generators $a, b, c, \ldots$. If $P, Q, R, \ldots$ is a set of defining relators for the group G on the generators $a, b, c, \ldots$, we call

(13) $\langle a, b, c, \ldots; \quad P(a, b, c, \ldots), Q(a, b, c, \ldots), R(a, b, c, \ldots), \ldots \rangle$

a *presentation of* G and write

$$G = \langle a, b, c, \ldots; \quad P, Q, R, \ldots \rangle.$$

We say a presentation is *finitely generated* (*finitely related*) if the number of generators (defining relators) in it is finite. If a presentation is both finitely generated and finitely related, we say the *presentation is finite.* (A presentation is often given using relations instead of relators; sometimes a hybrid using relators and relations is used; thus

$$\langle a, b; \quad a^2 = 1, b^2 = 1, ab = ba \rangle,$$

$$\langle a, b; \quad a^2, b^2, ab = ba \rangle,$$

and

$$\langle a, b; \quad a^2, b^2, aba^{-1}b^{-1} \rangle$$

are each interpreted as the last given presentation.)

Does every group have some presentation? Obviously, we can get a presentation of a group G by taking a distinct generating symbol for each element in the group, and using all relators on these generators as the set of defining relators.

The multiplication table of G provides another presentation; again take a distinct generating symbol for each element of G; the defining relators are all the words of length three of the form abc^{-1}, where a, b, and c are generating symbols such that the product of the elements defined by a and b is defined by c. To prove that every relator in G may be derived from these relators of length three, suppose that

$$W = f_1 f_2 \cdots f_n$$

is a relator in G. First we eliminate negative exponents occurring in W as follows:

Suppose b^{-1} occurs in W; there is a generating symbol, say, c, which defines the same element as b^{-1}; then bce^{-1}, where e defines the identity element of G, is one of our defining relators. Insert bce^{-1} to the right of b^{-1} in W obtaining $b^{-1}bce^{-1}$; then delete $b^{-1}b$ obtaining ce^{-1}; insert eee^{-1}

to the right of ce^{-1} obtaining $ce^{-1}eee^{-1}$, and now delete $e^{-1}e$ and ee^{-1} obtaining c. Thus in W, b^{-1} has been replaced by c, using (i) and (ii).

We next reduce the length of W. If a and d are consecutive in W, find the defining relator of the form adq^{-1}; insert $q^{-1}q$ to the right of ad in W obtaining $adq^{-1}q$; delete adq^{-1} obtaining q; thus ad has been replaced by q in W, using (i) and (ii).

Continuing in this way we may reduce the length of W until we obtain a word of length one, consisting of a single generating symbol. Since W is a relator, this generating symbol must define the identity of G; but distinct symbols were chosen for distinct elements. Thus the remaining generating symbol must be e; insert ee^{-1} to the right of e obtaining eee^{-1}, and then delete eee^{-1}. Thus, by (i) and (ii), using the multiplication table defining relators, W can be reduced to the empty word. Hence, the multiplication table gives a presentation for G.

The multiplication table presentation shows that every finite group has a finite presentation. But, in general, the multiplication table presentation is very large and gives much interdependent information.

In the next section we shall show that given an arbitrary set of symbols, and an arbitrarily prescribed set of words in these symbols, there is a unique group (up to isomorphism) with the symbols as generators and the set of prescribed words as defining relators. This provides us with a useful technique for constructing groups (the disadvantages will be discussed in Section 1.3).

Presentation theory also provides a method for establishing results about infinite groups. Whereas mathematical induction on the order of a group can be used to establish many theorems about finite groups, this method is not available for infinite groups. For infinite groups, counting arguments involving lengths of words often prove useful.

Finally, groups given by means of their presentations arise naturally in such areas as knot theory (see, e.g., Crowell and Fox, 1963), topology (see, e.g., Reidemeister, 1932a), automorphic functions (see, e.g., Lehner, 1964), and geometry (see, e.g., Coxeter, 1961).

Presentation theory attempts to derive information about a group from a presentation of it.

For presentations of many famous groups, we refer the reader to Coxeter and Moser, 1965.

Problems for Section 1.1

1. Show by induction on the length of the word $U(a, b, c, \ldots)$ that UU^{-1} is derivable from the trivial relators, and hence that the relation $U = U$ is

derivable from any set of relators. [*Hint:* If $U = Va^\epsilon$, where ϵ is 1 or -1 then $UU^{-1} = Va^\epsilon a^{-\epsilon}V^{-1}$.]

2. Show that insertion and deletion of the words P^{-1}, Q^{-1}, R^{-1}, . . . can be done by insertion and deletion of the words P, Q, R, . . . and the trivial relators. [*Hint:* To insert P^{-1}, use the trivial relators to insert PP^{-1} and then delete P.]

3. Let a, b, c, . . . be generating symbols for a group G: let P, Q, R, . . . be a set of relators. In the following, derivability is understood to be from P, Q, R, ...

(a) Show that if W is derivable then W^{-1} is derivable; hence if $U = V$ is derivable, so is $V = U$.

(b) Show that if UV and W are derivable, then so is UWV; in particular, the product of two derivable words is derivable.

(c) Show that if W is derivable, then so is KWK^{-1}.

(d) Show that if W_1, W_2, . . . , W_n is each derivable then so is the product $K_1W_1K_1^{-1} \ldots K_nW_nK_n^{-1}$.

(e) Show that if $U = V$ and $V = W$ are derivable, then so is $U = W$.

(f) Show that if $U = V$ and $K = M$ are derivable, then so is $UK = VM$.

(g) Show that if $U = V$ is derivable, then so is $KUM = KVM$.

4. Show that if the word K is derivable from the relators M, N, . . . , and M, N, . . . are derivable from the relators P, Q, R, . . . , then K is derivable from P, Q, R, . . . ; hence if M, N, . . . is a set of defining relators, then so is P, Q, R, ...

5. (a) Show that the "multiplication table" presentation for the cyclic group of order four has four generators and sixteen defining relators, and specifically write down this presentation.

(b) Give a presentation for this group using one generator and one defining relator.

(c) Give a presentation for this group using one generating symbol b and two defining relators, neither of which is b^4 or b^{-4}.

(d) Give a presentation for this group using two generators and three defining relators.

[*Hint:* For (*c*) use b^8 and b^{12} as defining relators, and show that b^4 can be derived. Hence, by the preceding exercise, b^8 and b^{12} are a set of defining relators. For (*d*), if the group is $1, i, -1, -i$ under multiplication, let $a \to i$, $b \to i$ and show that a^4, b^4, ab^{-1} is a set of defining relators, by using ab^{-1} to reduce $W(a, b)$ to a^n.]

6. (a) Show that the group Σ_3 of permutations of the objects x, y, z is presented by

$$\langle a, b;\ \ a^3, b^2, ab = ba^2\rangle$$

under the mapping $a \to (xyz)$, $b \to (xz)$.

(b) Show that Σ_3 is presented by

$$\langle a, b;\ \ a^2, b^2, (ab)^3\rangle$$

under the mapping $a \to (xy)$, $b \to (xz)$.

[*Hint:* For (a), show that any word in a and b can be reduced by (i) and (ii) to one of the words $1, b, a, ba, a^2, ba^2$; since the empty word is the only relator among these, a relator can be reduced to the empty word. To reduce a word to one of the six forms, use a^3 and b^2 to make exponents positive and $ab = ba^2$ to move b to the left. For (b), show that using (i) and (ii) any word in a and b can be reduced to one of the words $1, a, b, ab, ba, aba$; since the empty word is the only relator among these, a relator can be reduced to the empty word. To reduce a word to one of the six forms, use a^2 and b^2 to make exponents all one and show that $baba$ can be reduced to ab, $abab$ to ba, and bab to aba.]

7. Show that under the mapping $a \to i$, $b \to j$ the *quaternion group* has the presentation

$$\langle a, b;\quad ab = b^{-1}a, ba = a^{-1}b\rangle.$$

[*Hint:* First show that $a^2 = (ab)^2 = b^2$ may be derived. Next, show that a^4 and b^4 can be derived as follows: $a^4 = a \cdot a^2 \cdot a = a \cdot b^2 \cdot a = ab \cdot ba = b^{-1}a \cdot a^{-1}b = 1$. Finally, show that any word W in a and b can be reduced to one of the words $1, a, a^2, a^3, b, ab, a^2b, a^3b$ using (i) and (ii).]

8. The *dihedral group* D_{2n} of order $2n$ may be defined as the group of two-by-two matrices, with entries from the ring of integers mod n, of the form

$$\begin{pmatrix} \epsilon & k \\ 0 & 1 \end{pmatrix}$$

where ϵ is 1 or -1, and k is any integer mod n, under matrix multiplication. Show that this group can be presented by

$$\langle a, b;\quad a^n, b^2, ba = a^{-1}b\rangle,$$

under the mapping

$$a \to \begin{pmatrix} 1 & 1 \\ 0 & 1 \end{pmatrix}, \qquad b \to \begin{pmatrix} -1 & 0 \\ 0 & 1 \end{pmatrix}.$$

[*Hint:* Show by induction that $a^k = \begin{pmatrix} 1 & k \\ 0 & 1 \end{pmatrix}$, and $a^kb = \begin{pmatrix} -1 & k \\ 0 & 1 \end{pmatrix}$; hence, a and b are generating symbols for D_{2n}. Next, show using (i) and (ii), that every word in a and b can be reduced to a^k or a^kb where $0 \leq k < n$.]

9. The *infinite dihedral group* D may be defined as the group of two-by-two matrices, with entries from the ring of integers, of the form

$$\begin{pmatrix} \epsilon & k \\ 0 & 1 \end{pmatrix}$$

where ϵ is 1 or -1, and k is any integer, under matrix multiplication. Show that this group can be presented by

$$\langle a, b;\quad b^2, ba = a^{-1}b\rangle$$

under the mapping

$$a \to \begin{pmatrix} 1 & 1 \\ 0 & 1 \end{pmatrix}, \qquad b \to \begin{pmatrix} -1 & 0 \\ 0 & 1 \end{pmatrix}.$$

[*Hint:* See the hint to the preceding exercise; also show that $ba^{-1} = ab$ can be derived, and used together with $ba = a^{-1}b$ to move b to the right.]

10. Let G be a group which has a presentation in which all defining relators have even length. Show that no element defined by a generating symbol can have odd order; hence, if G is finite, G has even order. [*Hint:* Since the defining relators and the trivial relators have even length, applying (i) and (ii) to a word W does not change the parity of the length of W. Thus, if W is a relator, and hence can be reduced to the empty word, the length of W is even. Hence, if a is a generating symbol, a^{2k+1} cannot be a relator. All elements have odd order in a group of odd order.]

11. Show that $a^{-1}a$ can never be reduced to the empty word by insertions and deletions of aa^{-1}. [*Hint:* To show that $a^{-1}a$ can never be reduced to the empty word by insertion and deletion of aa^{-1}, we shall map a into the function f on the integers defined by $n \to 2n$ for all n, and a^{-1} into the function g on the integers defined by $n \to n$ if n is odd, and $n \to n/2$ if n is even. A word $W(a)$, which is a sequence in the symbols a and a^{-1}, is mapped onto the resultant composition of the corresponding sequence in f and g; for example, $aa^{-1}a^{-1}aa$ is mapped onto the function $n \to (n)fggff$. Since resultant composition of functions is associative and fg is the identity function, it follows that the words $Kaa^{-1}M$ and KM are mapped into the same function. Hence, if $a^{-1}a$ could be reduced to the empty word by insertions and deletions of aa^{-1}, then $a^{-1}a$ would be mapped into the identity function. But $a^{-1}a$ is mapped into gf which maps an odd integer into twice itself.]

12. For each of the following groups, find a set of generating elements; find several relators.

(a) The permutation group consisting of the identity and all the permutations (1234), (13)(24), (1432), (13), (24), (12)(34), (14)(23).

(b) The functions $\omega^k x$ and ω^k/x, where ω is a complex primitive nth root of unity and $0 \leq k < n$, under resultant composition.

(c) Matrices of the form

$$\begin{pmatrix} \epsilon & k \\ 0 & \epsilon \end{pmatrix}$$

where ϵ is 1 or -1 and k is an integer, under matrix multiplication.

(d) The complex functions $f(z) = e^{2\pi ki/n}z$, where $0 \leq k < n$, under resultant composition.

(e) The seven reflections of Euclidean 3-space in the origin and each of the coordinate axes and planes, together with the identity under resultant composition.

13. Show that if a set G together with an operation $\cdot$ satisfies postulates (1), (2), and has a right identity 1 such that $a \cdot 1 = a$ for all a in G, and for each a, a right inverse a^{-1}, such that $a \cdot a^{-1} = 1$, then $(G, \cdot)$ is a group.

14. (a) Show that if a group G is finitely generated by a set of elements $g_1, \ldots, g_n$, then from any set of generating elements for G finitely many can be chosen to generate G.

(b) Show that if G has a finite set of defining relators $R_1(a_\nu)$, $R_2(a_\nu), \ldots, R_m(a_\nu)$ on the generators $a_1, a_2, \ldots, a_n$, then from any set of defining relators for G on the generators $a_1, a_2, \ldots, a_n$ a finite subset of defining relators for G can be obtained.

[*Hint:* For (a), use that $g_1, \ldots, g_n$ each require only finitely many generating elements for its expression. For (b), use that if a word is derivable from a set of defining relators, it is derivable from a finite subset of defining relators; also use Problem 1.1.4.]

1.2. Construction of Groups from Generators and Defining Relators

Given a set of distinct symbols $a, b, c, \ldots$ and a set (possibly empty) of words $P, Q, R, \ldots$ in $a, b, c, \ldots$ we wish to show now that there is a unique group (up to isomorphism) with the presentation

$$\langle a, b, c, \ldots; P, Q, R, \ldots \rangle.$$

We shall use the words in $a, b, c, \ldots$ to construct such a group. Words in $a, b, c, \ldots$ under the operation of juxtaposition do not form a group. Although they do satisfy axioms I, II (juxtaposition is clearly an associative operation), and III (the empty word is the identity element), no word other than the empty one has an inverse (length arguments show easily that if $WU = 1$ then W and U are both 1).

If W^{-1} is to be the inverse of W under juxtaposition, we must require that aa^{-1}, $a^{-1}a$, bb^{-1}, $b^{-1}b$, cc^{-1}, $c^{-1}c, \ldots$ all be identified with the empty word. Moreover, if $P, Q, R, \ldots$ are to be relators, they must all be identified with the empty word. These natural requirements suggest introducing the following equivalence relation:

The words W_1 and W_2 in $a, b, c, \ldots$ are called *equivalent*, and this equivalence is denoted by

$$W_1 \sim W_2, \tag{1}$$

if the following operations applied a finite number of times, change W_1 into W_2:

(i) *Insertion of one of the words* $P, P^{-1}, Q, Q^{-1}, R, R^{-1}, \ldots$ *or one of the trivial relators between any two consecutive symbols of* W_1, *or before* W_1, *or after* W_1.

(ii) *Deletion of one of the words* $P, P^{-1}, Q, Q^{-1}, R, R^{-1}, \ldots$ *or one of the trivial relators, if it forms a block of consecutive symbols in* W_1.

The relation $\sim$ is an equivalence relation. It is reflexive, i.e.,

$$W \sim W, \tag{2}$$

because W can be carried into itself by inserting the empty word, a trivial relator. The relation is symmetric, i.e.,

(3) $$W_1 \sim W_2 \text{ implies } W_2 \sim W_1,$$

because the operations in (i) and (ii) carrying W_1 into W_2 can be performed in reverse to take W_2 into W_1. The relation is transitive, i.e.,

(4) $$W_1 \sim W_2 \text{ and } W_2 \sim W_3 \text{ imply } W_1 \sim W_3,$$

because W_1 can be carried into W_2 and then W_2 into W_3, thus carrying W_1 into W_3, in a finite number of steps using (i) and (ii). Furthermore, $\sim$ is preserved under juxtaposition, i.e.,

(5) $$W_1 \sim W_2 \text{ and } W_3 \sim W_4 \text{ imply } W_1W_3 \sim W_2W_4,$$

because we can change W_1W_3 to W_2W_3 and this in turn to W_2W_4.

The class of all words in $a, b, c, \ldots$ equivalent to W will be denoted by $\{W\}$, and W or any other word contained in $\{W\}$ will be called a *representative* of $\{W\}$. We introduce *multiplication of equivalence classes* by:

(6) $$\{W_1\} \cdot \{W_2\} = \{W_1W_2\}.$$

An equivalence class is completely determined by any one of its representatives because of (2), (3), and (4); moreover, the product of two classes is independent of the particular choice of representatives which are used to obtain the product, because of (5).

The set of equivalence classes under the multiplication introduced is precisely the group we are seeking.

THEOREM 1.1. *The set G of equivalence classes of words in $a, b, c, \ldots$ defined by the relation $\sim$ in* (1) *is a group under the multiplication defined by* (6). *Moreover, under the mapping*

(7) $$a \to \{a\}, \quad b \to \{b\}, \quad c \to \{c\}, \quad \ldots,$$

G has the presentation

(8) $$\langle a, b, c, \ldots; \quad P(a, b, c, \ldots), Q(a, b, c, \ldots), R(a, b, c, \ldots), \ldots \rangle$$

Finally, if G' is a group having the presentation (8), *then G' is isomorphic to G.*

PROOF. Since the product of any two equivalence classes in G is a well-defined equivalence class in G, axiom I of Section I is satisfied. Axiom II follows from (6) and the associativity of juxtaposition of words.

The class $\{1\}$, of the empty word is the identity element in G by (6), and so axiom III is satisfied. The inverse of the class $\{W\}$ is the class $\{W^{-1}\}$, i.e.,

$$\{W\}^{-1} = \{W^{-1}\}. \tag{9}$$

For $\{W\} \cdot \{W^{-1}\} = \{WW^{-1}\}$, and by deletions of trivial relators WW^{-1} may be carried into 1; thus $WW^{-1} \sim 1$ and so $\{WW^{-1}\} = \{1\}$. Similarly, $\{W^{-1}\} \cdot \{W\} = \{1\}$. Axiom IV is therefore satisfied and consequently G is a group.

To show that G has the presentation (8) under the mapping (7), we first show that $a, b, c, \ldots$ are generators for G. For, by (6) and (9),

$$W(\{a\}, \{b\}, \{c\}, \ldots) = \{W(a, b, c, \ldots)\}; \tag{10}$$

for example, $\{a\}\{b\}\{a\}^{-1} = \{a\}\{b\}\{a^{-1}\} = \{aba^{-1}\}$. Hence, under (7), the word $W(a, b, c, \ldots)$ defines the class $\{W(a, b, c, \ldots)\}$, and so $a, b, c, \ldots$ are generating symbols for G.

We next show that $P, Q, R, \ldots$ is a set of defining relators for G. The class defined by $P(a, b, c, \ldots)$ under (7), is, by (10), the class $\{P(a, b, c, \ldots)\}$. But by deletion of P, P can be carried into the empty word; hence $P \sim 1$ and $\{P\} = \{1\}$, the identity element of G. Thus $P, Q, R, \ldots$ are relators. Suppose now that $S(a, b, c, \ldots)$ is any relator in G; then, by (10), S defines the class $\{S(a, b, c, \ldots)\}$, which must be $\{1\}$. Hence, $S \sim 1$, and so S can be carried into the empty word by the operations (i) and (ii). By definition then S is derivable from the relators $P, Q, R, \ldots$. Thus, $P, Q, R, \ldots$ is a set of defining relators for G.

To prove the last assertion in Theorem 1.1, suppose that G' has the presentation (8) under the mapping

$$a \to g', \quad b \to h', \quad c \to k', \ldots.$$

We show that the mapping

$$\{W(a, b, c, \ldots)\} \to W(g', h', k', \ldots) \tag{11}$$

is an isomorphism of G onto G'.

We first show that the mapping (11) is well defined, i.e., if $W_1 \sim W_2$ then $W_1(g', h', k', \ldots) = W_2(g', h', k', \ldots)$. For, if $W_1 \sim W_2$ then $W_1W_2^{-1}$ can be carried first into $W_2W_2^{-1}$ and then into the empty word, by (i) and (ii). Hence, $W_1W_2^{-1}$ is derivable from $P, Q, R, \ldots$ and so is a relator in G'. Thus $W_1(g', h', k', \ldots) = W_2(g', h', k', \ldots)$.

We next show that the mapping (11) is one-one. Suppose $W_1(g', h', k', \ldots) = W_2(g', h', k', \ldots)$. Then $W_1W_2^{-1}$ is a relator in G' and since G' has the presentation (8), $W_1W_2^{-1}$ is derivable from $P, Q, R, \ldots$. Thus, $W_1W_2^{-1}$ can be carried into the empty word by (i) and (ii), and so W_1 can be carried into $W_1W_2^{-1}W_2$ and then into W_2 by (i) and (ii). Hence, $W_1 \sim W_2$, and the mapping (11) is one-one.

Since $a, b, c, \ldots$ are generators for G' the mapping (11) is onto.

Finally, the mapping (11) is a homomorphism because of (6), and because the element defined by W_1W_2 is the product of the elements defined by W_1 and W_2. Hence, the mapping (11) is an isomorphism of G onto G'. ◀

COROLLARY 1.1.1. *Any two groups with the presentation* (8) *are isomorphic under the obvious isomorphism.*

PROOF. For, both groups are isomorphic to the group of equivalence classes G above, under the obvious isomorphism. ◀

COROLLARY 1.1.2. *If $g'', h'', k'' \ldots$ are elements of a group G'' such that*

$$(12) \quad P(g'', h'', k'', \ldots), \quad Q(g'', h'', k'', \ldots), \quad R(g'', h'', k'', \ldots), \quad \ldots$$

is each the identity element in G'', then the group of equivalence classes G above [and hence any group G' with the presentation (8)] *can be mapped homomorphically into G'' under the obvious homomorphism.*

PROOF. The mapping

$$(13) \qquad \{W(a, b, c, \ldots)\} \to W(g'', h'', k'', \ldots)$$

gives a homomorphism of G into G''. For, since each element in (12) is the identity element in G'', if $W_1(a, b, c, \ldots)$ can be carried into $W_2(a, b, c, \ldots)$ by (i) and (ii), then $W_1(g'', h'', k'', \ldots)$ and $W_2(g'', k'', k'', \ldots)$ are the same element of G''. Hence the mapping (13) is well-defined; it is a homomorphism for the same reason that the mapping (11) is a homomorphism. ◀

COROLLARY 1.1.3. *If G has the presentation*

$$\langle a, b, c, \ldots ; P, Q, R, \ldots \rangle,$$

and G' has the presentation

$$\langle a, b, c, \ldots ; P, Q, R, \ldots, S, T, \ldots \rangle,$$

then G' is a homomorphic image of G under the obvious homomorphism.

PROOF. If $a, b, c, \ldots$ define $g', h', k', \ldots$ respectively in G', then P, Q, R define the identity in G'. Hence by Corollary 1.1.2 we have our result. ◀

Theorem 1.1 guarantees that anything which looks like a presentation is in fact the presentation of some group. We shall now consider some illustrations of Theorem 1.1.

Let us consider the equivalence class group G given by the presentation

$$\langle a, b; a^3, b^2, ab = ba \rangle. \tag{14}$$

We shall show that this group is of order six (it is in fact the cyclic group of order six) by first showing that every class in G has a representative among the words

$$1, a, a^2, b, ab, a^2b. \tag{15}$$

For, given any word $W(a, b)$, we can arrive at an equivalent word W' of the form

$$a^{\alpha_0}b^{\beta_0}a^{\alpha_1}b^{\beta_1} \ldots a^{\alpha_n}b^{\beta_n}, \tag{16}$$

where none of the exponents is zero (except possibly α_0 and β_n), by deleting the trivial relators in W until no more such deletions are possible. Since $a^{-1} \sim a^2$ and $b^{-1} \sim b$ [$W_1 \sim W_2$ if and only if $W_1W_2^{-1}$ is derivable from $P, Q, R, \ldots$] by (5) we may replace a^{-1} by a^2 and b^{-1} by b in W' and arrive at an equivalent word W'' in which all exponents are positive integers. Since $ab \sim ba$, it follows that $a^\alpha b^\beta \sim b^\beta a^\alpha$ (see Problem 1) and so we may replace $b^\beta a^\alpha$ by $a^\alpha b^\beta$ and arrive at an equivalent word. Assuming that W'' has the form (16), and replacing $b^{\beta_0}a^{\alpha_1}$ by $a^{\alpha_1}b^{\beta_0}$, W'' is changed to an equivalent word W''' given by

$$a^{\alpha_0+\alpha_1}b^{\beta_0+\beta_1}a^{\alpha_2}b^{\beta_2} \ldots a^{\alpha_n}b^{\beta_n}. \tag{17}$$

Continuing in this way, we arrive after finitely many steps at a word equivalent to W and having the form

$$a^\alpha b^\beta, \tag{18}$$

where α and β are non-negative integers. Since $a^3 \sim 1$ and $b^2 \sim 1$, we may reduce α modulo 3 and β modulo 2. Hence, W is equivalent to one of the words in (15). Thus the group G has at most six elements.

Showing that G has at least six elements (i.e., that none of the words in (15) are equivalent) can be done in a variety of ways. One way is to produce, for each pair of words in (15), a homomorphic image of G in which the words go into different elements. For example, the cyclic group X of order three generated by x is a homomorphic image of G under the mapping determined by $\{a\} \to x$, $\{b\} \to 1$; this follows from Corollary 1.1.2, since the defining relators in (14) are all relators in X. Thus only those words in (15) which have the same a-exponent can be equivalent. Similarly, by using the cyclic group Y of order two generated by y as the homomorphic image of G under the mapping determined by $\{a\} \to 1$, $\{b\} \to y$, it follows that only those words in (15) with the same b-exponent can be equivalent. Thus no two words in (15) can be equivalent and G has exactly six elements.

Alternatively, we could have considered the homomorphism of G into the cyclic group Z of order six generated by z, under the mapping determined by $\{a\} \to z^2$, $\{b\} \to z^3$ [a homomorphism is determined by Corollary 1.1.2 since the defining relators in (14) are relators in Z]. Since the words in (15) go into the distinct elements $1, z^2, z^4, z^3, z^5, z$, the words in (15) are inequivalent. (Moreover, since G and Z both have order six, the homomorphism is one-one onto and so G is isomorphic to Z.)

A second method of showing that G has order six is to construct the graph of G (see Section 1.6).

As another illustration of Theorem 1.1, let G be the equivalence class group given by the presentation

$$\langle a, b, c;\quad a^3, b^3, c^4, ac = ca^{-1}, aba^{-1} = bcb^{-1}\rangle.$$

We shall show that this is merely a complicated way of defining the identity group. Indeed, since $aba^{-1} \sim bcb^{-1}$, it follows that $(aba^{-1})^3 \sim (bcb^{-1})^3$. Since, $a^{-1}a \sim 1$ and $b^{-1}b \sim 1$, it follows that $(aba^{-1})^3 \sim ab^3a^{-1}$ and $(bcb^{-1})^3 \sim bc^3b^{-1}$. But $b^3 \sim 1$ and so $ab^3a^{-1} \sim 1$; then $bc^3b^{-1} \sim 1$ and so $c^3 \sim 1$. Since $c^4 \sim 1$ and $c^3 \sim 1$, it follows that $c \sim 1$; since $aba^{-1} \sim bcb^{-1}$, it follows that $aba^{-1} \sim 1$ and so $b \sim 1$. From $ac \sim ca^{-1}$ and $c \sim 1$ it follows that $a \sim a^{-1}$; hence, $a^2 \sim 1$ which with $a^3 \sim 1$ yields $a \sim 1$. Since a, b, and c are all equivalent to 1, every word $W(a, b, c)$ is equivalent to 1, and consequently G is the identity group.

It should be noted that in the construction of the equivalence class group of Theorem 1.1, we did not assume that the number of generators or defining relators was finite or even denumerably infinite. To give an example of the equivalence class group defined by a non-denumerable presentation, we select generating symbols r_α and t_ρ for each positive real number α and each real number ρ; as defining relations for our presentation we take the equations

$$r_\alpha r_\beta = r_{\alpha\beta}, \tag{19}$$

$$t_\rho t_\sigma = t_{\rho+\sigma}, \tag{20}$$

$$r_\alpha t_\rho = t_{\alpha\rho} r_\alpha. \tag{21}$$

As an immediate consequence of the relations (19), letting $\alpha = \beta = 1$ and then $\beta = 1/\alpha$, we have $r_1 \sim 1$ and $r_\alpha^{-1} \sim r_{1/\alpha}$. Similarly, from (20), letting $\rho = \sigma = 0$ and then $\sigma = -\rho$, we have $t_0 \sim 1$ and $t_\rho^{-1} \sim t_{-\rho}$. Using these equivalences, every word in the generating symbols is equivalent to a word

$$t_{\rho_1} r_{\alpha_1} t_{\rho_2} r_{\alpha_2} \cdots t_{\rho_n} r_{\alpha_n}. \tag{22}$$

Moreover, using (21), the word (22) is equivalent to a word of the form

$$t_\mu r_\lambda, \tag{23}$$

where μ and λ are real numbers, λ positive.

No two of the words in (23) are equivalent and, in fact, the mapping determined by

$$\{r_\alpha\} \to \begin{pmatrix} \alpha & 0 \\ 0 & 1 \end{pmatrix},$$

$$\{t_\rho\} \to \begin{pmatrix} 1 & \rho \\ 0 & 1 \end{pmatrix},$$

is a homomorphism of G onto a group of two-by-two real matrices under matrix multiplication. To verify this, we use Corollary 1.1.2 and show that (19), (20), and (21) are relations in the matrix group. Under the homomorphism,

$$\{t_\mu r_\lambda\} \to \begin{pmatrix} \lambda & \mu \\ 0 & 1 \end{pmatrix};$$

and so, distinct words in (23) are inequivalent. (G is, of course, isomorphic to the matrix group.)

As an example of the use of a presentation to construct a group satisfying stipulated conditions, we shall construct a group in which some element ($\neq 1$) is conjugate to all positive powers of itself. For this purpose, consider the equivalence class group G which has the presentation

$$\langle x, a_2, a_3, \ldots;\ a_2 x a_2^{-1} = x^2, a_3 x a_3^{-1} = x^3, \ldots \rangle \tag{24}$$

Clearly $\{x\}$ is conjugate in G to each of its positive powers; it is not clear, however, that $\{x\}$ is not the identity element in G, i.e., that x is not equivalent to 1. To show this, consider the homomorphism of G into the two-by-two real matrix group under matrix multiplication determined by

$$\{x\} \to \begin{pmatrix} 1 & 1 \\ 0 & 1 \end{pmatrix}, \qquad \{a_n\} \to \begin{pmatrix} n & 0 \\ 0 & 1 \end{pmatrix}. \tag{25}$$

Since

$$\begin{pmatrix} 1 & 1 \\ 0 & 1 \end{pmatrix}^n = \begin{pmatrix} 1 & n \\ 0 & 1 \end{pmatrix}$$

and

$$\begin{pmatrix} n & 0 \\ 0 & 1 \end{pmatrix}\begin{pmatrix} 1 & 1 \\ 0 & 1 \end{pmatrix}\begin{pmatrix} 1/n & 0 \\ 0 & 1 \end{pmatrix} = \begin{pmatrix} n & 0 \\ 0 & 1 \end{pmatrix}\begin{pmatrix} 1/n & 1 \\ 0 & 1 \end{pmatrix} = \begin{pmatrix} 1 & n \\ 0 & 1 \end{pmatrix},$$

each of the defining relators in (4) is a relator in the matrix group, and (25) determines a homomorphism by Corollary 1.1.2. But $\{x\}$ does not go into the unit matrix and so $\{x\}$ is not the identity element in G.

An alternative method to show that $\{x\}$ is not the identity element is to use Theorem 4.10 and the theory of amalgamated groups (see Section 4.4).

As a final example of the equivalence class group G defined from a given presentation, we take two generators a and b, and no defining relators, i.e., G has the presentation

$$\langle a, b \rangle.$$

The group G is called the *free group* on a and b (see Section 1.4). Since trivial relators may be deleted, every class in G has as a representative a word of the form (16) in which each α_i and β_j is a non-zero integer (except possibly α_0 and β_n). Moreover, it can be shown that no two distinct words in (16) are equivalent in G (see Section 1.4).

In this section we have been careful to distinguish between notations for an equivalence class and a representative word of such a class; in practice, however, this distinction is not strictly adhered to. Indeed, it is customary to denote an equivalence class by one of its representatives. Thus the equation $W_1(a, b, c, \ldots) = W_2(a, b, c, \ldots)$ may mean that the two words are identical or merely that $W_1 \sim W_2$, i.e., $\{W_1\} = \{W_2\}$; the context or the authors should make the meaning clear. Similarly, a mapping of the equivalence class group G into a group G' is usually described by

$$a \to g', \quad b \to h', \quad c \to k', \ldots$$

and

$$W(a, b, c, \ldots) \to W(g', h', k', \ldots)$$

rather than the more technically correct

$$\{a\} \to g', \quad \{b\} \to h', \quad \{c\} \to k', \ldots$$

and

$$\{W(a, b, c, \ldots)\} \to W(g', h', k', \ldots).$$

[The reader should be accustomed to this notational duplicity from his previous mathematical education; thus, when we say that the fractions a/b and c/d are equal, it may mean that $a = c$ and $b = d$, or merely that a/b and c/d define the same ratio, i.e., equivalence class of fractions. Similarly, the mapping

$$\frac{a}{b} \to \frac{a^2 + b^2}{a^2 + 2b^2}$$

is usually interpreted as a mapping of ratios; technically speaking, the mapping is a mapping of fractions and to interpret it as a mapping of ratios, we must verify that if a/b and c/d define the same ratio, then $(a^2 + b^2)/(a^2 + 2b^2)$ and $(c^2 + d^2)/(c^2 + 2d^2)$ also define the same ratio.]

In Section 1 we showed that every group has a presentation; in this section we showed that every presentation defines a unique group. In the next section we shall consider some typical problems of presentation theory.

Problems for Section 1.2

1. Show that $a^\alpha b^\beta = b^\beta a^\alpha$ is derivable from $ab = ba$, where α and β are any integers. [*Hint:* If $aba^{-1} = b$, then $a^{-1}ba = b$, $a^\alpha ba^{-\alpha} = b$, and $a^\alpha b^\beta a^{-\alpha} = b^\beta$.]

2. Show that $\langle a, b, c;\ P, Q, R, \ldots, ab = ba, ac = ca, bc = cb\rangle$ is an Abelian group.

3. Suppose that $P, Q, R, \ldots, S, T, \ldots$ are words in $a, b, c, \ldots$ such that $S, T, \ldots$ are derivable from $P, Q, R, \ldots$, and conversely. Show that the equivalence class group

$$\langle a, b, c, \ldots;\ P, Q, R, \ldots\rangle$$

is the same as the equivalence class group

$$\langle a, b, c, \ldots;\ S, T, \ldots\rangle.$$

4. Show that the equivalence class group

$$\langle a, b, c, \ldots;\ P, Q, R, \ldots\rangle$$

is the same as the equivalence class group of each of the following:

$$\langle a, b, c, \ldots;\ P^{-1}, Q, R, \ldots\rangle;$$
$$\langle a, b, c, \ldots;\ RPQ, Q, R, \ldots\rangle;$$
$$\langle a, b, c, \ldots;\ WPW^{-1}, Q, R, \ldots\rangle;$$
$$\langle a, b, c, \ldots;\ P', Q, R, \ldots\rangle,$$

where P' is a cyclic permutation of P, i.e., if $P = f_1 f_2 \cdots f_n$, then $P' = f_i f_{i+1} \cdots f_n f_1 f_2 \cdots f_{i-1}$.

5. Suppose $G = \langle a, b;\ P(a, b), Q(a, b)\rangle$ and $H = \langle x, y;\ S(x, y), T(x, y)\rangle$. Then show

$$G \times H = \langle a, b, x, y;\ P(a, b), Q(a, b), S(x, y), T(x, y),$$
$$ax = xa, ay = ya, bx = xb, by = yb\rangle,$$

where $G \times H$ is the direct product of G and H, i.e., the set of elements (g, h), g in G, h in H under the product

$$(g, h) \cdot (g', h') = (gg', hh').$$

[*Hint:* If G is presented under the mapping $a \to g$, $b \to g'$, and H is presented under the mapping $x \to h$, $y \to h'$, then show that the combined mapping $a \to (g, 1)$, $b \to (g', 1)$, $x \to (1, h)$, $y \to (1, h')$ determines a homomorphism of the alleged presentation for $G \times H$ onto $G \times H$. Next show that each element of the alleged presentation can be defined by a word $U(a, b)V(x, y)$. Show that if $U(a, b)V(x, y) \sim U'(a, b)V'(x, y)$, then $U(g, g') = U'(g, g')$ and $V(h, h') = V'(h, h')$ by mapping the alleged presentation for $G \times H$ into G under $a \to g$, $b \to g'$, $x \to 1$, $y \to 1$, and into H under $a \to 1$, $b \to 1$, $x \to h$, $y \to h'$.]

6. Generalize Problem 5 to arbitrary presentations G and H. Generalize Problem 5 to an arbitrary number of groups $G, H, \ldots$.

7. Show that the group

$$\langle a, b; a^3, b^2, ab = ba^2 \rangle$$

has order six, and is isomorphic to the permutation group Σ_3 on three objects. [*Hint:* Reduce every word in a and b to one of the words $a^\alpha b^\beta$, where $\alpha = 0, 1, 2$ and $\beta = 0, 1$. Then use the mapping $a \to (123)$, $b \to (12)$.]

8. Show that the group

$$\langle a, b, c; a^3, b^2, ab = ba^2, c^2, ac = ca, bc = cb \rangle$$

has order 12 and find a permutation group isomorphic to it.

9. Show that the group

$$\langle a, b; a^n, b^2, ab = ba^{-1} \rangle$$

has order $2n$ and is isomorphic to the multiplicative group of matrices with entries from the ring of integers mod n of the form

$$\begin{pmatrix} \epsilon & k \\ 0 & 1 \end{pmatrix},$$

where $\epsilon = 1$ or -1, and k is an integer mod n.

10. Let $G = \langle a, b, c, \ldots; \ P, Q, R, \ldots \rangle$. Show that if every word of length $n + 1$ defines the same element as some word of length $\leq n$, then a representative system for G can be chosen from the words of length $\leq n$.

11. Find a representative system for the equivalence classes of

$$\langle a, b; a^4, a^2 = b^2, ab = ba^3 \rangle;$$

construct a multiplication table for these canonical forms.

12. Show that

$$G = \langle a, b; a^3, b^2, (ab)^3 \rangle$$

is a presentation for A_4, the alternating group on 1, 2, 3, 4, under the mapping $a \to (234)$, $b \to (12)(34)$. [*Hint:* Show that $1, a, a^2, b, ab, a^2b, ba, aba, a^2ba, ba^2, aba^2$, and a^2ba^2 are canonical forms.]

13. Show that

$$G = \langle x, y; x^3, y^4, (xy)^2 \rangle$$

presents Σ_4, under the mapping $x \to (234)$, $y \to (1324)$, as follows:

(a) Show that the subgroup N generated by x and y^2 is normal in the equivalence class group by showing yxy^{-1} is in N.

(b) Show that N has index at most 2 in G, since $xN = N$, and $y^\alpha N$ is yN or N; hence $W(x, y)N$ is N or yN.

(c) Show that $xy^2 = y^{-1}x^{-1}y$ and hence $(xy^2)^3 = 1$.

(d) Show that N has order at most 12 by mapping the group in Problem 12 onto N under $a \to x$, $b \to y^2$.

(e) Conclude from (b) and (d) that G has order at most 24. Hence, under $x \to (234)$, $y \to (1324)$, G is isomorphic to Σ_4.

14. Let $G = \langle a, b; a^5, b^4, ab = ba^2\rangle$. Show that G has order 20, and that G is isomorphic to the multiplicative group of two-by-two matrices with entries in the ring of integers modulo five, of the form

$$\begin{pmatrix} k & 0 \\ m & 1 \end{pmatrix},$$

where $0 < k < 5$ and $0 \leq m < 5$, under the mapping

$$a \to \begin{pmatrix} 1 & 0 \\ 1 & 1 \end{pmatrix}, \qquad b \to \begin{pmatrix} 2 & 0 \\ 0 & 1 \end{pmatrix}.$$

15. Show that the equivalence class group

$$\langle a, b; a^{15}, b^4, ab = ba^{12}\rangle$$

and

$$\langle a, b; a^{20}, b^4, ab = ba^2\rangle$$

are the same as the equivalence class group G in Problem 14. [*Hint:* From $b^{-1}ab = a^{12}$ it follows that $b^{-1}a^5b = 1$, and hence, $a^5 = 1$. From $b^{-1}ab = a^2$ and $b^4 = 1$ it follows that $b^{-4}ab^4 = a^{16}$, and hence, $a^{15} = 1$.]

16. Show that the equivalence class group

$$\langle a, b; a^r, b^s, ab = ba^t\rangle,$$

is the same as the equivalence class group

$$\langle a, b; a^d, b^s, ab = ba^t\rangle$$

where d is the greatest common divisor of r and $t^s - 1$. [*Hint:* From $b^s = 1$ and $b^{-1}ab = a^t$, it follows that $b^{-s}ab^s = a^{t^s}$, and so $a^{t^s-1} = 1$.]

17. Show that the equivalence class group

$$G = \langle a, b; a^r, b^s, ab = ba^t\rangle$$

where r divides $t^s - 1$, has the canonical forms $b^\beta a^\alpha$, where $0 \leq \alpha < r$ and $0 \leq \beta < s$. Show that the multiplication rule for canonical forms is given by $b^{\beta_1}a^{\alpha_1}b^{\beta_2}a^{\alpha_2} = b^\delta a^\gamma$ where $\delta = \beta_1 + \beta_2$ modulo s and $\gamma = \alpha_1 t^{\beta_2} + \alpha_2$ modulo r. [*Hint:* To show that $b^\beta a^\alpha$ represent distinct elements, first map G into the multiplicative group of two-by-two matrices with entries in the integers modulo r, under

$$a \to \begin{pmatrix} 1 & 0 \\ 1 & 1 \end{pmatrix}, \qquad b \to \begin{pmatrix} t & 0 \\ 0 & 1 \end{pmatrix};$$

in particular show that $b^\beta a^\alpha = 1$ implies $\alpha = 0$; hence, if $b^\beta a^\alpha = b^{\beta'} a^{\alpha'}$, then $\alpha = \alpha'$. Next, show by mapping G into the cyclic group of order s generated by x, under the mapping

$$a \to 1, \quad b \to x$$

that if $b^\beta = b^{\beta'}$, then $\beta = \beta'$.

To show the multiplication rule, note that $b^{-1}ab = a^t$, $b^{-\beta}ab^\beta = a^{t^\beta}$, $b^{-\beta}a^\alpha b^\beta = a^{\alpha t^\beta}$.]

18. Show that if r divides $t^s - 1$, then the elements $d^\beta c^\alpha$ where $0 \leq \alpha < r$ and $0 \leq \beta < s$ form a group under the multiplication rule $d^{\beta_1}c^{\alpha_1} \cdot d^{\beta_2}c^{\alpha_2} = d^\delta c^\gamma$ where $\delta = \beta_1 + \beta_2$ modulo s and $\gamma = \alpha_1 t^{\beta_2} + \alpha_2$ modulo r. Show that the group G in Problem 17 is isomorphic to this group under $a \to c$, $b \to d$. [*Hint:* Verify the group postulates I through IV for this system.]

19. Show that the group

$$G = \langle a, b; a^p, b^{p-1}, ab = ba^q \rangle,$$

where p is a prime and q has order $p - 1$ in the multiplicative group of integers modulo p, is isomorphic to the group of linear transformations

$$x \to kx + m,$$

where k is different from zero and where x, k, and m are integers modulo p, under resultant composition, i.e., $f(x) \cdot g(x) = g(f(x))$. [*Hint:* Use the canonical forms from Problem 17 to show G has order $p(p - 1)$; map a into $x \to x + 1$ and b into $x \to qx$.]

20. Let

$$G = \langle a, b, c; a^r, b^s, c = b^{-1}a^{-1}ba, ac = ca, bc = cb \rangle.$$

(a) Show that c^γ commutes with every element.
(b) Show that $a^{-1}ba = bc$, and hence, $aba^{-1} = bc^{-1}$; generalize this to $a^{-\alpha}b^\beta a^\alpha = b^\beta c^{\alpha\beta}$.
(c) Show that $c^r = 1$ and $c^s = 1$ by letting $\alpha = 1$, $\beta = s$, and $\alpha = r$, $\beta = 1$ in (b).
(d) Show that every element of G can be defined by a word $a^\alpha b^\beta c^\gamma$, where $0 \leq \alpha < r$, $0 \leq \beta < s$, and $0 \leq \gamma < d = gcd(r, s)$.
(e) Show the multiplication rule

$$a^{\alpha_1}b^{\beta_1}c^{\gamma_1}a^{\alpha_2}b^{\beta_2}c^{\gamma_2} = a^{\alpha_3}b^{\beta_3}c^{\gamma_3},$$

where $\alpha_3 = \alpha_1 + \alpha_2$ modulo r, $\beta_3 = \beta_1 + \beta_2$ modulo s, and $\gamma_3 = \alpha_2\beta_1 + \gamma_1 + \gamma_2$ modulo d, by using (b).
(f) Show that the set of words $a^\alpha b^\beta c^\gamma$, where $0 \leq \alpha < r$, $0 \leq \beta < s$, and $0 \leq \gamma < d$ form a group under the multiplication given in (e). Conclude that G is isomorphic to this group.

21. Show that the group

$$\langle a, b; a^3, b^3, c = b^{-1}a^{-1}ba, ac = ca, bc = cb \rangle$$

has order 27.

1.3. Dehn's Fundamental Problems

A presentation

$$(1) \qquad \langle a, b, c, \ldots ; P, Q, R, \ldots \rangle,$$

no matter how numerous nor how strange the defining relators, always determines a unique group (up to isomorphism). But there may arise great difficulties as soon as we wish some more specific information about the group G defined by (1); e.g., is G Abelian? finite?

Part of the trouble is that the definition of equivalence of words used to obtain G is non-constructive, i.e., we say $W_1 \sim W_2$ if there exists some way of going from W_1 to W_2 by applying (i) and (ii) of Section 1.2 a finite number of times, but we give no procedure for determining whether one can go from W_1 to W_2. For example, if (1) is finitely generated, then G is Abelian if and only if the words $aba^{-1}b^{-1}$, $aca^{-1}c^{-1}, \ldots, bcb^{-1}c^{-1}, \ldots$ all define the identity element in G; if we had a constructive procedure for determining whether or not a word defines the identity element in G, we would be able to decide whether G is Abelian.

The problem of deciding whether a word in G defines the identity element (or equivalently whether two words define the same element) is the first of the following three fundamental decision problems formulated by Max Dehn in 1911. These problems are important for presentation theory, as well as for its applications.

Let a group G be defined by means of a given presentation.

(I) For an arbitrary word W in the generators, decide in a finite number of steps whether W defines the identity element of G, or not.

(II) For two arbitrary words W_1, W_2 in the generators, decide in a finite number of steps, whether W_1 and W_2 define conjugate elements of G, or not.

(III) For an arbitrary group G' defined by means of another presentation, decide in a finite number of steps whether G is isomorphic to G', or not.

The problems (I), (II), (III) are called the *word problem*, the *conjugacy* or *transformation problem*, and the *isomorphism problem*, respectively, for the presentation defining G.

The word problem has been solved for many classes of presentations of one specialized form or another: e.g., presentations in which there is at most one defining relator (see Sections 1.4 and 4.4); presentations which are finitely generated and in which, for each pair of generating symbols a and b, the relation $ab = ba$ is included among the defining relations (see

Section 3.3); presentations in which each pair of defining relators have only small (compared to their lengths) blocks of consecutive symbols in common (see Tartakovski, 1949, and Greendlinger, 1960a); and presentations constructed in a very simple manner from presentations in which the word problem is solved (see Section 4.1). However, there is some finite presentation for which the word problem cannot be solved, i.e., there is some word W in the generating symbols for which it cannot be decided in a finite number of steps whether W defines the identity element or not (see Novikov, 1955, and Boone, 1955). Thus there is no procedure for solving the word problem which will work for every presentation; in a sense then, every solution of the word problem for a class of presentations is a triumph over nature.

The transformation problem is even more difficult than the word problem. Indeed, by selecting W_1 to be the empty word, the solution to the transformation problem yields a solution to the word problem. Thus the classes of groups for which the transformation problem has been solved are included among those for which the word problem has been solved: e.g., presentations in which there are no defining relators (see Section 1.4); presentations which are finitely generated and in which for each pair of generating symbols a and b, the relation $ab = ba$ is included among the defining relations (here the transformation problem is equivalent to the word problem); presentations in which each pair of defining relators have only very small (compared to their lengths) blocks of consecutive symbols in common (see Greendlinger, 1960b). Although the word problem has been solved for presentations with a single defining relator, the transformation problem has not been solved [in fact, it may be incapable of solution, unless the relator is somewhat restricted].

The isomorphism problem is the most difficult of the three problems of Dehn. It has been shown that even if the given presentation of G is such that G is obviously the identity, e.g., $\langle a; a\rangle$, then the isomorphism problem is unsolvable; i.e., there exists some finite presentation for which we cannot decide in a finite number of steps whether the group it defines is the identity or not (see Rabin, 1958). It is therefore customary to restrict the presentations for G and G' to some special class. If the presentations for G and G' have no defining relators, then the isomorphism problem can be solved (see Section 2.2); if the presentations for G and G' are finite and each contains a relation of the form $ab = ba$ for every pair of its generating symbols, then the isomorphism problem can be solved (see Section 3.2); if one of the presentations has no defining relators and the other has a single defining relator, the isomorphism problem can be solved (see Whitehead, 1936a, and Section 3.5). However, if both G and G' have a single defining relator, then the solution to even this restricted isomorphism problem is unknown, if indeed the problem is not unsolvable (see Section 6.1).

Although the general isomorphism problem is unsolvable, there are a number of tests which can be used to investigate whether two given presentations define isomorphic groups; these tests give necessary but not sufficient conditions (see Sections 3.2 and 3.3, and Section 1.5).

It should be noted that the problems (I), (II), (III) refer to a given presentation of a group, and not to the group itself; thus it is conceivable that in one presentation of a group the word problem is solvable and yet in another it is not. However, it is customary to talk about the problems (I), (II), (III) for a group, assuming that it is given by one of its "standard" presentations, e.g., the *free group* on two generators has the "standard" presentation $\langle a, b\rangle$, the *group of braids with four strings* has the "standard" presentation

$$\langle a, b, c; aba = bab, bcb = cbc, ac = ca\rangle;$$

it is always some such "standard" presentation that is understood when discussing problems (I), (II), (III) for a group without mention of a presentation for the group.

The solution of the word problem for a finite presentation of a group G allows us to express an element of G in a *canonical form*, i.e., we may constructively select a unique representative from each equivalence class of words; moreover, by using the solution to the word problem we may construct a multiplication table for these canonical forms. Thus a solution of the word problem for a presentation of the group G allows us to view G in the usual way a group is given, i.e., in terms of a set of distinct elements together with a multiplication rule for these elements.

One method for constructing a representative system S for the equivalence classes G of

$$\langle a_1, a_2, \ldots, a_n; \quad R_1, R_2, \ldots, R_m\rangle, \tag{2}$$

assuming a solution to its word problem, is to select a "least" word from each equivalence class. Specifically, define an *order relation* $<$ (read "precedes") among the words $W(a_\nu)$ [i.e., the words $W(a_1, a_2, \ldots, a_n)$] as follows:

If $L(W_1) < L(W_2)$, then $W_1 < W_2$;

$$a_1 < a_1^{-1} < a_2 < a_2^{-1} < \ldots < a_n < a_n^{-1}. \tag{3}$$

If $L(W_1) = L(W_2)$, and W_1 and W_2 first differ in their kth terms, then order W_1 and W_2 according to their kth terms. For example,

$$1 < a_1 < a_2a_n < a_2a_n^{-1} < a_1^3.$$

A given word W has finitely many words preceding it, since the words preceding W have length $\leq L(W)$, and the number of generators in (2) is finite. Hence, any non-empty set of words has a least word. Let the representative system S consist of the least word in each equivalence class of (2). If we have a solution of the word problem for (2), we can constructively find the representative of any word W. For, list the finitely many words W_i that precede W; use the solution to the word problem to select those W_i which define the same element as W, and select the least such W_i. Moreover, given two representatives U and V in S, to constructively find the product of U and V in S, find the representative of UV in S by the above method. The set S clearly forms a group, under this multiplication, which is isomorphic to G; thus we can view G as the set of canonical forms under this multiplication rule.

Conversely, given (2), if we can find a set of words S such that every word is equivalent to one of the words in S and distinct words in S are inequivalent, and if we have a constructive process for obtaining in a finite number of steps the word in S equivalent to any given word W, then the word problem for (2) is solvable. For, given W, compute in a finite number of steps the two words in S equivalent to W and to the empty word 1, respectively; if these words in S coincide then W defines the identity element, and otherwise not.

Similarly, given (2), if we can obtain a set of words T such that every word has a conjugate equivalent to a word in T, and distinct words in T have inequivalent conjugates, and if we have a constructive process for obtaining in a finite number of steps the word in T equivalent to a conjugate of any given word W, then the transformation problem for (2) can be solved.

We illustrate these last remarks on the solvability of the word and transformation problem by solving the word and transformation problem for the group

$$(4) \quad G = \langle a, b, c; a^{-1}ba = c, a^{-1}ca = b, b^{-1}ab = c, b^{-1}cb = a, \\ c^{-1}ac = b, c^{-1}bc = a\rangle,$$

in which the conjugate of one generator by another is the remaining generator.

We first solve the word problem. For this purpose we note that $a^{-2}ba^2 = a^{-1}(a^{-1}ba)a = a^{-1}ca = b$. Hence, a^2 commutes with b; similarly, a^2 commutes with c and so is in the center of G. Thus $(c^{-1}ac)^2 = b^2 = c^{-1}a^2c = a^2$, and so $a^2 = b^2$. By symmetry of the relators, $a^2 = b^2 = c^2$. Also, from the relators in (4), it follows immediately that $ab = bc = ca$ and $ba = cb = ac$. We now show that the words of the form

$$(5) \qquad a^{2k}, a^{2k}a, a^{2k}b, a^{2k}c, a^{2k}ab, a^{2k}ba,$$

where k is any integer, are a representative system for the equivalence classes of (4); moreover, we give a process for reducing any word to its representative in a finite number of steps.

Every word of length zero or one can be reduced to one of the words in (5); indeed, $1 = a^0$, $a = a^0a$, $b = a^0b$, $c = a^0c$, $a^{-1} = a^{-2}a$, $b^{-1} = b^{-2}b = a^{-2}b$, and $c^{-1} = c^{-2}c = a^{-2}c$. Moreover, if one of the words in (5) is multiplied by a generator or its inverse, we obtain a word equivalent to one of the words in (5). For example,

$$(a^{2k}b)c^{-1} = a^{2k}b \cdot c^{-2}c = a^{2k}b \cdot a^{-2}c = a^{2(k-1)}bc.$$

Hence, if every word of length n is equivalent to one of the words in (5), then so is every word of length $n + 1$. Thus every word can be reduced in a finite number of steps to an equivalent word in (5).

Moreover, no two of the canonical words in (5) are equivalent. We use homomorphisms of (4) into known groups to show this. Consider the homomorphism determined by

$$a \to (12), \quad b \to (23), \quad c \to (31), \tag{6}$$

of G into the symmetric group of order six; (6) determines a homomorphism by Corollary 1.1.2. Under (6), a^{2k} is mapped into the identity and 1, a, b, c, ab, ba go into distinct elements; hence, equivalent words in (5) can differ only on k. But considering the homomorphism of G into the infinite cyclic group generated by x, given by

$$a \to x, \quad b \to x, \quad c \to x, \tag{7}$$

we see that equivalent words in (5) must have the same k. Therefore, distinct words in (5) are inequivalent. We have thus shown that the words in (5) are a representative system. Since we also showed how to reduce a word to its canonical form in (5) in a finite number of steps, we have solved the word problem for (4). [Instead of using homomorphisms to show the inequivalence of the canonical forms, conceivably we might have directly used the definition of a set of defining relators. However, even for presentations with only trivial relators this is by no means easy to do (see Section 1.4).]

To solve the transformation problem for (4), we make use of some of the canonical forms in (5), namely,

$$a^{2k}, a^{2k}a, a^{2k}ab \tag{8}$$

where k is any integer. Clearly, every word in (5) has a conjugate equivalent to some word in (8); for example, $b^{-1}(a^{2k}c)b = b^{-1}a^{2k}b \cdot b^{-1}cb = a^{2k} \cdot a$. Hence, every word has a conjugate equivalent to one of the words in (8). Moreover, using the homomorphism (6), if two words in (8) have equivalent conjugates they must both end in 1 or both in a or both in ab; using the homomorphism (7), they must both have the same word length. Thus no two distinct words in (8) can have equivalent conjugates. This solves the transformation problem for (4).

Although the problems of Dehn are unsolvable in general, for many presentations important for applications problems (I) and (II) can be solved. For more details the reader is referred to Section 6.1.

Problems for Section 1.3

1. Solve the word and transformation problem for

$$\langle a, b; a^5, b^2, ab = ba^{-1} \rangle.$$

[*Hint:* See Problem 1.2.17.]

2. Solve the word and transformation problem for

$$\langle a, b; a^5, b^4, ab = ba^{-1} \rangle.$$

[*Hint:* See Problem 1.2.17.]

3. Show that a finite multiplication table presentation has a solvable word problem and transformation problem. [*Hint:* See the end of Section 1.1 for the word problem. Conjugate each generating symbol by all the others to find its conjugates.]

4. Let $G = \langle a_1, a_2, \ldots, a_n; R_1, R_2, \ldots, R_m \rangle$, and let $<$ be the relation of precedes defined in this section. Show that for any words W_1, W_2 one and only one of the following holds: either $W_1 = W_2$ (identically) or $W_1 < W_2$ or $W_2 < W_1$. Show that if $W_1 < W_2$ and $W_2 < W_3$, then $W_1 < W_3$. Show that if $W_1 < W_2$, then $WW_1 < WW_2$ and $W_1W < W_2W$. Suppose $W = f_1 f_2 \cdots f_r$ where $f_i = a_{\nu_i}^{\epsilon_i}$, $\epsilon_i = 1$ or -1 and $1 \leq \nu_i \leq n$; then show that the number of words of length r which precede W is

$$p(f_1)(2n)^{r-1} + p(f_2)(2n)^{r-2} + \ldots + p(f_r),$$

where $p(f_i)$ is the number of generating symbols or their inverses preceding f_i.

5. Define the relation $<$ (called "precedes") for words in a countable number of symbols $a_1, a_2, \ldots, a_n, \ldots$ as follows: let $L(W)$ be the length of W and $H(W)$ be the highest subscript on any generating symbol in W; then

$$a_1 < a_1^{-1} < a_2 < a_2^{-1} < \ldots < a_n < a_n^{-1} < \ldots;$$

if $L(W_1) + H(W_1) < L(W_2) + H(W_2)$, then $W_1 < W_2$; if $L(W_1) + H(W_1) = L(W_2) + H(W_2)$ and $H(W_1) < H(W_2)$, then $W_1 < W_2$; if $H(W_1) = H(W_2)$ and $L(W_1) = L(W_2)$, and W_1 and W_2 first differ on their kth terms, then order them according to their kth terms. Show that one and only one of the following holds for any words W_1, W_2: either $W_1 = W_2$ or $W_1 < W_2$ or $W_2 < W_1$. Show that if $W_1 < W_2$ and $W_2 < W_3$, then $W_1 < W_3$. Show that $W_1 < W_2$ need not imply $WW_1 < WW_2$ nor $W_1W < W_2W$. Show that any word W has at most finitely many words preceding it. Use the relation $<$ to show that if the word problem is solvable for a countable presentation

$$\langle a_1, a_2, \ldots, a_n, \ldots; \quad R_1, R_2, \ldots \rangle,$$

then every word can be reduced to a canonical form in finitely many steps. Show that the product of two canonical forms can be reduced to canonical form in finitely many steps.

6. Consider a finite presentation

$$G = \langle a_1, a_2, \ldots, a_n; \quad R_1, R_2, \ldots, R_m \rangle.$$

Show that we can produce a countable sequence of relators of G such that every relator of G occurs in the sequence. [*Hint:* Show that the words that can be obtained from a given word $W(a_\nu)$ by insertion or deletion of one of the relators $R_1, R_2, \ldots, R_m$ or one of the trivial relators is finite. Hence, the number of words obtainable from the empty word by a finite number of such insertions or deletions is finite. Arrange the words that can be obtained by precisely k such insertions and deletions into a finite sequence, and juxtapose these sequences for $k = 0, 1, 2, \ldots$.]

7. Show that if Σ_k is the symmetric group of degree k, and

$$G = \langle a_1, a_2, \ldots, a_n; \quad R_1, R_2, \ldots, R_m \rangle$$

is a finitely presented group, then all homomorphisms of G into Σ_k can be determined in finitely many steps. Show that we can produce a countable sequence of homomorphisms of G into each Σ_k such that every homomorphism of G into any Σ_r occurs. [*Hint:* A homomorphism of G into Σ_k is determined by a mapping of the generating symbols into Σ_k, such that $R_1, R_2, \ldots, R_m$ are relators in Σ_k under that mapping.]

8. Show that if the finitely presented group

$$G = \langle a_1, a_2, \ldots, a_n; \quad R_1, R_2, \ldots, R_m \rangle$$

is known to be a finite group, then the word problem for the presentation can be solved. Moreover, the transformation problem for the presentation can be solved. [*Hint:* Let W be any word in $a_1, a_2, \ldots, a_n$. Alternately construct terms of the sequence of relators in Problem 6, and construct terms of the sequence of homomorphisms of G into Σ_k in Problem 7. After finitely many steps, W must appear in the sequence of relators and so W defines 1, or W must be mapped into an element not the identity in some Σ_k and hence W does not define 1. The solution of the word problem allows us to construct a multiplication table of canonical forms. Hence, by Problem 3, the transformation problem can be solved.]

9. Let $Q(x)$ be the field of rational functions in the variable x with rational coefficients, and let M be the multiplicative group of non-singular two-by-two matrices with entries in $Q(x)$.

(a) Show that

$$a \to \begin{pmatrix} 1 & 0 \\ x & 1 \end{pmatrix}, \qquad b \to \begin{pmatrix} 1 & x \\ 0 & 1 \end{pmatrix}$$

determines a homomorphism of the free group $\langle a, b \rangle$ into M.

(b) Show that

$$a^n \to \begin{pmatrix} 1 & 0 \\ nx & 1 \end{pmatrix}, \qquad b^n \to \begin{pmatrix} 1 & nx \\ 0 & 1 \end{pmatrix},$$

where n is any integer.

(c) Show that

$$a^\alpha b^\beta \to \begin{pmatrix} P_{11}(x) & P_{12}(x) \\ P_{21}(x) & P_{22}(x) \end{pmatrix}$$

where $P_{ij}(x)$ is a polynomial in x, and $P_{22}(x)$ has a larger degree than any other $P_{ij}(x)$ provided α and β are non-zero integers.

(d) Show that

$$b^\beta a^\alpha \to \begin{pmatrix} R_{11}(x) & R_{12}(x) \\ R_{21}(x) & R_{22}(x) \end{pmatrix}$$

where $R_{ij}(x)$ is a polynomial in x, and $R_{11}(x)$ has a larger degree than any other $R_{ij}(x)$ provided α and β are non-zero integers.

(e) Show that matrices with polynomial entries in which the lower right polynomial has larger degree than any other entry is closed under multiplication.

(f) Show that a^α, b^β, $b^\beta a^\alpha$ and $a^{\alpha_1} b^{\beta_1} \ldots a^{\alpha_r} b^{\beta_r}$, where $\alpha, \beta, \alpha_1, \beta_1, \ldots, \alpha_r, \beta_r$ are non-zero integers, cannot define the identity matrix.

(g) Show that $a^{\alpha_1} b^{\beta_1} \ldots a^{\alpha_r} b^{\beta_r}$ cannot define the same matrix as $a^{-\alpha_{r+1}}$, $b^{-\beta_0}$, or $b^{-\beta_0} a^{-\alpha_{r+1}}$, where $\beta_0, \alpha_1, \beta_1, \ldots, \alpha_r, \beta_r, \alpha_{r+1}$ are non-zero integers. Hence, none of the words

$$a^{\alpha_1} b^{\beta_1} \ldots a^{\alpha_r} b^{\beta_r} a^{\alpha_{r+1}},$$

$$b^{\beta_0} a^{\alpha_1} b^{\beta_1} \ldots a^{\alpha_r} b^{\beta_r},$$

and

$$b^{\beta_0} a^{\alpha_1} b^{\beta_1} \ldots a^{\alpha_r} b^{\beta_r} a^{\alpha_{r+1}}$$

can define the identity element.

(h) Conclude that one set of canonical forms for $\langle a, b \rangle$ consists of the empty word, together with the words of the form

$$a^{\alpha_1} b^{\beta_1} \ldots a^{\alpha_r} b^{\beta_r},$$

where α_1 or β_r may be zero, but otherwise all α_i and β_j are non-zero integers. This solves the word problem for the free group $\langle a, b \rangle$.

(i) Conclude also that $\langle a, b \rangle$ is a presentation for the group generated by

$$\begin{pmatrix} 1 & 0 \\ x & 1 \end{pmatrix} \quad \text{and} \quad \begin{pmatrix} 1 & x \\ 0 & 1 \end{pmatrix}.$$

10. Show that canonical forms for

$$\langle a, b; a^p, b^p\rangle,$$

where p is a positive prime, are

$$a^{\alpha_1}b^{\beta_1}\ldots a^{\alpha_r}b^{\beta_r},$$

where α_1 or β_r may be zero, but otherwise $0 < \alpha_i < p$, $0 < \beta_j < p$. [*Hint:* Use the method of Problem 9, replacing $Q(x)$ by the rational functions in x with coefficients from the field of integers modulo p.]

11. Show that canonical forms for

$$H = \langle a, b; a^p = b^p\rangle,$$

where p is a positive prime, are the words $a^{kp}W$, where k is an integer and W is a canonical form for a word of $\langle a, b; a^p, b^p\rangle$. [*Hint:* Map H into $\langle a, b; a^p, b^p\rangle$ under $a \to a$, $b \to b$; also map H into the infinite cyclic group generated by z under $a \to z$, $b \to z$.]

12. Solve the word problem for the presentation

$$G = \langle a_1, a_2, a_3, \ldots, a_n, \ldots; a_1 = a_2{}^2, a_2 = a_3{}^2, \ldots, a_n = a_{n+1}^2, \ldots\rangle.$$

Show that G is Abelian and, hence, solve the transformation problem. [*Hint:* Show that if k is the highest subscript on a generating symbol occurring in $W(a_\nu)$, then $W(a_\nu)$ can be reduced to a word $a_k{}^r$; show that if $r \neq 0$, then $a_k{}^r$ does not define the identity under the homomorphism $a_\nu \to \dfrac{1}{2^\nu}$ of G into the additive group Q of rational numbers.]

13. Find a set of canonical forms for the words of

$$G = \langle a_1, a_2, a_3, \ldots, a_n, \ldots; a_1 = a_2{}^2, a_2 = a_3{}^2, \ldots, a_n = a_{n+1}^2, \ldots\rangle.$$

Show that this group is isomorphic to the group of rational numbers with denominator a non-negative power of 2, under addition. [*Hint:* See the hint to Problem 12. Show that the words $a_1{}^r$ and $a_k{}^s$, where r is any integer, $k > 1$ and s is any odd integer are canonical forms.]

14. Solve the word problem for

$$G = \langle a_1, a_2, a_3, \ldots, a_n, \ldots; a_1 = a_2{}^t, a_2 = a_3{}^t, \ldots, a_n = a_{n+1}^t, \ldots\rangle,$$

and find a group of rational numbers under addition isomorphic to G, where $t \geq 2$. Find a set of canonical forms for G. [*Hint:* See the hints to Problems 12 and 13.]

15. Solve the word problem for

$$G = \langle a_1, a_2, a_3, \ldots, a_n, \ldots; a_1 = a_2{}^2, a_2 = a_3{}^3, \ldots, a_n = a_{n+1}^{n+1}, \ldots\rangle.$$

Show that G is isomorphic to the additive group of rational numbers. Find canonical forms for G. [*Hint:* See the hint to Problem 12; use the mapping $a_\nu \to 1/\nu!$. For canonical forms, use the words $a_1{}^{\alpha_1}a_2{}^{\alpha_2}\ldots a_r{}^{\alpha_r}$, where α_1 is any integer, $0 \leq \alpha_\nu < \nu$ for $\nu > 1$, and $\alpha_r \neq 0$.]

16. Solve the word problem for

$$G = \langle a_1, a_2, a_3, \ldots, a_n, \ldots; a_1 = a_2^{r_2}, a_2 = a_3^{r_3}, \ldots, a_n = a_{n+1}^{r_{n+1}}, \ldots \rangle,$$

where r_ν is a non-zero integer. [*Hint:* See the hint to Problem 12; use the mapping $a_1 \to 1$, $a_\nu \to 1/r_2 r_3 \ldots r_\nu$.]

17. Solve the isomorphism problem when it is restricted to finite presentations on one generator.

18. Let $S_1 \subset S_2 \subset \ldots \subset S_n \subset \ldots$ be an increasing sequence of sets of symbols; let $T_1 \subset T_2 \subset \ldots \subset T_n \subset \ldots$ be an increasing sequence of sets of words, where T_n is a set of words in the symbols of S_n. Let $G_n = \langle S_n; T_n \rangle$, and suppose that for each n, the subgroup of G_n generated by the symbols of S_{n-1} has the presentation $\langle S_{n-1}; T_{n-1} \rangle$, under the obvious mapping. Show that if $S = \bigcup_{i=1}^{\infty} S_n$, $T = \bigcup_{i=1}^{\infty} T_n$, and $G = \langle S; T \rangle$, then the subgroup of G generated by the symbols of S_n is G_n; hence, $G = \bigcup_{i=1}^{\infty} G_n$. [*Hint:* Consider the equivalence class group G_n. Select one word from each equivalence class, so that the canonical forms contain a representative system for the equivalence classes of G_{n-1}. If W_n is the set of representative words for G_n then W_n forms a group under the appropriate multiplication. Moreover, W_{n-1} is a subgroup of W_n. If $W = \bigcup_{i=1}^{\infty} W_n$, then S is a set of generating symbols for W under the obvious mapping. The set T contains relators of W; and if any word in the symbols of S is a relator in W, then it is a word in the symbols of some S_n, and so is a relator in G_n. But then the relator is derivable from the words in T_n and, hence, is derivable from the words in T. Thus, $W \simeq \langle S; T \rangle$, and we have our result.]

1.4. Definition and Elementary Properties of Free Groups

We now consider an important class of groups called *free groups*.

The *free group* F_n *on the n free generators* $x_1, x_2, \ldots, x_n$ is the group with generators $x_1, x_2, \ldots, x_n$ and the empty set of defining relators. Thus,

$$F_n = \langle x_1, x_2, \ldots, x_n \rangle.$$

It follows from Corollary 1.1.3 that every group G on n generators is a homomorphic image of F_n. In this respect, the theory of free groups precedes the general theory of groups with generators and defining relators, in the same way that the algebra of polynomials precedes the general theory of algebraic numbers.

The word problem and transformation problem are easily solved in a free group by using the concepts of *freely reduced* and *cyclically reduced* words.

A *freely reduced word in* $x_1, x_2, \ldots, x_n$ is a word in which the symbols $x_\nu^\epsilon, x_\nu^{-\epsilon}$ ($\epsilon = \pm 1$; $\nu = 1, 2, \ldots, n$) do not occur consecutively. Thus the words $x_1^2 x_2^5$ and $x_1 x_2 x_3 x_2^{-1} x_1^{-1}$ are freely reduced but the words $x_1^3 x_2^5 x_2^{-2} x_3$ and $x_1 x_2^{-1} x_2 x_3$ are not.

A *cyclically reduced word in* $x_1, x_2, \ldots, x_n$ is a freely reduced word which does not simultaneously begin with x_ν^ϵ and end with $x_\nu^{-\epsilon}$ ($\epsilon = \pm 1$; $\nu = 1, 2, \ldots, n$). Thus $x_1^2 x_2^5 x_3^{-1}$ is cyclically reduced but $x_1 x_2 x_3 x_2^{-1} x_1^{-1}$ is not. Clearly, a word is cyclically reduced if and only if all its cyclic permutations are freely reduced.

Two words $W_1(x_\nu)$ and $W_2(x_\nu)$ are called *freely equal* (denoted $W_1 \approx W_2$) if they determine the same element of F_n, the free group on $x_1, x_2, \ldots, x_n$. Thus $W_1 \approx W_2$ if and only if W_1 can be transformed into W_2 by insertions and deletions of the words $x_\nu x_\nu^{-1}$ or $x_\nu^{-1} x_\nu$ ($\nu = 1, 2, \ldots, n$). For example,

$$x_1 x_2^2 x_2^{-1} x_3^2 x_1^{-1} x_1 x_3^{-1} x_1^{-1} x_3 \approx x_1 x_2 x_3 x_1^{-1} x_3.$$

It is clear that every word in $x_1, x_2, \ldots, x_n$ is freely equal to some freely reduced word. For we may delete the words $x_\nu^\epsilon x_\nu^{-\epsilon}$ ($\epsilon = \pm 1$; $\nu = 1, 2, \ldots, n$) until no such word remains. Indeed, free reduction of a word such as

$$x_1 x_2^2 x_2^{-1} x_3^2 x_1^{-1} x_1 x_3^{-1} x_1^{-1} x_3$$

may be carried out in many ways. However, each of these ways leads to the same freely reduced word, namely, $x_1 x_2 x_3 x_1^{-1} x_3$.

THEOREM 1.2. *Every element of the free group F_n on the free generators $x_1, x_2, \ldots, x_n$ is defined by a unique freely reduced word, i.e., every word in $x_1, x_2, \ldots, x_n$ is freely equal to a unique freely reduced word.*

PROOF. We shall prove Theorem 1.2 by giving a specific process ρ for freely reducing a word. The process ρ freely reduces a word by successively reducing its initial segments. Thus, for example, to compute

$$\rho(x_1 x_2^{-1} x_3 x_3^{-1} x_2 x_2^{-1}),$$

we compute

$$\rho(x_1) = x_1, \qquad \rho(x_1 x_2^{-1}) = x_1 x_2^{-1}, \qquad \rho(x_1 x_2^{-1} x_3) = x_1 x_2^{-1} x_3,$$
$$\rho(x_1 x_2^{-1} x_3 x_3^{-1}) = x_1 x_2^{-1}, \qquad \rho(x_1 x_2^{-1} x_3 x_3^{-1} x_2) = x_1,$$
$$\rho(x_1 x_2^{-1} x_3 x_3^{-1} x_2 x_2^{-1}) = x_1 x_2^{-1}.$$

In general, ρ is defined inductively as follows:

$$\rho(1) = 1, \qquad \rho(x_\nu^\epsilon) = x_\nu^\epsilon \qquad (\epsilon = \pm 1; \nu = 1, 2, \ldots, n)$$

and if

$$\rho(U) = x_{\mu_1}^{\eta_1} \ldots x_{\mu_q}^{\eta_q} \qquad (\eta_i = \pm 1;\ \mu_i = 1, 2, \ldots, n),$$

then

$$\rho(Ux_\nu^{\epsilon}) = x_{\mu_1}^{\eta_1} \cdots x_{\mu_q}^{\eta_q} x_\nu^{\epsilon}, \qquad \text{if } \mu_q \neq \nu \text{ or } \eta_q \neq -\epsilon$$

$$= x_{\mu_1}^{\eta_1} \cdots x_{\mu_{q-1}}^{\eta_{q-1}}, \qquad \text{if } \mu_q = \nu \text{ and } \eta_q = -\epsilon.$$

We first establish some properties of ρ from which the theorem follows easily.

(a) $\rho(W)$ is freely reduced.
(b) $\rho(W) \approx W$.
(c) If V is freely reduced, then $\rho(V) = V$.
(d) $\rho(W_1 \cdot W_2) = \rho(\rho(W_1) \cdot W_2)$.
(e) $\rho(Wx_\nu^{\epsilon}x_\nu^{-\epsilon}) = \rho(W)$, $(\epsilon = \pm 1; \nu = 1, 2, \ldots, n)$.
(f) $\rho(W_1x_\nu^{\epsilon}x_\nu^{-\epsilon}W_2) = \rho(W_1W_2)$, $(\epsilon = \pm 1; \nu = 1, 2, \ldots, n)$.

The properties (a), (b), and (e) follow immediately from the definition of ρ by using induction on the length of W; property (c) follows by using induction on the length of V, and property (d) follows by using induction on the length of W_2; property (f) follows from (d) and (e).

We show next that if two words are freely equal, then the process ρ leads to the same reduced word when applied to each of them; for suppose $U \approx T$. Then we can find a sequence of words $U = U_1, U_2, \ldots, U_k = T$ such that consecutive words can be obtained from one another by a single insertion or deletion of $x_\nu^{\epsilon}x_\nu^{-\epsilon}$, $(\epsilon = \pm 1; \nu = 1, 2, \ldots, n)$. Hence, by (f), $\rho(U_i) = \rho(U_{i+1})$ and so $\rho(U) = \rho(T)$.

Finally, we show that any method for freely reducing a word U must result in $\rho(U)$; for suppose $U \approx V$, and V is freely reduced. Then $\rho(U) = \rho(V) = V$ by the above and (c). Thus every word U is freely equal to a unique freely reduced word, namely, $\rho(U)$. ◄

COROLLARY 1.2.1. *If F_n is the free group on the free generators $x_1, x_2, \ldots, x_n$, then the word problem can be solved.*

PROOF. To decide if a given word $W(x_1, x_2, \ldots, x_n)$ defines the identity in F_n, freely reduce W by using ρ. If $\rho(W)$ is not the empty word, then W does not define the identity, and conversely. ◄

COROLLARY 1.2.2. *If F_n is a free group on $x_1, x_2, \ldots, x_n$, then every element $\neq 1$ has infinite order.*

PROOF. Let $W = x_{\nu_1}^{\epsilon_1} \cdots x_{\nu_p}^{\epsilon_p}$ $(\epsilon_i = \pm 1; \nu_i = 1, 2, \ldots, n)$ be a freely reduced word defining an element $\neq 1$ of F_n. If W is also cyclically reduced, then $\nu_1 \neq \nu_p$ or $\epsilon_1 \neq -\epsilon_p$. Hence, if k is a positive integer,

$$W^k = x_{\nu_1}^{\epsilon_1} \cdots x_{\nu_p}^{\epsilon_p} x_{\nu_1}^{\epsilon_1} \cdots x_{\nu_p}^{\epsilon_p} \cdots x_{\nu_1}^{\epsilon_1} \cdots x_{\nu_p}^{\epsilon_p},$$

where the right-hand side consists of k factors, each equal to W. Now W^k is freely reduced and is not the empty word. Thus W^k cannot define 1, and so W has infinite order.

On the other hand, if W is not cyclically reduced, then W is a conjugate of some cyclically reduced word V. For if

$$W = x_{\nu_1}^{\epsilon_1} \ldots x_{\nu_r}^{\epsilon_r} x_{\mu_1}^{\eta_1} \ldots x_{\mu_q}^{\eta_q} x_{\nu_r}^{-\epsilon_r} \ldots x_{\nu_1}^{-\epsilon_1},$$

where

$$\mu_1 \neq \mu_q \quad \text{or} \quad \eta_1 \neq -\eta_q \qquad (\epsilon_i, \eta_j = \pm 1; \nu_i, \mu_j = 1, 2, \ldots, n),$$

then $W = UVU^{-1}$ where $U = x_{\nu_1}^{\epsilon_1} \ldots x_{\nu_r}^{\epsilon_r}$, $V = x_{\mu_1}^{\eta_1} \ldots x_{\mu_q}^{\eta_q}$, and V is cyclically reduced and $\neq 1$. Since V is cyclically reduced, by the above V defines an element in F_n of infinite order. Hence W, which defines a conjugate of this element, also defines an element of infinite order ◀

Theorem 1.2 allows us to give an alternative description of a free group, which is often taken as its definition. The free group F_n may be viewed as the set of freely reduced words in $x_1, x_2, \ldots, x_n$ (i.e., the canonical forms used in our solution of the word problem) with multiplication defined as juxtaposition followed by the reduction of this preliminary product to a freely reduced word.

To solve the transformation problem for the free group F_n on the free generators $x_1, x_2, \ldots, x_n$, we introduce a specific process σ for cyclically reducing a word. Roughly speaking, σ cyclically reduces a word by first freely reducing it and then cancelling first and last symbols (if necessary). For example,

$$\begin{aligned}\sigma(x_1x_2x_3x_3^{-1}x_2x_1x_2^{-1}x_1^{-1}) &= \sigma(x_1x_2^2x_1x_2^{-1}x_1^{-1}) \\ &= \sigma(x_2^2x_1x_2^{-1}) = \sigma(x_2x_1) = x_2x_1.\end{aligned}$$

In general, we define σ inductively for freely reduced words as follows:

$$\begin{aligned}\sigma(1) = 1, \quad \sigma(x_\nu^\epsilon) &= x_\nu^\epsilon \qquad (\epsilon = \pm 1, \nu = 1, 2, \ldots, n) \\ \sigma(x_\nu^\epsilon U x_\mu^\eta) &= x_\nu^\epsilon U x_\mu^\eta \qquad \text{if } \nu \neq \mu \quad \text{or} \quad \epsilon \neq -\eta, \\ &= \sigma(U) \qquad \text{if } \nu = \mu \quad \text{and} \quad \epsilon = -\eta,\end{aligned}$$

where $\epsilon, \eta = \pm 1, \nu, \mu = 1, 2, \ldots, n$. Finally, σ is defined for any word W by $\sigma(W) = \sigma(\rho(W))$.

It is easy to show from the definition of σ (by using induction on the length of W) that $\sigma(W)$ is a cyclically reduced word, and that W and $\sigma(W)$ define conjugate elements of F_n.

THEOREM 1.3. *If F_n is the free group on the free generators x_1, $x_2, \ldots, x_n$, then $W_1(x_\nu)$, $W_2(x_\nu)$ define conjugate elements of F_n if and only if the cyclic reduction $\sigma(W_1)$ of W_1 is a cyclic permutation of the cyclic reduction $\sigma(W_2)$ of W_2.*

PROOF. Suppose first that $\sigma(W_2)$ is a cyclic permutation of $\sigma(W_1)$, i.e.,

$$\sigma(W_1) = x_{\nu_1}^{\epsilon_1} \ldots x_{\nu_s}^{\epsilon_s} x_{\nu_{s+1}}^{\epsilon_{s+1}} \ldots x_{\nu_p}^{\epsilon_p}$$

and

$$\sigma(W_2) = x_{\nu_{s+1}}^{\epsilon_{s+1}} \ldots x_{\nu_p}^{\epsilon_p} x_{\nu_1}^{\epsilon_1} \ldots x_{\nu_s}^{\epsilon_s}$$

where $\epsilon_i = \pm 1$; $\nu_i = 1, 2, \ldots, n$. Then $\sigma(W_1) \approx K\sigma(W_2)K^{-1}$ where $K = x_{\nu_1}^{\epsilon_1} \ldots x_{\nu_s}^{\epsilon_s}$. Since $\sigma(W_1)$ and W_1, and also $\sigma(W_2)$ and W_2, define conjugate elements of F_n, it follows that W_1 and W_2 define conjugate elements of F_n.

Next, suppose that W_1 and W_2 define conjugate elements of F_n, i.e., $W_1 \approx TW_2T^{-1}$. We wish to show that $\sigma(W_1)$ and $\sigma(W_2)$ are cyclic permutations of one another. Since

$$\rho(W_1) = \rho(TW_2T^{-1}),$$

it follows that

$$\sigma(W_1) = \sigma(\rho(W_1)) = \sigma(\rho(TW_2T^{-1})) = \sigma(TW_2T^{-1}),$$

and so it suffices to show that $\sigma(TW_2T^{-1})$ is a cyclic permutation of $\sigma(W_2)$. We prove this by using induction on the length of T. If

$$T = x_\nu^{\epsilon} \qquad (\epsilon = \pm 1; \quad \nu = 1, 2, \ldots, n),$$

we must compare $\sigma(x_\nu^{\epsilon} W_2 x_\nu^{-\epsilon})$ with $\sigma(W_2)$. Now $\sigma(W_2) = \sigma(\rho(W_2))$; also since

$$x_\nu^{\epsilon} W_2 x_\nu^{-\epsilon} \approx x_\nu^{\epsilon} \rho(W_2) x_\nu^{-\epsilon},$$

it follows that

$$\sigma(x_\nu^{\epsilon} W_2 x_\nu^{-\epsilon}) = \sigma(\rho(x_\nu^{\epsilon} W_2 x_\nu^{-\epsilon})) = \sigma(\rho(x_\nu^{\epsilon} \rho(W_2) x_\nu^{-\epsilon})).$$

Suppose that

$$\rho(W_2) = x_{\nu_1}^{\epsilon_1} \ldots x_{\nu_r}^{\epsilon_r} \qquad (\epsilon_i = \pm 1; \quad \nu_i = 1, 2, \ldots, n);$$

then computation of

$$\rho(x_\nu^{\epsilon} \rho(W_2) x_\nu^{-\epsilon}) = \rho(x_\nu^{\epsilon} x_{\nu_1}^{\epsilon_1} \ldots x_{\nu_r}^{\epsilon_r} x_\nu^{-\epsilon})$$

suggests the consideration of four cases.

CASE 1. No cancellation takes place; i.e., $\nu \neq \nu_1$ or $\epsilon \neq -\epsilon_1$, and also $\nu \neq \nu_r$ or $\epsilon \neq \epsilon_r$. Then

$$\rho(x_\nu^{\epsilon} \rho(W_2) x_\nu^{-\epsilon}) = x_\nu^{\epsilon} x_{\nu_1}^{\epsilon_1} \ldots x_{\nu_r}^{\epsilon_r} x_\nu^{-\epsilon}.$$

Hence, in this case,

$$\sigma(x_\nu^{\epsilon} W_2 x_\nu^{-\epsilon}) = \sigma(x_\nu^{\epsilon} x_{\nu_1}^{\epsilon_1} \ldots x_{\nu_r}^{\epsilon_r} x_\nu^{-\epsilon}) = \sigma(x_{\nu_1}^{\epsilon_1} \ldots x_{\nu_r}^{\epsilon_r}) = \sigma(W_2).$$

Case 2. Cancellation takes place at both ends; i.e., $\nu = \nu_1 = \nu_r$ and $\epsilon = -\epsilon_1 = \epsilon_r$. Then

$$\rho(x_\nu^{\ \epsilon}\rho(w_2)x_\nu^{\ -\epsilon}) = x_{\nu_2}^{\epsilon_2} \ldots x_{\nu_{r-1}}^{\epsilon_{r-1}}.$$

Hence, in this case,

$$\sigma(x_\nu^{\ \epsilon}W_2x_\nu^{\ -\epsilon}) = \sigma(x_{\nu_2}^{\epsilon_2} \ldots x_{\nu_{r-1}}^{\epsilon_{r-1}})$$

while

$$\sigma(W_2) = \sigma(x_{\nu_1}^{\epsilon_1}x_{\nu_2}^{\epsilon_2} \ldots x_{\nu_{r-1}}^{\epsilon_{r-1}}x_{\nu_1}^{-\epsilon_1}) = \sigma(x_{\nu_2}^{\epsilon_2} \ldots x_{\nu_{r-1}}^{\epsilon_{r-1}}),$$

and again

$$\sigma(x_\nu^{\ \epsilon}W_2x_\nu^{\ -\epsilon}) = \sigma(W_2).$$

Case 3. Cancellation takes place only at the left end; i.e., $\nu = \nu_1$ and $\epsilon = -\epsilon_1$, but $\nu_1 \neq \nu_r$ or $\epsilon_r \neq \epsilon_1$. Then

$$\rho(x_\nu^{\ \epsilon}\rho(W_2)x_\nu^{\ -\epsilon}) = x_{\nu_2}^{\epsilon_2} \ldots x_{\nu_r}^{\epsilon_r}x_{\nu_1}^{\epsilon_1}.$$

Hence, since both $x_{\nu_1}^{\epsilon_1}x_{\nu_2}^{\epsilon_2} \ldots x_{\nu_r}^{\epsilon_r}$ and $x_{\nu_2}^{\epsilon_2} \ldots x_{\nu_r}^{\epsilon_r}x_{\nu_1}^{\epsilon_1}$ are freely reduced, they are both cyclically reduced. Therefore,

$$\sigma(x_\nu^{\ \epsilon}W_2x_\nu^{\ -\epsilon}) = \sigma(x_{\nu_2}^{\epsilon_2} \ldots x_{\nu_r}^{\epsilon_r}x_{\nu_1}^{\epsilon_1}) = x_{\nu_2}^{\epsilon_2} \ldots x_{\nu_r}^{\epsilon_r}x_{\nu_1}^{\epsilon_1}$$

while

$$\sigma(W_2) = \sigma(x_{\nu_1}^{\epsilon_1}x_{\nu_2}^{\epsilon_2} \ldots x_{\nu_r}^{\epsilon_r}) = x_{\nu_1}^{\epsilon_1}x_{\nu_2}^{\epsilon_2} \ldots x_{\nu_r}^{\epsilon_r}.$$

In this case, $\sigma(x_\nu^{\ \epsilon}W_2x_\nu^{\ -\epsilon})$ is a cyclic permutation of $\sigma(W_2)$.

Case 4. Cancellation takes place only at the right end; i.e., $\nu \neq \nu_1$ or $\epsilon \neq -\epsilon_1$, but $\nu = \nu_r$ and $\epsilon = \epsilon_r$. Then

$$\rho(x_\nu^{\ \epsilon}\rho(W_2)x_\nu^{\ -\epsilon}) = x_\nu^{\ \epsilon}x_{\nu_1}^{\epsilon_1} \ldots x_{\nu_{r-1}}^{\epsilon_{r-1}} = x_{\nu_r}^{\epsilon_r}x_{\nu_1}^{\epsilon_1} \ldots x_{\nu_{r-1}}^{\epsilon_{r-1}}.$$

Hence,

$$\sigma(x_\nu^{\ \epsilon}W_2x_\nu^{\ -\epsilon}) = \sigma(x_{\nu_r}^{\epsilon_r}x_{\nu_1}^{\epsilon_1} \ldots x_{\nu_{r-1}}^{\epsilon_{r-1}}) = x_{\nu_r}^{\epsilon_r}x_{\nu_1}^{\epsilon_1} \ldots x_{\nu_{r-1}}^{\epsilon_{r-1}}$$

which is a cyclic permutation of

$$\sigma(W_2) = \sigma(x_{\nu_1}^{\epsilon_1} \ldots x_{\nu_{r-1}}^{\epsilon_{r-1}}x_{\nu_r}^{\epsilon_r}) = x_{\nu_1}^{\epsilon_1} \ldots x_{\nu_{r-1}}^{\epsilon_{r-1}}x_{\nu_r}^{\epsilon_r}.$$

Thus in all four cases $\sigma(x_\nu^{\ \epsilon}W_2x_\nu^{\ -\epsilon})$ is a cyclic permutation of $\sigma(W_2)$. This shows that $\sigma(TW_2T^{-1})$ is a cyclic permutation of $\sigma(W_2)$ when T has length one.

Assume now as the inductive hypothesis that $\sigma(KW_2K^{-1})$ is a cyclic permutation of $\sigma(W_2)$. Then by the above

$$\sigma(x_\nu^{\ \epsilon}KW_2K^{-1}x_\nu^{\ -\epsilon})$$

is a cyclic permutation of $\sigma(KW_2K^{-1})$, and so $\sigma(x_\nu^{\ \epsilon}KW_2K^{-1}x_\nu^{\ -\epsilon})$ is a cyclic permutation of $\sigma(W_2)$.

This completes the proof of Theorem 1.3. ◀

It should be remarked that Theorems 1.2 and 1.3 give a constructive solution to the word and transformation problem for a free group only when the free group is presented on free generators. Note, however, that a free group may be presented on other than free generators. For example, the group

$$\langle a, b, c; ab^{-2} \rangle$$

is a free group on the free generators b and c. A less obvious example is the group

$$\langle a, b, c; a^2bacacab \rangle$$

which is a free group on the free generators ab and ac.

Moreover, there cannot be an algorithm for deciding whether a presentation defines a free group or not (see M. O. Rabin, 1958 and Section 6.1).

The methods used in establishing the solution of the word and transformation problems for free groups can be used for a larger class of groups called the *free products* (see Section 4.1). We now consider a special case of these groups (which includes the free groups), namely the *free products of cyclic groups*, i.e., groups with a presentation of the form

$$\langle x_1, x_2, \ldots x_n; x_1^{r_1}, x_2^{r_2}, \ldots, x_n^{r_n} \rangle$$

where r_i is a non-negative integer (if $r_j = 0$ we usually omit the defining relator $x_j^{r_j}$).

An important group having a presentation of this type is the *group of unimodular linear fractional transformations* (of the complex plane) *with integral coefficients*, i.e., the group of transformations

$$z \to \frac{az + b}{cz + d}$$

where a, b, c, d are integers and $ad - bc = 1$. This group can be shown to have the presentation

$$\langle x, y; x^2, y^3 \rangle$$

where $x = (z \to -1/z)$ and $y = (z \to 1/(-z + 1))$ [see Problem 19].

Moreover, it can be shown that the *reduced group of unimodular two-by-two matrices with integer entries*, in which we identify a matrix with its negative, also has the presentation

$$\langle x, y; x^2, y^3 \rangle$$

where $x = \begin{pmatrix} 0 & 1 \\ -1 & 0 \end{pmatrix}$ and $y = \begin{pmatrix} 0 & 1 \\ -1 & 1 \end{pmatrix}$ [see Problem 24(e)].

To solve the word and transformation problem for the group

$$G = \langle x_1, x_2, \ldots, x_n; x_1^{r_1}, x_2^{r_2}, \ldots, x_n^{r_n} \rangle$$

we again make use of notions of a "reduced" word and "cyclically reduced" word. However, the presence of the defining relators causes us to modify the meaning of these notions. Thus, in the group $\langle a, b; a^2, b^3 \rangle$, the words ab^4 and $a^{-5}b$ define the same elements although they are different freely reduced words. Similarly, b and aba define conjugate elements although both words are cyclically reduced in the free group sense.

If $G = \langle x_1, x_2, \ldots, x_n; x_1^{r_1}, x_2^{r_2}, \ldots, x_n^{r_n} \rangle$, then we define a word

$$W = x_{\nu_1}^{\alpha_1} x_{\nu_2}^{\alpha_2} \ldots x_{\nu_p}^{\alpha_p} \qquad (\alpha_i \text{ an integer}; \quad \nu_i = 1, 2, \ldots, n)$$

to be *reduced* (*in* G) if $\nu_i \neq \nu_{i+1}$ and α_i is a non-zero integral remainder modulo r_i; i.e., $0 < \alpha_i < r_i$ if $r_i \neq 0$, and $\alpha_i \neq 0$ if $r_i = 0$.

Thus if $G = \langle a, b; a^2, b^3 \rangle$ and $H = \langle a, b; a^2, b^2 \rangle$, then ab^2aba and ab^2a are reduced in G but not in H, while aba is reduced in both G and H.

If $G = \langle x_1, x_2, \ldots, x_n; x_1^{r_1}, x_2^{r_2}, \ldots, x_n^{r_n} \rangle$, then we define a word

$$W = x_{\nu_1}^{\alpha_1} x_{\nu_2}^{\alpha_2} \ldots x_{\nu_p}^{\alpha_p} \qquad (\alpha_i \text{ an integer}; \quad \nu_i = 1, 2, \ldots, n)$$

to be *cyclically reduced* (*in* G) if W is reduced in G and $\nu_1 \neq \nu_p$ if $p \neq 1$.

Thus, if $G = \langle a, b; a^2, b^3 \rangle$ and $H = \langle a, b; a^2, b^2 \rangle$, then ab^2ab is cyclically reduced in G but not in H, while $abab$ is cyclically reduced in both; $ababa$ is cyclically reduced in neither G nor H, although it is cyclically reduced in the free group on a and b.

As in the case of the free group, any word may be changed into a reduced word in G, although in this case we are allowed to delete the defining relators of G as well as $x_\nu x_\nu^{-1}$ or $x_\nu^{-1} x_\nu$. Similarly, by allowing cyclic permutations of a word and deletions, any word may be changed to a cyclically reduced word in G.

THEOREM 1.4. *If* $G = \langle x_1, \ldots, x_n; x_1^{r_1}, \ldots, x_n^{r_n} \rangle$, *then every element of* G *is defined by a unique word which is reduced in* G. *Moreover, two words define conjugate elements of* G *if and only if their cyclic reductions in* G *are cyclic permutations of one another.*

PROOF. The proof of this theorem is very similar to that given for Theorems 1.2 and 1.3, and is left for the exercises (see Problem 15). ◀

COROLLARY 1.4. *Let* $G = \langle x_1, \ldots, x_n; x_1^{r_1}, \ldots, x_n^{r_n} \rangle$. *Then every element* $\neq 1$ *of finite order in* G *is the conjugate of a power of some* x_j *such that* $r_j \neq 0$.

PROOF. Let $W(x_\nu)$ define an element of finite order in G. Then W is a conjugate of a word V which is cyclically reduced in G. Let

$$V = x_{\nu_1}^{\alpha_1} \ldots x_{\nu_p}^{\alpha_p}$$

where α_i is an integer, $\nu_i = 1, 2, \ldots, n$. If $p > 1$, then $x_{\nu_1} \neq x_{\nu_p}$ and

$$x_{\nu_1}^{\alpha_1} \ldots x_{\nu_p}^{\alpha_p} x_{\nu_1}^{\alpha_1} \ldots x_{\nu_p}^{\alpha_p} \ldots x_{\nu_1}^{\alpha_1} \ldots x_{\nu_p}^{\alpha_p}$$

is reduced in G and so cannot define 1. But V has the same order as W. Thus $p = 1$; then x_{ν_1} has finite order in G and so $r_{\nu_1} \neq 0$ (see Problem 16). ◀

As an application of this last corollary, consider the presentation

$$\langle x, y; x^2, y^3 \rangle$$

for the group of unimodular linear fractional transformations with integral coefficients. We can conclude that such a transformation can only have 2 or 3 for its order, if its order is finite. Similarly, there is no two-by-two matrix

$$\begin{pmatrix} a & b \\ c & d \end{pmatrix}, \qquad (a, b, c, d \text{ integers}, \quad ad - bc = 1)$$

which has a finite order other than 2, 3, 4, or 6 (since, if we do not identify a matrix with its negative, $\begin{pmatrix} -1 & 0 \\ 0 & -1 \end{pmatrix}$ has order 2).

Problems for Section 1.4

1. Show that if F is the free group on $x_1, x_2, \ldots, x_n$ and U is cyclically reduced, then U^k is also cyclically reduced. Show that if $T = RUR^{-1}$ is freely reduced, and U is cyclically reduced, then $\rho(T^k) = RU^kR^{-1}$.

2. Show that if U and V are freely reduced and $U^k \approx V^k$, k an integer $\neq 0$, then $U = V$. [*Hint:* If U is cyclically reduced but V is not, then, by Problem 1, $\rho(U^k) = U^k$ is cyclically reduced but $\rho(V^k)$ is not. Hence, if U is cyclically reduced, then V is cyclically reduced and $U^k = V^k$; thus $U = V$. On the other hand, if U is not cyclically reduced, then apply conjugation so as to cyclically reduce U.]

3. Let F be the free group on $x_1, x_2, \ldots, x_n$. Show that any element of F has at most finitely many roots. [*Hint:* Let W be freely reduced and $\neq 1$, $W = RUR^{-1}$, where U is cyclically reduced and $\neq 1$. Show that $L(W^k) = 2L(R) + kL(U)$, where $L(V)$ denotes the length of the word V. Show that if V is fixed, $V = W^k$ is only possible for finitely many k. Use Problem 2 to show each k contributes at most one W.]

4. Show that if F is the free group on $x_1, x_2, \ldots, x_n$, and if $U^kV^m \approx V^mU^k$, it follows that $UV \approx VU$. [*Hint:* Consider $V^{-m}U^kV^m$ and $U^{-1}V^mU$ and use Problem 2.]

5. Let F be the free group on $x_1, x_2, \ldots, x_n$. Suppose U, V are freely reduced words and U begins with x_μ^ϵ and V ends with x_ν^η (where $\epsilon, \eta = \pm 1$; $\mu, \nu = 1, 2, \ldots, n$). Show that if $U \neq RV^{-1}$ and $V \neq U^{-1}S$ (where RV^{-1} and $U^{-1}S$ are freely reduced), then $\rho(UV)$ begins with x_μ^ϵ and ends with x_ν^η. [*Hint:* Use induction on the sum of the lengths of U and V; the result is obvious if UV is freely reduced. Otherwise, $U = Kx_\lambda^\theta$, $V = x_\lambda^{-\theta}L$ (where $\theta = \pm 1$); if K or L is empty, then $U = RV^{-1}$ or $V = U^{-1}S$. Hence, we can assume K begins in x_μ^ϵ and L ends in x_ν^η; thus $\rho(UV) = \rho(KL)$ begins in x_μ^ϵ and ends in x_ν^η or $K = RL^{-1}$ or $L = K^{-1}S$. But then $U = Kx_\lambda^\theta = RL^{-1}x_\lambda^\theta = RV^{-1}$ or $V = x_\lambda^{-\theta}L = x_\lambda^{-\theta}K^{-1}S = U^{-1}S$.]

6. Let F be the free group on $x_1, x_2, \ldots, x_n$. Show that if $UV \approx VU$, then $U \approx W^k$, $V \approx W^m$ for some W, and integers k, m, i.e., two elements of F commute if and only if they are powers of the same element. [*Hint:* Use induction on the sum of the lengths $L(U)$ and $L(V)$ of U and V, respectively. By symmetry we may assume $L(U) \leq L(V)$. Moreover, we may assume $U \neq 1$, $V \neq 1$, and U, V are freely reduced. Assume first that UV is also freely reduced. Then $UV = VU$ and so U must be an initial segment of V, i.e., $V = UR$. Then $UV = UUR = VU = URU$. Thus $UR = RU$, and by inductive hypothesis $R \approx W^r$, $U \approx W^s$; hence, $V = UR \approx W^{r+s}$ and we have our result. Suppose, therefore, that UV is not freely reduced; suppose in addition, that U is cyclically reduced. Since UV is not freely reduced, but U and V are, $U = Kx_\nu^\epsilon$, $V = x_\nu^{-\epsilon}M$ (where $\epsilon = \pm 1$, $\nu = 1, 2, \ldots, n$). If both K and M are 1, then $U = V^{-1}$, and we have our result. If only one of K and M is 1, then $V = U^{-1}M$ or $U = KV^{-1}$; hence, $UM \approx MU$ or $VK \approx KV$, so that as in the preceding case, using the inductive hypothesis, we easily obtain our result. Suppose $K \neq 1$, $M \neq 1$. Now, by Problem 5, $\rho(UV)$ begins in the first symbol of U and ends in the last symbol of V (unless $U = RV^{-1}$ or $V = U^{-1}R$, in which case we again have our result, using the inductive hypothesis). Thus, if U begins with x_λ^η (since U is cyclically reduced), $\lambda \neq \nu$ or $\eta \neq -\epsilon$. Now $UV \approx VU$, and so $\rho(UV) = \rho(VU)$. But (unless $V = RU^{-1}$ or $U = V^{-1}S$, in which case we are finished by inductive hypothesis) $\rho(VU)$ begins with the first symbol of V, which is $x_\nu^{-\epsilon}$. Hence, $x_\lambda^\eta = x_\nu^{-\epsilon}$ contrary to U being cyclically reduced. Thus $U = RV^{-1}$ or $V = U^{-1}R$, and by using the inductive hypothesis we have our result. If now U is not cyclically reduced, we can write $U = T^{-1}QT$ where Q is cyclically reduced. Then $Q \approx TUT^{-1}$ and TVT^{-1} commute. Moreover, $L(Q) + L(TVT^{-1}) \leq L(U) + L(V)$. Hence, we may apply the above cases to show that Q and TVT^{-1} are powers of the same element; and hence, U and V are powers of the same element.]

7. Let F be the free group on $x_1, x_2, \ldots, x_n$. Show that the normalizer of any element $\neq 1$ is a cyclic group. [*Hint:* Use Problems 3, 4, and 6.]

8. State a reason or reasons why each of the following groups is not a free group:

(a) The positive real numbers under multiplication.
(b) The permutations of the integers.
(c) A group which contains the rationals under addition as a subgroup.

(d) The group of non-singular two-by-two matrices with integer entries and determinant ± 1.
(e) The group $\langle x, y, z; xz = zx\rangle$.
(f) The group $\langle x, y; x^2 = y^2\rangle$.
(g) The group $\langle x_1, x_2, x_3, \ldots; x_1 = x_3{}^2, x_3 = x_5{}^2, \ldots\rangle$.
(h) A group in which an element $\neq 1$ has two cube roots.

9. Let F be the free group on x, y, z. Show that the subgroup of F generated by x^r, y^s, z^t where r, s, $t \neq 0$, is a free group freely generated by x^r, y^s, z^t. [*Hint:* Map the free group on a, b, c into F by mapping $a \to x^r$, $b \to y^s$, $c \to z^t$. Show that if $W(a, b, c)$ is a freely reduced word $\neq 1$, in a, b, c, then $W(x^r, y^s, z^t)$ is a freely reduced word $\neq 1$, in x, y, z.]

10. Let F be the free group on x, y, z. Show that the subgroup of F generated by xyx, yzy, zxz is freely generated by them. [*Hint:* See the hint in Problem 9.]

11. Let F be the free group on x, y. Show that the subgroup of F generated by xy, yx is freely generated by them.

12. Let F be the free group on x, y. Show that the elements $p_n = x^n y^n$, $n = 1, 2, 3, \ldots$ freely generate a subgroup of F. [*Hint:* Show by induction on q that

$$p_{\nu_1}^{\alpha_1} p_{\nu_2}^{\alpha_2} \cdots p_{\nu_r}^{\alpha_r}$$

(where $\nu_i \neq \nu_{i+1}$, α_i is an integer $\neq 0$ and $\nu_i = 1, 2, 3, \ldots$) defines an element of F whose freely reduced form ends with y^{ν_q} if $\alpha_q > 0$ and ends with $x^{-\nu_q}$ if $\alpha_q < 0$.]

13. Show that if F is the free group on x, y, and H is the subgroup generated by x^2, y^2, xy, yx, then H is not freely generated by x^2, y^2, xy, yx. Show that H is the subgroup of F of words of even length. [*Hint:* Let $a = x^2$, $b = y^2$, $c = xy$, $d = yx$. Show $da^{-1}c = b$.]

14. Let F and H be as in Problem 13. Show H is freely generated by x^2, y^2, xy. [*Hint:* Let $W(a, b, c)$ be freely reduced in a, b, c. Show that

$$W(x^2, y^2, xy) = W_1(x^2, y^2)(xy)^{\alpha_1} \cdot W_2(x^2, y^2)(xy)^{\alpha_2} \cdots W_r(x^2, y^2)(xy)^{\alpha_r},$$

where $W_i(a, b)$ is freely reduced in a, b. Show that if $W_i(a, b)$ is freely reduced, $\neq 1$, then $W_i(x^2, y^2)$ is freely reduced in x, y and begins and ends in x^2, y^2, x^{-2}, or y^{-2}. Show that if $\alpha_i \neq 0$, then $(xy)^{\alpha_i}$ is freely reduced and begins and ends in xy or $y^{-1}x^{-1}$. Show that if $W_i(a, b)$ is freely reduced in a, b, $\neq 1$, and $\alpha_i \neq 0$, then $\rho(W_i(x^2, y^2) \cdot (xy)^{\alpha_i})$ begins in x^2, y^2, x^{-2}, y^{-2}, $x^{-1}y$, or yx^{-1}; moreover, $\rho(W_i(x^2, y^2) \cdot (xy)^{\alpha_i})$ ends in xy, $y^{-1}x^{-1}$, $x^{-1}y$, or yx^{-1}. Finally, show that if $W(a, b, c)$ is freely reduced, $\neq 1$, then $W(x^2, y^2, xy)$ cannot be freely reduced to 1 in x, y.]

15. Let $G = \langle x_1, x_2, \ldots, x_n; x_1{}^{r_1}, x_2{}^{r_2}, \ldots, x_n{}^{r_n}\rangle$. Write out the proof of Theorem 1.4 in detail. [*Hint:* See the proofs of Theorems 1.2 and 1.3.]

16. Let G be as in Problem 15. Show that x_i has order $|r_i|$ if $r_i \neq 0$, and infinite order if $r_i = 0$. Show that if $r_i, r_j \neq 1$, then x_i and x_j do not commute. [*Hint:* Use the word problem solution.]

17. State a reason or reasons why each of the following groups is not a free product of finitely many cyclic groups:

(a) Polynomials of degree less than 3, with integer coefficients, under addition.

(b) A group which has as a subgroup all the nth roots of unity, for all n, under multiplication.

(c) The group $\langle x, y; x^4, x^2 = y^2\rangle$.

(d) The group of n-by-n matrices with integer entries and determinant ± 1 (under multiplication).

(e) The group of linear fractional transformations $z \to \dfrac{pz + q}{rz + s}$ where p, q, r, s are real numbers, $ps - qr \neq 0$.

(f) The group $\langle x_1, x_2, x_3, \ldots; x_1{}^2, x_2{}^2, x_3{}^2, \ldots\rangle$.

(g) The group $\langle x_1, x_2, x_3, \ldots; x_1{}^r, x_2{}^r, x_3{}^r, \ldots\rangle$, $r \neq 0$.

(h) A finite non-cyclic group.

18. Let L be the group of unimodular linear fractional transformations with integer coefficients, i.e., the group

$$z \to \frac{az + b}{cz + d}$$

where a, b, c, d, are integers, $ad - bc = 1$. Moreover, let $x = (z \to 1/-z)$, $y = (z \to 1/(-z + 1))$.

(a) Show that a, b, c, d, and $-a, -b, -c, -d$ define the same linear transformation.

(b) Show that if

$$p = \left(z \to \frac{az + b}{cz + d}\right),$$

then

$$xp = \left(z \to \frac{cz + d}{-az - b}\right) = \left(z \to \frac{-cz - d}{az + b}\right)$$

(c) Show that

$$xy = (z \to z - 1), \qquad (xy)^{-1} = (z \to z + 1),$$

and

$$(xy)^k = (z \to z - k).$$

(d) Show that if

$$p = \left(z \to \frac{az + b}{cz + d}\right),$$

then

$$(xy)^k p = \left(z \to \frac{(a - kc)z + (b - kd)}{cz + d}\right).$$

(e) Verify the following computation: if

$$q = \left(z \to \frac{5z - 1}{-4z + 1}\right),$$

then

$$(xy)^{-1}q = \left(z \to \frac{z}{-4z + 1}\right); \; x(xy)^{-1}q = \left(z \to \frac{4z - 1}{z}\right),$$

$$(xy)^4x(xy)^{-1}q = \left(z \to \frac{-1}{z}\right); \; x(xy)^4x(xy)^{-1}q = (z \to z).$$

Conclude that

$$q = (xy)x^{-1}(xy)^{-4}x^{-1}.$$

(f) Write the following transformations as words in x and y:

$$z \to \frac{2z + 1}{5z + 3}; \qquad z \to \frac{7z - 3}{-9z + 4}; \qquad z \to \frac{1045z + 2}{522z + 1}.$$

(g) Show that if $p = \left(z \to \dfrac{az + b}{cz + d}\right)$, then by multiplying p on the left sufficiently often by x and $(xy)^k$ we can obtain a transformation

$$r = \left(z \to \frac{mz + j}{n}\right).$$

Show that since $mn = 1$, $r = (z \to z + nj) = (xy)^{-nj}$.

(h) Show that x and y generate L.

19. Show that the group L in Problem 18 has the presentation $\langle x, y; x^2, y^3\rangle$ by showing the following:

(a) the mapping $x \to (z \to 1/-z)$ and $y \to (z \to 1/(-z + 1))$ induces a homomorphism of $\langle x, y; x^2, y^3\rangle$ onto L;

(b) every element $\neq 1$ of $\langle x, y; x^2, y^3\rangle$ is defined by a word

$$x^{\alpha_1}y^{\beta_1}x^{\alpha_2}y^{\beta_2} \ldots x^{\alpha_k}y^{\beta_k}$$

where $\alpha_i = 0$ or 1, $\beta_i = 0, 1$ or -1;

(c) every element $\neq 1$ of $\langle x, y, x^2, y^3\rangle$ is conjugate to an element defined by a word of the type x, y, y^{-1}, xy, xy^{-1} or

$$(xy)^{\alpha}(xy^{-1})^{\gamma}(xy)^{\delta} \ldots (xy^{-1})^{\beta}$$

where $\alpha, \beta \geq 0$ but $\gamma, \delta, \ldots$ are ≥ 1;

(d) if some element $\neq 1$ goes into 1 under the homomorphism in (a), then an element of the type given in (c) must go into 1;

(e) x, y, y^{-1} do not go into 1;

(f) xy and xy^{-1} do not go into 1;

(g) $(xy)^{\gamma}$ goes into $z \to z - \gamma$;

(h) $(xy^{-1})^{\delta}$ goes into $z \to \dfrac{z}{-\delta z + 1}$;

(i) $(xy)^{\gamma}(xy^{-1})^{\delta}$ goes into $z \to \dfrac{(1 + \gamma\delta)z - \gamma}{-\delta z + 1}$;

(j) $(xy^{-1})^{\delta}(xy)^{\gamma}$ goes into $z \to \dfrac{z - \gamma}{-\delta z + (1 + \gamma\delta)}$;

(k) the set of linear fractional transformations

$$z \to \frac{az + b}{cz + d}$$

where $a, d > 0$; $b, c \leq 0$; and $ad > 1$ is closed under multiplication and does not contain $z \to z$;

(l) if $\alpha, \beta, \ldots, \gamma, \delta \geq 1$, then

$$(xy)^\alpha(xy^{-1})^\beta \ldots (xy)^\gamma(xy^{-1})^\delta$$

and

$$(xy^{-1})^\alpha(xy)^\beta \ldots (xy^{-1})^\delta(xy)^\gamma$$

cannot go into $z \to z$, since they go into elements in the subset in (k);

(m) the subset in (k) is closed under left multiplication by

$$z \to z - \gamma \quad \text{and} \quad z \to \frac{z}{-\delta z + 1} \qquad \text{if } \gamma, \delta > 0;$$

(n) no word in (c) can go into $z \to z$;

(o) $\langle x, y; x^2, y^3 \rangle$ is isomorphic to L.

20. Let $L = \langle x, y; x^2, y^3 \rangle$. Show each of the following:

(a) the subgroup of L generated by yxy and $xyxyx$ is a free group, freely generated by them;

(b) the subgroup of L generated by yxy^{-1} and $xyxy^{-1}x$ has the presentation $\langle r, s; r^2, s^2 \rangle$;

(c) the subgroup of L generated by yxy^{-1} and xyx has the presentation $\langle r, s; r^2, s^3 \rangle$ but is not L;

(d) the subgroup of L generated by yxy^{-1} and $xyxyx$ has the presentation $\langle r, s; r^2 \rangle$.

21. Show that $\langle x, y; x^2, y^3 \rangle$ has free subgroups with any finite number of free generators. [*Hint:* Use Problems 20(a) and 12.]

22. Let $G = \langle x, y; x^2 = y^3 \rangle$. Show that every element $\neq 1$ of G is defined by a unique word of the form

$$x^{2\alpha} \cdot x^{\alpha_1}y^{\beta_1}x^{\alpha_2}y^{\beta_2} \ldots x^{\alpha_k}y^{\beta_k}$$

where α is an integer; $\alpha_1 = 0$ or 1; $\alpha_2, \ldots, \alpha_k = 1$ (if they occur); $\beta_1, \beta_2, \ldots, \beta_{k-1} = 1$ or 2 (if they occur); $\beta_k = 0$, 1 or 2. [*Hint:* Consider the homomorphism of G into $\langle a, b; a^2, b^3 \rangle$ with $x \to a$, $y \to b$; then consider the homomorphism of G into $\langle a \rangle$ given by $x \to a^3$, $y \to a^2$.]

23. Let $G = \langle x, y; x^4, x^2 = y^3 \rangle$. Show that every element $\neq 1$ of G is defined by a unique word of the form in Problem 22, except that α is 0 or 1. [*Hint:* Consider homomorphisms of G into $\langle a, b; a^2, b^3 \rangle$ and $\langle a; a^4 \rangle$.]

24. Let M be the group of unimodular matrices

$$\begin{pmatrix} a & b \\ c & d \end{pmatrix}$$

where a, b, c, d are integers and $ad - bc = 1$.

(a) Show that the mapping

$$\begin{pmatrix} a & b \\ c & d \end{pmatrix} \to \left(z \to \frac{az + b}{cz + d}\right)$$

is a homomorphism of M onto the group L of Problem 18.

(b) Show that the kernel of the homomorphism in (a) is the center of M, i.e., the set of matrices

$$\begin{pmatrix} 1 & 0 \\ 0 & 1 \end{pmatrix}, \quad \begin{pmatrix} -1 & 0 \\ 0 & -1 \end{pmatrix}$$

(c) Show that M is generated by

$$w = \begin{pmatrix} -1 & 0 \\ 0 & -1 \end{pmatrix}, \quad x = \begin{pmatrix} 0 & 1 \\ -1 & 0 \end{pmatrix}, \quad \text{and} \quad y = \begin{pmatrix} 0 & 1 \\ -1 & 1 \end{pmatrix};$$

note that x and y are pre-images of $(z \to 1/(-z))$ and

$$(z \to 1/(-z + 1)).$$

(d) Show that every element of M can be written uniquely as $U(x, y)$ or $wU(x, y)$ where $U(a, b)$ is a word reduced in the group

$$\langle a, b; a^2, b^3 \rangle.$$

(e) Show that

$$x^2 = w, \qquad y^3 = w, \qquad w^2 = \begin{pmatrix} 1 & 0 \\ 0 & 1 \end{pmatrix}$$

and finally that

$$M = \langle x, y; x^4, x^2 = y^3 \rangle.$$

25. Solve the word problem for $G = \langle x, y; x^{rs}, x^s = y^t \rangle$ where r and t are coprime, by showing that every element of G is defined by a unique word of the form

$$x^{ks} \cdot U(x, y)$$

where $0 \leq k < r$ and $U(a, b)$ is reduced in the group

$$\langle a, b; a^s, b^t \rangle.$$

[*Hint:* Consider the homomorphism of G onto $\langle a, b; a^s, b^t \rangle$ in which $x \to a$, $y \to b$; then consider the homomorphism of G into the cyclic group $\langle a; a^{rs} \rangle$ given by $x \to a$, $y \to a^{ns}$ where $tn = 1$ modulo r, and so $tns = s$ modulo rs.]

26. Show that $\langle x, y; x^{12}, x^2 = y^6 \rangle$ cannot be mapped homomorphically into $\langle a; a^{12} \rangle$ by sending $x \to a$. [*Hint:* Every sixth power in $\langle a; a^{12} \rangle$ has order 1 or 2.]

27. Show that $G = \langle x, y; x^{rs}, x^s = y^t \rangle$ has the word problem solution given in Problem 25, even if r and t are not coprime. [*Hint:* Consider the homomorphism of G into $\langle a; a^{rst} \rangle$ in which $x \to a^t$, $y \to a^s$.]

28. Show that the group $\langle a, b; a^m, a^p = b^q \rangle$ can be presented as a group

$$\langle x, y; x^{rs}, y^{rq}, x^s = y^{nq} \rangle$$

where n and r are coprime. [*Hint:* Let s denote the greatest common divisor of p and m. Then $m = sr$, $p = sv$ where r and v are coprime. If now $nv = 1$ modulo r, then $a^{sr} = 1$, $a^{sv} = b^q$ and $a^{nsv} = a^s = b^{nq}$. Therefore,

$$G = \langle a, b; a^{sr}, a^{sv} = b^q, a^s = b^{nq} \rangle.$$

On the other hand, we also have $b^{rq} = a^{rsv} = 1$; hence,

$$G = \langle a, b; a^{sr}, b^{rq}, a^{sv} = b^q, a^s = b^{nq} \rangle.$$

Now $a^{sv} = b^q$ can be derived from $b^{rq} = 1$ and $a^s = b^{nq}$. For, $a^{sv} = b^{nqv} = b^q$ since $nv = 1$ modulo r, and so $nqv = q$ modulo rq. Thus

$$G = \langle a, b; a^{rs}, b^{rq}, a^s = b^{nq} \rangle$$

where n and r are coprime.]

29. Solve the word problem for

$$G = \langle x, y; x^{rs}, y^{rq}\, x^s, = y^{nq} \rangle$$

where n and r are coprime. [*Hint:* Show that every element of G is defined by some word $x^{ks} \cdot U(x, y)$ where $U(a, b)$ is reduced in $\langle a, b; a^s, b^q \rangle$, and $0 \leq k < r$. Show the word is unique by considering homomorphisms of G into

$$\langle a, b; a^s, b^q \rangle$$

and into $\langle a; a^{rsq} \rangle$ in which $x \to a$, $y \to b$, and $x \to a^{nq}$, $y \to a^s$, respectively.]

30. Show that $G = \langle a, b; a^m, b^k, a^p = b^t \rangle$ has a presentation as in Problem 29. [*Hint:* See hint to Problem 28, and use the same method.]

31. Show that F is a free group on the n generators $a_1, a_2, \ldots, a_n$ if and only if every mapping

$$a_1 \to g_1, \quad a_2 \to g_2, \quad \ldots, \quad a_n \to g_n$$

into a group G can be extended to a homomorphism of F into G. (This characterization is often used to define the term "free group.") [*Hint:* Given F is free, use Corollary 1.1.2. To show that F is free, map F into the free group $\langle b_1, b_2, \ldots, b_n \rangle$ under $a_\nu \to b_\nu$; conclude that any relator $R(a_\nu)$ is freely equal to the empty word.]

1.5. Tietze Transformations

A group G can have many presentations; for, given a set of generating elements for G (and the corresponding generating symbols) there are many possible sets of defining relators.

As an illustration, let P be the group of permutations on 1, 2, 3; the three cycle (123) and the two cycle (12) form a set of generating elements for P. Under the mapping $a \to (123)$, $b \to (12)$, P has the presentation

(1) $$\langle a, b; a^3, b^2, ab = ba^2 \rangle$$

(see Problem 1.6(a)). P may also be presented by

(2) $$\langle a, b; a^3, b^2, ab = ba^{-1} \rangle$$

under the same mapping; for, since the defining relators (and relations) in (1) are derivable from those in (2) and conversely, (1) and (2) define the same equivalence class group. It is less obvious that P can be presented by

(3) $$\langle a, b; ab^2a^2, a^2(b^2a^3)^4a, b^3a^4ba \rangle$$

under the same mapping; the reader may verify that the defining relators in (3) are derivable from those in (1) and conversely.

In general, if G has two presentations,

(4) $$G = \langle a_1, a_2, \ldots; R_1(a_\nu), R_2(a_\nu), \ldots \rangle$$

and

(5) $$G = \langle a_1, a_2, \ldots; S_1(a_\nu), S_2(a_\nu), \ldots \rangle,$$

under the same mapping, then each of the defining relators in (5) are derivable from those in (4) and conversely; for $S_1(a_\nu)$, $S_2(a_\nu)$, ... are relators and so are derivable from the defining relators $R_1(a_\nu)$, $R_2(a_\nu)$, ... and conversely.

Moreover, other presentations for G can be obtained by using other sets of generating elements for G.

For example, the permutation group P is generated by the two-cycles (13) and (23). Under the mapping $c \to (13)$, $d \to (23)$, P has the presentation

(6) $$\langle c, d; c^2, d^2, (cd)^3 \rangle$$

(see Problem 1.6(b)). Is there any method for changing the presentation (1) into (6)?

In 1908, H. Tietze showed that given a presentation

(7) $$\langle a, b, c, \ldots; P, Q, R, \ldots \rangle$$

for a group G, any other presentation for G can be obtained by a repeated application of the following transformations to (7):

(T1) *If the words S, T, ... are derivable from P, Q, R, ..., then add S, T, ... to the defining relators in* (7).

(T2) *If some of the relators, say, S, T, ..., listed among the defining relators P, Q, R, ... are derivable from the others, delete S, T, ... from the defining relators in* (7).

(T3) *If $K, M, \ldots$ are any words in $a, b, c, \ldots$, then adjoin the symbols $x, y, \ldots$ to the generating symbols in* (7) *and adjoin the relations $x = K$, $y = M, \ldots$ to the defining relators in* (7).

(T4) *If some of the defining relations in* (7) *take the form $p = V$, $q = W, \ldots$ where $p, q, \ldots$ are generators in* (7) *and $V, W, \ldots$ are words in the generators other than $p, q, \ldots$, then delete $p, q, \ldots$ from the generators, delete $p = V$, $q = W, \ldots$ from the defining relations, and replace $p, q, \ldots$ by $V, W, \ldots$ respectively, in the remaining defining relators in* (7).

The transformations (T1), (T2), (T3), and (T4) are called *Tietze transformations*; a Tietze transformation is called *elementary* if it involves the insertion or deletion of one defining relator, or the insertion or deletion of one generator and the corresponding defining relation.

The Tietze transformations do not change the group defined by a presentation. Indeed, suppose (7) presents G under the mapping

$$a \to g,\ b \to h,\ c \to k,\ \ldots. \tag{8}$$

Then applying (T1) or (T2) to (7) yields a presentation for G under the same mapping (8). Applying (T3) to (7) yields a presentation for G under the mapping (8) supplemented by

$$x \to K(g, h, k, \ldots), \quad y \to M(g, h, k, \ldots), \quad \ldots; \tag{9}$$

for, if $N(a, b, c, \ldots, x, y, \ldots)$ is a relator in G under the mapping determined by (8) and (9), then using the relations $x = K$, $y = M, \ldots$, the word N may be changed to a word in $a, b, c, \ldots$ alone which is a relator under (8), and hence can be derived from $P, Q, R, \ldots$; therefore, $P, Q, R, x = K, y = M, \ldots$ is a set of defining relators for G under (8) and (9). Applying (T4) to (7) yields a presentation for G under the restriction of (8) to the generators of (7) remaining in the new presentation; for using (T3) to insert the deleted generators and their corresponding defining relations, we arrive back at (7) (after replacing $V, W, \ldots$ by $p, q, \ldots$ wherever necessary using (T1) and (T2)); since (T3) does not change the group defined by a presentation, neither does (T4).

As an illustration of the use of Tietze transformations, we show that the group

$$\langle a, b, c;\ (ab)^2ab^2\rangle$$

is a free group on two generators. To do this, we introduce the new generators ab and ab^2 by (T3), and obtain

$$\langle a, b, c, x, y;\ (ab)^2ab^2,\ x = ab,\ y = ab^2\rangle.$$

Next we apply (T1) to obtain

$$\langle a, b, c, x, y;\ (ab)^2ab^2,\ x^2y,\ x = ab,\ y = ab^2\rangle$$

and then (T2) to obtain

$$\langle a, b, c, x, y; x^2y, x = ab, y = ab^2\rangle.$$

We now "solve" for a and b, and use (T1) to obtain

$$\langle a, b, c, x, y; x^2y, x = ab, y = ab^2, b = x^{-1}y, a = xy^{-1}x\rangle.$$

Applying (T4) to eliminate a and b, we obtain

$$\langle c, x, y; x^2y, x = (xy^{-1}x)(x^{-1}y), y = (xy^{-1}x)(x^{-1}y)^2\rangle$$

which by (T2) is

$$\langle c, x, y; x^2y\rangle$$

This presentation can be changed by (T1) and (T2) to

$$\langle c, x, y; y = x^{-2}\rangle$$

and hence, applying (T4) again, we have the presentation

$$\langle c, x\rangle,$$

a free group on two generators.

THEOREM 1.5. *Given two presentations for a group* G,

(10) $$G = \langle a_1, a_2, \ldots; R_1(a_\nu), R_2(a_\nu), \ldots\rangle$$

and

(11) $$G = \langle b_1, b_2, \ldots; S_1(b_\mu), S_2(b_\mu), \ldots\rangle,$$

then (11) *can be obtained from* (10) *by a repeated application of the Tietze transformations* (T1), (T2), (T3), *and* (T4).

PROOF. Let (10) be a presentation of G under the mapping

(12) $$a_1 \to g_1, a_2 \to g_2, \ldots,$$

and let (11) be a presentation of G under the mapping

(13) $$b_1 \to h_1, b_2 \to h_2, \ldots.$$

We shall first change (10) by Tietze transformations so that the symbols $b_1, b_2, \ldots$ of (11) appear as generating symbols; for this purpose we wish to express $h_1, h_2, \ldots$ in terms of $g_1, g_2, \ldots$. Since $g_1, g_2, \ldots$ is a set of generating elements for G,

(14) $$h_1 = B_1(g_1, g_2, \ldots), \qquad h_2 = B_2(g_1, g_2, \ldots), \qquad \ldots.$$

Then by (T3), we adjoin the new symbols $b_1, b_2, \ldots$ to the generating symbols in (10), and adjoin the corresponding relations

(15) $$b_1 = B_1(a_1, a_2, \ldots), \qquad b_2 = B_2(a_1, a_2, \ldots), \qquad \ldots,$$

obtaining the presentation

(16) $\langle a_1, a_2, \ldots, b_1, b_2, \ldots, R_1(a_\nu), R_2(a_\nu), \ldots,$

$$b_1 = B_1(a_\nu), b_2 = B_2(a_\nu), \ldots\rangle.$$

Moreover, G is presented by (16) under the mapping

(17) $$a_1 \to g_1, a_2 \to g_2, \ldots, \qquad b_1 \to h_1, b_2 \to h_2, \ldots,$$

determined by (12) and (13).

We wish now to bring the defining relators of (11) into (16). For this purpose we note that

(18) $$S_1(b_1, b_2, \ldots), \quad S_2(b_1, b_2, \ldots), \quad \ldots$$

are relators under (17) since they are relators under (13); hence, (18) may be adjoined to the defining relators in (16) by T1, obtaining the presentation

(19) $\langle a_1, a_2, \ldots, b_1, b_2, \ldots; R_1(a_\nu), R_2(a_\nu), \ldots,$

$$b_1 = B_1(a_\nu), b_2 = B_2(a_\nu), \ldots, S_1(b_\mu), S_2(b_\mu), \ldots\rangle,$$

which presents G under (17).

We wish now to express $a_1, a_2, \ldots$ in terms of $b_1, b_2, \ldots$ so that we can delete $a_1, a_2, \ldots$ from (19); for this purpose we express $g_1, g_2, \ldots$ as words in $h_1, h_2, \ldots$. Since $h_1, h_2, \ldots$ is a set of generating elements of G,

$$g_1 = A_1(h_1, h_2, \ldots), \qquad g_2 = A_2(h_1, h_2, \ldots), \qquad \ldots.$$

Hence, under the mapping (17),

(20) $$a_1 = A_1(b_1, b_2, \ldots), \qquad a_2 = A_2(b_1, b_2, \ldots), \qquad \ldots$$

are relations in G, and hence derivable from the defining relators in (19). Thus, by (T1) we may adjoin the relations (20) to the defining relators in (19), obtaining the presentation

(21) $\langle a_1, a_2, \ldots, b_1, b_2, \ldots; R_1(a_\nu), R_2(a_\nu), \ldots, b_1 = B_1(a_\nu),$

$$b_2 = B_2(a_\nu), \ldots, S_1(b_\mu), S_2(b_\mu), \ldots, a_1 = A_1(b_\mu), a_2 = A_2(b_\mu), \ldots\rangle.$$

Instead of deleting $a_1, a_2, \ldots$ as planned, we observe that (21) is symmetric; hence, we can obtain (21) from (11) by Tietze transformations. Since the inverse of a Tietze transformation is a sequence of Tietze transformations (see Problem 1), (11) can be obtained from (21) by a sequence of Tietze transformations. Thus (11) can be obtained from (10) by a repeated application of Tietze transformations. ◀

COROLLARY 1.5. *If the presentations* (10) *and* (11) *in Theorem* 1.5 *are finite, then* (10) *may be changed into* (11) *by a finite sequence of elementary Tietze transformations.*

PROOF. If the presentations (10) and (11) are finite, then the relations (15), (18), and (20) are finite in number and so may be added one at a time to (10), by elementary Tietze transformations, to obtain (21). Similarly, (11) may be obtained from (21) by a finite number of elementary Tietze transformations. ◀

Although Theorem 1.5 states that any presentation for G can be obtained by Tietze transformations from any other presentation for G, Theorem 1.5 does not give any constructive procedure for deciding in a finite number of steps whether one presentation can be obtained from another by Tietze transformations. Thus Theorem 1.5 does not yield a solution to the isomorphism problem. However, Corollary 1.5 has been used to derive tests for isomorphism (see R. H. Fox, 1954 and Section 3.4). The tests are made by constructing certain mathematical objects (such as polynomials and ideals) associated with a finite presentation; by showing that these objects remain invariant under elementary Tietze transformations, it follows that two presentations defining the same group must have the same objects associated with them.

The Corollary has also been used by M. O. Rabin to show that if two finite presentations define the same group, then the word (transformation) problem is solvable for one of the presentations if and only if it is solvable for the other (see Problems 10 and 11).

Problems for Section 1.5

1. Show that the inverse of a Tietze transformation (T1) is a Tietze transformation (T2), and conversely. Show that the inverse of a Tietze transformation (T3) is a Tietze transformation (T4). Show that the inverse of a Tietze transformation (T4) is a Tietze transformation (T3) followed by a (T1) and then a (T2).

2. Show by means of Tietze transformations that the presentations

$$\langle a, b, c; b^2, (bc)^2 \rangle$$

and

$$\langle x, y, z; y^2, z^2 \rangle$$

define isomorphic groups.

3. Show by means of Tietze transformations that the cyclic group of order mn, where m and n are coprime, has the presentation

$$\langle b, c; b^m, c^n, bc = cb \rangle.$$

[*Hint:* Start with the presentation $\langle a; a^{mn} \rangle$ and adjoin new generators $b = a^n$, $c = a^m$; adjoin b^m, c^n, and $bc = cb$ to the defining relators; since m and n are coprime, $rm + sn = 1$; hence, $a = b^s c^r$ may be adjoined to the defining relators.]

4. Show that the group

$$\langle a, b; aba = bab \rangle$$

is isomorphic to $\langle c, d; c^3 = d^2 \rangle$. [*Hint:* Start with the presentation

$$\langle a, b; aba = bab \rangle$$

and adjoin new generators $c = ab$, $d = aba$; solve for a and b in terms of c and d; apply (T4) to eliminate a and b.]

5. Show by means of Tietze transformations that

$$\langle a, b, c; b(abc^{-1})^2a, c(abc^{-1})^3 \rangle$$

is a free group on two generators. [*Hint:* Let $\langle x, y \rangle$ be the free group on two generators; adjoin the new generators $a = xy$, $b = y^{-1}x$, $c = x^3$; solve for x and y in terms of a, b, and c, and use (T4).]

6. Show how to Tietze transform

$$\langle a, b; a^3, b^2, ab = ba^2 \rangle$$

into

$$\langle c, d; c^2, d^2, (cd)^3 \rangle$$

by adjoining $c = b$, $d = ab$. Also show how to Tietze transform the first presentation into the second by adjoining $c = ab$, $d = ba$.

7. Show how to Tietze transform

$$\langle a, b; P(a, b), Q(a, b), R(a, b) \rangle$$

into

$$\langle x, y; P(x, y^{-1}), Q(x, y^{-1}), R(x, y^{-1}) \rangle$$

as well as

$$\langle x, y; P(xy^k, y), Q(xy^k, y), R(xy^k, y) \rangle.$$

[*Hint:* If a is to be xy^k and b is to be y, then $x = ab^{-k}$, $y = b$.]

8. Show how to Tietze transform

$$G = \langle a, b; a^4, b^4, ba = a^2b^2 \rangle$$

to

$$\langle c, d; c^5, d^4, cd = dc^2 \rangle.$$

In particular, show that G has order twenty. [*Hint:* Let $c = ba$ and $d = b$. To show $(ba)^5 = 1$, first show

$$(ba)^2 = ba \cdot ba = a^2b^2 \cdot ba = a^2b^{-1}a = a^3 \cdot a^{-1}b^{-1}a = a^3 \cdot b^2a^3$$
$$= a \cdot a^2b^2 \cdot a^3 = ab \cdot a^4 = ab.$$

Hence,

$$(ba)^6 = (ab)^3 = a(ba)^2b = a^2b^2 = ba.$$

That G has order twenty follows from Problem 2.17.]

9. Show how to Tietze transform

$$G = \langle a, b; a^5, b^5, ba = a^2b^2 \rangle$$

to

$$\langle c, d; c^{11}, d^5, cd = dc^3 \rangle.$$

In particular, show that G has order fifty-five. [*Hint:* Show $(ba)^{11} = 1$ by first showing $(ba)^3 = ab$ and then $(ba)^{12} = (ab)^4 = a(ba)^3b = ba$. See hint to Problem 8.]

10. Show that if the word (transformation) problem is solvable for the finite presentation

$$G = \langle a_1, a_2, \ldots, a_n; R_1, R_2, \ldots, R_m \rangle,$$

then it is solvable on any presentation which can be obtained from this one by a single elementary Tietze transformation. [*Hint:* Given another presentation for G which can be obtained from the given one by one elementary Tietze transformation, then we can determine by examining its generating symbols whether (T1), (T2), (T3), or (T4) was used. If (T1), (T2), or (T4) was used, then the new generating symbols are included in $a_1, a_2, \ldots a_n$; a word in the new generating symbols determines the identity if and only if it does in the original presentation. If (T3) was used and b is the new generating symbol and $b = U(a_\nu)$ is the corresponding defining relation, then $W(a_\nu, b)$ defines the identity in the new presentation if and only if $W(a_\nu, U(a_\nu))$ does in the original presentation.]

11. Show that if the word (transformation) problem is solvable for the finite presentation

$$G = \langle a_1, a_2, \ldots, a_n; R_1, R_2, \ldots, R_m \rangle$$

and a specific finite sequence of elementary Tietze transformations is given, then the presentation resulting from applying this sequence to the original presentation also has a solvable word (transformation) problem. [*Hint:* See Problem 10.]

(We assume in the following that the generating symbols for all presentations are chosen from the infinite sequence $a_1, a_2, \ldots, a_n, \ldots$.)

12. A relator R is *derivable in r steps* from $R_1, \ldots, R_m$ if R can be obtained from the empty word by a sequence of r steps each of which consists of the insertion or deletion of one of the relators $R_1, \ldots, R_m$ or one of the trivial relators. Show that only finitely many relators are derivable in no more than r steps from $R_1, \ldots, R_m$.

13. The *rank* of an elementary Tietze transformation T of the finite presentation

$$\langle a_{i_1}, a_{i_2}, \ldots, a_{i_k}; R_1, \ldots, R_m \rangle$$

is r under the following conditions: T is a (T1) transformation and the relator adjoined is derivable in no more than r steps from $R_1, \ldots, R_m$; T is a (T2) transformation and the relator deleted is derivable in no more than r steps from the remaining defining relators; T is a (T3) transformation and the generator a_ν and corresponding relation $a_\nu = V(a_{i_j})$ satisfy $\nu \leq r$ and $L(V) \leq r$; T is any (T4) transformation. Show that a given finite presentation has finitely many elementary Tietze transformations of rank r, each of which may be constructed from the presentation. Show that every elementary Tietze transformation of a finite presentation has some rank.

14. Show that the presentations that can be obtained from a given finite presentation by a sequence of r elementary Tietze transformations each of rank r are finite in number, and can be constructed from the presentation.

15. Construct a sequence of finite presentations from a given one, such that every finite presentation which defines the same group as the given one occurs after finitely many steps. [*Hint:* List those presentations which can be obtained from the given one by a sequence of r elementary Tietze transformations each of rank r. Then juxtapose these into a sequence for $r = 1, 2, \ldots$. By Corollary 1.5, every finite presentation for the same group must occur.]

16. Show that if the word (transformation) problem is solvable on one finite presentation for a group G, then it is solvable for any other given finite presentation for G. [*Hint:* Use Problems 15 and 11.]

1.6 Graph of a Group

This section is an introduction to the concept of the graph of a group, and consists largely of definitions. The terminology will not be needed for the later chapters. However, graphical methods have been used explicitly in important papers on the theory of finitely generated groups: see, for example M. Dehn, 1911, 1914; O. Schreier, 1927; R. Baer and F. Levi, 1936; J. H. C. Whitehead 1936a, 1936b; O. H. Keller, 1954; and Howson, 1954. Moreover, they are, in algebraic disguise, the supporting element in proofs given by R. C. Lyndon, 1950, and H. W. Kuhn, 1952. (See especially the discussion in the last section of Kuhn's paper.)

The graph of a group also provides a method by which a group can be visualized; in many cases it suggests an economical algebraic proof for a result. For finite groups of small order the graph of a group can be used in place of its multiplication table; it gives the same information but in a much more efficient way.

We shall first give an intuitive description of the graph of a group in order to elucidate the more abstract definition which is to follow.

Let G be a group which, for the sake of simplicity, we assume to have a presentation with two generators a and b. For each element g_ν of G select a point P_ν (for instance, in a plane or in three space) so that the elements of G are in one-one correspondence with the points selected. We shall join the points P_ν by oriented edges of two different types; (in the drawing of the graph) orientation is indicated by an arrow and the two different types of edges are indicated by two different colors, C_1 and C_2, corresponding to the two generators a and b.

Specifically, suppose

(1) $$g_\nu a = g_\mu. \qquad g_\nu a^{-1} = g_\mu', \qquad g_\nu b = g_\lambda, \qquad g_\nu b^{-1} = g_\lambda'.$$

Then we join P_ν to P_μ by an edge of color C_1 beginning at P_ν and ending at P_μ; we join P_μ' to P_ν by an edge of color C_1, beginning at P_μ' and ending at P_ν; we join P_ν to P_λ by an edge of color C_2 beginning at P_ν and ending at P_λ; we join P_λ' to P_ν by an edge of color C_2 begining at P_λ' and ending at P_ν. Thus exactly one positively oriented edge of each color begins at every point P_ν, and exactly one positively oriented edge of each color ends at every point P_ν. Moreover, the edges do not intersect except in the points P_ν.

This system of points and edges is usually called the *graph* of the group G on the generators a, b. Other names used in the literature are "group diagram" (translated from Dehn's term "Gruppenbild"), Cayley diagram, and color group (this term was introduced by Cayley, and is used in Burnside, 1911).

If, in the graph of the group G, P_0 is the point corresponding to the identity element 1 of G, then any word $W(a, b)$ is uniquely represented by a path of oriented edges starting from P_0. For example, if $W(a, b) = aba^{-1}$, then the path corresponding to $W(a, b)$ consists of the positively oriented C_1-edge going out of P_0 to its other endpoint, say, P_1, followed by the positively oriented C_2-edge going out of P_1 to its endpoint, say, P_2, and finally followed by the negatively oriented C_1-edge going out of P_2 to its other endpoint, say, P_3. Clearly, from the construction of the edges, P_3 is the point corresponding to the element of G defined by $W(a, b)$. In particular, $W(a, b)$ is a relator if and only if the corresponding path is closed.

The problem of constructing the graph of G from a presentation

$$G = \langle a, b; P, Q, R \rangle$$

is obviously equivalent to the word problem for the presentation.

Figure 1 shows the graph of the group

$$G = \langle a, b; a^2, b^3, (ab)^2 \rangle,$$

which is the symmetric group of degree three. Since G has six elements, the graph has six points. Instead of colors, solid and broken lines are used: the solid lines are edges corresponding to a; the broken lines are edges corresponding to b. The elements of G are defined by the words

$$1, a, b, ab, b^2, ab^2.$$

From the graph we see readily that G is not Abelian; for the paths corresponding to ab and ba lead from 1 to different points. The graph may be used to find the element of G defined by a given word; for example,

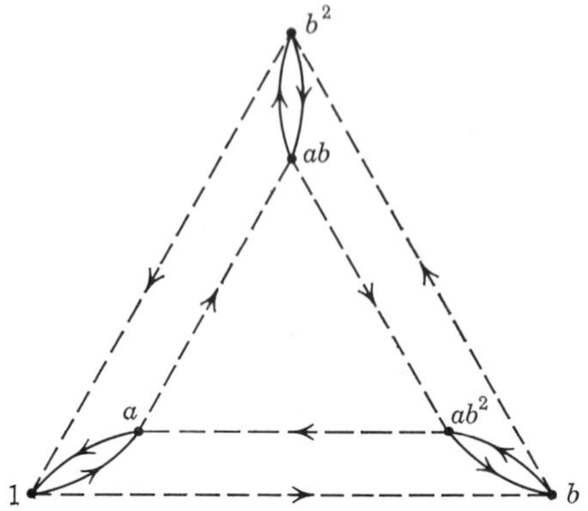

Figure 1

since the path corresponding to $aba^{-1}b^{-1}$ leads from 1 to b, it follows that $aba^{-1}b^{-1} = b$, i.e., $aba^{-1}b^{-1}$ and b define the same element of G.

Figure 2 shows the singular graph S_2 with one point and four edges; this is the graph of the identity group presented by $\langle a, b; a, b\rangle$. Every path in this graph is closed, corresponding to the fact that every word in a and b is a relator.

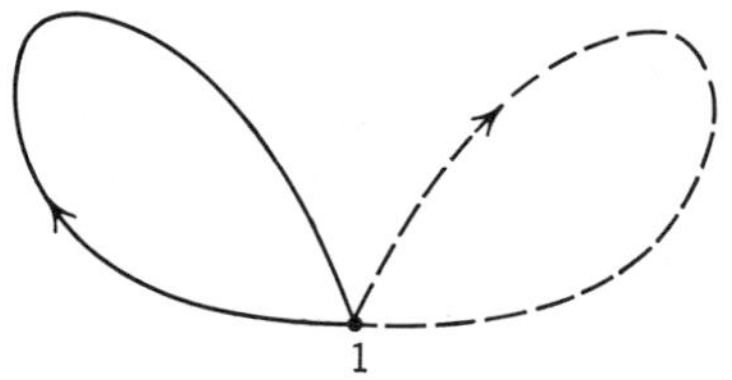

Figure 2

Figure 3 is the graph of the free group $F_2 = \langle a, b\rangle$. Its points are denoted by the freely reduced words which uniquely define the elements of F_2. The graph has infinitely many points. It is a so-called *tree*, which means that it does not contain any non-trivial closed paths. (Here the term "non-trivial" is used in the sense of "reduced" as defined below.)

We shall now give some definitions needed for a more precise discussion of the graph of a group.

A *graph* or *one-dimensional complex* is a set of two types of elements called *points* and *edges* which satisfy the following postulates:

(G1) *To every edge E, there is associated a uniquely determined ordered pair P, P' of points (not necessarily distinct) which are called the boundary points of E. P is called the initial point and P' is called the end point of E.*

(G2) *To every edge E, there is associated a unique edge $E^{-1} \neq E$ which is called the inverse of E and for which $(E^{-1})^{-1} = E$.*

(G3) *If E begins at P and ends at P', then E^{-1} begins at P' and ends at P.*

(In the usual interpretations, of course, E^{-1} is simply the edge which differs from E only in its orientation. However, see Problem 2.)

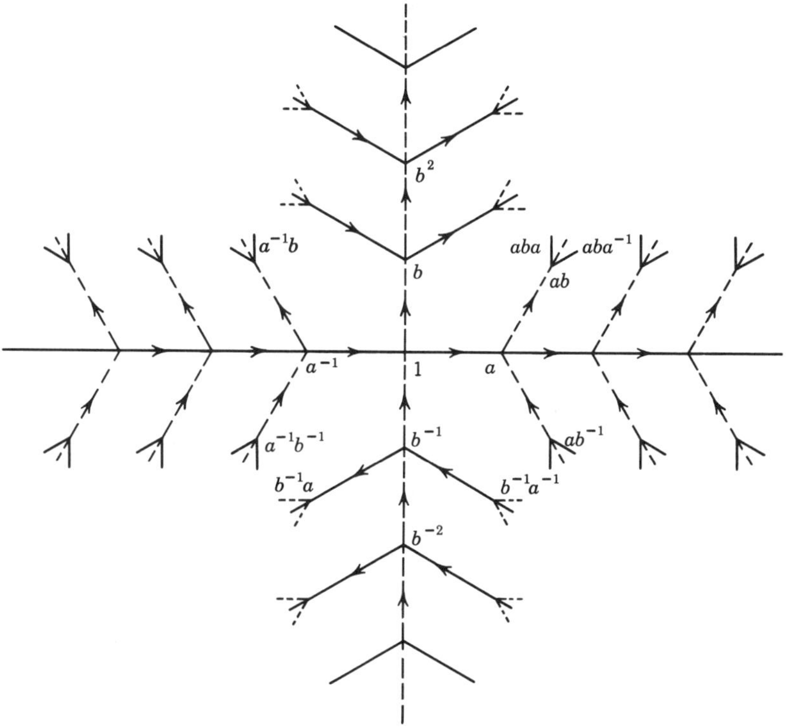

Figure 3

It should be noted that in a graph the points P, P' may be the initial and end points of many different edges. In the special case where the graph consists of a single point which is the initial and end point of all edges, the graph is called *singular*.

Two *graphs*, Γ and Γ^*, are called *isomorphic* if there exists a one-one mapping T of the points and edges of Γ onto the points and edges respectively of Γ^*, which preserves the relations "is the initial point of," "is the end point of," and "is the inverse of."

Any two singular graphs with the same number of edges are isomorphic (see Problem 3). S_n denotes a *singular graph with* $2n$ edges, $s_1, s_1^{-1}, s_2, s_2^{-1}, \ldots, s_n, s_n^{-1}$.

A *path* π_0 in a graph is defined as a sequence of edges

$$\pi_0 = E_1 E_2 \ldots E_m$$

such that, for $\mu = 1, 2, \ldots, m - 1$, $E_{\mu+1}$ begins at the end point of E_μ. The *initial point* of π_0 is the *initial point* of E_1, and the *end point* of π_0 is the end point of E_m. We admit, for each point, a path of length zero, called an *empty path*, which has that point as both its initial and end point.

Since the singular graph S_n has only a single point and has the $2n$ edges $s_1, s_1^{-1}, \ldots, s_n, s_n^{-1}$, a path in S_n is simply a word in $s_1, \ldots, s_n$.

If

$$\pi_0 = E_1 \ldots E_m, \qquad \pi_1 = E_{m+1} \ldots E_n,$$

and the end point of π_0 is the initial point of π_1, we define their product to be the path

$$\pi_0 \pi_1 = E_1 \ldots E_m E_{m+1} \ldots E_n.$$

We define the *inverse* π_0^{-1} of the path π_0 to be the path $E_m^{-1} \ldots E_1^{-1}$, and the *inverse of an empty path* to be the path itself.

A path is *closed* if its initial point coincides with its end point.

We call a path *reduced* if no two consecutive edges of it are inverse to one another. It is clear that if $\pi_0 = \pi_1 E E^{-1} \pi_2$, then $\pi_1 \pi_2$ is a path with the same initial and end points as π_0. Therefore, starting with any path, we can always arrive at a reduced path by a finite number of deletions of consecutive edges which are inverse to one another.

A graph Γ is called *connected* if, for any two of its points, P, Q, there exists a path whose initial point is P and whose end point is Q.

We now define the *graph* Γ *of a group* G *presented on a given set of generators* a_ν. We take as *points of* Γ, the elements of G.

It is clear from the intuitive discussion earlier in the section that an edge of Γ is determined by its initial and end point, its color (i.e., the generator it is associated with), and its orientation (i.e., whether its arrow is directed toward its initial or end point). Hence, an edge is described completely by the triple $(g_1, g_2; a_\nu^\epsilon)$, where g_1 and g_2 (which equals $g_1 a_\nu^\epsilon$) are its initial and end point respectively, a_ν is its associated generator, and $\epsilon = \pm 1$ specifies its orientation.

We therefore take as the *edges of* Γ the set of triples $(g_1, g_2; a_\nu^\epsilon)$ where g_1, g_2 are in G, a_ν is one of the given generating symbols, $\epsilon = \pm 1$, and $g_2 = g_1 a_\nu^\epsilon$. If $E = (g_1, g_2; a_\nu^\epsilon)$, then the *initial point* of E is defined as g_1, the *end point* as g_2, and the *inverse* of E as $(g_2, g_1; a_\nu^{-\epsilon})$.

If we wish to embed Γ in a particular space, for example, the Euclidean plane or three space or the non-Euclidean hyperbolic plane, we represent the points of Γ by points in the space, and the edges of Γ by Jordan curves.

It is easily seen that the graph Γ of a group satisfies the postulates for a graph, (G1), (G2), and (G3). Two questions naturally arise: When is a graph isomorphic to the graph of a group? Given a graph isomorphic to the graph of a group, how can we recover the group?

Let us consider some necessary conditions that a graph Γ be isomorphic to the graph of a group. The graph of a group G is connected; for, any element g of G is defined by a product $g = a_{\nu_1}^{\epsilon_1}a_{\nu_2}^{\epsilon_2} \ldots a_{\nu_r}^{\epsilon_r}$. If $g_j = a_{\nu_1}^{\epsilon_1}a_{\nu_2}^{\epsilon_2} \ldots a_{\nu_j}^{\epsilon_j}$ and $E_j = (g_{j-1}, g_j; a_{\nu_j}^{\epsilon_j})$, then the path $E_1E_2 \ldots E_r$ joins 1 to g. Thus a graph Γ isomorphic to the graph of a group must be connected.

The graph of a group has "colored and oriented" edges; hence, the graph Γ should be capable of having its edges "colored and oriented." To make this notion precise, we make use of singular graphs.

A singular graph S_n with $2n$ edges.

$$s_1, s_1^{-1}, s_2, s_2^{-1}, \ldots, s_n, s_n^{-1},$$

is the simplest example of a graph whose edges may be "colored with n colors and oriented"; namely, we interpret s_ν as the positively oriented edge of a certain color and s_ν^{-1} as the negatively oriented edge of the same color. In general, a graph can be "colored with n colors and oriented" by associating with each of its edges, an edge of S_n. This intuitive idea leads to the following definition.

A *coloring (with n colors) and orientation* of a graph Γ is a mapping M of the edges of Γ into the edges of S_n with the following properties:

(C1) *For each point P in Γ, the edges of Γ with initial point P are mapped one-one onto all edges of S_n.*

(C2) *For each edge E in Γ, $M(E^{-1}) = [M(E)]^{-1}$.*

Condition (C1) says that one edge of every color and orientation begins at each point P of Γ. Condition (C2) says that the color of E^{-1} is the same as that of E, but the orientation is reversed.

Clearly, the graph of a group on n generators $a_1, a_2, \ldots, a_n$ has a coloring and orientation; namely, map the edge $(g_\mu, g_\lambda; a_\nu^\epsilon)$ into s_ν^ϵ. However, not every graph which admits a coloring and orientation is the graph of a group. For example, Figure 4 shows a graph "colored and oriented" which cannot be isomorphic to the graph of a group. To show this we use the following definitions:

If M is a coloring and orientation of Γ and

$$\pi = E_1 \ldots E_r$$

is a path in Γ, we define $M(\pi) = M(E_1) \ldots M(E_n)$, and we say that the path π *covers* the path $M(\pi)$.

A coloring and orientation M of the graph Γ is called *regular* if for any two paths π and π' of Γ such that $M(\pi) = M(\pi')$, π is closed if and only if π' is closed.

The coloring and orientation of the graph Γ^* given in Figure 4 is not regular; for the paths consisting of the positively oriented solid edge with

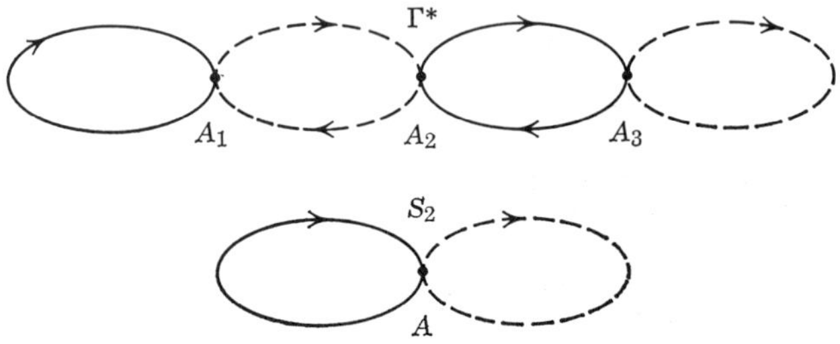

Figure 4

initial point A_1 and the positively oriented solid edge with initial point A_2 cover the same path in S_2, although the first is closed and the second is not.

On the other hand, the coloring and orientation of the graph of a group G on the generators $a_1, a_2, \ldots, a_n$ is regular. For if

$$\pi = E_1 \ldots E_r, \qquad \pi' = E_1' \ldots E_r',$$

where $E_j = (g_{j-1}, g_j; a_{\nu_j}^{\epsilon_j})$ and $E_j' = (h_{j-1}, h_j; a_{\mu_j}^{\delta_j})$, and $M(\pi) = M(\pi')$, then $s_{\nu_j}^{\epsilon_j} = s_{\mu_j}^{\delta_j}$, and so $\nu_j = \mu_j$ and $\epsilon_j = \delta_j$. Now π is closed if and only if $g_0 = g_r = g_0 a_{\nu_1}^{\epsilon_1} \ldots a_{\nu_r}^{\epsilon_r}$, i.e., if and only if $a_{\nu_1}^{\epsilon_1} \ldots a_{\nu_r}^{\epsilon_r}$ is a relator. Similarly, π' is closed if and only if $a_{\mu_1}^{\delta_1} \ldots a_{\mu_r}^{\delta_r}$ is a relator. Thus π is closed if and only if π' is.

The connectedness of a graph Γ and the existence of a regular coloring and orientation of Γ are necessary and sufficient conditions for Γ to be isomorphic to the graph of a group. To prove this we require the following lemma:

Lemma 1.1. *If M is a regular coloring (with n colors) and orientation of Γ then for each point P in Γ, M is a one-one correspondence between the paths in Γ with initial point P and all paths in S_n (i.e., words in $s_1, \ldots, s_n$).*

Proof. Since M preserves the number of edges in a path, it suffices to show the following: if

$$\pi = E_1 \ldots E_r, \qquad \pi' = E_1' \ldots E_r',$$

π and π' have initial point P, and $M(\pi) = M(\pi')$, then $\pi = \pi'$; and if $\sigma = s_{\nu_1}^{\epsilon_1} \ldots s_{\nu_r}^{\epsilon_r}$ is any path in S_n, then there is some path $\pi = E_1 \ldots E_r$ in Γ with initial point P such that $M(\pi) = \sigma$.

We obtain both results by induction on r. For $r = 1$, the result follows from (C1). Assume both results hold for r.

Suppose $M(E_1 \ldots E_r E) = M(E_1' \ldots E_r' E')$. Then $M(E_1) = M(E_1')$ and so $E_1 = E_1'$. Moreover, $M(E_2 \ldots E_r E) = M(E_2' \ldots E_r' E')$ and the initial points of $E_2 \ldots E_r E$ and $E_2' \ldots E_r' E'$, which are the end points E_1 and E_1', respectively, coincide. Hence, by inductive assumption, $E_2 \ldots E_r E = E_2' \ldots E_r' E'$ and so $E_1 E_2 \ldots E_r E = E_1' E_2' \ldots E_r' E'$.

Suppose now that $\sigma = s_{\nu_1}^{\epsilon_1} \ldots s_{\nu_r}^{\epsilon_r} s_\nu^{\epsilon}$. By inductive hypothesis there exists a path $\pi = E_1 \ldots E_r$ in Γ with initial point P such that $M(\pi) = s_{\nu_1}^{\epsilon_1} \ldots s_{\nu_r}^{\epsilon_r}$. If Q is the end point of π, let E be the edge with initial point Q which covers s_ν^{ϵ}. Then $E_1 \ldots E_r E$ is a path in Γ with initial point P such that $M(E_1 \ldots E_r E) = \sigma$. ◄

THEOREM 1.6. *Let Γ be a connected graph with a regular coloring of n colors and orientation M. Then Γ is isomorphic to the graph of a group G on n generators $a_1, \ldots, a_n$. The words $W(a_\nu)$ in $a_1, \ldots, a_n$ are in one-one correspondence with the paths π in Γ having a fixed point P_0 as initial point, under the mapping $W(a_\nu) \to \pi$, where π covers $W(s_\nu)$. Moreover, the relators of G are precisely those words $R(a_\nu)$ which correspond to closed paths in Γ.*

PROOF. To construct the group G, we give a presentation using $a_1, \ldots, a_n$ as the generating symbols; all the words $R(a_\nu)$, whose corresponding path in Γ (under $W(a_\nu) \to \pi$, where π covers $W(s_\nu)$) is closed, constitute the set of defining relators. G is then the equivalence class group of the given presentation.

To show that the graph of G on $a_1, \ldots, a_n$ is isomorphic to Γ (in such a way that coloring and orientation is preserved) we must map the elements of G onto the points of Γ and the edges $(g_\mu, g_\lambda; a_\nu^{\epsilon})$ onto the edges of Γ.

For this purpose, let g_μ be the element of G defined by a word $W_\mu(a_\nu)$, and let π_μ be the path in Γ with initial point P_0 corresponding to W_μ; then we map g_μ into P_μ, the end point of π_μ.

This mapping is well-defined; for suppose that $W_\mu'(a_\nu)$ is another word defining g_μ. Then $W_\mu'(a_\nu)$ can be obtained from $W_\mu(a_\nu)$ by a finite number of insertions or deletions of the defining relators $R(a_\nu)$ or the trivial relators. We shall show that the paths corresponding to the words

(1) $$K(a_\nu)T(a_\nu),\ K(a_\nu)a_\lambda^{\epsilon}a_\lambda^{-\epsilon}T(a_\nu),\ K(a_\nu)R(a_\nu)T(a_\nu)$$

have the same end point.

Now the path π corresponding to a product $U(a_\nu)V(a_\nu)$ is obtained as follows: let π_1 be the path with initial point P_0 covering $U(s_\nu)$ and let π_2 be the path with initial point at the end of π_1 and covering $V(s_\nu)$; then $\pi = \pi_1\pi_2$.

Suppose, then, that $K(a_\nu) \to \pi_1'$. If the end point of π_1' is Q,

$$K(a_\nu)T(a_\nu) \to \pi_1'\pi_2',$$

where π_2' is the unique path with initial point Q covering $T(s_\nu)$. Since $s_\lambda^\epsilon s_\lambda^{-\epsilon}$ and $R(s_\nu)$ are covered by closed paths with initial point P_0, and Γ has a regular coloring and orientation, the paths π', π'' with initial point Q that cover $s_\lambda^\epsilon s_\lambda^{-\epsilon}$ and $R(s_\nu)$, respectively, are closed. Hence,

$$K(a_\nu)a_\lambda^\epsilon a_\lambda^{-\epsilon}T(a_\nu) \to \pi_1'\pi'\pi_2'$$

and

$$K(a_\nu)R(a_\nu)T(a_\nu) \to \pi_1'\pi''\pi_2'.$$

Thus the words (1) correspond to paths with the same end point. Hence, the mapping g_μ into P_μ is well defined.

In addition to mapping g_μ into P_μ, we map the edge $(g_\mu, g_\lambda; a_\nu^\epsilon)$ in the graph of G into the edge E of Γ with initial point P_μ which covers s_ν^ϵ.

We show now that the mapping $g_\mu \to P_\mu$ and $(g_\mu, g_\lambda; a_\nu^\epsilon) \to E$ is an isomorphism between the graph of G and the graph Γ.

Since Γ is connected there is a path π from P_0 to any point P of Γ; if π covers $W(s_\nu)$, then the element g of G defined by $W(a_\nu)$ is mapped into P. Hence, the mapping of points is onto.

Let g_μ and g_μ' be mapped into the same point P of Γ. If $W_\mu(a_\nu)$, $W_\mu'(a_\nu)$ are words defining g_μ, g_μ', respectively, with corresponding paths π_μ, π_μ', then $\pi_\mu'\pi_\mu^{-1}$ is a closed path with initial point P_0. Hence, $W_\mu'W_\mu^{-1}$ is a relator of G, and so $g_\mu = g_\mu'$. Thus the mapping of points is one-one.

Showing that the mapping of edges is one-one onto, and that "initial point," "end point," "inverse," and "color and orientation" are preserved, is straightforward, and is left to the reader (see Problem 5). ◀

In the remainder of this section we briefly discuss the connection between the graph of a group and the fundamental group of certain topological spaces. For the definition of a fundamental group we refer the reader to any standard topology text. (See, for example, Hu, 1959, and Chapter 8 in L. Pontrajagin, 1946.)

The free group F_n is the fundamental group of the one-dimensional complex S_n, the singular graph with n edges. Moreover, the group G of Theorem 1.6 can be considered as the fundamental group of a two-dimensional complex C_2 which is derived from S_n by spanning two-cells into all paths of S_n which are covered by closed paths in Γ. (It suffices to span two-cells into those paths of S_n which correspond to a set of defining relators for G on the generators $a_1, \ldots, a_n$; in many cases this reduces the number of two-cells that are needed, to finitely many.)

The fundamental groups of two-dimensional manifolds, especially orientable ones, have been studied in great detail. (For a review of some results in this connection, see Sections 3.7 and 6.1.) These groups also provide examples of discontinuous groups of motions in a space S. For groups of this type the graph of the group can be easily constructed if the covering of the space S by the maps of the fundamental region of the group is given.

For the general theory of discontinuous groups of motions in the non-Euclidean hyperbolic plane, see F. Klein and R. Fricke, 1890, R. Fricke and F. Klein, 1897 (in which many illustrations appear), and L. R. Ford, 1929. We shall confine ourselves to giving an example of such a discontinuous group.

Figure 5 shows part of the hyperbolic plane which is represented as the interior of the unit circle. The dotted lines are circles orthogonal to the unit circle and represent non-Euclidean straight lines. Figure 5 shows a triangle with angles $2\pi/7$, $\pi/3$, $\pi/3$ and some of its images under a group M of non-Euclidean motions. M can be presented by means of two generators a and b and the defining relators.

$$a^7,\ b^2,\ (ab)^3.$$

(M is the non-Euclidean rotation group generated by $a = xy$, $b = yz$, where x is reflection in one of the legs, y is reflection in the altitude to the base, and z is reflection in the base of the given triangle. Now, the product of two reflections in lines intersecting at an angle α is a rotation through an angle 2α. Since the angle between a leg and altitude is $\pi/7$, between altitude and base is $\pi/2$, and between leg and base is $\pi/3$, we have $a^7 = 1$, $b^2 = 1$ and $(ab)^3 = (xz)^3 = 1$.)

If we mark the midpoint of the altitude in one triangle and its images under M within each of the other triangles, and join the points thus obtained across the boundary lines of neighboring triangles we obtain the graph of M represented in the non-Euclidean plane. Since Figure 5 has so many lines in it, we have represented the edges corresponding to the generator a by a solid line with an arrow, and the edges corresponding to the generator b by a solid line without an arrow. Moreover, since $b^2 = 1$ we have coalesced the edges corresponding to b and b^{-1}. Since M is of infinite order only part of its graph is drawn.

The definitions given in this section are essentially special cases of those given by Reidemeister, 1932a. Condition (C1) is sometimes broken up into several weaker conditions. What is called a "colored and oriented graph" here would then have to be called an unramified unbounded covering graph of S_n. Reidemeister, 1932a, also discusses a generalization of the graph of a group (the graph of the set of cosets of a subgroup), and the relation between the fundamental group of a two dimensional complex and its covering complexes.

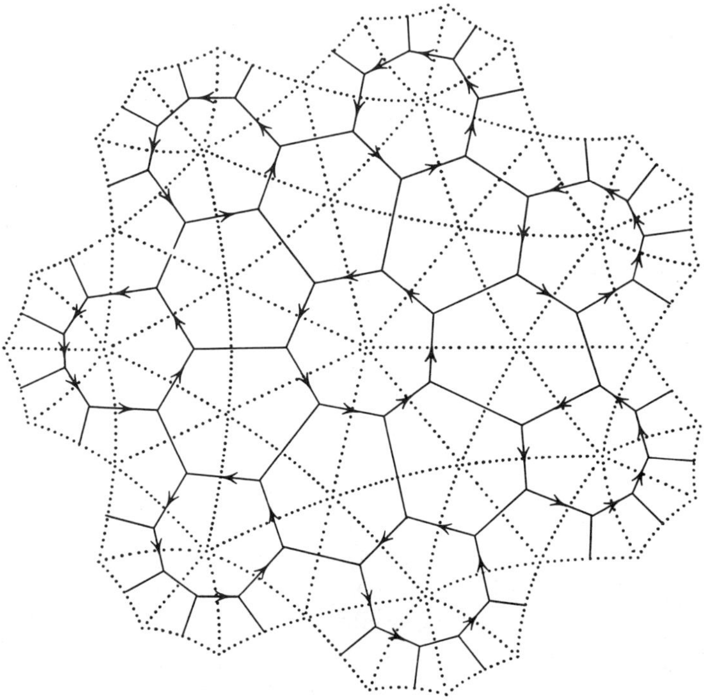

Figure 5

For the general concept of the fundamental group of an n-dimensional manifold see textbooks and monographs on topology; for the special case of the group of a knot or link see Reidemeister, 1932b, and Crowell and Fox, 1963.

The methods used by J. H. C. Whitehead, 1936a and 1936b, for group theoretical investigations involve high powered topological theorems which cannot even be indicated here. A review of some of his results is given in Section 3.5.

For additional examples of graphs of groups and for various problems connected with them see M. Dehn 1911, 1912, and 1914; H. Maschke, 1896; A. Hurwitz, 1893; W. Threlfall, 1932; R. P. Baker, 1931; H. R. Brahana, 1926; and Coxeter and Moser, 1965.

Problems for Section 1.6

1. Let Γ be a graph.
   (a) Show that if the number of edges is finite and Γ is connected, then the number of points is finite.
   (b) Show that if the number of edges is finite, then the number of edges is even.
   (c) Show that if the number of edges is finite, then there is an even number of points which are initial points of exactly an odd number of edges.

[*Hint:* For (b), partition the edges into sets consisting of an edge and its inverse. For (c), count the number of ordered pairs (P, E), where P is the initial point of edge E, in two ways: first by collecting according to the point, and then by collecting according to the edge.]

2. Let K be a circle. Let P and Q be diametrically opposite points of K. Let R and S be a different pair of diametrically opposite points of K. Show that the following points and edges form a graph, i.e., satisfy (G1), (G2), and (G3): the points are P, Q; the edges are $E_1 = \text{arc }(PRQ)$, $E_2 = \text{arc }(QSP)$, $E_3 = \text{arc }(PSQ)$, $E_4 = \text{arc }(QRP)$; P is the initial point of E_1, E_3, and P is the end point of E_2, E_4, and contrariwise for Q; $E_1^{-1} = E_2$, $E_2^{-1} = E_1$, $E_3^{-1} = E_4$, $E_4^{-1} = E_3$.

3. Show that any two singular graphs with the same number of edges are isomorphic.

4. Let Γ and Γ^* be graphs. Show that if T is a mapping of the points of Γ one-one onto the points of Γ^*, and of the edges of Γ one-one onto the edges of Γ^* such that the relations "is the initial point of" and "is the inverse of" are preserved, then T is an isomorphism of Γ onto Γ^*.

5. Let G, Γ, and the mapping $g_\mu \to P_\mu$ and $(g_\mu, g_\lambda; a_\nu^\epsilon) \to E$, be as in the proof of Theorem 1.6. Show that the mapping preserves the relations "is the initial point of," "is the inverse of," and moreover, show that $(g_\mu, g_\lambda; a_\nu^\epsilon)$ and E have the same "color and orientation." Show the mapping is one-one onto the edges.

6. Show that the graph of the symmetric group of degree four presented by $\langle a, b; a^3, b^4, (ab)^2\rangle$ can be represented in the plane by the colored oriented graph in Figure 6. Use the graph to show that $aba^2b^2a = ba^2b^2$, ba^2bab^2a and $a^2bab^{-2}ab$ are relators; find the orders of aba^2b^2, a^2b^2, and ab^3a^2.

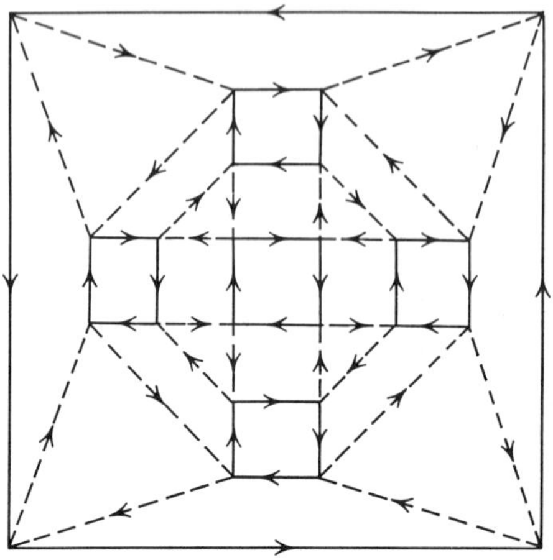

Figure 6

7. Show that the direct product of the group of symmetries of the square and the cyclic group of order two has the presentation

$$G = \langle a, b, c; a^2, b^2, (ab)^4, c^2, ac = ca, bc = cb\rangle.$$

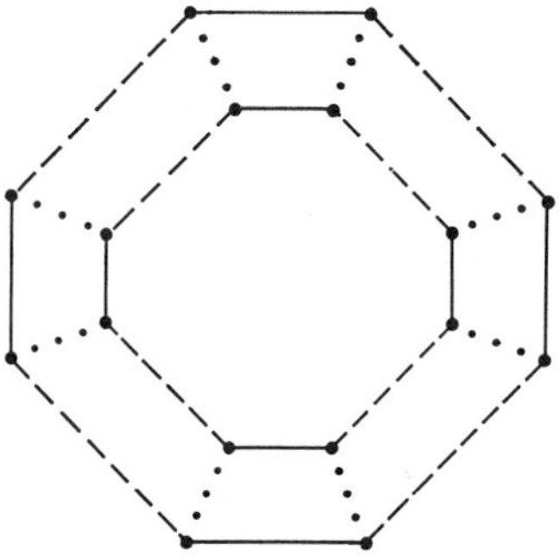

Figure 7

Show that the graph of G can be represented in the plane by the color graph in Figure 7, where the solid, broken, and dotted lines are for a, b, and c, respectively. (Note that since a has order two, the edges representing a and a^{-1} have been coalesced, and similarly, for b and c.) Find the orders of bac, $abca$, and bab, using the graph.

8. Represent the graph of the dihedral groups on the presentations $\langle a, b; a^2, b^n, (ab)^2\rangle$ and $\langle c, d; c^2, d^2, (cd)^n\rangle$ in the plane. Picture the edges corresponding to x and x^{-1}, where x is a generator of order two, uncoalesced; then picture them coalesced.

9. Represent the graph of the cyclic group of order six on the presentations $\langle a; a^6\rangle$ and $\langle b, c; b^3, c^2, bc = cb\rangle$ in the plane.

10. Represent the graph of the following presented groups in the stated space:

(a) $\langle a, b; ab = ba\rangle$, in the Euclidean plane;
(b) $\langle a, b; b^n, ab = ba\rangle$, on the surface of an infinite circular cylinder;
(c) $\langle a, b; a^m, b^n, ab = ba\rangle$, on the surface of a torus;
(d) $\langle a, b, c; ab = ba, ac = ca, bc = cb\rangle$, in Euclidean 3-space;
(e) $\langle a, b; a^2, b^3\rangle$, in the Euclidean plane;
(f) $\langle a, b; a^2, b^n\rangle$, in the Euclidean plane;

(g) $\langle a, b; a^3, b^3\rangle$, in the Euclidean plane;
(h) $\langle a, b; a^3, b^n\rangle$, in the Euclidean plane;
(i) $\langle a, b; a^m, b^n\rangle$, in the Euclidean plane;
(j) $\langle a, b, c; a^4, b^3, c^2\rangle$, in Euclidean 3-space.

11. Represent the graph of the group

$$\langle a, b, c; a^3, b^3, c = b^{-1}a^{-1}ba, ac = ca, bc = cb\rangle$$

in the plane. (*Note:* It may be impossible to draw a graph in the plane without crossing of edges which may have no common points in the graph of the group.) [*Hint:* See Problem 2.21.]

12. Let Γ be the graph of a group G on generators $a_1, a_2, \ldots, a_n$. Show that if the edges in Γ corresponding to a_1 are deleted, then Γ decomposes into disjoint isomorphic connected subgraphs. Show that each of the subgraphs consists of the elements in a left coset of H, the subgroup of G generated by $a_2, \ldots, a_n$. Show that H is a normal subgroup of G if and only if every positively oriented edge corresponding to a_1 in Γ leads from points of one subgraph precisely to the points of one other subgraph. Verify these results in the graphs of Figures 1, 2, 3, 5, 6, and 7.

13. Let Γ be the graph of a group G. Show that left translation L_g by the element g, i.e.,

$$g_\mu \to gg_\mu \quad \text{and} \quad (g_\mu, g_\lambda; a_\nu^\epsilon) \to (gg_\mu, gg_\lambda; a_\nu^\epsilon),$$

is an isomorphism of Γ onto itself which preserves color and orientation. Show that any isomorphism T of Γ onto itself which preserves color and orientation coincides with some left translation. Show that G is isomorphic to the isomorphisms of Γ that preserve color and orientation, under the mapping $g \to L_{g^{-1}}$. Conclude that if Γ^* is a connected regular colored and oriented graph, then the isomorphisms of Γ^* which preserve color and orientation form a group whose graph is isomorphic to Γ^*.

14. Let Γ be a graph of the group G on the generators $a_1, a_2, \ldots, a_n$, and let H be a subgroup of G. Form the graph Γ^* whose points are Hg, the right cosets of G mod H, and whose edges are the triples $(Hg_\mu, Hg_\lambda; a_\nu^\epsilon)$, where $Hg_\mu a_\nu^\epsilon = Hg_\lambda$; the initial point of a triple is the first entry and the end point is the second entry; the inverse of $(Hg_\mu, Hg_\lambda; a_\nu^\epsilon)$ is $(Hg_\lambda, Hg_\mu; a_\nu^{-\epsilon})$. Show that Γ^* satisfies (G1), (G2), and (G3). Show that the mapping

$$(Hg_\mu, Hg_\lambda; a_\nu^\epsilon) \to s_\nu^\epsilon$$

is a coloring with n colors and orientation of Γ^*. Show that the coloring and orientation of Γ^* is regular if and only if H is a normal subgroup of G. (Γ^* is called the coset graph of G mod H.)

15. Let Γ be a colored and oriented graph with n colors, which is connected. Show Γ is regular if and only if there is an isomorphism of Γ into itself, which preserves color and orientation, mapping any point P of Γ into any other point Q of Γ. [*Hint:* If α is an isomorphism of Γ, and π is a path, then $\alpha(\pi)$ is closed if and only if π is. Also use Theorem 1.6 and Problem 13.]

16. Let the group G be generated by $a_1, \ldots, a_n$ and let the mapping $a_\nu \to b_\nu$ induce a homomorphism α of G onto a group H. Show that the graph of G on $a_1, \ldots, a_n$ covers the graph of H on $b_1, \ldots, b_n$, i.e., there is a mapping of points onto points and edges onto edges which preserves the relations "is the initial point of," "is the end point of," and "is the inverse of"; moreover, color and orientation are preserved.

Chapter 2

Factor Groups and Subgroups

2.1. Factor Groups

In this section we shall show how to obtain a presentation for a factor group G/N from a presentation for G,

(1) $$G = \langle a, b, c, \ldots ; P, Q, R, \ldots \rangle.$$

In order to do this we must have the normal subgroup N given in terms of the presentation (1). Specifically, we shall assume that N is defined as the normal subgroup of G generated by the words $S(a, b, c, \ldots)$, $T(a, b, c, \ldots)$, $\ldots$ (or, more properly, by the elements of G defined by these words). (Recall that the *normal subgroup of a group generated by a set of elements* is the smallest normal subgroup containing these elements, or equivalently, it is the subgroup generated by the set of elements and their conjugates.) We may then assert the following:

THEOREM 2.1. *Let G have the presentation (1) under the mapping*

(2) $$a \to g, \quad b \to h, \quad c \to k, \quad \ldots,$$

and let N be the normal subgroup of G generated by $S(g, h, k, \ldots)$, $T(g, h, k, \ldots)$, ... Then the factor group G/N has the presentation

(3) $$\langle a, b, c, \ldots ; P, Q, R, \ldots, S, T, \ldots \rangle$$

under the mapping

(4) $$a \to gN, \quad b \to hN, \quad c \to kN, \quad \ldots .$$

PROOF. Under the mapping (4), the word $W(a, b, c, \ldots)$ is sent into $W(gN, hN, kN, \ldots) = W(g, h, k, \ldots)N$. Since G has the presentation (1) and N is the normal subgroup generated by $S(g, h, k, \ldots)$, $T(g, h, k, \ldots)$, $\ldots$, the elements of G defined by $P, Q, R, \ldots, S, T, \ldots$ are all in N. Thus $P, Q, R, \ldots, S, T, \ldots$ are relators in G/N under (4), and so by Corollary 1.1.3, (4) induces a homomorphism into G/N.

Since G is generated by $g, h, k, \ldots, G/N$ is generated by $gN, hN, kN, \ldots$ and so the homomorphism induced by (4) is onto.

To show that this homomorphism is one-one, suppose $W(a, b, c, \ldots)$ is mapped into the identity in G/N. Then $W(g, h, k, \ldots)$ is in N. Since N is the normal subgroup of G generated by $S, T, \ldots,$

$$W(g, h, k, \ldots) = U_1V_1U_1^{-1} \cdot \ldots \cdot U_rV_rU_r^{-1} \tag{5}$$

where U_i are elements of G and V_i is one of $S, S^{-1}, T, T^{-1}, \ldots$. Hence,

$$W(a, b, c, \ldots) \sim U_1V_1U_1^{-1} \cdot \ldots \cdot U_rV_rU_r^{-1}, \tag{6}$$

with respect to the presentation (1) for G. Thus $W(a, b, c, \ldots)$ may be changed to the right-hand side of (6) by insertions and deletions of the defining relators in (1) and trivial relators. But the right-hand side of (6) may be reduced to the empty word by insertions and deletions of $S, T, \ldots$ and trivial relators. Thus $W(a, b, c, \ldots) \sim 1$ in the presentation (3), and hence, the homomorphism induced by (4) is one-one from (3) to G/N. ◀

COROLLARY 2.1. *If F is the free group on $a, b, c, \ldots$ and N is the normal subgroup of F generated by $P(a, b, c, \ldots)$, $Q(a, b, c, \ldots)$, $R(a, b, c, \ldots)$, $\ldots$, then*

$$F/N = \langle a, b, c, \ldots; P, Q, R, \ldots\rangle. \tag{7}$$

PROOF. This follows immediately from Theorem 2.1. ◀

[Some authors define the free group without first defining the general presentation (1); they then take (7) as the definition of the presentation (1).]

Problems for Section 2.1

1. Show that if two words in a and b are freely equal, their lengths have the same parity. Let F be the free group on a and b, and let N be the set of words of even length. Show that N is a normal subgroup of F. Show that N is the normal subgroup of F generated by a^2 and ab. Express ba and b^2 as a product of conjugates of a^2 and ab and their inverses. Show that N is not the normal subgroup of F generated by any single word $R(a, b)$. [*Hint:* Let M be the normal subgroup of F generated by a^2 and ab. Show that M has index two in F by showing F/M has order two. Show F/N has index two by showing its cosets are N and aN. Conclude since $M \subset N$, that $M = N$. To show that N is not the normal subgroup generated by a single word $R(a, b)$, make $\langle a, b; R(a, b)\rangle$ Abelian, i.e., consider $G = \langle a, b; R(a, b), ab = ba\rangle$. Using $ab = ba$, $R(a, b)$ is equivalent to a relator of the form $a^m = b^n$. Hence, $G = \langle a, b; a^m = b^n, ab = ba\rangle$. Show G is infinite by mapping G into an infinite cyclic group; hence, $\langle a, b; R(a, b)\rangle$ is infinite; but F/N has order two.]

2. Let F be the free group on $a_1, a_2, \ldots, a_n$ and let N be the normal subgroup in F consisting of all words of even length. Show that $a_1^2, a_1a_2, \ldots, a_1a_n$ normally generate N. [*Hint:* See the hint to Problem 1.]

3. Show directly without using Corollary 2.1 that $W(a, b, c, \ldots)$ is derivable from $P(a, b, c, \ldots)$, $Q(a, b, c, \ldots)$, $R(a, b, c, \ldots), \ldots$ if and only if W is freely equal to a product of conjugates of $P, Q, R, \ldots$ and their inverses. [*Hint:* To show that a word W derivable from $P, Q, R, \ldots$ is a product of conjugates, use induction on the number of insertions and deletions of $P, Q, R, \ldots$ and trivial relators needed to obtain W from the empty word. Note that if $W' = UV$ and $W = UPV$, then $W \approx W' \cdot V^{-1}PV$; and if $W' = UPV$ and $W = UV$, then $W \approx W' \cdot V^{-1}P^{-1}V$.]

4. Let $F = \langle a, b\rangle$. If N is the normal subgroup of F generated by each of the following sets of words, find the index of N in F:

(a) b^2, ab
(b) ab, ab^{-1}
(c) a^2, b^2
(d) $a^3, b^2, aba^{-1}b^{-1}$
(e) $a^3, b^2, (ab)^2$
(f) $aba^{-1}b, bab^{-1}a$.

5. For each of the following groups G, let H be the subgroup generated by the given elements. Show that H is normal and find the order of G/H.

(a) $G = \langle a, b; a^4, a^2 = b^2 = (ab)^2\rangle$; a^2.
(b) $G = \langle a, b, c; a^{2r}, b^2, (ab)^2, c^2\rangle$; a^r, c.
(c) $G = \langle a, b; a^{22}, b^{15}, ab = ba^3\rangle$; a^2.
(d) $G = \langle a, b; a^{22}, b^{15}, ab = ba^3\rangle$; a^{11}.
(e) $G = \langle a, b; a^{22}, b^{15}, ab = ba^3\rangle$; b^5.

6. Let F be a free group on $a_1, \ldots, a_n$. Show that N, the set of words W in F such that the sum of all exponents occurring on a_1 in W is a multiple of a fixed positive integer d, is a normal subgroup of F of index d. Show that N is the normal subgroup of F generated by $a_1^d, a_2, \ldots, a_n$. Show that N is the subgroup generated by a_1^d and all conjugates $a_1^k a_\nu a_1^{-k}$, where $\nu = 2, \ldots, n$ and $0 \leq k < d$. [*Hint:* As a subgroup, N is generated by $a_1^d, a_2, \ldots, a_n$ and their conjugates.]

7. Let $G = \langle a_1, \ldots, a_n; R_\mu(a_\nu), \ldots\rangle$, where the sum of all the exponents on a_1 occurring in the relator R_μ is a multiple of a fixed positive integer d, for all μ. Show that G has a normal subgroup of index d. [*Hint:* Adjoin $a_1^d, a_2, \ldots, a_n$ to the defining relators of the presentation.]

8. The *commutator subgroup* G' of a group

$$G = \langle a, b, c, \ldots; P, Q, R, \ldots\rangle$$

is the normal subgroup of G generated by the *commutators*

$$aba^{-1}b^{-1}, aca^{-1}c^{-1}, \ldots, bcb^{-1}c^{-1}, \ldots.$$

Show that G/N is Abelian if and only if $N \supset G'$. [*Hint:* Use Theorem 2.1.]

2.2. Verbal Subgroups and Reduced Free Groups

A normal or invariant subgroup N of a group G is closed under inner automorphisms, i.e., conjugations. A subgroup K of G is called *characteristic* if K is closed under all automorphisms of G. A characteristic subgroup is necessarily normal, but not conversely (see Problem 15). An even stronger property of a subgroup than being characteristic is being *fully invariant*, i.e., closed under all *endomorphisms* of G (homomorphisms of G into itself). The concept of a "verbal subgroup" will allow us to construct many examples of fully invariant subgroups of a group G.

Let $W_\mu(X_\lambda)$, where $\mu = 1, 2, \ldots$, be a set of words in the symbols X_λ, where $\lambda = 1, 2, \ldots$. Then the *$\{W_\mu\}$-verbal subgroup $G(W_\mu, \ldots)$ of a group* G is the subgroup of G generated by all elements of the form $W_\mu(g_\lambda)$, where g_λ ranges over G.

For example, $G(X^2)$ is the subgroup of G generated by the squares g^2 of all elements g of G. The verbal subgroup $G(X^4, X^6)$ is the same as $G(X^2)$. The *commutator subgroup* of G is the verbal subgroup $G(X_1X_2X_1^{-1}X_2^{-1})$ generated by all the elements $g_1g_2g_1^{-1}g_2^{-1}$ of G. The verbal subgroup $G(X)$ is, of course, just G. The verbal subgroup $G(X_1X_2)$ is also G; for $G(X_1X_2) \subset G$ and substituting g for X_1 and 1 for X_2 we obtain all of G. If Σ_n is the symmetric group on $1, 2, \ldots, n$ then $\Sigma_n(X^3)$ is Σ_n; for Σ_n is generated by two-cycles and the cube of a two-cycle is itself. On the other hand $\Sigma_n(X^2)$ is A_n, the alternating group on $1, 2, \ldots, n$; for the square of any permutation is even, and A_n is generated by three-cycles, which are squares.

The verbal subgroup $G(W_\mu, \ldots)$ is a fully invariant subgroup of G. For let α be an endomorphism of G. Then $\alpha(W_\mu(g_\lambda)) = W_\mu(\alpha(g_\lambda))$. Hence, $\alpha(W_\mu(g_\lambda))$ can be obtained by substituting $\alpha(g_\lambda)$ for X_λ in $W_\mu(X_\lambda)$, and so is in $G(W_\mu, \ldots)$. Thus, $G(W_\mu, \ldots)$ is fully invariant.

Not every fully invariant subgroup of a group G need be verbal (see Problem 16). However, there is an important class of groups for which this is true.

A *reduced free group R of rank n with the identities $W_\mu(X_\lambda) = 1$* is the factor group of the free group F_n of rank n by the verbal subgroup $F_n(W_\mu, \ldots)$. Equivalently, by the corollary to Theorem 1, the reduced free group R could be defined as the group

$$(1) \qquad \langle a_1, a_2, \ldots, a_n; W_\mu(U_\lambda(a_1, a_2, \ldots, a_n)), \ldots\rangle,$$

where U_λ ranges over all words in $a_1, a_2, \ldots, a_n$. Note that, in general, we cannot restrict the U_λ to range only over the generators $a_1, a_2, \ldots, a_n$

(see Problem 18). We often denote R simply by

$$\langle a_1, a_2, \ldots, a_n; W_\mu(X_\lambda), \ldots\rangle. \tag{2}$$

It is clear, from the presentation (1), that any group G on n generators, in which the *identities* $W_\mu(X_\lambda) = 1$ *hold*, i.e., in which $W_\mu(g_\lambda) = 1$ for any g_λ in G, is a homomorphic image of the reduced free group of rank n with the identities $W_\mu(X_\lambda) = 1$.

If F_n is the free group on $a_1, a_2, \ldots, a_n$, then F_n is characterized by the fact that any mapping of these generators into any group G can be extended to a homomorphism of F_n into G (see Problem 1.4.31). The following lemma generalizes this type of characterization to reduced free groups.

LEMMA 2.1. *Let R be a group in which the identities $W_\mu(X_\lambda) = 1$ hold. Then R is the reduced free group of rank n with these identities if and only if there exist n generators $a_1, \ldots, a_n$ of R such that any mapping of these generators into a group G, in which the given identities hold, can be extended to a homomorphism of R into G.*

PROOF. Let R be a reduced free group with the presentation (1). Then by Corollary 2 of Theorem 1.1, any mapping of $a_1, \ldots, a_n$ into a group G, in which the identities $W_\mu(X_\lambda) = 1$ hold, can be extended to a homomorphism of R into G.

Conversely, suppose that the identities $W_\mu(X_\lambda) = 1$ hold in R, and R has n generators $a_1, a_2, \ldots, a_n$ such that any mapping of $a_1, a_2, \ldots, a_n$ into a group G in which the identities $W_\mu(X_\lambda) = 1$ hold, can be extended to a homomorphism of R into G. Let

$$G = \langle b_1, b_2, \ldots, b_n; W_\mu(X_\lambda), \ldots\rangle. \tag{3}$$

Then the identities $W_\mu(X_\lambda) = 1$ hold in G, and the mapping $a_\nu \to b_\nu$ can be extended to a homomorphism of R into G. Therefore, if $P(a_\nu)$ is a relator in R, then $P(b_\nu)$ is a relator in (3). Hence, $P(b_\nu)$ is derivable from the words $W_\mu(U_\lambda(b_1, b_2, \ldots, b_n))$ and so $P(a_\nu)$ is derivable from the words $W_\mu(U_\lambda(a_1, a_2, \ldots, a_n))$. Thus R has the presentation

$$\langle a_1, a_2, \ldots, a_n; \quad W_\mu(U_\lambda(a_1, a_2, \ldots, a_n)), \ldots\rangle,$$

and so is a reduced free group of rank n with the identities $W_\mu(X_\lambda) = 1$. ◀

THEOREM 2.2. *Every fully invariant subgroup J of a reduced free group R of rank n (and, in particular, of a free group) is a verbal subgroup $R(V_\rho, \ldots)$. Moreover, we may choose the words $V_\rho(Y_\nu)$ defining J so that the number of symbols Y_ν used is n.*

PROOF. Let R have the presentation

$$\langle a_1, \ldots, a_n; W_\mu(X_\lambda), \ldots \rangle,$$

and consider the set of all words $V_\rho(Y_1, \ldots, Y_n)$ such that

$$V_\rho(a_1, \ldots, a_n) \tag{4}$$

is in J. We show that $J = R(V_\rho, \ldots)$.

For clearly, $V_\rho(a_1, \ldots, a_n)$ ranges over all elements of J, and so $J \subset R(V_\rho, \ldots)$.

Conversely, consider any element

$$V_\rho(U_1(a_\nu), \ldots, U_n(a_\nu)). \tag{5}$$

Now the mapping $a_i \to U_i(a_\nu)$ can be extended to an endomophism of R by the lemma. Hence, since (4) is in the fully invariant subgroup J, and (4) goes into (5) under the endomorphism, (5) must be in J. But the elements in (5) generate $R(V_\rho, \ldots)$. Thus $R(V_\rho, \ldots) \subset J$, and so $J = R(V_\rho, \ldots)$. ◀

Further insight into reduced free groups is given by the following theorem, which is due to F. Levi, 1933. In order to prove the theorem, we need the following definition:

If W is a word in $a_1, a_2, \ldots, a_n$, and

$$W = a_{\nu_1}^{\alpha_1} a_{\nu_2}^{\alpha_2} \ldots a_{\nu_r}^{\alpha_r},$$

where the α_i are integers and $\nu_i = 1, 2, \ldots, n$, then the *exponent sum of W on a_ν* is the integer

$$\sigma_\nu(W) = \sum_{\nu_i = \nu} \alpha_i.$$

For example, if $W = a_1^2 a_2 a_1^{-3} a_2^{-1} a_1^{-1}$, then $\sigma_1(W) = -2$ and $\sigma_2(W) = 0$.

Clearly, if $W_1 \approx W_2$, then $\sigma_\nu(W_1) = \sigma_\nu(W_2)$; moreover, $\sigma_\nu(UV) = \sigma_\nu(U) + \sigma_\nu(V)$. Hence, $W \to \sigma_\nu(W)$ is a homomorphism of F_n, the free group on $a_1, \ldots, a_n$, onto the additive group of integers. Which elements of F_n are in the kernel of the homomorphism σ_ν? Those elements with zero exponent sum on a_ν; this is the same as the normal subgroup of F_n generated by those a_λ with $\lambda \neq \nu$ (see Problem 1). Which elements of F_n are in the kernel of all the homomorphisms σ_ν? Those elements with zero exponent sum on all a_ν; this is the same as the commutator subgroup of F_n (see Problem 2).

THEOREM 2.3. *Let $W_\mu(X_\lambda)$ be a set of elements in the free group F on $X_1, X_2, \ldots, X_n$. Then there exists a non-negative integer d and a set of words $V_\mu(X_\lambda)$ in the commutator subgroup of F such that for any group G,*

$$G(W_\mu, \ldots) = G(X_1^d, V_\mu, \ldots).$$

PROOF. Let d be the greatest common divisor of the set of integers $\sigma_\lambda(W_\mu)$ where λ and μ range over all their possible values. Let

$$V_\mu = W_\mu \cdot X_1^{-\beta_1} \cdots X_n^{-\beta_n}, \tag{6}$$

where $\beta_\lambda = \sigma_\lambda(W_\mu)$.

Clearly $\sigma_\lambda(V_\mu) = \sigma_\lambda(W_\mu) - \beta_\lambda = 0$. Hence V_μ is in the commutator subgroup of F, the free group on the X_λ.

We show now that if G is any group then $G(W_\mu, \ldots) = G(X_1{}^d, V_\mu, \ldots)$. For, since $d \mid \beta_\lambda$, clearly if g_λ is in G, then $g_\lambda{}^{\beta_\lambda}$ is in $G(X_1{}^d, V_\mu, \ldots)$. Hence,

$$W_\mu(g_\lambda) = V_\mu(g_\lambda) g_n{}^{\beta_n} \cdots g_1{}^{\beta_1}$$

is in $G(X_1{}^d, V_\mu, \ldots)$, and so $G(W_\mu, \ldots) \subset G(X_1{}^d, V_\mu, \ldots)$.

On the other hand, in $W_\mu(X_\lambda)$, substitute g for X_ν and 1 for X_λ, $\lambda \neq \nu$; then the element g^{β_ν} results. Since g^{β_ν} is in $G(W_\mu, \ldots)$ for all β_ν, it follows that g^d is in $G(W_\mu, \ldots)$. But then from (6), $V_\mu(g_\lambda)$ must be in $G(W_\mu, \ldots)$. Hence $G(X_1{}^d, V_\mu, \ldots) \subset G(W_\mu, \ldots)$, and the two verbal subgroups must be equal. ◀

COROLLARY 2.3.1. *Every reduced free group of rank n has a presentation*

$$\langle a_1, \ldots, a_n; X_1{}^d, W_\mu(X_\lambda), \ldots \rangle, \tag{7}$$

where d is a non-negative integer, $\lambda = 1, \ldots, n$, and the W_μ are in the commutator subgroup of F, the free group on $X_1, \ldots, X_n$.

PROOF. The proof is immediate from Theorems 2.2 and 2.3. ▶

COROLLARY 2.3.2. *The only verbal subgroups of an Abelian group G are the "power subgroups" $G(X_1{}^d)$.*

PROOF. Any element of the commutator subgroup of an Abelian group is the identity. Hence, $G(X_1{}^d, V_\mu, \ldots)$ reduces to $G(X_1{}^d)$. ◀

Using Theorem 2.3, we may classify all Abelian reduced free groups.

The *free Abelian group A_n of rank n* (n possibly infinite) is the reduced free group

$$\langle a_1, \ldots, a_n; X_1X_2X_1{}^{-1}X_2{}^{-1} \rangle. \tag{8}$$

It is the quotient group of the free group F_n by its commutator subgroup. The group A_n is isomorphic to the additive group of all "vectors"

$$(\alpha_1, \ldots, \alpha_n),$$

with entries from the ring of integers (and all but finitely many entries equal to zero) under the mapping

$$W(a_1, \ldots, a_n) \rightarrow (\sigma_1(W), \ldots, \sigma_n(W))$$

(see Problem 3).

The *free Abelian group* $A_{n,d}$ *of rank* n *and exponent* d (where n is possibly infinite and d is an integer greater than one) is the reduced free group

$$\langle a_1, \ldots, a_n; X_1{}^d, X_1X_2X_1{}^{-1}X_2{}^{-1}\rangle.$$

The group $A_{n,d}$ is isomorphic to the additive group of all "vectors"

$$(\alpha_1, \ldots, \alpha_n)$$

with entries from the ring of integers mod d (and all but finitely many entries equal to zero) under the mapping

$$W(a_1, \ldots, a_n) \rightarrow (\alpha_1, \ldots, \alpha_n),$$

where $\alpha_\nu = \sigma_\nu(W)$ mod d (see Problem 5).

It is easy to see from this representation of $A_{n,d}$ that its order is d^n if n is finite, and n if n is infinite; moreover, d is the maximum order of any element. Hence, if $A_{n,d}$ is isomorphic to $A_{m,c}$, then $d = c$ and $n = m$.

Clearly, the smallest number of generators of $A_{n,d}$ is n. For, if $A_{n,d}$ has p generators, it is a homomorphic image of $A_{p,d}$ (since $X_1{}^d = 1$ and $X_1X_2X_1{}^{-1}X_2{}^{-1} = 1$ are identities that hold in $A_{n,d}$); hence, the order of $A_{n,d}$ must be less than or equal to the order of $A_{p,d}$. This implies $n \leq p$.

These remarks enable us to prove that n is the minimum number of generators for any reduced free group of rank n.

THEOREM 2.4. *A reduced free group R of rank n ($R \neq 1$) cannot be generated by fewer than n elements. Hence, reduced free groups, and, in particular, free groups and free Abelian groups, of different ranks cannot be isomorphic.*

PROOF. By Corollary 2.3.1, we may assume that

$$R = \langle a_1, \ldots, a_n; X_1{}^d, W_\mu(X_\lambda), \ldots\rangle,$$

where $d > 1$ or $d = 0$ (since $R \neq 1$), and W_μ is in the commutator subgroup of F, the free group on the X_λ. If p is any divisor of d, $p > 1$, then

$$A_{n,p} = \langle a_1, \ldots, a_n; X_1{}^p, X_1X_2X_1{}^{-1}X_2{}^{-1}\rangle$$

is a homomorphic image of R; for clearly the identities $X_1{}^d = 1$ and $W_\mu(X_\lambda) = 1$ hold in $A_{n,p}$. Hence, if R has fewer than n generators, so does $A_{n,p}$; but $A_{n,p}$ cannot have fewer than n generators. Thus R cannot have fewer than n generators. ◀

Theorem 2.4 solves the isomorphism problem for two freely presented free groups, as well as for two free Abelian groups presented as such. Moreover, the concept of verbal subgroups can be used to provide tests for isomorphism.

THEOREM 2.5. *Let* $\{W_\mu(X_\lambda)\}$ *be a set of words. Then if two groups* G_1 *and* G_2 *are isomorphic, the groups* Γ_1 *and* Γ_2, *where*

$$\Gamma_i = G_i/G_i(W_\mu, \ldots)$$

are also isomorphic.

PROOF. For, under an isomorphism of G_1 onto G_2, $G_1(W_\mu, \ldots)$ is mapped onto $G_2(W_\mu, \ldots)$. Moreover, the cosets of Γ_1 are mapped isomorphically onto the cosets of Γ_2. ◀

In view of the infinite variety of verbal subgroups, it might seem that Theorem 2.5 provides an infinite number of useful tests for isomorphism. However, all Theorem 2.5 does is reduce the question of the isomorphism of G_1 and G_2 to that of Γ_1 and Γ_2. If Γ_1 and Γ_2 are finitely generated Abelian groups, or even Abelian groups finitely generated over a (principal ideal) ring of operators, then the isomorphism of Γ_1 and Γ_2 can be decided (see Chapter 3). If Γ_1 and Γ_2 are finitely generated and the words W_μ are *simple commutators* in the sense defined in Section 5.3, then the isomorphism of Γ_1 and Γ_2 can be decided. In general, however, there are no tests for deciding whether Γ_1 and Γ_2 are isomorphic.

The study of the innocent-looking reduced free groups with the identity $X^d = 1$, leads to rather difficult problems (except for some special values of d), most of which are still unsolved (see Section 5.13).

The *variety of groups* corresponding to the identities $\{W_\mu(X_\lambda) = 1\}$ is just the set of groups in which the identities $W_\mu(X_\lambda) = 1$ hold. Varieties of groups are discussed in Section 6.3.

The terms "endomorphism" and "fully invariant" were introduced by F. Levi, 1933. Most of the results of this section are due to B. H. Neumann, 1937. For related results and generalizations, see also R. Baer, 1944, F. Wever, 1950, B. H. Neumann, 1951, and H. Neumann, 1962.

Problems for Section 2.2

1. Let F be the free group on $a_1, \ldots, a_n$ and let N be the set of words W in F such that $\sigma_1(W) = 0$. Show that N is the normal subgroup of F generated by $a_2, \ldots, a_n$. [*Hint:* Show that N is the kernel of the obvious homomorphism of F into $\langle a_1, a_2, \ldots, a_n; a_2, \ldots, a_n\rangle$.]

2. Let F be the free group on $a_1, \ldots, a_n$. Show that $W(a_\nu)$ is in the commutator subgroup $F(X_1X_2X_1^{-1}X_2^{-1})$ if and only if $\sigma_\nu(W) = 0$ for all ν. [*Hint:* To show that if $\sigma_\nu(W) = 0$ for all ν, then W is in $F(X_1X_2X_1^{-1}X_2^{-1})$, use induction on the "syllable length" of W (if $W = a_{\nu_1}^{\alpha_1} \ldots a_{\nu_r}^{\alpha_r}$, where $\nu_i = 1, \ldots, n$ and $\nu_i \neq \nu_{i+1}$ then r is the *syllable length* of W). Note that if $W = a_\nu^\alpha U a_\nu^\beta V$, then $W = a_\nu^\alpha U a_\nu^{-\alpha} U^{-1} \cdot U a_\nu^{\alpha+\beta} V$; and $U a_\nu^{\alpha+\beta} V$ has the same exponent sums as W but shorter syllable length.]

3. Show that if F is the free group on $a_1, \ldots, a_n$ then the mapping

$$W \to (\sigma_1(W), \ldots, \sigma_n(W))$$

has kernel $F(X_1X_2X_1^{-1}X_2^{-1})$. [*Hint:* Use Problem 2.]

4. Show that if F is the free group on $a_1, \ldots, a_n$ then the set J of words W such that $\sigma_\nu(W)$ is a multiple of a fixed positive integer d for all ν, is a fully invariant subgroup of F. Show that $J = F(X_1^d, X_1X_2X_1^{-1}X_2^{-1})$. [*Hint:* If $W(a_\nu)$ is any word in $a_1, \ldots, a_n$, and $U_1, \ldots, U_n$ are n words in $a_1, \ldots, a_n$, then $\sigma_\lambda(W(U_\nu)) = \sigma_1(W)\sigma_\lambda(U_1) + \ldots + \sigma_n(W)\sigma_\lambda(U_n)$. Hence, if $W(a_\nu)$ is in J, so is $W(U_\nu)$. Thus J is fully invariant. To show that

$$J = F(X_1^d, X_1X_2X_1^{-1}X_2^{-1}),$$

use induction on the "syllable length" of a word W in J and see the hint to Problem 2.]

5. Show that if F is the free group on $a_1, \ldots, a_n$, then the mapping

$$W \to (\beta_1, \ldots, \beta_n),$$

where $\beta_\nu = \sigma_\nu(W)$ modulo a fixed positive integer d, has kernel

$$F(X_1^d, X_1X_2X_1^{-1}X_2^{-1}).$$

[*Hint:* Use Problem 4.]

6. Show that the only reduced free groups which are Abelian are A_n, $A_{n,d}$, and 1.

7. Let F be the free group on $a_1, \ldots, a_n$. Show that $F(W_\mu(X_\nu), \ldots)$ is the smallest fully invariant subgroup of F containing $W_\mu(a_\nu), \ldots$.

8. Let F be the free group on X_λ, $\lambda = 1, \ldots, n, \ldots$. Show that a word $W(X_\lambda)$ defines an element of the verbal subgroup $F(W_\mu(X_\lambda), \ldots)$ if and only if $W(X_\lambda)$ is freely equal to a word obtained from the words $W_\mu, \ldots$ by a repeated application of the following operations:

(i) Replace each X_ν by a word $U_\nu(X_\lambda)$;
(ii) Take the inverse of a word;
(iii) Multiply several words together.

9. Let H be a verbal subgroup of a group G, and let N be a verbal subgroup of H. Show that N is a verbal subgroup of G. [*Hint:* If

$$H = G(W_\mu(X_\lambda), \ldots) \quad \text{and} \quad N = H(V_\rho(X_\sigma), \ldots) \quad \text{then} \quad N = G(U_\tau(X_\lambda), \ldots),$$

where $U_\tau(X_\lambda), \ldots$ is the set of words obtained by replacing X_σ in $V_\rho(X_\sigma)$ by a word in $F(W_\mu(X_\lambda), \ldots)$, and F is the free group on the X_λ.]

10. Suppose that $W(X_\lambda) = 1$ is an identity which holds in every group G in which the identities $W_\mu(X_\lambda) = 1, \ldots$ hold. Show that $W(X_\lambda)$ can be obtained from $W_\mu(X_\lambda), \ldots$ by the operations (i), (ii), and (iii) of Problem 8. [*Hint:* Let F be the free group on the X_λ and consider $G = F/F(W_\mu(X_\lambda), \ldots)$. Use Problem 8.]

11. Let H be a factor group G/N of a group G. Show that $H(W_\mu, \ldots) = G(W_\mu, \ldots) \cdot N/N$, the coset of the verbal subgroup $G(W_\mu, \ldots)$. Show that if the set of words $V_\nu, \ldots$ contains the set of words $W_\mu, \ldots$, then

$$H(W_\mu, \ldots)/H(V_\nu, \ldots) \simeq G(W_\mu, \ldots) \cdot N/G(V_\nu, \ldots) \cdot N.$$

($\simeq$ means "is isomorphic to.") [*Hint:* Use the isomorphism theorem,

$$(K/N)/(M/N) \simeq K/M, \qquad \text{where } K \supset M \supset N.]$$

12. A subgroup H of a group G is a *retract* of G if there exists a homomorphism of G into H which is the identity on H. Show that if H is a retract of G then $H(W_\mu, \ldots) = G(W_\mu, \ldots) \cap H$. [*Hint:* To show $H(W_\mu, \ldots) \supset G(W_\mu, \ldots) \cap H$, let α be the *retractive* homomorphism of G onto H. Then, since α is the identity on H, α leaves $G(W_\mu, \ldots) \cap H$ fixed; since α is a homomorphism carrying G into H, α carries $G(W_\mu, \ldots)$ into $H(W_\mu, \ldots)$.]

13. Show that R is a reduced free group of rank n if and only if there exist n generators $a_1, \ldots, a_n$ for R such that any mapping of $a_1, \ldots, a_n$ into R can be extended to an endomorphism of R. [*Hint:* Let F be the free group on $a_1, \ldots, a_n$ and let $R = F/N$. Then any mapping $a_\lambda \to U_\lambda(a_\nu)$ defines an endomorphism of F/N if and only if N is mapped into itself, and so, if and only if N is a fully invariant subgroup of F. Now use Theorem 2.2.]

14. Let Q be the additive group of rational numbers. Show that Q and $\{0\}$ are the only verbal subgroups of Q. Show that an endomorphism of Q can be found carrying any non-zero element q into any other element p of Q. Conclude that the only fully invariant subgroups of Q are Q and $\{0\}$. Show that Q is not a reduced free group, and hence, the converse of Theorem 2 is false. [*Hint:* The mapping $x \to xp/q$ is an endomorphism of Q. Since Q is Abelian, if it is a reduced free group it must be A_n or $A_{n,d}$. Since the elements of Q do not have finite order, Q must be A_n. But not every element in A_n is a square, though every element of Q is a square, i.e., a double.]

15. Let K be the quaternion group of order 8. Show that every subgroup of K is normal, but that the only characteristic (and fully invariant) subgroups of K are K, $\{1, -1\}$, $\{1\}$. Show that K is not a reduced free group. [*Hint:* To show that K is not a reduced free group, suppose it were. Since the minimum number of generators of K is two, K would be of rank two. Now, for any two generators a and b of K, the word $aba^{-1}b$ is a relator. Hence, $X_1X_2X_1^{-1}X_2 = 1$ must hold in K. But letting $X_1 = 1$ and $X_2 = a$, we get $a^2 = 1$, although $a^2 = -1$.]

16. Let G be the direct product of the cyclic group of order two and the cyclic group of order four. Show that the elements of G whose orders divide two form a fully invariant but non-verbal subgroup of G. [*Hint:* Since G is Abelian, its only verbal subgroups are powers of it.]

17. Show that if G is any group, then

$$G(X^2) \supset G(X_1X_2X_1^{-1}X_2^{-1}).$$

[*Hint:* $G/G(X^2)$ is Abelian; for, $ab = (ab)^{-1} = b^{-1}a^{-1} = ba$, if every square is the identity.]

18. Show that $\langle a, b; a^2, b^2\rangle$ and $\langle a, b; X^2\rangle$ are not isomorphic. [*Hint:* $\langle a, b; a^2, b^2\rangle$ has the non-Abelian group of order six as a homomorphic image, while $\langle a, b; X^2\rangle$ is Abelian.]

19. Show that the *Burnside group of exponent d and rank n*,

$$B(d, n) = \langle a_1, \ldots, a_n; X^d\rangle,$$

where $n > 1$ and $d > 2$, is non-Abelian. [*Hint:* Consider the non-Abelian group

$$G_p = \langle a, b, c; a^p, b^p, c = b^{-1}a^{-1}ba, ac = ca, bc = cb\rangle$$

(here $r = s = p > 1$) of Problem 1.2.20. Using the multiplication rule in (e), show by induction on k that

$$(a^\alpha b^\beta c^\gamma)^k = a^{k\alpha}b^{k\beta}c^{m\alpha\beta+k\gamma},$$

where $m = k(k - 1)/2$. Conclude from (c) that if p is odd, then the identity $X^p = 1$ holds in G_p; conclude from (c) that if p is even, then the identity $X^{2p} = 1$ holds in G_p. Map $B(d, n)$ onto G_d if d is odd, and onto $G_{d/2}$ if d is even, under the mapping $a_1 \to a$, $a_2 \to b$, $a_3 \to 1, \ldots, a_n \to 1$.]

20. Show that if G is generated by finitely many elements, each of finite order and having finitely many conjugates, then G is finite. [*Hint:* Let $a_1, \ldots, a_n$ be the finitely many generators and their finitely many conjugates, each of which is of finite order. We show by induction on the syllable length of a word $W(a_\nu)$ that $W(a_\nu)$ defines the same element of G as a word

$$a_{\nu_1}^{\alpha_1} \cdots a_{\nu_r}^{\alpha_r}$$

where $\nu_i \neq \nu_j$ if $i \neq j$, $\nu = 1, \ldots, n$, and α_i is a non-negative integer less than the order of a_{ν_i}. For, if a_ν occurs twice, i.e., if $W = Ua_\nu{}^\alpha Va_\nu{}^\beta T$, then $W = U \cdot a_\nu^{\alpha+\beta} \cdot (a_\nu{}^{-\beta} Va_\nu{}^\beta) \cdot T$. Since the conjugate of a_i by $a_\nu{}^\beta$ is one of $a_1, \ldots, a_n$, the syllable length of W has been shortened.]

21. Show that an Abelian group G with finitely many generators, each of finite order, is a finite group. [*Hint:* Use Problem 20, or show that G is the homomorphic image of some reduced free group $A_{n,d}$.]

22. Show that

$$B(2, n) = \langle a_1, \ldots, a_n; X^2\rangle$$

is finite, if n is finite. [*Hint:* Use Problem 21.]

23. Let

$$B(3, n) = \langle a_1, \ldots, a_n; X^3\rangle.$$

Show that any element of $B(3, n)$ commutes with all its conjugates. [*Hint:* If x and y are any elements of $B(3, n)$ then $xyx = y^{-1}x^{-1}y^{-1}$. Hence, $s \cdot t^{-1}st = st^{-1}s \cdot t = ts^{-1}t \cdot t = ts^{-1}t^{-1} = t \cdot s^{-1}t^{-1}s^{-1} \cdot s = t \cdot tst \cdot s = t^{-1}st \cdot s$.]

24. It is known that a subgroup of finite index in a finitely generated group is itself finitely generated (see Corollary 2.7.1 in Section 2.3). Use this result to show that if a group G has generators $a_1, \ldots, a_n$, n finite, such that each generator a_ν is of finite order and commutes with all its conjugates, then G is a finite group. In particular, $B(3, n)$ is finite for finite n. [*Hint:* Use induction on n. If $n = 1$ the result is immediate. Suppose that the result holds for all groups with $n - 1$ generators, each of finite order and commuting with its conjugates. Let G be generated by $a_1, \ldots, a_n$, each of finite order and commuting with its conjugates. Consider the normal subgroup N of G generated by any generator, say, a_n. Then G/N is generated by $a_1, \ldots, a_{n-1}$ and satisfies the inductive hypothesis, and so is finite. Hence, N is of finite

index in G, and since G is finitely generated, so is N. But N is generated by the conjugates of a_n, and so is Abelian. By Problem 21, N is finite. Thus, a_n has finitely many conjugates. But this argument works for any a_ν. Hence, by Problem 20, G is finite. To show $B(3, n)$ finite, use Problem 23.]

25. Let

$$G_1 = \langle a_1, \ldots, a_n; P, Q, \ldots \rangle$$

and

$$G_2 = \langle b_1, \ldots, b_m; S, T, \ldots \rangle,$$

where $n \neq m$, and suppose there is an integer $d > 1$ which divides the exponent sum of each defining relator of G_1 and G_2 on any generator. Show G_1 is not isomorphic to G_2. [*Hint:* Use Theorem 2.5 with

$$\{W_\mu\} = \{X_1{}^d, X_1X_2X_1{}^{-1}X_2{}^{-1}\},$$

and show that $\Gamma_1 = A_{n,d}$, whereas $\Gamma_2 = A_{m,d}$.]

26. Let

$$G_1 = \langle a_1, \ldots, a_n; P, Q, \ldots \rangle$$

and

$$G_2 = \langle b_1, \ldots, b_n; S, T, \ldots \rangle,$$

where n is finite and there is an integer $d > 1$ which divides $\sigma_\nu(P)$, $\sigma_\nu(Q), \ldots$ for each ν, but which does not divide $\sigma_1(S)$. Show G_1 is not isomorphic to G_2. [*Hint:* Use Theorem 2.5 with $\{W_\mu\} = \{X^d, X_1X_2X_1{}^{-1}X_2{}^{-1}\}$, and show $\Gamma_1 = A_{n,d}$. Since Γ_2 is Abelian, and S is a relator in Γ_2, $b_1^{\sigma_1(S)}$ is a word in $b_2, \ldots, b_n$. Hence, the number of distinct elements $b_1{}^{\alpha_1}b_2{}^{\alpha_2} \ldots b_n{}^{\alpha_n}$ in Γ_2 is less than d^n.]

27. Construct specific presentations for groups G_1 and G_2 which satisfy the hypotheses for Problem 25; do the same for the hypotheses of Problem 26.

28. Let G_1 and G_2 be isomorphic groups. If $\{W_\mu(X_\lambda)\}$ is a set of words contained in the set of words $\{V_\rho(X_\lambda)\}$, and

$$\Gamma_i = G_i(W_\mu, \ldots)/G_i(V_\rho, \ldots),$$

show that Γ_1 and Γ_2 are isomorphic. [*Hint:* See the proof of Theorem 2.5.]

29. Show that if G is an Abelian group, then

$$G(X^d) \cap G(X^p) = G(X^q),$$

where q is the least common multiple of d and p. Show that if G is not Abelian, this result may not be true. [*Hint:* In an Abelian group, $G(X^r)$ is just the set of rth powers of all elements. If $d = st$ and $p = su$, where t and u are coprime, then $q = stu$. Moreover, if x is in $G(X^d) \cap G(X^p)$, then $x = y^d = z^p$. Hence, $x^u = y^{du} = y^q$ and $x^t = z^{pt} = z^q$. Thus x^u and x^t are in $G(X^q)$; since t and u are coprime, x is in $G(X^q)$. If F is the free group on a and b, then $F(X^2) \cap F(X^3) \neq F(X^6)$. For,

$$F(X^2) \cap F(X^3) \supset F(X_1X_2{}^3X_1{}^{-1} \cdot X_2{}^{-3}).$$

If $F(X^6) \supset F(X_1X_2{}^3X_1{}^{-1}X_2{}^{-3})$, then in any group on two generators in which the identity $X^6 = 1$ holds, the identity $X_1X_2{}^3X_1{}^{-1}X_2{}^{-3} = 1$ must also hold. But in the non-Abelian group of order six, $X^6 = 1$ holds but not $X_1X_2^3X_1^{-1}X_2^{-3} = 1$.]

30. Show that the identities $X^6 = 1$ and $X_1^2X_2^2X_1^{-2}X_2^{-2} = 1$ hold in the non-Abelian group G of order six. (It can be shown that any other identity which holds in the group G can be derived from these two identities; see Wever 1950.) Show that G is not a reduced free group. [*Hint:* To show that G is not a reduced free group, use Problem 13.]

31. Let R be a reduced free group of rank n. Show that some element of R has infinite order, or every element has a finite order which divides the maximum order of any element. [*Hint:* If R satisfies the identity $X^d = 1$ then by considering $R/R(X_1X_2X_1^{-1}X_2^{-1}) = A_{n,d}$, we see there is an element in R of order d, and the order of every element in R divides d. Otherwise, $R/R(X_1X_2X_1^{-1}X_2^{-1}) = A_n$ and R has an element of infinite order.]

32. Let G be the alternating group on 1, 2, 3, 4, 5. Show that the order of the elements of G are 1, 2, 3, and 5. Let $[U, V] = UVU^{-1}V^{-1}$. Show $X^{30} = 1$, $[[X_1, X_2^5], X_2^6] = 1$ are identities which hold in G. Find other identities that hold in G. Show that if neither p nor q is divisible by 30, then $X_1^pX_2^qX_1^{-p}X_2^{-q} = 1$ is not an identity in G. Show that G is not a reduced free group. [*Hint:* To show $X_1^pX_2^qX_1^{-p}X_2^{-q} = 1$ is not an identity note that $G(X^p) = G(X^q) = G$, since G is simple. To show that G is not a reduced free group use Problem 31.]

33. Show that if $\{W_\mu(X_\lambda)\}$ is a finite set of words then there is a single word $V(Y_\nu)$ such that for any group G,

$$G(W_\mu, \ldots) = G(V).$$

Show that even if $\{W_\mu(X_\lambda)\}$ is a non-countable set, there exists a finite or countable set of words $\{V_\rho(Y_\nu)\}$ such that for any group G,

$$G(W_\mu, \ldots) = G(V_\rho, \ldots).$$

[*Hint:* Use $G(W_1(X_\lambda), W_2(X_\lambda)) = G(W_1(Y_\lambda) \cdot W_2(Z_\lambda))$, for the first part.

For the second part, note that each $W_\mu(X_\lambda)$ involves only finitely many symbols X_λ. Hence, $W_\mu(X_\lambda)$ may be replaced by a word involving finitely many of the countable number of symbols. $Y_1, \ldots, Y_n, \ldots$.]

34. Let $\mathscr{V}$ be a collection of groups, each of which satisfies the identities $W_\mu(X_\lambda) = 1$. Show that any subgroup or factor group of a group in $\mathscr{V}$ also satisfies the identities $W_\mu(X_\lambda) = 1$. Show that the unrestricted direct product of any groups in $\mathscr{V}$ also satisfies the identities $W_\mu(X_\lambda) = 1$.

35. Let G be a finite group, and let $\{W_\mu(X_\lambda) = 1\}$ be the totality of identities that G satisfies, where λ ranges over the positive integers. Show that a reduced free group of finite rank with the identities $W_\mu(X_\lambda) = 1$, must be a finite group. [*Hint:* Suppose G has order r. We show that the reduced free group of rank n with the identities $W_\mu(X_\lambda) = 1$, is a subgroup of the direct product of r^n copies of G. In order to do this, make up a table with r^n columns, each column of which, is one of the possible n-tuples with elements from G. Define a_i as the r^n-tuple which is the ith row of the table. Then $a_1, \ldots, a_n$

generate a subgroup R of the direct product of r^n copies of G. Hence, R satisfies the identities $W_\mu(X_\lambda) = 1$. Moreover, if $P(a_1, \ldots, a_n)$ is a relator in R, then by construction of the a_i, $P(g_1, \ldots, g_n) = 1$ in G for any n-tuple $g_1, \ldots, g_n$ from G. Hence, $P(X_1, \ldots, X_n) = 1$ is an identity in G. Thus,

$$R = \langle a_1, \ldots, a_n; W_\mu(X_\lambda), \ldots \rangle .$$

If n is finite, then R is a subgroup of the finite group which is the direct product of r^n copies of G.]

36. Show that a non-empty collection $\mathscr{C}$ of groups is a variety of groups if and only if $\mathscr{C}$ contains the subgroups and factor groups of any group in it, and also the unrestricted direct product of any groups in it. [*Hint:* To show that any collection $\mathscr{C}$ of groups closed under subgroups, factor groups, and unrestricted direct products, is a variety of groups, let $W_\mu(X_\lambda) = 1, \ldots,$ where λ ranges over the positive integers, be the totality of identities satisfied by all groups in $\mathscr{C}$. We show that $\mathscr{C}$ is the variety of all groups which satisfy $W_\mu(X_\lambda) = 1, \ldots.$ To show that any group satisfying the given identities is in $\mathscr{C}$, it suffices to show that a reduced free group of arbitrary rank with the given identities is in $\mathscr{C}$; for $\mathscr{C}$ is closed under factor groups. We show that such a reduced free group is a subgroup of an unrestricted direct product of groups from $\mathscr{C}$. Let n be any cardinal number. Consider the words $V_\rho(X_\lambda)$, where $\lambda = 1, \ldots, r_\rho$, such that $V_\rho(X_\lambda) = 1$ is not an identity for all groups in $\mathscr{C}$. Hence, there exists a group G_ρ in $\mathscr{C}$ and a sequence $g_1, \ldots, g_{r_\rho}$ of elements from G_ρ such that $V_\rho(g_1, \ldots, g_{r_\rho}) \neq 1$. Construct a table whose columns are all the possible n-tuples with entries from $\{g_1, \ldots, g_{r_\rho}\}$. Put these tables next to one another for all ρ. Let a_i by the ith row in the juxtaposed tables. Then all the a_i generate a subgroup R of the unrestricted direct product of a certain number of copies of the groups G_ρ for all ρ. If now $U(a_{\nu_1}, \ldots, a_{\nu_p})$ is a relator in R, but $U(X_1, \ldots, X_p) = 1$ is not an identity for all groups in $\mathscr{C}$, we obtain a contradiction; for, if $U(X_1, \ldots, X_p) = V_\rho(X_\lambda)$ then $U(g_1, \ldots, g_p) \neq 1$, contrary to $U(a_{\nu_1}, \ldots, a_{\nu_p}) = 1$. Hence, R is the reduced free group of rank n with identities $W_\mu(X_\lambda) = 1$.]

37. Show that the intersection of varieties of groups is a variety of groups. Show that any non-empty collection of groups is contained in a smallest variety of groups. Show that the smallest variety containing an infinite cyclic group is the variety of Abelian groups.

38. Show that the collection $\mathscr{C}$ of groups, all of whose elements have finite order, is not a variety. Show that $\mathscr{C}$ is closed under subgroups, factor groups, and restricted direct product. [*Hint:* Use Problem 36.]

39. Show that the collection of groups whose elements $\neq 1$ have infinite order, is closed under subgroups and unrestricted direct product, but is not a variety.

40. Show that the collection of groups each of whose elements is a square, is closed under factor groups and unrestricted direct product, but is not a variety.

2.3. Presentations of Subgroups (The Reidemeister-Schreier Method)

Let G be a group with the presentation

$$G = \langle a_1, \ldots, a_n; R_\mu(a_\nu), \ldots \rangle. \tag{1}$$

In Section 2.1 we showed how to present a factor group G/N of G; in this section we show how to present a subgroup H of G.

To present the factor group G/N we required words in the a_ν, which generated the normal subgroup N; to present the subgroup H we require words in the a_ν, which generate the subgroup H. But in addition to these generators for H, we shall require a process for "rewriting" a word in the a_ν which defines an element of H, as a word in the generators of H.

Specifically, let G be presented as in (1), and let H be the subgroup of G generated by the words $J_i(a_\nu), \ldots$. Then a *rewriting process for H (with respect to the generators* $J_i(a_\nu)$*)* is a mapping

$$\tau\colon \quad U(a_\nu) \to V(s_i) \tag{2}$$

of words $U(a_\nu)$ which define elements of H, into words in the symbols s_i, such that the words

$$U(a_\nu), \qquad V(J_i(a_\nu))$$

define the same element of H. (The symbol s_i will be the generating symbol used for $J_i(a_\nu)$ in the presentation of H which we shall obtain.)

For example, let G be the free group on a and b, and let H be the normal subgroup of G generated by b. Then H is generated by b and its conjugates by powers of a, i.e., by a^kba^{-k}, where k is any integer. Let

$$J_k(a, b) = a^kba^{-k}.$$

A word $W(a, b)$ defines an element of H if and only if the exponent sum of W on a is zero. Indeed, if

$$U(a, b) = a^{\alpha_1}b^{\beta_1}a^{\alpha_2}b^{\beta_2} \cdots a^{\alpha_r}b^{\beta_r} \tag{3}$$

has zero exponent sum on a, then $U(a, b)$ and

$$(a^{\alpha_1}ba^{-\alpha_1})^{\beta_1} \cdot (a^{\alpha_1+\alpha_2}ba^{-\alpha_1-\alpha_2})^{\beta_2} \cdot \ldots \cdot (a^{\alpha_1+\alpha_2+\cdots+\alpha_r}ba^{-\alpha_1-\alpha_2-\cdots-\alpha_r})^{\beta_r},$$

define the same element of H. Hence, the mapping τ which sends the word in (3) into

$$s_{\alpha_1}^{\beta_1} \cdot s_{\alpha_1+\alpha_2}^{\beta_2} \cdot \ldots \cdot s_{\alpha_1+\alpha_2+\cdots+\alpha_r}^{\beta_r} \tag{4}$$

is a rewriting process for H.

We wish now to present a subgroup H of the group G given in (1), using generating symbols s_i for the generating elements $J_i(a_\nu)$.

THEOREM 2.6. *Let H be a subgroup of the group G presented in* (1). *If $J_i(a_\nu)$ are generators for H and the mapping τ is a rewriting process for H (with respect to the generators $J_i(a_\nu)$), then a presentation for H under the mapping $s_i \to J_i(a_\nu)$ is obtained by using the symbols s_i as generating symbols and using the following equations as defining relations:*

$$s_i = \tau(J_i(a_\nu)); \tag{5}$$

$$\tau(U) = \tau(U^*), \tag{6}$$

where $U(a_\nu)$ and $U^*(a_\nu)$ are freely equal words which define elements of H;

$$\tau(U_1 \cdot U_2) = \tau(U_1)\tau(U_2), \tag{7}$$

where $U_1(a_\nu)$ and $U_2(a_\nu)$ define elements of H; and

$$\tau(WR_\mu W^{-1}) = 1, \tag{8}$$

where $R_\mu(a_\nu)$ is a defining relator in (1) *and W is any word in a_ν.*

PROOF. We first show that (5), (6), (7), and (8) are relations. For, if $U(a_\nu)$ defines an element of H, and

$$\tau(U(a_\nu)) = V(s_i),$$

then $U(a_\nu)$ and $V(J_i(a_\nu))$ define the same element of H, by definition of a rewriting process. Hence, under the mapping $s_i \to J_i(a_\nu)$, the equations (5), (6), (7), and (8) are relations.

To show that (5), (6), (7), and (8) are defining relations, we must show that any relator

$$s_{i_1}^{\epsilon_1} \ldots s_{i_r}^{\epsilon_r} \qquad (\epsilon_{i_j} = \pm 1), \tag{9}$$

can be reduced to the empty word using the relations (5), (6), (7), and (8). It is convenient to first derive some relations from (5), (6), (7), and (8); the term "derivable" shall mean derivable with respect to (5), (6), (7), and (8).

In (7), if we replace U_1 and U_2 by the empty word 1, we obtain $\tau(1) = \tau(1) \cdot \tau(1)$, or equivalently $\tau(1) = 1$. Now, in (7) we replace U_2 by U_1^{-1}, and use (6) to obtain $\tau(U_1) \cdot \tau(U_1^{-1}) = 1$, or equivalently

$$\tau(U_1^{-1}) = \tau(U_1)^{-1}. \tag{10}$$

Hence, from (7) and (10) we may derive

$$\tau(U_1^{\epsilon_1} \ldots U_p^{\epsilon_p}) = \tau(U_1)^{\epsilon_1} \ldots \tau(U_p)^{\epsilon_p}. \tag{11}$$

We may now proceed to show that (9) is derivable. Using (5), (9) may be replaced by

$$\tau(J_{i_1}(a_\nu))^{\epsilon_1} \cdot \ldots \cdot \tau(J_{i_r}(a_\nu))^{\epsilon_r}$$

which by (11) may be replaced by

$$\tau(J_{i_1}^{\epsilon_1} \ldots J_{i_r}^{\epsilon_r}). \tag{12}$$

Now, since (9) is a relator under $s_i \to J_i(a_\nu)$,

$$J_{i_1}(a_\nu)^{\epsilon_1} \cdot \ldots \cdot J_{i_r}(a_\nu)^{\epsilon_r} \tag{13}$$

defines the identity in H, and hence, also in G. Since G has the presentation (1), it follows from the Corollary to Theorem 2.1 that (13) is freely equal to a product

$$(W_1 R_{\mu_1} W_1^{-1})^{\eta_1} \cdot \ldots \cdot (W_t R_{\mu_t} W_t^{-1})^{\eta_t}, \tag{14}$$

where $\eta_j = \pm 1$ and R_{μ_j} is a defining relator in (1). Hence, using (6), (12) can be replaced by

$$\tau((W_1 R_{\mu_1} W_1^{-1})^{\eta_1} \cdot \ldots \cdot (W_t R_{\mu_t} W_t^{-1})^{\eta_t}). \tag{15}$$

Using (11), (15) can be replaced by

$$\tau(W_1 R_{\mu_1} W_1^{-1})^{\eta_1} \cdot \ldots \cdot \tau(W_t R_{\mu_t} W_t^{-1})^{\eta_t}. \tag{16}$$

But then the relations in (8) allow us to reduce (16) to the empty word. ◀

The presentation for H obtained in Theorem 2.6 is very cumbersome. By a judicious choice of generators and rewriting process, the presentation can be greatly simplified. One such choice may be made by using a "right coset function" for G mod H, to obtain both generators and a rewriting process.

If G has the presentation (1), then a *right coset representative function for G (on the generators a_ν) modulo a subgroup H*, is a mapping of words in a_ν,

$$W(a_\nu) \to \overline{W}(a_\nu),$$

where the $\overline{W}(a_\nu)$ form a right coset representative system for G mod H, which contains the empty word, and where $\overline{W}(a_\nu)$ is the representative of the coset of $W(a_\nu)$.

THEOREM 2.7. *If $W \to \overline{W}$ is a right coset function for G mod H, then H is generated by the words*

(17) $$Ka_\nu \cdot \overline{Ka_\nu}^{-1},$$

where K is an arbitrary representative and a_ν is a generator for G.

PROOF. To show that the words in (17) are generators for H, it is convenient to use the following easily verified properties of a right coset representative function:

(i) $\overline{W} = 1$ if and only if W defines an element of H;
(ii) W is freely equal to V implies $\overline{W} = \overline{V}$;
(iii) $\overline{\overline{W}} = \overline{W}$;
(iv) $\overline{WV} = \overline{\overline{W}V}$.

The word $Ka_\nu\overline{Ka_\nu}^{-1}$ clearly defines an element of H; for, Ka_ν and $\overline{Ka_\nu}$ determine the same right coset of H.

We now show that every element of H can be expressed as a product of the words in (17) and their inverses.

First observe that the word

(18) $$Ka_\nu^{-1}\overline{Ka_\nu^{-1}}^{-1},$$

is the inverse of a word in (17), for since K is a representative, it follows that

$$\overline{\overline{Ka_\nu^{-1}}a_\nu} = \overline{Ka_\nu^{-1}a_\nu} = \overline{K} = K,$$

by using (iv), (ii), and (iii). Hence, (18) is the inverse of the word $Ma_\nu\overline{Ma_\nu}^{-1}$ where $M = \overline{Ka_\nu^{-1}}$, which is included in (17).

Suppose now that

(19) $$U = a_{\nu_1}^{\epsilon_1}a_{\nu_2}^{\epsilon_2} \ldots a_{\nu_r}^{\epsilon_r} \qquad (\epsilon_i = \pm 1)$$

defines an element of H; we must express U in terms of the words in (17) and (18). We therefore insert before and after each $a_{\nu_j}^{\epsilon_j}$ in U, the words $\overline{W}_j$ and $\overline{W_j a_{\nu_j}^{\epsilon_j}}^{-1}$, respectively, and try to choose the W_j so that our new product,

(20) $$\overline{W}_1 a_{\nu_1}^{\epsilon_1}\overline{W_1 a_{\nu_1}^{\epsilon_1}}^{-1} \cdot \overline{W}_2 a_{\nu_2}^{\epsilon_2}\overline{W_2 a_{\nu_2}^{\epsilon_2}}^{-1} \cdot \ldots \cdot \overline{W}_r a_{\nu_r}^{\epsilon_r}\overline{W_r a_{\nu_r}^{\epsilon_r}}^{-1}$$

defines the same element of H as (19). Now, (20) will certainly define the same element of H as (19) does, if we choose

$$W_1 = 1, \quad W_2 = W_1 a_{\nu_1}^{\epsilon_1}, \quad W_3 = W_2 a_{\nu_2}^{\epsilon_2}, \quad \ldots \quad, W_r = W_{r-1}a_{\nu_{r-1}}^{\epsilon_{r-1}},$$

i.e., if we choose

$$W_1 = 1,\ W_2 = a_{\nu_1}^{\epsilon_1},\ W_3 = a_{\nu_1}^{\epsilon_1}a_{\nu_2}^{\epsilon_2},\ \ldots,\ W_r = a_{\nu_1}^{\epsilon_1}a_{\nu_2}^{\epsilon_2}\ldots a_{\nu_{r-1}}^{\epsilon_{r-1}}$$

(W_j is called the $(j-1)$st *initial segment* of U); for if W_j is so chosen, then (20) is freely equal to

$$\overline{W}_1 U \overline{W_r a_{\nu_r}^{\epsilon_r}}^{-1} = \overline{1} U \overline{U}^{-1} = 1U1^{-1} = U$$

($\overline{U} = 1$, since U is an element of H).

Moreover, using (iv), it is clear that

$$\overline{W}_j a_{\nu_j} \overline{W_j a_{\nu_j}}^{-1} \tag{21}$$

is one of the words in (17), and

$$\overline{W}_j a_{\nu_j}^{-1} \overline{W_j a_{\nu_j}^{-1}}^{-1} \tag{22}$$

is the inverse of

$$\overline{W_j a_{\nu_j}^{-1}} a_{\nu_j} \overline{W}_j^{-1}$$

which is one of the words in (17).

Hence, every word $U(a_\nu)$ which defines an element of H is a product (20) of words in (17) and their inverses. Thus the words in (17) generate H. ◀

Corollary 2.7.1. *If G is finitely generated and H is a subgroup of finite index, then H is finitely generated.*

Proof. Indeed, if G has n generators and H has index j, then the generators (17) for H are nj in number (later on, we shall obtain a better bound for the number of generators needed for H). Hence, H is finitely generated. ◀

Corollary 2.7.2. *Let $W \to \overline{W}$ be a right coset representative function for G mod H. Introduce the generating symbol*

$$s_{K,a_\nu}$$

for the element of H defined by the word

$$Ka_\nu \overline{Ka_\nu}^{-1},$$

where K is an arbitrary right coset representative and a_ν is a generator in (1). *Define the mapping τ of the word U in* (19), *which defines an element of H, by*

$$\tau(U) = s_{K_1,a_{\nu_1}}^{\epsilon_1} s_{K_2,a_{\nu_2}}^{\epsilon_2} \cdots s_{K_r,a_{\nu_r}}^{\epsilon_r}, \tag{23}$$

where K_j is the representative of the initial segment of U preceding a_{ν_j} if

$\epsilon_j = 1$, *and* K_j *is the representative of the initial segment of* U *up to and including* $a_{\nu_j}^{-1}$ *if* $\epsilon_j = -1$. *Then* τ *is a rewriting process for* H.

PROOF. To show that τ is a rewriting process, we must show that if $s_{K_j,a_{\nu_j}}$ is replaced by $K_j a_{\nu_j} \overline{K_j a_{\nu_j}}^{-1}$ in (23) then a word in a_ν results which defines the same element of H as $U(a_\nu)$.

If W_j is the initial segment of U which precedes $a_{\nu_j}^{\epsilon_j}$, then by construction of K_j,

$$K_j = \overline{W}_j, \qquad \text{if } \epsilon_j = 1,$$

and

$$K_j = \overline{W_j a_{\nu_j}^{-1}}, \qquad \text{if } \epsilon_j = -1.$$

Hence, in any event, $s_{K_j,a_{\nu j}}^{\epsilon_j}$ is replaced by

$$\overline{W}_j a_{\nu_j}^{\epsilon_j} \overline{W_j a_{\nu_j}^{\epsilon_j}}^{-1}$$

But then (23) becomes (20), and this defines the same element as $U(a_\nu)$. Thus τ is a rewriting process. ◀

A rewriting process τ obtained from a right coset representative function, as in the last corollary, is called a *Reidemeister rewriting process.*

To illustrate the method of rewriting in a Reidemeister process, suppose $a_1{}^2 a_2{}^{-1} a_3$ defines an element of H. Then

$$\tau(a_1 a_1 a_2{}^{-1} a_3) = s_{\overline{1},a_1} \cdot s_{\overline{a}_1,a_1} \cdot s^{-1}_{\overline{a_1{}^2 a_2{}^{-1}},a_2} \cdot s_{\overline{a_2{}^2 a_2{}^{-1}},a_3}.$$

It should be noted that computation of $\tau(U)$ can be carried out by replacing a symbol $a_\nu{}^\epsilon$ of U by the appropriate s-symbol s_{K,a_ν}^ϵ.

Using a Reidemeister rewriting process τ in Theorem 2.6 greatly simplifies the presentation for H: the defining relations (6) and (7) can be eliminated, and (8) can be restricted.

THEOREM 2.8 (Reidemeister). *Let* τ *be the Reidemeister rewriting process given by* (23) *for a subgroup* H *of a group* G. *If* G *has the presentation* (1), *then* H *has the presentation*

$$(24) \qquad \langle s_{K,a_\nu}, \ldots ; s_{K,a_\nu} = \tau(Ka_\nu \overline{Ka_\nu}^{-1}), \ldots, \tau(KR_\mu K^{-1}), \ldots \rangle$$

under the mapping $s_{K,a_\nu} \to Ka_\nu \overline{Ka_\nu}^{-1}$, *where* K *is an arbitrary representative (used in the right coset function determining* τ), a_ν *is an arbitrary generator and* R_μ *is an arbitrary defining relator in* (1).

PROOF. It suffices to show that the relations (6), (7), and (8) can be derived from the defining relations in (24); for then by using a Tietze transformation we may delete the redundant defining relations.

To derive (6), (7), and (8), it is convenient to have the following properties of a Reidemeister rewriting process:

(v) If U and U^* are freely equal words (in the a_ν) which define elements of H then $\tau(U)$ and $\tau(U^*)$ are freely equal words (in the s-symbols).

(vi) If U_1 and U_2 define elements of H then $\tau(U_1 \cdot U_2)$ and $\tau(U_1) \cdot \tau(U_2)$ are identical words (in the s-symbols).

Clearly, to show (v), it suffices to show that $\tau(Va_\nu^{\epsilon}a_\nu^{-\epsilon}W)$ is freely equal to $\tau(VW)$, where $\epsilon = \pm 1$. Now, in computing τ of a word we may do so by replacing each a-symbol by the appropriate s-symbol. Since the s-symbol replacing a given a-symbol depends only upon the a-symbol and the representative of the segment preceding the a-symbol, the s-symbols replacing the a-symbols of V and W in computing $\tau(VW)$ and $\tau(Va_\nu^{\epsilon}a_\nu^{-\epsilon}W)$, will be the same. Hence, we need only consider the s-symbols which replace a_ν^{ϵ} and $a_\nu^{-\epsilon}$. These are

$$s_{\overline{V},a_\nu} \quad \text{and} \quad s^{-1}_{\overline{Va_\nu a_\nu^{-1}},a_\nu} = s^{-1}_{\overline{V},a_\nu},$$

if $\epsilon = 1$, and

$$s^{-1}_{\overline{Va_\nu^{-1}},a_\nu} \quad \text{and} \quad s_{\overline{Va_\nu^{-1}},a_\nu},$$

if $\epsilon = -1$. This proves (v).

To show (vi), we note that the s-symbols replacing the a-symbols of U_1 in computing $\tau(U_1U_2)$, are precisely the s-symbols of $\tau(U_1)$. Moreover, since U_1 is in H, $\overline{U_1W} = \overline{W}$ for any word $W(a_\nu)$. Hence, the s-symbols replacing the a-symbols of U_2 in computing $\tau(U_1U_2)$, are precisely the s-symbols of $\tau(U_2)$.

It is clear from properties (v) and (vi) that the relations (6) and (7) may be derived from trivial relators.

To simplify the relations in (8), we note any word W is freely equal to the word UK, where K is $\overline{W}$, and U is WK^{-1}, which defines an element of H. Thus

$$\begin{aligned} \tau(WR_\mu W^{-1}) &\approx \tau(U \cdot KR_\mu K^{-1} \cdot U^{-1}) \\ &= \tau(U) \cdot \tau(KR_\mu K^{-1})\tau(U^{-1}). \end{aligned}$$

Since (6) and (7) hold, $\tau(U^{-1}) = \tau(U)^{-1}$ holds, and so the relator $\tau(WR_\mu W^{-1})$ is derivable from the relator

$$\tau(U)\tau(KR_\mu K^{-1})\tau(U)^{-1},$$

and hence, from the relator

$$\tau(KR_\mu K^{-1}).$$

Thus (24) presents H. ◄

COROLLARY 2.8. *If G is finitely presented and H is of finite index in G, then H is finitely presented.*

PROOF. This is an immediate consequence of Theorem 2.8. ◀

To simplify the presentation (24) still further, we shall restrict ourselves to a special class of right coset functions.

A *Schreier right coset function* is one for which any initial segment of a representative is again a representative. The system of representatives is then called a *Schreier system*. A Reidemeister rewriting process using a Schreier system is called a *Reidemeister-Schreier rewriting process*.

As an illustration of a Schreier system, let G be the free group on a and b, and let H be the normal subgroup of G generated by a^2, b^2, and $aba^{-1}b^{-1}$. Then G/H has four cosets. The representative system $1, a, b, ab$ is Schreier, as is the representative system $1, a, b, ab^{-1}$. The representative system $1, a, b, a^{-1}b^{-1}$ is not Schreier; for, the initial segment a^{-1} of $a^{-1}b^{-1}$ is not a representative.

LEMMA 2.2. *Let G have the presentation* (1) *and let H be a subgroup of G. Then there is some Schreier system of representatives for G mod H.*

PROOF. We call the *length of a coset* of G mod H the length of the shortest word in it, and shall define our Schreier representatives inductively, using the length of a coset.

Choose the empty word as the representative of H, the coset of length zero. If S_1 is a coset of length one, choose any word of length one in S_1 as its representative. If S_2 has length two, select a word b_1b_2 of length two in S_2 (b_1 and b_2 are generators in (1) or their inverses). Now $\overline{b_1}b_2$ is in S_2 by (iv), and since $\overline{b_1}$ has length at most one, $\overline{b_1}b_2$ must have length two. We choose $\overline{b_1}b_2$ as the representative of S_2. In general, assuming we have chosen representatives for all cosets of length less than r, if S_r is a coset of length r and $b_1 \ldots b_{r-1}b_r$ is a word in S_r, we choose $\overline{b_1 \ldots b_{r-1}}b_r$ (which has length r) as the representative of S_r. Clearly, by construction, if the last symbol is deleted from a representative, another representative is obtained. Hence, any initial segment of a representative is a representative, and we have constructed a Schreier system for G mod H. ◀

It should be noted that in the proof of the lemma we constructed a *minimal Schreier system*, i.e., a Schreier system in which each representative has a length not exceeding the length of any word it represents. Not every Schreier system is minimal (see Problem 16).

Using a Reidemeister-Schreier rewriting process the relations (5) in a presentation for H can be greatly simplified.

THEOREM 2.9 (Schreier). *Let G be presented as in* (1) *and let H be a subgroup of G. If τ is a Reidemeister-Schreier rewriting process, then H can be presented as*

(25) $$\langle s_{K,a_\nu}, \ldots ; s_{M,a_\lambda}, \ldots , \tau(KR_\mu K^{-1}), \ldots \rangle,$$

where K is an arbitrary Schreier representative, a_ν is an arbitrary generator and R_μ is an arbitrary defining relator in (1), *and M is a Schreier representative and a_λ a generator such that*

(26) $$Ma_\lambda \approx \overline{Ma_\lambda}.$$

PROOF. If $Ma_\lambda \approx \overline{Ma_\lambda}$ then $Ma_\lambda\overline{Ma_\lambda}^{-1} \approx 1$. Hence, by (v), $\tau(Ma_\lambda\overline{Ma_\lambda}^{-1}) \approx \tau(1) = 1$. Thus the relation

$$s_{M,a_\lambda} = \tau(Ma_\lambda\overline{Ma_\lambda}^{-1})$$

is derivable from the relator

(27) $$s_{M,a_\lambda},$$

and conversely.

Moreover, using the Schreier property of the representative system, we shall show that

$$\tau(Ka_\nu\overline{Ka_\nu}^{-1})$$

can be reduced to s_{K,a_ν} by deleting s-symbols of the type (27) for which (26) holds. For, any s-symbol replacing an a-symbol of K in computing $\tau(Ka_\nu\overline{Ka_\nu}^{-1})$ will have the form

$$s_{N,a_\rho},$$

or

$$s^{-1}_{Na_\rho^{-1},a_\rho},$$

where N is the initial segment of K preceding the a-symbol replaced. In the first case, Na_ρ is an initial segment of K, and so

$$\overline{Na_\rho} = Na_\rho;$$

in the second case, N is an initial segment of K, and so

$$\overline{Na_\rho^{-1} \cdot a_\rho} = \overline{N} = N \approx Na_\rho^{-1} \cdot a_\rho.$$

Thus, for any s-symbol s_{M,a_λ} entering from K, we have $\overline{Ma_\lambda} \approx Ma_\lambda$. Using that $\tau(U^{-1})$ is the same word in the s-symbols as $\tau(U)^{-1}$, we can show that an s-symbol s_{M,a_λ} replacing an a-symbol from $\overline{Ka_\nu}^{-1}$ in computing $\tau(Ka_\nu\overline{Ka_\nu}^{-1})$, will satisfy $\overline{Ma_\lambda} \approx Ma_\lambda$ (see Problem 1).

Hence, in the face of the defining relators (27), the relation in (5)

$$s_{K,a_\nu} = \tau(Ka_\nu\overline{Ka_\nu}^{-1}),$$

is derivable from the relation

$$s_{K,a_\nu} = s_{K,a_\nu},$$

which may clearly be deleted. ◀.

Corollary 2.9 (Nielsen-Schreier). *A subgroup of a free group is free.*

Proof. If G is a free group, it can be presented with an empty set of defining relators. Hence, the defining relators in (25) are just certain of the generators of the subgroup. Using a Tietze transformation to delete the generators which occur as defining relators, together with the corresponding defining relators, the resulting presentation has an empty set of defining relators. Hence, the subgroup is free. ◀

It should be noted that not only have we shown that a subgroup of a free group is free, but we have also obtained a set of free generating elements for H; namely, those $Ka_\nu\overline{Ka_\nu}^{-1}$ such that Ka_ν is not freely equal to a representative. Schreier obtained this set of free generators for H in 1927. In 1921, Nielsen, using quite a different method, had constructed a set of free generators for H if F were finitely generated (see Section 3.2). The generators of Nielsen are precisely those Schreier generators obtained when a minimal Schreier system is used (see Theorem 3.4).

As an illustration of Theorem 2.9, we shall compute the presentation of the orientation-preserving subgroup P of a certain group R of non-Euclidean motions. Given a triangle in the non-Euclidean plane with angles $\pi/8$, $\pi/8$, and $\pi/2$, if r_1 and r_3 are the reflections in the equal sides of the triangle and r_2 is reflection in the base of the triangle, then it can be shown that the group R of non-Euclidean motions generated by the reflections r_1, r_2, and r_3 has the presentation

(28) $$R = \langle r_1, r_2, r_3; r_1^2, r_2^2, r_3^2, (r_1r_2)^8, (r_2r_3)^8, (r_1r_3)^2\rangle.$$

Now, a reflection reverses orientation; hence, the orientation-preserving subgroup P is just the set of words in (28) of even length. The words 1 and r_1 form a Schreier system for R mod P. Hence, a presentation for P may be obtained using the following generating symbols:

(29) $$s_{1,r_1};\quad s_{1,r_2};\quad s_{1,r_3};\quad s_{r_1,r_1};\quad s_{r_1,r_2};\quad s_{r_1,r_3}.$$

Since $1 \cdot r_1 \approx \overline{1 \cdot r_1}$, the generating symbol s_{1,r_1} is a defining relator; the other defining relators are the following [note that in computing

$\tau(KR_\mu K^{-1})$ only the s-symbols from R_μ need be considered (see Problem 7)]:

(30) $$\tau(1 \cdot r_1{}^2 \cdot 1^{-1}) = \tau(r_1{}^2) = s_{1,r_1} \cdot s_{r_1,r_1};$$

(31) $$\tau(r_1 \cdot r_1{}^2 \cdot r_1{}^{-1}) \approx \tau(r_1{}^2);$$

(32) $$\tau(1 \cdot r_2{}^2 \cdot 1^{-1}) = \tau(r_2{}^2) = s_{1,r_2} \cdot s_{r_1,r_2};$$

(33) $$\tau(r_1 \cdot r_2{}^2 \cdot r_1{}^{-1}) = s_{r_1,r_2} \cdot s_{1,r_2};$$

(34) $$\tau(1 \cdot r_3{}^2 \cdot 1^{-1}) = s_{1,r_3} \cdot s_{r_1,r_3};$$

(35) $$\tau(r_1 \cdot r_3{}^2 \cdot r_1{}^{-1}) = s_{r_1,r_3} \cdot s_{1,r_3};$$

(36) $$\tau(1 \cdot (r_1r_2)^8 \cdot 1^{-1}) = \tau(r_1r_2)^8 = (s_{1,r_1} \cdot s_{r_1,r_2})^8;$$

(37) $$\tau(r_1 \cdot (r_1r_2)^8 \cdot r_1{}^{-1}) \approx \tau(r_1 \cdot r_1r_2 \cdot r_1{}^{-1})^8 = (s_{r_1,r_1} \cdot s_{1,r_2})^8;$$

(38) $$\tau(1 \cdot (r_2r_3)^8 \cdot 1^{-1}) = \tau(r_2r_3)^8 = (s_{1,r_2} \cdot s_{r_1,r_3})^8;$$

(39) $$\tau(r_1 \cdot (r_2r_3)^8 \cdot r_1{}^{-1}) \approx \tau(r_1 \cdot r_2r_3 \cdot r_1{}^{-1})^8 = (s_{r_1,r_2} \cdot s_{1,r_3})^8;$$

(40) $$\tau(1 \cdot (r_1r_3)^2 \cdot 1^{-1}) = \tau(r_1r_3)^2 = (s_{1,r_1} \cdot s_{r_1,r_3})^2;$$

(41) $$\tau(r_1 \cdot (r_1r_3)^2 \cdot r_1{}^{-1}) = \tau(r_1 \cdot r_1r_3 \cdot r_1{}^{-1})^2 = (s_{r_1,r_1} \cdot s_{1,r_3})^2.$$

[In computing (36) through (41) we used that $\tau(U^n) = \tau(U)^n$, if U is in H.]

Thus P can be presented on the generating symbols (29) with the defining relators (30) through (41), and s_{1,r_1}. To simplify the presentation, note that (30) implies that s_{r_1,r_1} is a relator; (31) states the same; (32) and (33) state that s_{1,r_2} defines the same element as s_{r_1,r_2}^{-1}; (34) and (35) state that s_{1,r_3} defines the same element as s_{r_1,r_3}^{-1}; (36) and (37) state that s_{r_1,r_2}^8 is a relator; (38) and (39) state that $(s_{r_1,r_2} \cdot s_{r_1,r_3}^{-1})^8$ is a relator; (40) and (41) state that s_{r_1,r_3}^2 is a relator. Hence, using appropriate Tietze transformations, and denoting s_{r_1,r_2} by p_2 and s_{r_1,r_3} by p_3, P has the presentation

(42) $$\langle p_2, p_3; p_2{}^8, p_3{}^2, (p_2{}^{-1}p_3)^8\rangle,$$

under the mapping $p_2 \to r_1r_2$ (r_1r_2 is $r_1 \cdot r_2 \cdot \overline{r_1r_2}^{-1}$) and $p_3 \to r_1r_3$ (r_1r_3 is $r_1 \cdot r_3 \cdot \overline{r_1r_3}^{-1}$). Using Tietze transformations to introduce $p_1 = p_2{}^{-1}p_3$, and then to delete p_3, we may also present P by

(43) $$\langle p_1, p_2; p_1{}^8, p_2{}^8, (p_1p_2)^2\rangle,$$

under the mapping $p_1 \to r_2r_3$ and $p_2 \to r_1r_2$.

As another illustration of Theorem 2.9, let

$$L = \langle x, y; x^2, y^3\rangle$$

be the group of unimodular linear fractional transformations with integer coefficients (see Problem 1.4.18); let $H = L(X_1X_2X_1^{-1}X_2^{-1})$. We shall show that H is a free group; for we may choose

$$1,\ x,\ y,\ xy,\ y^2,\ xy^2,$$

as a Schreier representative system for L mod H. Hence, a presentation for H may be obtained using the following generating symbols:

$$(44)\quad s_{1,x};\ s_{1,y};\ s_{x,x};\ s_{x,y};\ s_{y,x};\ s_{y,y};\ s_{xy,x};\ s_{xy,y};\ s_{y^2,x};\ s_{y^2,y};\ s_{xy^2,x};\ s_{xy^2,y}.$$

The first, second, third, fifth, and seventh s-symbol have property (26), and hence are defining relators; the other defining relators are the following (where we omit any s-symbol which is a defining relator):

$$(45)\quad \tau(1 \cdot x^2 \cdot 1^{-1}) = \tau(x^2) = s_{x,x};$$

$$(46)\quad \tau(x \cdot x^2 \cdot x^{-1}) \approx \tau(x^2);$$

$$(47)\quad \tau(y \cdot x^2 \cdot y^{-1}) = s_{y,x} \cdot s_{xy,x};$$

$$(48)\quad \tau(xy \cdot x^2 \cdot y^{-1}x^{-1}) = s_{xy,x} \cdot s_{y,x};$$

$$(49)\quad \tau(y^2 \cdot x^2 \cdot y^{-2}) = s_{y^2,x} \cdot s_{xy^2,x};$$

$$(50)\quad \tau(xy^2 \cdot x^2 \cdot y^{-2}x^{-1}) = s_{xy^2,x} \cdot s_{y^2,x};$$

$$(51)\quad \tau(1 \cdot y^3 \cdot 1^{-1}) = \tau(y^3) = s_{y^2,y};$$

$$(52)\quad \tau(x \cdot y^3 \cdot x^{-1}) = s_{xy^2,y};$$

$$(53)\quad \tau(y \cdot y^3 \cdot y^{-1}) \approx \tau(y^3);$$

$$(54)\quad \tau(xy \cdot y^3 \cdot y^{-1}x^{-1}) \approx \tau(x \cdot y^3 \cdot x^{-1});$$

$$(55)\quad \tau(y^2 \cdot y^3 \cdot y^{-2}) \approx \tau(y^3);$$

$$(56)\quad \tau(xy^2 \cdot y^3 \cdot y^{-2}x^{-1}) \approx \tau(x \cdot y^3 \cdot x^{-1}).$$

We note that the defining relators (45), (51), and (52) are three more s-symbols. Moreover, the defining relators (47), (48), (49), and (50) state that

$$s_{xy,x} = s_{y,x}^{-1}, \qquad \text{and} \qquad s_{xy^2,x} = s_{y^2,x}^{-1}.$$

Hence, by Tietze transformations, all s-symbols except $s_{y,x}$ and $s_{y^2,x}$, and all defining relators may be deleted from the presentation for H. Thus H is a free group on the two generators

$$yxy^{-1}x^{-1}, \qquad y^2xy^{-2}x^{-1}.$$

We may use this to get a matrix representation of the free group on two generators; for the mapping onto L,

$$\begin{pmatrix} 0 & 1 \\ -1 & 0 \end{pmatrix} \to x, \qquad \begin{pmatrix} 0 & 1 \\ -1 & 1 \end{pmatrix} \to y,$$

is a homomorphism of a subgroup of the multiplicative group of two-by-two unimodular matrices with integer coefficients. Under this homomorphism,

$$\begin{pmatrix} 1 & 1 \\ 1 & 2 \end{pmatrix} \to yxy^{-1}x, \qquad \text{and} \qquad \begin{pmatrix} 2 & 1 \\ 1 & 1 \end{pmatrix} \to y^2xy^{-2}x^{-1}.$$

Since $yxy^{-1}x^{-1}$ and $y^2xy^{-2}x^{-1}$ freely generate a free group, so must the matrices mapped into them. Thus the two matrices

$$\begin{pmatrix} 1 & 1 \\ 1 & 2 \end{pmatrix}, \quad \begin{pmatrix} 2 & 1 \\ 1 & 1 \end{pmatrix}$$

freely generate a free group. Consequently, every free group of finite or countable rank can be represented faithfully by two-by-two unimodular matrices with integer coefficients (see Problem 1.4.12). For other matrix representations of the free group, see Problem 14.

Problems for Section 2.3

In the following problems, τ is a Reidemeister rewriting process for a subgroup H of a group G presented by (1).

1. Show that $\tau(U)$ has the same length in the s-symbols as U has in the generating symbols of G. Show that $\tau(U^{-1})$ and $\tau(U)^{-1}$ are identical words in the s-symbols, where U is in H. [*Hint:* Let $U = Va^\epsilon T^{-1}$ be in H where a is a generating symbol for G, and $\epsilon = \pm 1$. If U has length r and a^ϵ is the kth symbol in U, then U^{-1} has length r and $a^{-\epsilon}$ is the $(r - k + 1)$th symbol in U^{-1}. Hence, if $\epsilon = 1$, then $s_{\overline{V},a}$ is the kth symbol in $\tau(U)$, and

$$s^{-1}_{\overline{Ta^{-1}},a} = s^{-1}_{\overline{V},a}$$

is the $(r - k + 1)$th symbol in $\tau(U^{-1})$; if $\epsilon = -1$, then $s_{\overline{Va^{-1}},a}$ is the kth symbol in $\tau(U)$, and $s^{-1}_{\overline{T},a} = s^{-1}_{\overline{Va^{-1}},a}$ is the $(r - k + 1)$th symbol in $\tau(U^{-1})$.]

2. Let $U_1(a_\nu)$ and $U_2(a_\nu)$ define elements in H. Then U_1 and U_2 are identical words in a_ν if and only if $\tau(U_1)$ and $\tau(U_2)$ are identical words in the s-symbols. Show that U_1 is freely (cyclically) reduced in the free group on

the a_ν if and only if $\tau(U_1)$ is freely (cyclically) reduced in the s-symbols. [*Hint:* If

$$\tau(U) = s^{\epsilon_1}_{K_1,a_{\nu_1}} \cdots s^{\epsilon_r}_{K_r,a_{\nu_r}}, \quad \text{then} \quad U = a^{\epsilon_1}_{\nu_1} \cdots a^{\epsilon_r}_{\nu_r}.]$$

3. Let UV, US, and TV define elements of H. Show that the s-symbols which replace the symbols of U in computing $\tau(UV)$ are the same as the s-symbols which replace U in computing $\tau(US)$. Show that the s-symbols that replace the symbols of V when computing $\tau(UV)$ and $\tau(TV)$ are the same. [*Hint:* The s-symbol replacing a generating symbol a_ν^ϵ in computing τ of a word depends only upon a_ν^ϵ and the representative of the initial segment preceding a_ν^ϵ (for $\epsilon = -1$, use that if $\overline{P} = \overline{Q}$ then $\overline{Pa_\nu^{-1}} = \overline{Qa_\nu^{-1}}$). Moreover, if UV and TV define elements of H, then $\overline{U} = \overline{V^{-1}} = \overline{T}$ and $\overline{UP} = \overline{\overline{U}P} = \overline{\overline{T}P} = \overline{TP}$.]

4. Although the process τ has only been defined for a word $U(a_\nu)$ which defines an element of H, show that if W defines a word not in H, but nevertheless the replacement of an a-symbol in W by an s-symbol is carried out using (39), and then s_{K,a_ν} is replaced by $Ka_\nu\overline{Ka_\nu}^{-1}$, a word results which is freely equal to $W\overline{W}^{-1}$ in the a_ν. [*Hint:* Use the hint to Problem 3, together with the fact that $s^{\epsilon_j}_{K,a_{\nu_j}}$ is replaced by $\overline{a^{\epsilon_1}_{\nu_1} \cdots a^{\epsilon_{j-1}}_{\nu_{j-1}}}a^{\epsilon_j}_{\nu_j}\overline{a^{\epsilon_1}_{\nu_1} \cdots a^{\epsilon_j}_{\nu_j}}^{-1}$.]

5. Suppose that the Reidemeister rewriting process for H uses a right coset function which is *minimal*, i.e., every representative has the smallest length of any word in its coset. Show that if $U(a_\nu)$ defines an element of H, and s^ϵ_{K,a_ν} occurs in computing $\tau(U)$, then the length of $Ka_\nu\overline{Ka_\nu}^{-1}$ in the a_ν symbols is no larger than the length of $U(a_\nu)$. [*Hint:* If $U = Va_\nu T^{-1}$ and $\overline{V} = K$, then $\overline{T} = \overline{Va_\nu} = \overline{Ka_\nu}$. Hence, $L(K) \le L(V)$ and $L(T^{-1}) = L(T) \ge L(\overline{Ka_\nu})$. Similarly, if $\epsilon = -1$.]

In the following problems τ is a Reidemeister-Schreier rewriting process for a subgroup H of a group G presented by (1).

6. Show that if K is a Schreier representative, then K is freely reduced in the a_ν. Also show that $Ka_\nu\overline{Ka_\nu}^{-1}$ is freely reduced or freely equal to the empty word in the a_ν. [*Hint:* If $K = Ma_\nu^\epsilon a_\nu^{-\epsilon}T$, then M and $\overline{Ma_\nu^\epsilon a_\nu^{-\epsilon}}$ define the same coset. Also, if $K = Ma_\nu^{-1}$, then $\overline{Ka_\nu} = M$; and if $\overline{Ka_\nu} = Ma_\nu$, then $K = \overline{Ka_\nu \cdot a_\nu^{-1}} = \overline{Ma_\nu \cdot a_\nu^{-1}} = M$.]

7. Show that if K is a Schreier representative and KT defines an element of H, then the s-symbols replacing the symbols of K in computing $\tau(KT)$ are all of the form s^ϵ_{M,a_λ}, where $Ma_\lambda \approx \overline{Ma_\lambda}$. Conclude that if KVP^{-1} defines an element of H, where K and P are Schreier representatives, then the s-symbols replacing the symbols of K and P^{-1} in computing $\tau(KVP^{-1})$ all have the form s^ϵ_{M,a_λ}, where $Ma_\lambda \approx \overline{Ma_\lambda}$. [*Hint:* If $K = Qa_\nu S$, then $\overline{Qa_\nu} = Qa_\nu \approx Qa_\nu$; if $K = Qa_\nu^{-1}S$ then $\overline{\overline{Qa_\nu^{-1}} \cdot a_\nu} = \overline{Qa_\nu^{-1}a_\nu} = Q \approx Qa_\nu^{-1}a_\nu = \overline{Qa_\nu^{-1}} \cdot a_\nu$. To show that the s-symbols replacing the symbols of P^{-1} have the desired form, use $\tau(U^{-1}) = \tau(U)^{-1}$.]

8. If F is the free group on a and b, find a Schreier representative system for F mod H, where $H = F(X^2)$. Using this Schreier system, construct a set of free generators for H by means of Theorem 2.9. [*Hint:* For example, $1, a, b, ab$ is a Schreier system for F mod H. Generators for H are the elements corresponding to $s_{1,a}, s_{a,a}, s_{b,a}, s_{ab,a}, s_{1,b}, s_{a,b}, s_{b,b}$, and $s_{ab,b}$. The first, fourth, and fifth generating symbols are of the form s_{M,a_λ}, where $Ma_\lambda \approx \overline{Ma_\lambda}$. Hence, H is freely generated by

$$a \cdot a \cdot \overline{aa}^{-1} = a^2, \qquad b \cdot a \cdot \overline{ba}^{-1} = bab^{-1}a^{-1}, \qquad ab \cdot a \cdot \overline{ab \cdot a}^{-1} = abab^{-1},$$
$$b \cdot b \cdot \overline{b \cdot b}^{-1} = b^2, \qquad \text{and} \qquad ab \cdot b \cdot \overline{ab \cdot b}^{-1} = ab^2a^{-1}.]$$

9. If F is the free group on a, b, and c, find a Schreier representative system for F mod H, where $H = F(X^2)$. Using this Schreier system, construct a set of free generators for H by means of Theorem 2.9. [*Hint:* For example, $1, a, b, c, ab, bc, ac, abc$ is a Schreier system for F mod H.]

10. If F is the free group on a and b, and N is the normal subgroup of F generated by a^r and b, $r > 0$, show that N is freely generated by a^r and a^kba^{-k}, where $0 \le k < r$. [*Hint:* Use the Schreier system $1, a, \ldots, a^{r-1}$ for F mod N, and Theorem 2.9.]

11. If F is the free group on a, b, and c, and N is the normal subgroup of F generated by b and c, show that N is freely generated by a^kba^{-k}, and a^mca^{-m}, where k and m range over all integers. [*Hint:* Use the Schreier system a^k, k any integer, for F mod N, and Theorem 2.9.]

12. Let $L = \langle x, y; x^2, y^3 \rangle$, and let $H = L(X^2)$. Find a Schreier system for L mod H. Using this system and Theorem 2.9, find a presentation for H. [*Hint:* In computing $\tau(Kx^2K^{-1})$, by Problem 7, only the s-symbols that occur in replacing the symbols in x^2 need be computed. Similarly, for $\tau(Ky^3K^{-1})$.]

13. Let $L = \langle x, y; x^2, y^3 \rangle$ be the group of unimodular linear fractional transformations with integer coefficients (see Problem 1.4.18).

(a) Show that the mapping α of L into the multiplicative group T of two-by-two non-singular matrices with entries from the ring of integers modulo two, given by

$$z \to \frac{az + b}{cz + d}$$

goes into

$$\begin{pmatrix} a' & b' \\ c' & d' \end{pmatrix},$$

where a', b', c', d' are the remainders of a, b, c, d, respectively, modulo two, is a homomorphism of L onto T.

(b) Show that T has order six.

(c) Show that $(xy)^2$ goes into the identity in T, under α.

(d) Show that N, the normal subgroup of L generated by $(xy)^2$ has index six in L.

(e) Conclude that N is the kernel of α.

(f) Find a presentation for N using the Schreier system 1, x, y, xy, y^2, xy^2 for L mod N, and Theorem 2.9.

(g) Show that N can be presented as a free group on the generating symbols $s_{y,x}$ and $s_{y^2,x}$; hence, show that N is freely generated by $(yx)^2$ and $(y^{-1}x)^2$.

(h) Show that the two matrices

$$\begin{pmatrix} 1 & 0 \\ 2 & 1 \end{pmatrix}, \quad \begin{pmatrix} 1 & 2 \\ 0 & 1 \end{pmatrix}$$

generate a free group.

(i) Show that the set of matrices with odd entries on the main diagonal and even entries elsewhere is a free group on two generators, if we identify a matrix with its negative.

[*Hint:* For (h), see the last illustration at the end of this section. For (i), show that the matrices described in (i) define the linear fractional transformations of N.]

14. Let P be the group

$$\langle p_1, p_2; p_1{}^8, p_2{}^8, (p_1p_2)^2 \rangle$$

in the illustration at the end of the section. Let N be the normal subgroup of P generated by $p_1p_2{}^{-3}$.

(a) Show that P/N is a cyclic group of order eight.

(b) Show that $p_1{}^\nu$, $\nu = 0, 1, \ldots, 7$ is a Schreier system for P mod N, and that $\overline{W(p_1, p_2)} = p_1{}^\lambda$, where $\lambda = \sigma_1(W) + 3\sigma_2(W)$ modulo 8 (σ_i is exponent sum on p_i).

(c) Show that $s_{p_1{}^\nu, p_1}$ is a defining relator in the presentation of N in Theorem 2.9, for $\nu = 0, 1, \ldots, 6$.

(d) Show that $\tau(p_1{}^\nu \cdot p_1{}^8 \cdot p_1{}^{-\nu}) \approx \tau(p_1{}^8) = \tau(p_1{}^7 \cdot p_1)$, and hence that the defining relators $\tau(p_1{}^\nu \cdot p_1{}^8 \cdot p_1{}^{-\nu})$ can be replaced by the defining relator $s_{p_1{}^7, p_1}$.

(e) Conclude from (c) and (d) that in computing $\tau(U)$ only those s-symbols replacing p_2 or $p_2{}^{-1}$ need be considered.

(f) Using s_ν to abbreviate, $s_{p_1{}^\nu, p_2}$ show that the defining relator $\tau(p_1{}^\nu \cdot p_1p_2p_1p_2 \cdot p^{-\nu})$ can be replaced by the defining relation

$$s_{\nu+1} = s_{\nu+5}^{-1},$$

where $\nu + 1$ and $\nu + 5$ are computed modulo eight.

(g) Show that the defining relator $\tau(p_2{}^8)$ can be replaced by the defining relator

$$S = s_0 \cdot s_3 \cdot s_6 \cdot s_1 \cdot s_4 \cdot s_7 \cdot s_2 \cdot s_5,$$

and that the defining relator $\tau(p_1{}^\nu \cdot p_2{}^8 \cdot p_1{}^{-\nu})$ can be replaced by a cyclic permutation of S.

(h) Conclude from (f) and (g) that the presentation of N given in Theorem 2.9 can be simplified to

$$\langle x, y, z, w; xyzwx^{-1}y^{-1}z^{-1}w^{-1}\rangle.$$

(i) Show by introducing the new generators

$$a_1 = y^{-1}, \qquad b_1 = x, \qquad a_2 = xzy, \qquad \text{and} \qquad b_2 = y^{-1}wx^{-1},$$

by Tietze transformations, and then deleting x, y, z, w, that N has the presentation

$$\langle a_1, b_1, a_2, b_2; a_1b_1a_1^{-1}b_1^{-1}a_2b_2a_2^{-1}b_2^{-1}\rangle.$$

(Note that this is the fundamental group of the sphere with two handles.)

15. Let G have the presentation

$$\langle a, b, c, \ldots ; R_\mu(a, b, c, \ldots), \ldots\rangle,$$

where each R_μ has zero exponent sum on a. Let N be the normal subgroup of G generated by $b, c, \ldots$. Show that $W(a, b, c, \ldots)$ defines an element of N if and only if W has zero exponent sum on a. Show that the words a^ν, ν any integer, form a Schreier representative system for G mod N, and that $b_\nu = a^\nu ba^{-\nu}$, $c_\nu = a^\nu ca^{-\nu}, \ldots$, where ν is any integer, generate N.

16. Let $G = \langle a, b; a^2, b^2\rangle$. Find two minimal Schreier systems for the verbal subgroup $H = G(X_1X_2X_1^{-1}X_2^{-1})$. Find a minimal system of representatives which is not a Schreier system. Show that $1, a, ab, aba$ is a Schreier system which is not minimal.

17. Let F be the free group on a and b. Let N be the normal subgroup of F generated by a^2, b^3, and $(ab)^2$, and let H be the normal subgroup of F generated by a^2, b^3, and $aba^{-1}b^{-1}$. Show that $N \neq H$, although there is a set of words which is a Schreier representative system for N and H simultaneously.

18. Let G be generated by $a_1, \ldots, a_n$, and let H be a subgroup of G. Show that a right coset representative function for H is determined by the set of words used for representatives of G mod H together with the mapping which assigns to each pair K, a_ν^ϵ, the representative $\overline{Ka_\nu^\epsilon}$, where K is a representative, a_ν is a generator of G, and $\epsilon = \pm 1$. [*Hint:* Use induction on the length of W to show that $\overline{W}$ is determined.]

19. Show that if H has index j in G, then any Schreier representative of G mod H has length at most $j - 1$. Hence, if G is finitely generated, there are only finitely many right coset Schreier representative functions which can correspond to some subgroup H of index j in G. Conclude that a finitely generated group has only finitely many subgroups of a given finite index.

20. Let F be the free group on $a_1, \ldots, a_n$, and let $J = j_1, \ldots, j_r$ be a non-empty subset of $1, \ldots, n$. Let F_J be the subgroup of F which consists of all $W(a_\nu)$ such that the sum of the exponents of W on all the generators a_ν with ν in J is even. Show that F_J has index two in F. Conversely, show that every subgroup of index two in F is some F_J. Hence, conclude that there

are $2^n - 1$ subgroups of index two in F. [*Hint:* Let H be of index two in F. Then H is normal and there is some generator, say a_1, such that a_1 is not in H and so $1, a_1$ is a Schreier system for F mod H. Let $a_{j_1}, \ldots, a_{j_r}$ be the subset of generators whose representative is a_1. Since H is normal and of index two,

$$\overline{W} = \overline{a_{\nu_1}^{\epsilon_1} \cdot \ldots \cdot a_{\nu_k}^{\epsilon_k}} = \overline{\overline{a_{\nu_1}} \cdot \ldots \cdot \overline{a_{\nu_k}}} = \overline{a_1{}^q},$$

where q is the sum of the exponents of W on $a_{j_1}, \ldots, a_{j_r}$. Hence, W is in H if and only if q is even, i.e., $H = F_J$, where J consists of $j_1, \ldots, j_r$.]

21. Let H be a subgroup of finite index in G. Then H is finitely generated if and only if G is. [*Hint:* Use Corollary 2.7.1; use also that generators for H and representatives for G mod H together generate G.]

22. Let F be the free group on $a_1, a_2, \ldots, a_n, \ldots$. Show that if H is a subgroup containing $a_1, \ldots, a_r$, then $a_1, \ldots, a_r$ generate a *free factor* of H, i.e., $a_1, \ldots, a_r$ are part of a set of free generators for H. [*Hint:* Use Theorem 2.9 and note that $\overline{a_\rho} = 1$ so that s_{1,a_ρ} defines a_ρ.]

23. Let G be a group generated by $a_1, \ldots, a_n$, and let H be a subgroup of G. Suppose that a set of words S in the a_ν is closed under taking initial segments, and that different words of S define different cosets of G mod H. Then S may be extended to a Schreier system for G mod H. Moreover, if the words in S have minimal length in their cosets, S may be extended to a minimal Schreier system. [*Hint:* Modify the proof of Lemma 2.2 preceding Theorem 2.9 so that cosets containing a word of S have that word as representative. Then proceed to define representatives for the other cosets inductively.]

24. Let $G = \langle a_1, \ldots, a_n; R(B(a_\nu), C(a_\nu))\rangle$, where $B(a_\nu)$ and $C(a_\nu)$ are two words in the a_ν. Show that the subgroup H of G, generated by $B(a_\nu)$ and $C(a_\nu)$ cannot, in general, be presented simply by

$$\langle b, c; R(b, c)\rangle,$$

under the mapping $b \to B(a_\nu)$ and $c \to C(a_\nu)$. [*Hint:* Let $G = \langle a_1, a_2; (a_1a_2)^3\rangle$, and let H be the subgroup generated by a_1a_2 and a_2a_1. Then, since in G, $a_2a_1 = a_1^{-1}(a_1a_2)a_1$, it follows that a_2a_1 has order three, while in $\langle b, c; b^3\rangle$, c has infinite order.]

25. Let $G = \langle a_1, \ldots, a_n; R_\mu(a_\nu), \ldots\rangle$ be a splitting extension of its normal subgroup N by its subgroup H, i.e., $G = HN$ and $H \cap N = 1$. Choose a set of representative words $K(a_\nu)$ for the cosets of G mod N, where $K(a_\nu)$ defines an element of H. Show that if U and V are any words in a_ν, then $\overline{UV} = \overline{U} \cdot \overline{V} \cdot T$, where T is a product of conjugates of the $R_\mu{}^\epsilon$, with $\epsilon = \pm 1$. Show that N is generated by the elements of G defined by $a_\nu\overline{a_\nu}^{-1}$ and their conjugates. Conclude from Theorem 2.1 that $H = G/N = \langle a_1, \ldots, a_n; R_\mu(a_\nu), \ldots, a_\nu\overline{a_\nu}^{-1}, \ldots\rangle$. [*Hint:* Since N is normal $\overline{UV}$ and $\overline{U}\overline{V}$ are in the same coset of N. Moreover, since representatives define elements of H, and each coset has a unique element of H in it, $\overline{UV}$ and $\overline{U}\overline{V}$ define the same element of H, and so $\overline{UV} = \overline{U} \cdot \overline{V} \cdot T$.

Since N is generated by $Ka_\nu \overline{Ka_\nu}^{-1}$, and $\overline{Ka_\nu} = \overline{\overline{K}a_\nu T} = \overline{Ka_\nu}T$, the element of G defined by $Ka_\nu \overline{Ka_\nu}^{-1}$ is the same as the element of G defined by $Ka_\nu \overline{a_\nu}^{-1} K^{-1}$.]

26. Let $G = \langle a_1, \ldots, a_n; R_\mu(a_\nu), \ldots \rangle$ and let H be a subgroup of G. Suppose that n is finite, and that it can be decided in a finite number of steps whether a word $U(a_\nu)$ defines an element of H or not. Show that there is a Schreier representative system for G mod H such that the representative of any word $V(a_\nu)$ can be obtained in finitely many steps. Show that if H has finite index in G, and G has finitely many defining relators, then a presentation for H can be constructed in finitely many steps. [*Hint:* Order words in the a_ν as in (3) of Section 1.3. Then choose the representative of a coset of G mod H to be the smallest word in the coset. Since $W_1 < W_2$ implies $W_1 W < W_2 W$, the initial segment of a representative is a representative. To compute the representative of a word $V(a_\nu)$, examine the finite number of words V_i less than V, and take the least such for which $V_i V^{-1}$ determines an element of H. If H has finite index, keep listing words in order, until all those of some length have representatives of smaller length. Then all representatives for G mod H can be obtained. Moreover, the rewriting process τ can be effectively computed; and so the presentation of Theorem 2.9 can be obtained in finitely many steps, if G has finitely many defining relators.]

2.4. Subgroups of Free Groups

In this section, we apply the Reidemeister-Schreier rewriting process to subgroups of free groups.

We have already shown that a subgroup of a free group is free; we now establish a formula for the rank of a subgroup of a free group.

THEOREM 2.10. *Let F be the free group on $a_1, \ldots, a_n$, and let j be the index of the subgroup H in F. If both n and j are finite, then H is a free group on $j(n-1)+1$ generators. If n is infinite and j is finite, then H is a free group on infinitely many generators. Finally, if j is infinite, then H may be finitely or infinitely generated; however, if H contains a normal subgroup N of F, $N \neq 1$, then H is a free group on infinitely many generators.*

PROOF. The presentation in Theorem 2.9 for the subgroup H of a free group F has as generators the s-symbols s_{K,a_ν}; moreover, the defining relators are the s-symbols

$$s_{M,a_\lambda} \tag{1}$$

such that

$$Ma_\lambda \approx \overline{Ma_\lambda}. \tag{2}$$

The rank of H will be the number of generators minus the number of defining relators.

We shall show that there are as many s-symbols which are defining relators as there are representatives other than the empty word in the Schreier system for F mod H. To do this associate an s-symbol to every Schreier representative $K \neq 1$ as follows:

If $K = Ma_\lambda$, then associate s_{M,a_λ}, and if $K = Ma_\lambda^{-1}$, then associate $s_{Ma_\lambda^{-1},a_\lambda}$. Clearly, the s-symbol associated with a representative occurs as a defining relator in the presentation for H. Conversely, suppose

$$s_{P,a_\mu} \tag{3}$$

occurs as a defining relator in the presentation for H. Then

$$Pa_\mu \approx \overline{Pa_\mu}.$$

Now, if P does not end in a_μ^{-1}, then Pa_μ and $\overline{Pa_\mu}$ are freely reduced and so $K = \overline{Pa_\mu} = Pa_\mu$ is the Schreier representative with which (3) is associated; if P ends in a_μ^{-1}, then $K = P$ is the Schreier representative with which (3) is associated. Thus, if H has index j in F, then there are $j - 1$ s-symbols (1) which satisfy (2).

These s-symbols may be deleted from the presentation for H by Tietze transformations; since there were jn generators originally, H has $jn - (j - 1) = j(n - 1) + 1$ free generators.

If j is finite and n infinite, then infinitely many generators remain in the presentation for H, and so H has infinitely many free generators.

Finally, suppose j is infinite and H contains N, a normal subgroup of F such that $N \neq 1$. We shall show that there are infinitely many s-symbols

$$s_{Q,a_\sigma} \tag{4}$$

such that Qa_σ is not freely equal to $\overline{Qa_\sigma}$. In order to construct such representatives Q, let U be a freely reduced word $\neq 1$, all of whose conjugates are in H, and let K be any representative. Consider the initial segments KV of KU. Now, KUK^{-1} is in H, and so

$$\overline{KU} = \overline{KUK^{-1} \cdot K} = \bar{K} = K.$$

Hence, KU is not freely equal to a representative; for otherwise, $KU \approx \overline{KU} = K$, and so $U \approx 1$. Since K is a representative, there must be an initial segment V of U, $V = Wa_\sigma^\epsilon$, $\epsilon = \pm 1$, of smallest length such that KV is not freely equal to a representative. By construction, KW is freely equal to a representative, and hence, to $\overline{KW}$. Hence, if $\epsilon = 1$,

$\overline{KWa_\nu}$ is freely equal to $KWa_\nu = KV$ which is not freely equal to a representative; therefore, in this case, if $Q = \overline{KW}$ and $a_\sigma = a_\nu$, then the corresponding s-symbol (4) is of the type we seek. On the other hand, if $\epsilon = -1$, then for $Q = \overline{KWa_\nu^{-1}}$ and $a_\sigma = a_\nu$, the s-symbol in (4) is of the type we seek; for, if $\overline{KWa_\nu^{-1}} \cdot a_\nu$ is freely equal to a representative, it must be freely equal to $\overline{\overline{KWa_\nu^{-1}} \cdot a_\nu} = \overline{KW}$, which is freely equal to KW; hence, $\overline{KWa_\nu^{-1}} \cdot a_\nu \approx KW$, and so $KWa_\nu^{-1} = KV$ is freely equal to a representative, contrary to the construction of KV.

Thus for each representative K there is an initial segment P of U such that if $Q = \overline{KP}$, then there is an a_σ such that

$$s_{Q,a_\sigma}$$

is of the type where Qa_σ is not freely equal to $\overline{Qa_\sigma}$.

To show that there are infinitely many such s-symbols, suppose $\overline{KP_1} = \overline{MP_2}$ where P_1 and P_2 are initial segments of U, and K and M are representatives. Then

$$M = \overline{M} = \overline{MP_2 \cdot P_2^{-1}} = \overline{\overline{MP_2} \cdot P_2^{-1}} = \overline{\overline{KP_1}P_2^{-1}} = \overline{KP_1P_2^{-1}}.$$

Since P_1 and P_2 are initial segments of U, there are only finitely many M such that for a given representative K, the s-symbols (4) constructed from K and M and initial segments of U coincide. Since there are infinitely many representatives, there are infinitely many s-symbols (4) which are not defining relators in the presentation for H. ◀

Corollary 2.10. *A subgroup H of a finitely generated free group F is of finite index in F if and only if H is finitely generated and contains the verbal subgroup $F(X^d)$, for some positive integer d.*

Proof. If H is finitely generated and contains $F(X^d)$, then Theorem 2.10 applies, and H is of finite index in F.

On the other hand, if H is of finite index in F and F is finitely generated, then H is finitely generated. Moreover, if j is the index of H in F, then for any word W,

$$1, W, W^2, \ldots, W^j.$$

cannot all determine different cosets of H. Hence, for each W, W^m is in H for some m with $1 \leq m \leq j$; thus $W^{j!}$ is in H, and so $F(X^{j!})$ is in H. ◀

If F is the free group on a and b, then it can be shown that a^nb^2 and $a^{n+1}b^2$ are free generators of the subgroup H of F generated by a and b^2 (see Corollary 2.13.1); note that a^nb^2 and $a^{n+1}b^2$ have length as large as we please, and yet H has an element of length one. However, there are free generators for H of "minimal" length, namely, a and b^2. The existence of free generators with "minimal length" is guaranteed for a subgroup H of a free group F by the following:

THEOREM 2.11(Levi). *Let F be the free group on $a_1, \ldots, a_n$, and let H be a subgroup of F. Then there exists a set of free generators $b_\mu = B_\mu(a_\nu)$ for H, such that if $U(a_\nu)$ is in H and involves b_λ (when expressed as a freely reduced word in the b_μ), then the length of $B_\lambda(a_\nu)$ does not exceed the length of $U(a_\nu)$ (length with respect to the a_ν generators).*

PROOF. We show that a system of free generators for H arising from a minimal Schreier system has the above property. For suppose that in applying a Reidemeister-Schreier rewriting process to $U(a_\nu)$, the s-symbol s_{K,a_λ} occurs and is not a defining relator. Then

$$U = Va_\lambda W^{-1},$$

where $\overline{V} = K$. Now $\overline{W} = \overline{Va_\lambda} = \overline{Ka_\lambda}$. Hence,

$$\begin{aligned} L(Ka_\lambda\overline{Ka_\lambda}^{-1}) &= L(K) + 1 + L(\overline{Ka_\lambda}) \\ &\leq L(V) + 1 + L(W) = L(U), \end{aligned}$$

since in a minimal Schreier system a representative has length not exceeding that of any word it represents.

Similarly, if the s-symbol s_{K,a_λ}^{-1} occurs in computing τ, then

$$U = Va_\lambda^{-1}W^{-1},$$

where $\overline{Va_\lambda^{-1}} = K$. Now $\overline{W} = \overline{Va_\lambda^{-1}} = K$, and so

$$\begin{aligned} L(Ka_\lambda\overline{Ka_\lambda}^{-1}) &= L(K) + 1 + L(\overline{Ka_\lambda}) \\ &\leq L(W) + 1 + L(\overline{\overline{Va_\lambda^{-1}} \cdot a_\lambda}) \\ &\leq L(W) + 1 + L(V) = L(U). \end{aligned}$$

Since the s_{K,a_λ} with Ka_λ not freely equal to $\overline{Ka_\lambda}$ form a free set of generators for H, if some s_{K,a_λ} did not occur in computing $\tau(U)$, then U would not involve $Ka_\lambda\overline{Ka_\lambda}^{-1}$. Hence, $L(U)$ is at least as large as $L(Ka_\lambda\overline{Ka_\lambda}^{-1})$ for each such $Ka_\lambda\overline{Ka_\lambda}^{-1}$ involved in U. ◀

COROLLARY 2.11. *If a subgroup H of the free group F on $a_1, \ldots, a_n$ is generated by a set of words of length $\leq r$ then H has a set of free generators of length $\leq r$.*

PROOF. Choose a set of free generators $B_\mu(a_\nu)$ as described in Theorem 2.11, for H. Then those $B_\mu(a_\nu)$ involved in any of the given generators must have length $\leq r$. Since the $B_\mu(a_\nu)$ are a free set of generators for H, each $B_\mu(a_\nu)$ must be involved in at least one of the given generators of H. Hence, we have our results. ◀

A *set of primitive elements of a free group* F is a set of elements which can be completed to a set of free generators for F. As an illustration, let F be the free group on a, b, and c. Then ac and bc are a set of primitive elements of F, since ac, bc, and c freely generate F (see Corollary 2.1 of Theorem 2.13). On the other hand, a^2 and b^2 are not a set of primitive elements of F. [For example, if a^2, b^2, and $W(a, b, c)$ were to freely generate F, then $F/F(X^2)$ would have order two at most though, in fact, it has order eight.] Similarly, abc is a primitive element of F, although $abca^{-1}b^{-1}c^{-1}$ is not.

As a further application of the Reidemeister-Schreier rewriting process we have the following:

THEOREM 2.12 (Takahasi). *Let F be a free group and let*

$$F = F_1 \supset F_2 \supset \ldots \supset F_n \supset \ldots \tag{5}$$

be a descending chain of subgroups such that F_{i+1} contains no primitive element of F_i. Then

$$\bigcap_{i=1}^{\infty} F_i = 1.$$

PROOF. We use induction on k to prove the following statement. In a free group, any word W which has length k in some system of free generators for the free group, and which is in a descending chain of the type (5), defines the identity element.

The statement is obviously true if $k = 0$. Suppose now that the statement holds for all words of length less than k, and suppose W has length k on some set of free generators a_ν for F, and is in the intersection of the chain (5). Replacing a generator of F by its inverse if necessary, we may assume that

$$W = a_{\nu_1} a_{\nu_2}^{\epsilon_2} \ldots a_{\nu_k}^{\epsilon_k},$$

where $\epsilon_i = \pm 1$. Since a_{ν_1} is not in F_2 (for F_2 contains no primitive element of F_1), we may choose a_{ν_1} as its own representative in the construction of a Schreier system for F mod F_2 (see the proof of Lemma 2.2 preceding Theorem 2.9, in the previous section). If now we apply τ to W, then although the apparent length of $\tau(W)$ is k, the first s-symbol s_{1,a_ν} is a defining relator in the presentation of F_2 given by Theorem 2.9.

Hence, W has length less than k on a set of free generators for F_2. Since W is in

$$\bigcap_{i=2}^{\infty} F_i,$$

W defines the identity by inductive hypothesis. ◀

COROLLARY 2.12 (Levi). *Let F be a free group and let*

$$F = F_1 \supset F_2 \supset \dots \supset F_n \supset \dots$$

be a descending chain of distinct subgroups such that F_{i+1} is characteristic in F_i. Then

$$\bigcap_{i=1}^{\infty} F_i = 1.$$

PROOF. Since F is a free group, each subgroup F_i is a free group, and hence there is an automorphism of F_i carrying any primitive element of F_i into any other primitive element of F_i. Since F_{i+1} is characteristic in F_i if F_{i+1} were to contain a primitive element of F_i, then $F_i = F_{i+1}$. Hence, Theorem 2.12 applies. ◀

This last corollary leads to an interesting property of finitely generated free groups, the so-called "Hopfian" property.

THEOREM 2.13. *A free group of finite rank cannot be isomorphic to one of its proper factor groups.*

PROOF. Let F be a free group. Consider the chain of verbal subgroups

$$(6) \quad V_0 = F, \qquad V_1 = V_0(X^2), \qquad V_2 = V_1(X^2), \dots, V_{i+1} = V_i(X^2), \dots.$$

Since a verbal subgroup is a fortiori characteristic, and the X^2-subgroup of a free group is different from the free group, the hypothesis of the last corollary is satisfied. Hence,

$$(7) \qquad \bigcap_{i=0}^{\infty} V_i = 1.$$

Suppose that $N \neq 1$ is a normal subgroup of F such that

$$F/N \simeq F.$$

Now, under the isomorphism of F onto F/N, the chain (6) must correspond to the chain

$$(8) \qquad V_0' = F/N, \qquad V_1' = V_0'(X^2), \dots, V_{i+1}' = V_i'(X^2), \dots.$$

Moreover, because of the construction of the V_i as verbal subgroups,

(9) $$V_0' = V_0N/N, \qquad V_1' = V_1N/N, \ldots, V_i' = V_iN/N, \ldots$$

(see Problem 2.11).

Hence, if V_j is the last V_i containing N [not all V_i can contain N because of (7)], then from the isomorphic correspondence between the chains (6) and (8), and by (9), we have

$$\frac{V_j}{V_{j+1}} \simeq \frac{V_j'}{V_{j+1}'} = \frac{V_jN/N}{V_{j+1}N/N} \simeq \frac{V_jN}{V_{j+1}N} = \frac{V_j}{V_{j+1}N} \simeq \frac{V_j/V_{j+1}}{V_{j+1}N/V_{j+1}}.$$

Since $V_{j+1}N \neq V_{j+1}$, V_j/V_{j+1} is isomorphic to a proper factor group of itself. But by induction on n it is easily shown that V_n is finitely generated and that V_{n+1} has finite index in V_n. Hence, V_j/V_{j+1} is finite, and so cannot be isomorphic to a proper factor group of itself. This contradiction shows that F is not isomorphic to F/N. ◀

Corollary 2.13.1. *Any n generators of a free group of finite rank n are free generators.*

Proof. For if F has the free generators $a_1, \ldots, a_n$, and $b_1, \ldots, b_n$ generate F, the mapping

(10) $$a_\nu \to b_\nu$$

is a homomorphism of F onto itself. Since the image F is isomorphic to F/N, where N is the kernel of (10), $N = 1$ and (10) is an isomorphism. Hence, $b_1, \ldots, b_n$, freely generate F. ◀

Corollary 2.13.2. *Let*

(11) $$F_1 \subset F_2 \subset \ldots \subset F_n \subset \ldots$$

be an ascending chain of free groups, each of rank not exceeding some fixed number r. If

(12) $$G = \bigcup_{i=1}^{\infty} F_i,$$

then G is not isomorphic to any proper factor group G/N, $N \neq 1$.

Proof. We prove this result by induction on r. If $r = 0$, the result is trivially true.

Suppose that the result holds for all chains (11) of free groups F_i of rank less than r. We show it holds if the free groups F_i have rank not exceeding r.

For, suppose G is isomorphic to G/N, $N \neq 1$. Any finitely generated subgroup of G is a free group; for from (12) it follows that any finitely generated subgroup of G is contained in some free group F_i. Hence, G/N has the same property. Now

$$G/N = \bigcup_{i=1}^{\infty} (F_iN/N). \tag{13}$$

Since F_iN/N is isomorphic to $F_i/F_i \cap N$ (by a standard isomorphism theorem), it is finitely generated, and hence free. Moreover, $F_i/F_i \cap N$ has at most rank r. But from (11) and (12), $F_i \cap N \neq 1$, for all i large enough. Thus $F_i/F_i \cap N$ cannot be isomorphic to F_i by the preceding theorem; hence, F_iN/N has rank less than r for large enough values of i. It then follows from (13) that G/N, and so G, is the union of an ascending chain of free groups of rank less than r. But then the inductive hypothesis yields the required result. ◀

This last corollary provides an example of a non-Abelian group which is not free although every finitely generated subgroup of it is; for if all the F_i are distinct, then G is infinitely generated. But an infinitely generated free group is not "Hopfian"; e.g., if $a_1, a_2, \ldots, a_n, \ldots$ freely generate a group F and N is the normal subgroup of F generated by a_1 then $F/N \simeq F$.

We conclude this section with some references and remarks about the results proved in this and the previous section. The method of Schreier was simplified by Hurewicz, 1931; see also Kuhn, 1952. Schreier systems were studied and applied by M. Hall and T. Rado, 1948, by K. Iwasawa, 1948, and by M. Hall, 1949b. The last paper contains a refinement of the method of Nielsen, 1921, which enables us to obtain, among other results, the following:

Let $U_1, \ldots, U_n$ and $V_1, \ldots, V_m$ be two finite sets of elements of a free group F such that the subgroup H generated by the U_ν does not contain any V_μ. Then there exists a subgroup H' of finite index in F such that H' contains all U_ν but no V_μ.

Nielsen, 1921, proved Theorem 2.13 in an entirely different manner (see Section 3.2).

H. Hopf, in 1932, raised the question as to whether a finitely generated group can be isomorphic to a proper factor of itself. This was answered in the affirmative: by B. H. Neumann, 1950, with a two-generator group with infinitely many defining relators; by G. Higman, 1951c, with a three-generator group with two defining relators; and by Baumslag and Solitar, 1962, with a two-generator group with one defining relator (see Section 4.4). A group which cannot be isomorphic to a proper factor group of itself is called *Hopfian*.

For other proofs of Theorems 2.11 and 2.12, see Levi, 1930 and 1931, and Takahasi, 1944, respectively. For a generalization of Theorem 2.12 to arbitrary descending chains of free groups, see Problem 36.

Problems for Section 2.4

1. Use the Schreier formula in Theorem 2.10 to show that a subgroup of an infinite or finite cyclic group is cyclic. [*Hint:* If F is infinite cyclic use that any subgroup has finite index.]

2. Show that if F is a free group of rank two or more, then the commutator subgroup of F is an infinitely generated free group. [*Hint:* Use Theorem 2.10.]

3. Let F be a free group of rank two or more, and let $F = F_1 \supset F_2 \supset \ldots \supset F_n \supset \ldots$, be a descending chain of different subgroups, F_{i+1} of finite index in F_i. Show that the rank of F_n tends to infinity monotonically. [*Hint:* Use the Schreier formula in Theorem 2.10]

4. Let F be the free group on a and b, and let N_k be the normal subgroup of F generated by a^{2^k} and b. Show that N_k has finite index in F if $k \geq 0$, and hence, N_k is finitely generated. Show that $\bigcap_{k=0}^{\infty} N_k = N$ is the normal subgroup of F generated by b, and that N has infinite index in F. Conclude that the intersection of infinitely many finitely generated subgroups of a free group need not be finitely generated (Howson, 1954, has shown that the intersection of finitely many finitely generated subgroups of a free group is finitely generated.) [*Hint:* Use Theorem 2.10, and that if $W(a, b)$ is in N_k then $\sigma_a(W)$ is divisible by 2^k, and so, if W is in N then $\sigma_a(W) = 0$.]

5. Let H be a finitely generated subgroup of a free group F. Show that the *normalizer* of H (the set of elements x in F such that $xHx^{-1} = H$) is finitely generated, and that H is of finite index in its normalizer. Show that if H is of infinite index in F, then H has infinitely many conjugates, whose intersection is the identity. [*Hint:* Use Theorem 2.10, and the facts that H is normal in its normalizer, and the number of conjugates of H is the index of its normalizer in F.]

6. Show that the Burnside group

$$B_{d,n} = \langle a_1, \ldots, a_n; X^d \rangle$$

is finite if and only if $F(X^d)$ is finitely generated, where F is the free group on $a_1, \ldots, a_n$. [*Hint:* Use Theorem 2.10.]

7. (Takahasi). Let H be a finitely generated subgroup of the free group F. Show that there are only finitely many subgroups of F in which H can be of finite index. [*Hint:* Let F be freely generated by $a_1, \ldots, a_n$, n finite or infinite. If we choose generators for H, $U_1, \ldots, U_r$, then the U_ρ will involve altogether finitely many a_ν, say, $a_1, \ldots, a_k$. If a word $W(a_\nu)$ involves a_λ, $\lambda > k$, then $W^p(a_\nu)$ involves a_λ for positive p, and so W^p is not in H. Hence, the powers of W define infinitely many cosets of H. Thus any subgroup

Q of F in which H is of finite index involves only $a_1, \ldots, a_k$. Moreover, if $B_1(a_\nu), \ldots, B_m(a_\nu)$, are generators of Q chosen in accordance with Theorem 2.11, then each B_μ must be involved in some U_ρ (since $[Q:H]$ is finite). Hence, the maximum length of $B_\mu(a_\nu)$ does not exceed the maximum length of $U_\rho(a_\nu)$. Since only finitely many words whose lengths have a given bound can be constructed from $a_1, \ldots, a_k$, we have our result.]

8. (Takahasi). A subgroup J of a free group is called a *free factor*, if a set of generators for J is a primitive set of elements for the free group. Show that if H is a finitely generated subgroup of a finitely generated free group F, then there are only finitely many subgroups Q of F containing H, such that no proper free factor of Q contains H. [*Hint:* See the hint to the previous problem.]

9. Let $F = F_1 \supset F_2 \supset \ldots \supset F_k \supset \ldots$, be a descending chain of free subgroups such that no primitive element of F_i is in F_{i+1}. Prove that any element of F, $\neq 1$, which is in F_k, cannot be defined by a word of length less than k in any system of free generators for F. [*Hint:* Use induction on k. Note that if $W \neq 1$ is defined by a word of length p in F, then using the method in the proof of Theorem 2.12, show that W can be defined by a word of length $\leq p - 1$ in F_2; hence $k - 1 \leq p - 1$, and so $k \leq p$.]

10. Let F be the free group on a and b. Let N_1 be the normal subgroup of F generated by a, and, in general, let N_{i+1} be the normal subgroup of N_i generated by a. Show that a is a primitive element of the free group N_i. Show that $\bigcap_{i=1}^{\infty} N_i$ is the cyclic group of F generated by a. [*Hint:* Suppose $\bigcap_{i=1}^{\infty} N_i$ is not just the set of powers of a. Consider all words W in a set of free generators containing a for the free group N_i, such that W is not a power of a but W is in $\bigcap_{i=1}^{\infty} N_i$. Choose such a word V of minimum length. Since every power of a is in $\bigcap_{i=1}^{\infty} N_i$, V cannot begin with a or a^{-1}. If V begins with x^ϵ, $\epsilon = \pm 1$, where x is a free generator of N_j, then choose a Schreier system for $N_j \bmod N_{j+1}$ in which x^ϵ is a representative. Then the free generators obtained by Theorem 2.9 include $1 \cdot a1 \cdot \overline{a}^{-1} = a$, but V has smaller length in these free generators for N_{j+1}.]

11. Let $G = \langle a, b; a^n, b^m \rangle$, n and $m > 0$, and let $H = G(X_1X_2X_1^{-1}X_2^{-1})$. Find a Schreier system for $G \bmod H$. Use Theorem 2.9 to obtain a presentation for H, and simplify this presentation to show that H is a free group on $(n - 1)(m - 1)$ generators.

12. Show that the elements $a^\nu b^\mu a^{-\nu} b^{-\mu}$, $1 \leq \nu \leq n - 1$, and $1 \leq \mu \leq m - 1$ generate the subgroup H of the preceding exercise. Conclude from Corollary 2.13.1 that these are free generators for H. [*Hint:* Use induction on the syllable length of a word W in H to show that W is a product of words $a^\nu b^\mu a^{-\nu} b^{-\mu}$ and their inverses. To do this, note that

$$a^{\alpha_1}b^{\beta_1}a^{\alpha_2}b^{\beta_2} \ldots a^{\alpha_k}b^{\beta_k} \approx a^{\alpha_1}b^{\beta_1}a^{-\alpha_1}b^{-\beta_1} \cdot b^{\beta_1}a^{\alpha_1+\alpha_2}b^{\beta_2} \ldots a^{\alpha_k}b^{\beta_k}.]$$

13. Let F be the free group on a and b. Show that $H = F(X_1X_2X_1^{-1}X_2^{-1})$ is freely generated by $a^{\nu}b^{\mu}a^{-\nu}b^{-\mu}$, where ν and μ are integers not zero. [*Hint:* Use the hint of the preceding excercise to show that $a^{\nu}b^{\mu}a^{-\nu}b^{-\mu}$ generate H. Next, given any subset of generators $a^{\alpha_i}b^{\beta_i}a^{-\alpha_i}b^{-\beta_i}$, $i = 1, \ldots, k$, choose an integer M so large that $\alpha_i = \alpha_j \bmod M$ implies that $\beta_i \neq \beta_j \bmod M$. Consider the obvious homomorphism of F onto $G = \langle a, b; a^M, b^M\rangle$. Show that the subset of generators go into a subset of free generators for G, and hence, are free.]

14. Let F be a free group on $a_1, \ldots, a_n$, n finite or infinite, and let $W \neq 1$ be in F. Show that there is a normal subgroup N of finite index excluding W. [*Hint:* If n is finite then use one of the subgroups V_i in the proof of Theorem 2.13, for N. If n is infinite and W involves $a_1, \ldots, a_k$, k finite, then map F into the free group on $a_1, \ldots, a_k$, sending $a_i \to a_i$ if $i \leq k$ and $a_i \to 1$ if $i > k$. Use the first case and take preimages.]

15. Show that there is no non-trivial identity which holds in all finite groups. [*Hint:* Given a freely reduced word $W(X_\lambda)$, $\neq 1$, the word $W(a_\lambda)$ is excluded from some normal subgroup N of finite index in the free group F on the a_λ. Hence, F/N does not satisfy the identity $W(X_\lambda) = 1$.]

16. Let Σ be the group of all permutations of the integers, which leave all but finitely many integers fixed. Show that Σ satisfies no non-trivial identity. [*Hint:* Σ contains every finite group as a subgroup.]

17 (Takahasi). If $H_1 \subset H_2 \subset \ldots \subset H_n \subset \ldots$ is an ascending chain of subgroups each of rank $\leq r$ in a free group F then all H_i coincide for i sufficiently large. [*Hint:* If infinitely many H_i are distinct, then by Corollary 2.13.2, $H = \bigcup_{i=1}^{\infty} H_i$ is Hopfian. But H is infinitely generated (any finite set of words is contained in some H_j) and as a subgroup of F, H is an infinitely generated free group, and so not Hopfian.]

18. Show that if G satisfies the *ascending chain condition for normal subgroups* (i.e., any ascending chain of distinct normal subgroups is finite), then G is Hopfian. [*Hint:* If $G \simeq G/N_1$, then $G/N_1 \simeq (G/N_1)/(N_2/N_1)$, and so $G/N_1 \simeq G/N_2$.]

19. Show that G is Hopfian if and only if every endomorphism of G onto itself is an automorphism. [*Hint:* If G is Hopfian and α is an endomorphism of G onto G then $G/N \simeq G$, where N is the kernel of α, and hence, $N = 1$. If every endomorphism of G onto G is an automorphism, then the natural mapping $G \to G/N$, of G onto the cosets of N, followed by an isomorphism of G/N onto G, is an endomorphism of G onto G. Hence, $N = 1$.]

20. A subgroup D of G is called *distinguished* if D is mapped into itself by every endomorphism of G onto itself.

(a) Show that a distinguished subgroup is characteristic.

(b) Show that if G is Hopfian, then every characteristic subgroup of G is distinguished.

(c) Show that the center of a group is distinguished.

(d) Show that the intersection of any number of distinguished subgroups is distinguished

(e) Show that a fully invariant subgroup is distinguished, and give an example of a distinguished subgroup which is not fully invariant.

[*Hint:* For (b), use Problem 19. For (e), let $G = S_8 \times Z_2$, where S_8 is the symmetries of a square and Z_2 is a cyclic group of order two. Map G into its subgroup H generated by the permutation (13) by sending the normal subgroup generated by the permutation (1234) into the identity and sending all other elements of S_8 into the permutation (13), and also sending Z_2 onto H. The center of G is not mapped into itself.]

21. A subgroup C of G is called *completely distinguished* if C is its own complete pre-image under any endomorphism of G onto itself.

(a) Show that a completely distinguished subgroup is distinguished.

(b) Show that if C is distinguished then C is completely distinguished if and only if G/C is Hopfian. In particular, G is Hopfian if and only if 1 is completely distinguished.

(c) Show that the intersection of any number of completely distinguished subgroups is completely distinguished.

(d) Show that G is Hopfian if and only if the intersection of some set of completely distinguished subgroups of G is the identity.

[*Hint:* For (b), use that if α is any endomorphism of G onto itself and C is distinguished, then $gC \to \alpha(g)C$ is an endomorphism of G/C onto itself. Hence, if G/C is Hopfian, $\alpha(g)$ is in C implies g is in C. Moreover, if there is an $\alpha(g)$ in C with g not in C, then G/C is not Hopfian. For (d), if $G/1 = G$ is Hopfian, then 1 is completely distinguished. If the intersection of some set of completely distinguished subgroup is 1, then use (c) and (b).]

22. Let P be a property of subgroups of a group which is preserved under taking complete pre-images under a homomorphism onto. Show that if only finitely many subgroups of a group G have the property P then the intersection of the subgroups of G having property P is completely distinguished. [*Hint:* Let α be an endomorphism of G onto G. If $H_1, \ldots, H_r$ are the subgroups of G with property P, then $\alpha^{-1}(H_1), \ldots, \alpha^{-1}(H_r)$ also have property P and hence, $\alpha^{-1}(H_i) = H_{j_i}$. Since

$$\alpha^{-1}\left(\bigcap_{i=1}^{r} H_i\right) = \bigcap_{i=1}^{r} \alpha^{-1}(H_i) = \bigcap_{i=1}^{r} H_{j_i} = \bigcap_{i=1}^{r} H_j,$$

we have our result.]

23. Let G be a finitely generated group. Show that the intersection of the following sets of subgroups are completely distinguished:

(a) all subgroups of a given finite index j;

(b) all normal subgroups of index j;

(c) all subgroups with index $\leq j$;

(d) all normal subgroups with index $\leq j$;

(e) all normal subgroups whose factor groups are cyclic groups of order j;

(f) all normal subgroups whose factor groups are Abelian of order j;

(g) all normal subgroups whose factor groups are in a given finite set of finite groups;

(h) all normalizers of subgroups of index j.

[*Hint:* Show that each listed property satisfies the condition of property P in the preceding exercise, and use Problem 3.19.]

24. A group G is called *residually finite* if the intersection of all normal subgroups with finite index is the identity.

(a) Show that G is residually finite if and only if for each element $g \neq 1$ of G, there is a finite homomorphic image of G in which g is not mapped into the identity.

(b) Show that G is residually finite if every $g \neq 1$ is excluded from some subgroup of finite index.

(c) Show that a free group is residually finite.

(d) Show that a finitely generated residually finite group is Hopfian.

(e) Show that an infinitely generated free group is residually finite but not Hopfian.

[*Hint:* For (b) use that the intersection of the conjugates of a subgroup of finite index is a normal subgroup of finite index. For (c), use Problem 14. For (d), use Problems 21(d) and 23(b).]

25 (Baumslag). Show that if G is a finitely generated residually finite group then any group of automorphisms A of G is residually finite. Hence, if A is finitely generated, then A is Hopfian. [*Hint:* Let α be an automorphism, not the identity of the group A of automorphisms of G. If g is in G and $\alpha(g) \neq g$, then $\alpha(g)g^{-1}$ is not in some subgroup N of finite index in G. Since G is finitely generated, there are only finitely many subgroups in G of the same index as N, say, $N_1, \ldots, N_r$. If $M = \bigcap_{i=1}^{r} N_i$ then $\alpha(g)g^{-1}$ is not in M, M is characteristic in G, and M has finite index in G. The automorphisms of A permute the right cosets of M. The set of automorphisms of A which leave all cosets of M fixed, has finite index in A and does not contain α.]

26. (a) Show that if $G_1 \subset G_2 \subset \ldots \subset G_n \subset \ldots$ is a chain of simple groups then $\bigcup_{i=1}^{\infty} G_i$ is a simple group. In particular, the group of all permutations of the integers which affect only finitely many integers, and on these are even permutations, is an infinite simple group.

(b) Show that an infinite simple group is not residually finite but is Hopfian.

27. Show that if n is finite, then the free Abelian group A_n on $a_1, \ldots, a_n$ is residually finite and hence, Hopfian. [*Hint:* If $W(a_\nu)$ has non-zero exponent on a_i and $\sigma_i(W) = d$, map A_n onto the cyclic group on x of order $d + 1$ by sending $a_i \to x$ and $a_j \to 1$, $j \neq i$.]

28. It is known that a subgroup of a free Abelian group is free Abelian (see Section 3.3). Use this to show that the union of a chain of free Abelian groups each of rank $\leq r$ is Hopfian. [*Hint:* See the proof of Corollary 2.13.2.]

29. Let F be a free group and let A and B be subgroups of F such that B is finitely generated and contains a normal subgroup of A. Show that $A \cap B$ has finite index in A. [*Hint:* The index of $A \cap B$ in A is also the number of cosets of B determined by elements of A. Now, if there are infinitely many cosets of B determined by elements of A, then a proof similar to the one for the last assertion in Theorem 2.10, will show the required result; the element U should be chosen from the normal subgroup of A in B, and the representative K as a representative of an element of A.]

The definition of *free factor* given in Problem 8 is needed for the remaining problems.

30. Show that if H is a free factor of F, then any set of free generators for H can be extended to a free set of generators for F. Hence, if G is a free factor of H, and H is a free factor of F, then G is a free factor of F. [*Hint:* Suppose F is freely generated by $a_1, \ldots, a_r, a_{r+1}, \ldots, a_n$, where $a_1, \ldots, a_r$ generate H. Let $W_1(a_\rho), \ldots, W_r(a_\rho)$ freely generate H. Show that

$$W_1, \ldots, W_r, a_{r+1}, \ldots, a_n$$

freely generate F.]

31. Show that if G is a free factor of F, and H is a subgroup of F containing G, then G is a free factor of H. [*Hint:* If F is freely generated by $a_1, \ldots, a_r, a_{r+1}, \ldots, a_n$, where $a_1, \ldots, a_r$ generate G, then the elements of H defined by s_{1,a_ρ} are just a_ρ.]

32. Generalize the preceding problem and show that if G is a free factor of F and H is any subgroup of F, then $G \cap H$ is a free factor of H. Show that this is equivalent to the following: If F is freely generated by $a_1, \ldots, a_r, a_{r+1}, \ldots, a_n$ and H is a subgroup of F then the words in H which involve only $a_1, \ldots, a_r$ constitute a free factor of H. [*Hint:* Clearly the words in F which involve only $a_1, \ldots, a_r$ constitute a free factor G of F. To show that $G \cap H$ is a free factor of H, we shall construct a Schreier representative system for F mod H in which the representative of an element of G is again in G. For then, the s-symbols s_{M,a_ρ}, where M is a representative of an element in G and $\rho \leq r$, define the elements $Ma_\rho\overline{Ma_\rho}^{-1}$ which are in G; moreover, $\tau(W(a_\rho))$ involves only such s-symbols. Hence, such s-symbols generate $G \cap H$, and can be extended to a set of free generators for H (namely, all s-symbols which are not defining relators). To obtain a Schreier system in which representatives of G are again in G, well-order the symbols a_ρ^{ϵ}, e.g.,

$$a_1 < a_1^{-1} < a_2 < a_2^{-1} < \ldots < a_r < a_r^{-1}.$$

Next, order words $W(a_\rho)$ first by length and then lexicographically, e.g., $a_1^3 < a_1a_2a_1^{-1}a_3 < a_1a_2a_2a_1$. Under this ordering, the words in G are well-ordered, and moreover, if $W_1 < W_2$ then $W_1W < W_2W$. Choose from each coset of H containing an element of G, the smallest element of G. An initial segment of such a smallest element is again a smallest element; for if W_2W is smallest in its coset but W_2 is not, then $W_1 < W_2$ and $HW_1 = HW_2$, so that $W_1W < W_2W$ and $HW_1W = HW_2W$. Since the smallest element of G in a given H coset is unique, these smallest elements define different cosets of H. Hence, they may be extended to a Schreier system for $F \bmod H$ (see Problem 2.3.23).]

33. Let $H_1 \supset H_2 \supset \ldots \supset H_n \supset \ldots$ be a descending chain of free groups, where H_1 has finite or countable rank. Show that if $H = \bigcap_{i=1}^{\infty} H_i$ then H has a free set of generators $U_1, U_2, \ldots, U_n, \ldots$ such that any finite set $U_1, \ldots, U_r$ freely generate a free factor of all but finitely many H_i. [*Hint:* Since H_1 is countably generated, H is countable, and so we may write the elements of H in a sequence $1, W_1, W_2, \ldots$. Let H_{j_1} be the first term of the chain in which W_1 has smallest possible length as a word in a set of free generators for H_{j_1}. If

$$W_1 = a_{i_1}^{\epsilon_1} \ldots a_{i_p}^{\epsilon_p},$$

then each a_{i_k} is in H; for if a_{i_k} is in H for $k < q$ but a_{i_q} is not in H, then a_{i_q} is not in some H_t, $t > j_1$. We may choose $a_{i_q}^{\epsilon_q}$ as its own representative in a Schreier system for $H_{j_1} \bmod H_t$. In rewriting W_1 in H_t using s-symbols, $a_{i_q}^{\epsilon_q}$ will be replaced by $s_{1,a_{i_q}}$ if $\epsilon_q = 1$ and by $s^{-1}_{a_{i_q}^{-1},a_{i_q}}$ if $\epsilon_q = -1$; in any event the s-symbol will be a defining relator. Hence W_1 has shorter length in H_t, which is impossible. Thus $a_{i_1}, \ldots, a_{i_p}$ is a free factor of H, and of all but finitely many H_i. Go to the first W_ν not in this free factor of H, say, W_2. Let H_{j_2} be the first term of the chain in which W_2 has smallest possible length as a word in a set of free generators for H_{j_2} containing $a_{i_1}, \ldots, a_{i_p}$. If

$$W_2 = b_{k_1}^{\eta_1} \ldots b_{k_x}^{\eta_x},$$

then each b_{k_i} is in H; for otherwise, the length of W_2 could be shortened. Hence, $a_{i_1}, \ldots, a_{i_p}, b_{k_1}, \ldots, b_{k_x}$ is a free factor of H, and of all but finitely many H_i; moreover, W_1 and W_2 are in this free factor. Continuing in this way we obtain the desired set of free generators for H.]

34. Let $H_1 \supset H_2 \supset \ldots \supset H_n \supset \ldots$ be a descending chain of free groups, where H_1 has finite or countable rank. If $H = \bigcap_{i=1}^{\infty} H_i$, then any finitely generated free factor of H is a free factor of all but finitely many H_i. [*Hint:* Let $U_1, U_2, \ldots$ be the set of generators constructed in Problem 33. If $V_1, \ldots, V_t$ freely generate a free factor of H, then $V_1, \ldots, V_t$ are contained in a free factor of H generated by finitely many U_i, say, $U_1, \ldots, U_n$. Then $U_1, \ldots, U_n$ generate a free factor of all but finitely many H_i. Since $V_1, \ldots, V_t$ generate a free factor of H, by Problem 31, they generate a free factor of the subgroup generated by $U_1, \ldots, U_n$, and hence, of all but finitely many H_i.]

35. Let $H_1 \supset H_2 \supset \ldots \supset H_n \supset \ldots$ and let $H = \bigcap_{i=1}^{\infty} H_i$, where H_1 is a free group. Show that any countable subgroup G of H is contained in a descending chain $H_1^* \supset H_2^* \supset \ldots \supset H_n^* \supset \ldots$ where H_1^* is countable and H_i^* is a free factor of H_i. [*Hint:* For each H_i, choose a fixed set of free generators. Since G is countable, it is contained in a subgroup $H_{1,i}$ of H_i,

generated by countably many free generators of H_i. Since $H_{1,i}$ and $H_{1,i+1}$ are countable, they are contained in a subgroup $H_{2,i}$ of H_i, generated by countably many of the free generators of H_i. In general, if $H_{j,i}$ and $H_{j,i+1}$ are countable, then they are contained in a subgroup $H_{j+1,i}$ of H_i, generated by countably many of the free generators of H_i. Then if $H_i^* = \bigcup_{j=1}^{\infty} H_{j,i}$, it follows that H_i^* is generated by countably many free generators of H_i and hence, that H_i^* is a countable free factor of H_i. Moreover, $H_1^* \supset H_2^* \supset \ldots$ and each H_i^* contains G.]

36. Show that if $H_1 \supset H_2 \supset \ldots \supset H_n \supset \ldots$ is a descending chain of free groups and $H = \bigcap_{i=1}^{\infty} H_i$, then every finitely generated free factor of H is a free factor of all but finitely many H_i. [*Hint:* Let G be a finitely generated free factor of H. Then by Problem 35, there is a descending chain $H_1^* \supset H_2^* \supset \ldots \supset H_n^* \supset \ldots$ such that H_i^* is a countable free factor of H_i and each H_i^* contains G. Hence, $G \subset \bigcap_{i=1}^{\infty} H_i^* \subset H$ and so, G is a finitely generated free factor of $\bigcap_{i=1}^{\infty} H_i^*$. Thus, by Problem 34, G is a free factor of all but finitely many H_i^*, and hence, of all but finitely many H_i.]

37. Show that if H is a subgroup of the free group on $a_1, \ldots, a_n$, n finite or infinite, then the subgroup H_r of H generated by all words $U(a_\nu)$ in H of length $\leq r$ (in the a_ν) is a free factor of H. [*Hint:* Choose a set of generators $B_\mu(a_\nu)$ for H, as in Theorem 2.11. Then show that H_r is generated by those $B_\mu(a_\nu)$ whose length is $\leq r$.]

Chapter 3

Nielsen Transformations

3.1. Introduction

Presenting a group G in terms of generators and defining relations is as arbitrary a procedure as choosing a coordinate system to describe a geometric configuration. Naturally, we try to find a system which leads to a simple description. But even before we can make a choice, we have to know the set of systems from which we can choose.

In order to be specific, let a_ν, $\nu = 1, 2, \ldots, n$ be a set of generators of G. Let b_μ, $\mu = 1, 2, \ldots, n$, be another n-tuple of elements of G. If G is defined solely in terms of generators and defining relations, we can define the b_μ only by giving them as words $W_\mu(a_\nu)$ in the a_ν. The possible choices for the n-tuples W_μ will depend strongly on the structure of the group G. For instance, in a cyclic group of order q defined by one generator a_1 and one defining relator $a_1{}^q$, the word $a_1{}^2$ will be a generator if and only if q is odd. To free ourselves from the influence of the particular nature of G, we shall restrict ourselves to the problem:

Which n-tuples of words W_μ in the generators a_ν of a group G will generate G, no matter what defining relations G has?

It is clear that we have to solve this problem only in the case where G is the free group F_n on n free generators a_ν; for if the $W_\mu(a_\nu)$ generate F_n, they will generate every quotient group of F_n. Thus let us confine ourselves to the case $G = F_n$. This has the additional advantage that now our words W_μ, if freely reduced, uniquely define elements of F_n. In particular, two different n-tuples of freely reduced words $W_\mu(a_\nu)$ will define different sets of generators of F_n. (Incidentally, we know from Corollary 2.13.1 that these generators are free again, and therefore that the mapping $a_\nu \to W_\nu$ can be extended to an automorphism of F_n.)

We shall call the replacing of the a_ν by the words W_ν a *free substitution*. There are obvious ways of constructing such substitutions.

We can

(I) Exchange two of the a_ν.
(II) Replace an a_ν by a_ν^{-1}.
(III) Replace an a_ν by $a_\nu a_\mu$, $\mu \neq \nu$.
(IV) Carry out substitutions of types (I), (II), (III) repeatedly a finite number of times.

One of the results of this chapter will be a proof of the fact that we can obtain all free substitutions from (IV) (see Theorem 3.2). If we define a *primitive element* as an element of a free group which is in a set of free generators for the free group, then Theorem 3.2 will enable us to effectively construct a sequence of primitive elements which exhausts all primitive elements. Nevertheless, it is exceedingly difficult to decide whether a given word $W(a_\nu)$ is primitive; we shall state (but not prove) a method for making this decision in Section 3.5.

The process described above for obtaining new generators of a group is particularly important if the group is a subgroup of a free group F^*, and therefore, the elements a_ν are themselves words in the free generators of F^*. In Section 3.2, we shall obtain the above result for subgroups of a free group and shall relate this result to previous ones on subgroups of a free group. Although the underlying ideas are very simple, the proofs are complicated. They are of a purely "combinatorial" nature, involving case distinctions in fairly large numbers. Still, they are simple enough to be presented in a textbook; even this is no longer true for several results stated but not proved in the last three sections of this chapter (Sections 3.5, 3.6, and 3.7). In particular, Section 3.7 merely reports results for some groups important in number theory or topology, and connected with the group of automorphisms of a free group. We do not go into any applications, but the references given do explain the significance of these groups. In Section 3.3, we develop in detail the effects of free substitution on the commutator quotient group of a group. In Section 3.4, we present (with full proofs) the theory developed by Alexander for the investigation of fundamental groups of three dimensional knots. In both of these sections, the methods of ordinary (commutative) algebra will be used when needed.

3.2. A Reduction Process

In the previous chapter, we proved, using Schreier systems, that a subgroup of a free group is free (Corollary 2.9) and that a finitely generated free group is Hopfian (Theorem 2.13). (Schreier proved Corollary 2.9 in

1927. Theorem 2.13 was stated as a problem by H. Hopf 1932, and was subsequently proved in several ways.) These significant properties of a free group were first obtained (for finitely generated free groups) by Nielsen in 1921 using a different approach. Moreover, his method yields important information about the automorphsisms of a free group.

The method is essentially a finite reduction process for transforming a finite system of generators of a subgroup of a free group into a free system of generators for that subgroup. Moreover, the free system arrived at enjoys certain properties which we now describe.

Let F be a free group on the free generators $\{x_\nu\}$ and let $\{W_i(x_\nu)\}$ be freely reduced words $\neq 1$, which generate the subgroup H. Now every element V of H can be expressed as a word $V(W_i)$ in the symbols W_i as well as a word $V(W_i(x_\nu))$ in the x_ν. We may therefore distinguish two "lengths" for each word $V(W_i)$, namely, its W-length (when freely reduced as a word in the W-symbols) $L_W(V)$, and its x-length (when freely reduced in the x-symbols) $L_x(V)$. Thus, for example, $L_W(W_i) = 1$ while $L_x(W_i) = L(W_i(x_\nu))$. We call the set of freely reduced (non-empty) words $\{W_i(x_\nu)\}$ *Nielsen reduced* if it satisfies the following conditions:

(i) *Each W-symbol occurring in $V(W_i)$ contributes at least one x-symbol to the freely reduced form of $V(W_i(x_\nu))$. More precisely, let*

$$V(W_i) = W_{i_1}^{\epsilon_1} W_{i_2}^{\epsilon_2} \cdots W_{i_r}^{\epsilon_r}, \qquad \epsilon_j = \pm 1,$$

be freely reduced in the W_i. Then in the free reduction of $V(W_i(x_\nu))$ at least one symbol $x_{\nu_j}^{\eta_j}$, $\eta_j = \pm 1$, is not deleted from $W_{i_j}^{\epsilon_j}$, for each j. (Thus in particular, $L_W(V) \leq L_x(V)$.)

(ii) *The x-length of $V(W_i(x_\nu))$ is at least as large as the x-length of any W-symbol occurring in $V(W_i)$.*

(*Note that either* (*i*) *or* (*ii*) *implies that a relator $R(W_i)$ in the generators W_i of H must* (*when freely reduced in the W_i*) *be the empty word, so that the W_i freely generate H.*)

As a simple illustration, let F be the free group on $x_1, x_2, \ldots$. Then $x_1, x_2, \ldots$ generate F and are Nielsen reduced. Conversely if $\{W_i\}$ is a Nielsen reduced set of generators for F, then $W_i = x_{\nu_i}^{\epsilon_i}$, $\epsilon_i = \pm 1$; for (i) or (ii) implies that the W_i freely generate F and hence no proper subset of them can. Therefore each W_i must occur in rewriting some x_ν in terms of the W-symbols, so that $L_x(W_i) \leq 1$, whence $W_i = x_{\nu_i}^{\epsilon_i}$. Since no proper subset of the x_ν can generate F, ν_i ranges over $1, 2, \ldots$ in some order.

As a second illustration, take $x_1 = a$, $x_2 = b$, $W_1 = a^2$, $W_2 = ab$, and $W_3 = ba$. Then the W_i freely generate the subgroup H consisting of all words in F of even length (see Problem 1.4.14 and Theorem 2.13). If $V(W_i) = W_2^{-1}W_1W_3^{-1}$, then

$$V(W_i(x_\nu)) = b^{-1}a^{-1} \cdot a^2 \cdot a^{-1}b^{-1} \approx b^{-2},$$

so that all x-symbols from W_1 are deleted. Thus (i) is violated. However, (ii) holds, since for any word $V(W_i)$, $L_x(V)$ is even, and $L_x(V) = 0$ implies $V(W_i) \approx 1$ because the W_i are free. On the other hand, H is also freely generated by W_1, W_2, and $W_4 = ba^{-1}$. As before (ii) holds. This time $W_2^{-1}W_1W_4^{-1} = b^{-1}a^{-1} \cdot a^2 \cdot ab^{-1} \approx b^{-1}a^2b^{-1}$, so that each W_i does contribute at least one x-symbol. That (i) holds for these W_i follows easily from Lemma 3.1.

Evidently a Nielsen reduced set $\{W_i(x_\nu)\}$ is such that in reducing any $V(W_i(x_\nu))$, not "too much" cancellation can take place. Conversely, if not "too much" cancellation can take place, then the set of words will be Nielsen reduced. To make this precise and also to give more convenient criteria for determining whether $\{W_i(x_\nu)\}$ is Nielsen reduced, we introduce some terminology and establish Lemma 3.1.

Let $\{W_i(x_\nu)\}$ be a set of freely reduced words ($\neq 1$). An *initial segment* of a W-symbol (i.e., of either $W_i(x_\nu)$ or $W_i^{-1}(x_\nu)$) is called *isolated* if it does not occur as an initial segment of any other W-symbol. (Note that $W_i(x_\nu)$ and $W_i^{-1}(x_\nu)$ are different W-symbols.) Similarly, a *terminal segment is isolated* if it is a terminal segment of a unique W-symbol.

Now if an initial segment S of W_i^ϵ ($\epsilon = \pm 1$) is isolated, then in freely reducing $W_j^\eta W_i^\epsilon$ ($W_j^\eta \neq W_i^{-\epsilon}$), not all the symbols of S can be deleted. For otherwise, S^{-1} is a terminal segment of W_j^η, and so S is an initial segment of $W_j^{-\eta}$, contrary to the isolation of S. Thus an isolated initial segment cannot be absorbed or completely cancelled from the left. Analogous remarks clearly apply to isolated terminal segments.

Let $W(x_\nu)$ be a freely reduced word $\neq 1$. That initial segment S of W which is "a little more than half" of W (i.e., $\frac{1}{2}L_x(W) < L_x(S) \leq \frac{1}{2}L_x(W) + 1$) is called the *major initial segment* of W. The *minor initial segment of* W is that initial segment S' which is "a little less than half" of W (i.e., $\frac{1}{2}L_x(W) - 1 \leq L_x(S') < \frac{1}{2}L_x(W)$). Similarly, we define *major* and *minor terminal segments*. (Thus, if $W_1 = aba^{-1}$ and $W_2 = a^2b^2$, then the major, minor initial and major, minor terminal segments of W_1 are ab, a, ba^{-1}, a^{-1} and of W_2 are a^2b, a, ab^2, b, respectively.) If the length of W is even we define the *left half* and the *right half* of W in the obvious manner.

LEMMA 3.1. *Let $\{W_i(x_\nu)\}$ be a set of freely reduced words ($\neq 1$). Then $\{W_i(x_\nu)\}$ is Nielsen reduced if and only if the following conditions are satisfied:*

(i*) *Both the major initial and major terminal segments of each W_i are isolated.*

(ii*) *For each W_i of even length, either its left half or its right half is isolated.*

PROOF. Suppose $\{W_i(x_\nu)\}$ is Nielsen reduced. Let $W_i = S_iT$ where S_i is the major initial segment of W_i. If S_i is not isolated then there exists a $W_j^\epsilon \neq W_i$ such that $W_j^\epsilon = S_iU$. Then

$$\begin{aligned} L_x(W_j^{-\epsilon}W_i) &= L_x(U^{-1}T) \leq L_x(U) + L_x(T) < L_x(U) + L_x(S_i) \\ &= L_x(W_j) \end{aligned}$$

But since $W_j^{-\epsilon}W_i$ is freely reduced in the W-symbols, (ii) is contradicted. Similarly, we show major terminal segments are isolated.

To show (ii*), let $W_i = U_iV_i$, where U_i, V_i are the halves of W_i, a word of even length. If neither U_i nor V_i is isolated then there exists W_j^ϵ, W_k^η (both different from W_i) such that $W_j^\epsilon = U_iT$ and $W_k^\eta = SV_i$. Then $W_j^{-\epsilon}(x_\nu)W_i(x_\nu)W_k^{-\eta}(x_\nu) \approx T^{-1}S^{-1}$ and so $W_i(x_\nu)$ contributes no x-symbol, contrary to (i).

Conversely, let $\{W_i(x_\nu)\}$ satisfy (i*) and (ii*). It is easily shown that (i*) and (ii*) hold for W_i^{-1} as well as W_i. For example, if $W_i^{-1} = S_iT$, S_i the major initial segment of W_i^{-1} and $W_j^\epsilon = S_iU$, then $W_i = T^{-1}S_i^{-1}$, $W_j^{-\epsilon} = U^{-1}S_i^{-1}$. Since S_i^{-1}, the major terminal segment of W_i, is isolated, we have $W_j^\epsilon = W_i^{-1}$.

We shall prove that the $\{W_i(x_\nu)\}$ satisfy not only (i) and (ii) but also the following condition:

(iii) *Let* $V(W_i) = W_{i_1}^{\epsilon_1}W_{i_2}^{\epsilon_2} \ldots W_{i_r}^{\epsilon_r}$ *be freely reduced in the* W*-symbols. Then after freely reducing* $V(W_i(x_\nu))$ *either at least the major terminal segment of* $W_{i_r}^{\epsilon_r}$ *remains or precisely the right half of* $W_{i_r}^{\epsilon_r}$ *remains. In either case the terminal segment of* $W_{i_r}^{\epsilon_r}$ *which remains is isolated.*

It is clear, by taking inverses, that (iii) implies an analogous statement for $W_{i_1}^{\epsilon_1}$.

Induction on r [the length of $V(W_i)$] will be used to establish (i), (ii), and (iii). The case $r = 1$ is immediate. Assume that (i), (ii), and (iii) hold for $r = n$. Consider

$$V(W_i) = W_{i_1}^{\epsilon_1} \ldots W_{i_n}^{\epsilon_n}W_{i_{n+1}}^{\epsilon_{n+1}}.$$

Let ST be the freely reduced form of

$$V' = W_{i_1}^{\epsilon_1}(x_\nu) \ldots W_{i_n}^{\epsilon_n}(x_\nu)$$

where T is the isolated major terminal segment or isolated right half of $W_{i_n}^{\epsilon_n}$ guaranteed by the inductive hypothesis. Now

$$(ST) \cdot W_{i_{n+1}}^{\epsilon_{n+1}}(x_\nu) = S(T \cdot W_{i_{n+1}}^{\epsilon_{n+1}}(x_\nu)).$$

Since T is isolated,

$$TW_{i_{n+1}}^{\epsilon_{n+1}}(x_\nu) = T_1Y \cdot Y^{-1}U,$$

where $T_1 \neq 1$ and $T_1 U$ is freely reduced. Moreover, since the major initial segment of $W_{i_{n+1}}^{\epsilon_{n+1}}$ is isolated,

$$L_x(Y) \leq \tfrac{1}{2} L_x(W_{i_{n+1}}).$$

If strict inequality holds, then U contains the major terminal segment of $W_{i_{n+1}}^{\epsilon_{n+1}}$, so that U is isolated. Otherwise, Y^{-1} is the left half of $W_{i_{n+1}}^{\epsilon_{n+1}}$, and is not isolated, so that the right half of U must be isolated. Consequently, the freely reduced form of $V(W_i(x_\nu))$ is ST_1U, and so (i) and (iii) hold. Also

$$\begin{aligned} L_x(V(W_i)) = L_x(ST_1U) &= L_x(ST_1) + L_x(U) \\ &= L_x(ST_1Y) + L_x(Y^{-1}U) - 2L_x(Y) \\ &= L_x(V') + L_x(W_{i_{n+1}}) - 2L_x(Y). \end{aligned}$$

Therefore, since $L_x(Y) \leq \frac{1}{2}L_x(W_{i_n})$, $\frac{1}{2}L_x(W_{i_{n+1}})$, and since $L_x(W_{i_n}) \leq L_x(V')$, we have $L_x(V) \geq L_x(V')$, $L_x(W_{i_{n+1}})$. Now by applying the inductive hypothesis we obtain (ii). ◀

Note that the major initial and major terminal segments as well as the left and right halves (if any) of W_i cannot be initial or terminal segments of W_i^{-1} (see Problem 3.9). Hence, in testing a particular W_i relative to (i*) and (ii*), we need only compare W_i with W_j and W_j^{-1}, $j \neq i$.

We return to the illustrations above. The set $W_1 = a^2$, $W_2 = ab$, $W_3 = ba$ is not Nielsen-reduced, since neither the left half nor right half of W_1 is isolated. On the other hand, the set W_1, W_2, $W_4 = ba^{-1}$ is Nielsen-reduced, since (i*) obviously holds, and (ii*) holds because the right halves of W_1 and W_2, and the left half of W_4, are isolated. Observe that W_1, W_2, W_4 can be obtained from W_1, W_2, W_3 by replacing W_3 by $W_3W_1^{-1}$. In general, starting with a finite sequence of words $W_1, \ldots W_m$, we can Nielsen-reduce them by applying a finite succession of such simple operations; specifically, these operations are as follows.

Let $\{x_\nu\}$ be free generators of F. The following transformations of the set of m-tuples $(W_1, \ldots, W_m)$ of freely reduced words in the x_ν are called *elementary Nielsen transformations of rank m*.

Permuting the W_μ and taking inverses of some of them, i.e., replacing each W_μ by

(1) $$W_{i_\mu}^{\epsilon_\mu}, \qquad \epsilon_\mu = \pm 1,$$

where i_μ runs through $1, \ldots, m$ in some order.

Leaving W_μ fixed, $\mu \neq r$, and replacing W_r by the freely reduced form of any one of the following:

(2) $$W_r W_s^\eta,$$

(3) $$W_s^\eta W_r,$$

(4) $$W_s^\eta W_r W_s^{-\eta},$$

where $1 \leq r, s, \mu \leq m$ (r, s fixed and distinct), $\eta = \pm 1$. It is clear that if an elementary Nielsen transformation is applied to a sequence of words, the resulting sequence generates the same subgroup as the original set.

By the *total x-length*, $L_x(W)$, of a finite m-tuple $W = (W_1, \ldots, W_m)$ we mean $L_x(W_1) + \ldots + L_x(W_m)$.

THEOREM 3.1. *Let $W = (W_1, \ldots, W_m)$ be a finite m-tuple of freely reduced words in the free generators x_ν of F. Then we can find a sequence $\tau_1, \ldots, \tau_k$ of elementary Nielsen transformations of rank m such that:*

$$L_x(W) \geq L_x(\tau_1 W) \geq \ldots \geq L_x(\tau_k \ldots \tau_1 W), \tag{5}$$

and

$$\tau_k \ldots \tau_1 W = (\tilde{W}_1, \ldots, \tilde{W}_m), \tag{6}$$

where $(\tilde{W}_1, \ldots, \tilde{W}_t)$ is Nielsen reduced and $\tilde{W}_{t+1} = 1, \ldots, \tilde{W}_m = 1$.

PROOF. We want to obtain from W a Nielsen reduced set, i.e., a set of words satisfying (i*) and (ii*). Suppose W does not satisfy (i*). This means that there exists a W_i such that its major initial segment S or major terminal segment S' is not isolated. Suppose $W_i = ST$, $W_j^\epsilon = SU$, $\epsilon = \pm 1, j \neq i$. Then $L_x(W_j^{-\epsilon} W_i) = L_x(W_i^{-1} W_j^\epsilon) = L_x(T^{-1}U) < L_x(W_j)$. Similarly, if S' is not isolated, we have $L_x(W_i W_j^{-\epsilon}) = L_x(W_j^\epsilon W_i^{-1}) < L_x(W_j)$. Hence, by applying a transformation of type (2) or (3) with $r = j$, $s = i$, we can reduce the total x-length of W.

In general, we see that if a set of words does not satisfy (i*) [or more accurately, if its subset of words different from 1 does not satisfy (i*)], we can by applying an elementary Nielsen transformation, reduce its total x-length. Let W^* be an m-tuple of shortest x-length obtainable from W by a sequence $\tau_1, \ldots, \tau_p$ of elementary Nielsen transformations *descending from* W, i.e., satisfying (5). Then W^* and any m-tuple obtained by a sequence descending from W^* must satisfy (i*). Moreover, clearly (i*) holds for any subset of such an m-tuple.

This last remark suggests the use of an inductive process on the words of W^* to obtain a set satisfying (ii*). Let $W_1^*, \ldots, W_t^*$ be the words of W^* different from 1, so arranged that $L_x(W_\lambda^*) \leq L_x(W_{\lambda+1}^*)$, $1 \leq \lambda \leq t - 1$. Assume that the set $\overline{W} = \{W_1^*, \ldots, W_{t-1}^*\}$ satisfies (ii*). We wish to modify $\overline{W}$ so that all right halves of the words in it are isolated. If the right half of W_λ^* is not isolated in $\overline{W}$, then replace W_λ^* by its inverse. Thus we can obtain from $\overline{W}$ the set $W' = \{W_1', \ldots, W_{t-1}'\}$ in which W_i' is W_i^* or its inverse and the right half of every word of even length is isolated in W'.

Next we isolate the right halves of words in W' from W_t^* and its inverse. Suppose that $W_\lambda' = U_\lambda V_\lambda$ and $W_t^* = S_1 V_\lambda$, where U_λ, V_λ are the left and right halves of W_λ'. Then replace W_t^* by $W_t' = W_t^* W_\lambda'^{-1} = S_1 U_\lambda^{-1}$. Now $S_1 U_\lambda^{-1}$ is freely reduced (for otherwise we could shorten the total x-length of W^*). If W_t' contains as a terminal segment the right half V_μ of W_μ' in W', then V_μ properly contains U_λ^{-1}; for V_μ cannot be a terminal segment of W_λ^{-1}, and not of U_λ^{-1}. Therefore, $W_t' = S_2 V_\mu$, where $L_x(S_2) < L_x(S_1)$. Now replace W_t' by $W_t'' = W_t' W_\mu'^{-1}$. Continuing in this way we arrive at a $\overline{W}_t$ which does not terminate with a right half of a word in W'. However, $\overline{W}_t^{-1}$ may end with such a right half. If it does, repeat the above procedure with $\overline{W}_t^{-1}$ in place of W_t^*. We finally arrive at a word $\tilde{W}_t^{-1}$ which does not end with a right half from W'; for in going from $\overline{W}_t^{-1}$ to ${}_t\tilde{W}^{-1}$, only right halves of words in W' are deleted. Therefore, since $L_x(\overline{W}_t) \geq L_x(W_\lambda')$, every initial segment of $\overline{W}_t^{-1}$ of length $\leq \frac{1}{2}L_x(\overline{W}_t)$ is unaltered. Hence, the terminal segments of $\tilde{W}_t$ with length $\leq \frac{1}{2}L_x(\tilde{W}_t)$ are the same as those of $\overline{W}_t$. Consequently, neither $\tilde{W}_t$ nor $\tilde{W}_t^{-1}$ ends with a right half from W'.

If $\tilde{W}_t$ has odd length, then W' together with $\tilde{W}_t$ satisfies (ii*). Suppose then $\overline{W}_t$ has even length and let U_t, V_t be the left and right halves of it. If U_t or V_t is isolated, then again W' together with $\tilde{W}_t$ satisfies (ii*). Assume neither U_t nor V_t is isolated. Then U_t and V_t must be halves of W-symbols $(W_\lambda')^{\pm 1}$ with W_λ' in W'. Now, V_t cannot be a right half of a W_λ' since a right half of a W_λ' cannot be a terminal segment of $\tilde{W}_t$. Similarly, U_t^{-1} cannot be the right half of a W_λ', so that U_t cannot be the left half of a $W_\lambda'^{-1}$. Therefore, $U_t = U_\rho$ and $V_t = U_\sigma^{-1}$ where U_ρ, U_σ are the left halves of W_ρ', W_σ'. We now modify W' so as to isolate the right half U_σ^{-1} of $\tilde{W}_t$ from those $W_\lambda'^{-1}$ which end with U_σ^{-1}. If $W_\lambda' = U_\sigma V_\lambda$, replace W_λ' by $\tilde{W}_\lambda = \tilde{W}_t W_\lambda' = U_\rho V_\lambda$, and otherwise let $\tilde{W}_\lambda = W_\lambda'$. This gives us a set $\tilde{W} = \{\tilde{W}_1, \ldots, \tilde{W}_t\}$ which satisfies (ii*). For, let V_λ, $\lambda < t$, be the right half of W_λ' and therefore of $\tilde{W}_\lambda$. Then V_λ cannot be the terminal segment of a word of odd length (nor its inverse) in $\tilde{W}$, since words of odd length were not altered in going from W' to $\tilde{W}$. Also V_λ is not a terminal segment of $\tilde{W}_t$ or $\tilde{W}_t^{-1}$ (by construction of $\tilde{W}_t$). If now V_λ is a terminal segment of $\tilde{W}_\mu^\epsilon$, $\epsilon = \pm 1$, $\mu < t$, $\mu \neq \lambda$, then V_λ is a terminal segment of the right half of W_μ^ϵ (since (i*) holds). Therefore, V_λ must be a terminal segment of V_μ or U_μ^{-1} or U_ρ^{-1} (since $\tilde{W}_\mu$ is either $U_\mu V_\mu$ or $U_\rho V_\mu$). But then V_λ is a terminal segment of W_μ' or $W_\mu'^{-1}$ or $W_\rho'^{-1}$, contrary to W' satisfying (ii*). Thus the right half of $\tilde{W}_\lambda$, $\lambda < t$ is isolated in $\tilde{W}$. Similarly, the right half U_σ^{-1} of $\tilde{W}_t$ cannot be a terminal segment of a word of odd length (nor of its inverse) in $\tilde{W}$. If U_σ^{-1} is a terminal segment of $\tilde{W}_\lambda^\epsilon$, $\epsilon = \pm 1$, $\lambda < t$, then U_σ^{-1} must be the right half of W_λ^ϵ. Now $\epsilon \neq 1$, since we have just shown that the right half of $\tilde{W}_\lambda$ $(\lambda < t)$ cannot be a terminal segment of $\tilde{W}_t$. Moreover, $\epsilon \neq -1$, since otherwise U_σ is the left half of $\tilde{W}_\lambda$, contrary to the construction of $\tilde{W}_\lambda$, $\lambda < t$.

Consequently, $\tilde{W}$ satisfies (ii*) as well as (i*) and is therefore Nielsen-reduced.

We have thus shown how to go from W^* to $\tilde{W}$, assuming that the set $\{W_1^*, \ldots, W_{t-1}^*\}$ is Nielsen reduced. But the set $\{W_1^*\}$ is Nielsen reduced, and so using the above process we can Nielsen-reduce $\{W_1^*, W_2^*\}$. Continuing in this manner, we can Nielsen-reduce $W_1^*, \ldots, W_{t-1}^*$ and, therefore, reduce W^*.

Finally, we can give an explicit (although perhaps tedious) method for descending from W to a W^*. Apply each elementary Nielsen transformation of type (2) and (3) to W. If each of these properly increases the total x-length of W, take $W^* = W$. Otherwise, let $W_{11}, W_{12}, \ldots, W_{1k_1}$ be the sets of shortest x-length obtained, different from W. Again apply each elementary Nielsen transformation of type (2) and (3) to each of the W_{1i}, $1 \leq i \leq k_1$. If for some W_{1j} each such transformation properly increases the total x-length of W_{1j}, let $W^* = W_{1j}$. Otherwise, let $W_{21}, \ldots, W_{2k_2}$ be the sets of shortest total x-length obtained, which are not among $W, W_{11}, \ldots, W_{1k_1}$. Repeat the procedure on $W_{21}, \ldots, W_{2k_2}$, and continue as before. The list $W, W_{11}, \ldots, W_{21}, \ldots, W_{31}, \ldots$ must be finite, since its terms are distinct, have total x-length not exceeding that of W, and involve only the finite number of x-symbols which occur in the words of W. Consequently, if we do not arrive at a W_{ij} whose total x-length is strictly increased by the elementary Nielsen transformations which we are considering, we must reach a set of m tuples $W_{q1}, \ldots, W_{qk_q}$ such that applying our transformations to them only yields m tuples $W_{q+1,j}$ already listed in $W, W_{11}, \ldots, W_{q1}, \ldots, W_{qk_q}$. Then we may take W^* equal to any of these W_{qj}. The justification of this method is not difficult and is left as an exercise. ◀

Corollary 3.1. *Let H be a finitely generated subgroup of the free group F on $\{x_\nu\}$ and let $\tilde{W}_1, \ldots, \tilde{W}_t$ be a Nielsen reduced set of generators for H. Then, of all systems of generators for H, $\{\tilde{W}_i\}$ has the shortest total x-length.*

Proof. Let $V_1, \ldots, V_m$ be a set of generators for H. By the above theorem there exists a descending sequence of elementary Nielsen transformations carrying $V_1, \ldots, V_m$ into $\tilde{V}_1, \ldots, \tilde{V}_t, 1, \ldots, 1$ where $\{\tilde{V}_i\}$ is Nielsen-reduced. Clearly, then, it suffices to show $L_x(\{\tilde{V}_i\}) \geq L_x(\{\tilde{W}_i\})$. We show that $\{\tilde{V}_i\}$ and $\{\tilde{W}_i\}$ have the same number of words of length k. By condition (ii), the words of x-length $\leq k$ in $\{\tilde{V}_i\}$ generate H_k, the subgroup of H generated by all words in H of x-length $\leq k$. Since the rank of H_k is unique, it follows that $\{\tilde{V}_i\}$ and $\{\tilde{W}_i\}$ have the same number of words of length $\leq k$, and hence the same number of length k. Thus, $L_x(\{\tilde{V}_i\}) = L_x(\{\tilde{W}_i\})$. ◀

Theorem 3.1 implies immediately that a finitely generated subgroup of a free group is free. Moreover, let F_n be a finitely generated free group on $x_1, \ldots, x_n$ and let H be an arbitrary subgroup. We show H has a Nielsen-reduced system of generators, and is therefore free. Let H_k denote the subgroup of H generated by all words in H of x-length $\leq k$. Since F is finitely generated, H_k is finitely generated. Given a Nielsen-reduced set of generators $W_1, \ldots, W_t$ for H_k, we can extend it to a Nielsen-reduced set of generators for H_{k+1}; for let $V_1, \ldots, V_s$ be a Nielsen-reduced set for H_{k+1}. Now, by (ii), those V_j with x-length $\leq k$, say, $V_1, \ldots, V_t$, generate H_k. Hence,

$$W_1, \ldots, W_t, \qquad V_{t+1}, \ldots, V_s \tag{7}$$

has the same total x-length as $V_1, \ldots, V_s$, and so may serve as W^* in the reduction process given in the proof of Theorem 3.1. Since each $L_x(W_i) \leq k < L_x(V_j)$, $j > t$, and $W_1, \ldots, W_t$ is Nielsen-reduced, we note that in reducing (7) by the process given in Theorem 3.1, $W_1, \ldots, W_t$ will (up to inverses) remain unaltered. Consequently, we can construct a sequence of sets, $W^{(k)}$, $k = 1, 2, \ldots$, such that

$$W^{(1)} \subset W^{(2)} \subset \ldots,$$

where $W^{(k)}$ is a Nielsen-reduced set of generators for H_k. It then follows that

$$W = \bigcup_{k=1}^{\infty} W^{(k)},$$

is a Nielsen-reduced set of generators for H. (For the case that F is countably generated, see Problem 5.)

To connect the preceding theory to automorphisms of a free group, we require some simple observations. Let $\{x_\nu\}$, $\nu = 1, \ldots, n$, freely generate F_n. An endomorphism α of F_n is determined by its effect on the x_ν. If $\alpha(x_\nu) = Y_\nu(x_\lambda)$ and α is an automorphism, then $\{Y_\nu\}$ freely generates F_n. Conversely, if $\{Y_\nu(x_\lambda)\}$ is a set of words, then the mapping $\bar{\alpha}$,

$$\bar{\alpha}: \quad x_\nu \to Y_\nu(x_\lambda), \tag{8}$$

determines a unique endomorphism α of F_n, namely,

$$\alpha: \quad U(x_\nu) \to U(Y_\nu(x_\lambda)).$$

Moreover, if the set $\{Y_\nu\}$ freely generates F_n, then α is an automorphism. The mapping (8) is called a *substitution on the* x_ν, and if $\{Y_\nu\}$ freely generates F_n, the substitution is called a *free substitution on the* x_ν. If we have a second substitution $\bar{\beta}$,

$$\bar{\beta}: \quad x_\nu \to Z_\nu(x_\lambda), \tag{9}$$

then the *product*, $\bar{\beta}\bar{\alpha}$ is the substitution,

(10) $$\bar{\beta}\bar{\alpha}: \quad x_\nu \to Y_\nu(Z_\lambda(x_\mu)).$$

It is clear then that $\bar{\beta}\bar{\alpha} = \overline{\beta\alpha}$. In particular, the group A_n of automorphisms of F_n is isomorphic to the group of free substitutions on $x_1, \ldots, x_n$.

With each free substitution on $x_1, \ldots, x_n$ we may associate not only an automorphism of F_n, but also a transformation of n-tuples of freely reduced words in the x_ν. For example, let $n = 2$ and consider the following substitutions $\bar{\alpha}_0, \bar{\beta}_0$:

$$\bar{\alpha}_0: \quad x_1 \to x_1x_2, \qquad x_2 \to x_2$$
$$\bar{\beta}_0: \quad x_1 \to x_1, \qquad x_2 \to x_1^{-1}x_2.$$

Since $\bar{\alpha}_0, \bar{\beta}_0$ have inverses, namely,

$$\bar{\alpha}_0^{-1}: \quad x_1 \to x_1x_2^{-1}, \qquad x_2 \to x_2$$
$$\bar{\beta}_0^{-1}: \quad x_1 \to x_1, \qquad x_2 \to x_1x_2,$$

$\bar{\alpha}_0, \bar{\beta}_0$ determine automorphisms α_0, β_0 of F_n, and hence are free substitutions. We now associate to $\bar{\alpha}_0, \bar{\beta}_0$ the following transformations of ordered pairs $(W_1(x_\nu), W_2(x_\nu))$ of freely reduced words:

$$N_{\alpha_0}: \quad (W_1, W_2) \to (W_1W_2, W_2)$$
$$N_{\beta_0}: \quad (W_1, W_2) \to (W_1, W_1^{-1}W_2).$$

Of course, in this case $N_{\alpha_0}, N_{\beta_0}$ are just elementary Nielsen transformations. Notice also that $N_{\alpha_0} \cdot N_{\beta_0}$ is the transformation

$$N_{\alpha_0}N_{\beta_0}: \quad (W_1, W_2) \to (W_2, W_1^{-1}W_2),$$

whereas $N_{\alpha_0\beta_0}$ is the transformation,

$$N_{\alpha_0\beta_0}: \quad (W_1, W_2) \to (W_1W_2, W_2^{-1}W_1^{-1}W_2).$$

However, $N_{\alpha_0}N_{\beta_0}$ is the same transformation as $N_{\beta_0\alpha_0}$. In general, if $\bar{\alpha}$ is the free substitution given by (8), we define the *Nielsen transformation* N_α *of rank* n by

$$N_\alpha: \quad (\ldots, W_\nu(x_\mu), \ldots) \to (\ldots, Y_\nu(W_\lambda(x_\mu)), \ldots),$$

that is, replace the νth entry by the Y_ν-combination of the other entries. Moreover, if $\bar{\beta}$ is the free substitution given by (9), then $N_\alpha N_\beta = N_{\beta\alpha}$. It then follows that the correspondence

(11) $$\alpha \to N_\alpha,$$

is an anti-isomorphism of A_n, the group of automorphisms of F_n, onto N_n, the group of Nielsen transformations of rank n. The *elementary automorphisms of* F_n *on the free generators* $x_1, \ldots, x_n$ are those which correspond [under (11)] to the elementary Nielsen transformations (1) through (4). For example, the automorphisms α_0, β_0 determined by $\bar{\alpha}_0, \bar{\beta}_0$ above are elementary automorphisms of F_2 on x_1, x_2.

THEOREM 3.2. *The elementary automorphisms of* F_n *on the free generators* $x_1, \ldots, x_n$, *n finite, are a finite set of generators for* A_n, *the group of automorphisms of* F_n.

PROOF. Let α be an automorphism of F_n and let N_α be the corresponding Nielsen transformation. Suppose

$$N_\alpha(x_1, \ldots, x_n) = (Y_1(x_\lambda), \ldots, Y_n(x_\lambda)).$$

Using Theorem 3.1, we can find a sequence $N_{\alpha_1}, \ldots, N_{\alpha_k}$ of elementary Nielsen transformation such that

$$N_{\alpha_k} \ldots N_{\alpha_1}(Y_1, \ldots, Y_n) \tag{12}$$

is Nielsen reduced. No one's can occur in (12), since no fewer than n elements can generate F_n (see Theorem 2.4). But we have already observed that a Nielsen-reduced set of generators for F_n can be obtained from $(x_1, \ldots, x_n)$ by an elementary Nielsen transformation $N_{\alpha_{k+1}}^{-1}$ of type (1). Therefore

$$N_{\alpha_{k+1}} N_{\alpha_k} \ldots N_{\alpha_1}(Y_1, \ldots, Y_n) = (x_1, \ldots, x_n),$$

so that

$$N_\alpha = N_{\alpha_1}^{-1} \ldots N_{\alpha_k}^{-1} N_{\alpha_{k+1}}^{-1}.$$

Consequently,

$$\alpha = \alpha_{k+1}^{-1} \alpha_k^{-1} \ldots \alpha_1^{-1}. \quad ◀$$

Clearly, Theorem 3.1 implies that a finitely generated subgroup of a free group is free. Moreover, we can now easily derive the Hopfian property for finitely generated free groups; for let $W_1(x_\lambda), \ldots, W_n(x_\lambda)$ be a set of generators of F_n, n finite. As in the proof of Theorem 3.2, there exists a sequence of elementary Nielsen transformations carrying the n-tuple of W's into $(x_1, \ldots, x_n)$. Hence, there exists an automorphism mapping x_ν into $W_\nu(x_\lambda)$, $\nu = 1, \ldots, n$, so that the W's are free.

As a further application of Nielsen transformations we solve the "generalized word problem" for finitely generated subgroups of a free group.

Let

$$G = \langle x_1, \ldots, x_n; R_1(x_\nu), \ldots \rangle,$$

and let $W_1(x_\nu), \ldots, W_m(x_\nu)$ be a set of words in the x_ν. The following problem is called the *generalized word problem for G relative to its subgroup*

H generated by $W_1, \ldots, W_m$: Find a method by which it can be decided in a finite number of steps whether a given word $V(x_\nu)$ defines an element of H or not.

Let F be the free group on $x_1, \ldots, x_n$, and let $W_1(x_\nu), \ldots, W_m(x_\nu)$, m finite, be freely reduced words which generate a subgroup H. We may first Nielsen-transform (by the finite reduction process of Theorem 3.1) $(W_1, \ldots, W_m)$ into $(\tilde{W}_1, \ldots, \tilde{W}_t, 1, \ldots, 1)$, where $\{\tilde{W}_1, \ldots, \tilde{W}_t\}$ is Nielsen-reduced. Then H is generated by $\tilde{W}_1, \ldots, \tilde{W}_t$. Suppose $V(x_\nu)$ is the freely reduced form of a product. $\Pi \tilde{W}_{i_r}^{\epsilon_r}$, $\epsilon_r = \pm 1$; then either the major initial segment or the left half of $\tilde{W}_{i_1}^{\epsilon_1}$ is an initial segment of $V(x_\nu)$ [by condition (iii)]. Make a table with rows labeled $\tilde{W}_1, \tilde{W}_1^{-1}, \ldots, \tilde{W}_t$, $\tilde{W}_t^{-1}$, consisting of the major initial segments in the first column and left halves (if any) in the second column. Run down the first column to see if a major initial segment S of some $\tilde{W}_j^{\epsilon}$ is an initial segment of $V(x_\nu)$. If so, $\tilde{W}_{i_1}^{\epsilon_1} = \tilde{W}_j^{\epsilon}$. For, an isolated initial segment S' of $\tilde{W}_{i_1}^{\epsilon_1}$ is an initial segment of $V(x_\nu)$. Either S is an initial segment of S' or vice versa. Since both S and S' are isolated, $\tilde{W}_{i_1}^{\epsilon_1} = \tilde{W}_j^{\epsilon}$. On the other hand, if no major initial segment of a $\tilde{W}$-symbol is an initial segment of $V(x_\nu)$, then run down the second column to find the left half U of maximum x-length which is an initial segment of $V(x_\nu)$. Let $\tilde{W}_k^{\eta}$ have left half U. Then since U', the left half of $\tilde{W}_{i_1}^{\epsilon_1}$, is an initial segment of $V(x_\nu)$, U' must be an initial segment of U. But U' is isolated, so that $\tilde{W}_{i_1}^{\epsilon_1} = \tilde{W}_k^{\eta}$. Thus we can recover $\tilde{W}_{i_1}^{\epsilon_1}$. Repeating the above process on the freely reduced form of $\tilde{W}_{i_1}^{-\epsilon_1} V(x_\nu)$, we can recover $\tilde{W}_{i_2}^{\epsilon_2}$. Continuing in this way we finally obtain the empty word and hence have expressed $V(x_\nu)$ as a word in the $\tilde{W}$-symbols. The number of steps required to obtain the empty word is the same as the $\tilde{W}$-length of V which $\leq L_x(V)$ [by condition (i)]. To decide now whether an arbitrary word $V'(x_\nu)$ defines an element of H, we start to apply the above procedure. If we reach the empty word in no more then $L_x(V')$ steps, $V'(x_\nu)$ is in H. Otherwise, it is not.

Theorem 3.1 can be applied to determine the normal subgroups of a free group whose quotient groups are also free. Namely, if $Y_1(x_\nu), \ldots, Y_n(x_\nu)$ are free generators for the free group F_n on $x_1, \ldots, x_n$, then clearly the group

$$G = \langle x_1, \ldots, x_n; Y_1(x_\nu), \ldots, Y_k(x_\nu) \rangle, \qquad k \leq n,$$

is a free group. Conversely, we have

THEOREM 3.3. *Let F_n, n finite, be the free group on $x_1, \ldots, x_n$. If F_n/H is a free group, then there exists a set of free generators, $Y_1(x_\nu), \ldots, Y_n(x_\nu)$, for F_n such that H is the normal subgroup generated by $Y_1, \ldots, Y_k$, $(k \leq n)$.*

PROOF. Let $a_1, \ldots, a_m$ freely generate F_n/H and let η be the natural homomorphism of F_n onto F_n/H and suppose $\eta(x_\nu) = W_\nu(a_\mu)$. Then $W_1(a_\mu), \ldots, W_n(a_\mu)$ generate F_n/H. Since F_n/H is a free group, there exists a Nielsen transformation N_α, of rank n, such that

$$N_\alpha(W_1(a_\mu), \ldots, W_n(a_\mu)) = (1, \ldots, 1, a_1, \ldots, a_m).$$

If now

$$N_\alpha(x_1, \ldots, x_n) = (Y_1(x_\nu), \ldots, Y_n(x_\nu)),$$

then

$$\eta(Y_j(x_\nu)) = Y_j(\eta(x_\nu)) = Y_j(W_\nu(a_\mu)) = 1, \qquad j = 1, \ldots, k,$$

and

$$\eta(Y_{k+\lambda}(x_\nu)) = Y_{k+\lambda}(W_\nu(a_\mu)) = a_\lambda, \qquad \lambda = 1, \ldots, m,$$

where $k = n - m$. Clearly, then, $Y_1, \ldots, Y_n$ freely generate F_n and $Y_j \in H$, $j = 1, \ldots, k$. Suppose $U(Y_\nu) \in H$, $\nu = 1, 2, \ldots, n$. Then, if N is the normal subgroup of F_n generated by $Y_1(x_\nu), \ldots, Y_k(x_\nu)$, it follows that the cosets of F_n/N are generated by $Y_{k+1}(x_\nu), \ldots, Y_n(x_\nu)$. Hence,

$$U(Y_\nu) = S(Y_{k+1}, \ldots, Y_n) \cdot R(Y_\nu),$$

where S is a freely reduced word in Y_λ, $k < \lambda \leq n$, and $R(Y_\nu)$ is in N. Thus

$$\eta(U(Y_\nu)) = 1 = S(\eta(Y_{k+1}), \ldots, \eta(Y_n)) \cdot \eta(R(Y_\nu)) = S(a_1, \ldots, a_m).$$

Since $a_1, \ldots, a_m$ are free generators, S is the empty word and $U(Y_\nu)$ is in N, and so $H = N$. ◀

If a group G is given by means of a presentation, then we may sometimes find a more convenient one by using a Nielsen transformation. Specifically, we have

LEMMA 3.2. *Let*

$$G = \langle x_1, \ldots, x_n; R_\mu(x_\nu)\rangle, \tag{13}$$

let N_α be a Nielsen transformation of rank n such that

$$N_\alpha(x_1, \ldots, x_n) = (Y_1(x_\lambda), \ldots, Y_n(x_\lambda)), \tag{14}$$

and suppose

$$N_\alpha^{-1}(x_1, \ldots, x_n) = (Z_1(x_\lambda), \ldots, Z_n(x_\lambda)). \tag{15}$$

Then G can be presented by

$$G = \langle Y_1, \ldots, Y_n; R_\mu(Z_\nu(Y_\lambda))\rangle.$$

(In other words, to get the new defining relators in the Y-symbols, solve for the x's in terms of the Y's and substitute in the old defining relators.)

PROOF. Let F_n be the free group on $x_1, \ldots, x_n$ and let H be the normal subgroup of F_n generated by $R_\mu(x_\nu)$. Clearly, if we rewrite the $R_\mu(x_\nu)$ in terms of the new set of free generators $\{Y_\nu(x_\lambda)\}$ of F_n, the rewritten words will "normally generate" the same subgroup H. From (14), $(x_1, \ldots, x_n)$ equals

$$N_\alpha^{-1}(Y_1(x_\lambda), \ldots, Y_n(x_\lambda)),$$

which by (15) is

$$(Z_1(Y_\lambda(x_\rho)), \ldots, Z_n(Y_\lambda(x_\rho))),$$

that is, the νth entry is the Z_ν-combination of the Y_λ. Therefore $R_\mu(x_\nu)$, when rewritten in terms of the Y-symbols is $R_\mu(Z_\nu(Y_\lambda))$. ◀

The Nielsen transformation (14) determines a free substitution $\bar{\alpha}$, and hence, an automorphism of the free group on $x_1, \ldots, x_n$. Under what conditions will (14) induce an automorphism of G?

LEMMA 3.3. *The mapping defined by*

$$x_\nu \to Y_\nu(x_\lambda)$$

can be extended to an automorphism of

$$G = \langle x_1, \ldots, x_n; R_\mu(x_\nu)\rangle$$

if and only if for all μ *we have*

(a) $R_\mu(Y_\nu(x_\lambda)) = 1,$

and

(b) $R_\mu(Z_\nu(x_\lambda)) = 1,$

as equalities in G, *where* Y_ν *and* Z_ν *are as in* (14) *and* (15), *respectively.*

PROOF. Suppose (14) induces an automorphism α of G. Then, since $R_\mu(x_\nu) = 1$ in G, $\alpha(R_\mu(x_\nu)) = 1 = R_\mu(\alpha(x_\nu)) = R_\mu(Y_\nu(x_\lambda))$. Moreover, the inverse α^{-1} of α is an automorphism. Now the Nielsen transformation corresponding to α^{-1} is the inverse of the Nielsen transformation corresponding to α, i.e., $N_{\alpha^{-1}} = N_\alpha^{-1}$. Hence, $\alpha^{-1}(x_\nu) = Z_\nu(x_\lambda)$, and so, $\alpha^{-1}(R_\mu(x_\nu)) = 1 = R_\mu(\alpha^{-1}(x_\nu)) = R_\mu(Z_\nu(x_\lambda))$. Thus (a) and (b) hold.

Conversely, if (a) and (b) hold, then by Corollary 1.1.2, the mappings

$$x_\nu \to Y_\nu(x_\lambda)$$

and

$$x_\nu \to Z_\nu(x_\lambda)$$

can be extended to homomorphisms α and β of G. Since (14) and (15) are inverse Nielsen transformations, $\alpha\beta$ and $\beta\alpha$ map x_ν into itself, and so α and β are automorphisms of G. ◀

G. Higman, 1951, discovered a group G and a mapping α arising from a Nielsen transformation and satisfying (a). Yet α does not define an automorphism of G. Of course, G is a non-Hopfian group; for Hopfian groups, condition (b) will follow from (a).

Lemma 3.2 will be used in later sections.

The significance of Lemma 3.3 will be illustrated by the results of Section 3.6.

The results obtained thus far deal primarily with finitely generated free groups. Problem 10 of this section indicates that there are infinite sets of words which cannot be Nielsen reduced by a finite sequence of Nielsen transformations. But although there exists no general finite constructive procedure for obtaining a Nielsen-reduced set from a given infinite set of generators, nevertheless every subgroup H of a free group F possesses a Nielsen-reduced set of generators. In fact, any *minimal Schreier system of generators* for H, i.e., a set of generators for H arising from a minimal Schreier representative system for F mod H as in Section 2.3, is Nielsen-reduced.

THEOREM 3.4. *Let $\{x_\nu\}$ freely generate F, and let H be a subgroup of F. Then any minimal Schreier system of generators for H is Nielsen reduced. Conversely, every Nielsen reduced set of generators for H is (up to inverses) a minimal Schreier system of generators for H.*

PROOF. We have already shown in the proof of Theorem 2.11 that a minimal Schreier generating system for H has property (ii). To show (i), let $\{K_i\}$ be a Schreier representative system for F mod H (not necessarily minimal). Then the corresponding Schreier generators for H are $\{Kx_\nu M^{-1}\}$ where M is the representative of Kx_ν, and Kx_ν is not freely equal to a representative. Therefore, both Schreier generators and their inverses have the form $Kx_\nu{}^\epsilon M^{-1}$, $\epsilon = \pm 1$, K, M representatives and $Kx_\nu{}^\epsilon$ not freely equal to a representative. Let

$$(16) \qquad U = (K_1 x_{\nu_1}^{\epsilon_1} M_1{}^{-1}) \ldots (K_r x_{\nu_r}^{\epsilon_r} M_r{}^{-1}),$$

be a word freely reduced in the Schreier generators. We shall show by use of induction on r that in the freely reduced form of U relative to the x_ν, there remain the $x_{\nu_j}^{\epsilon_j}$, $j = 1, \ldots, r$, as well as K_1, $M_r{}^{-1}$. For $r = 1$, the result is immediate since Schreier generators are already freely reduced in the x_ν. Consider the word of length $r + 1$ in the Schreier generators,

$$(17) \qquad U \cdot (K_{r+1} x_{\nu_{r+1}}^{\epsilon_{r+1}} M_{r+1}^{-1}).$$

Suppose $x_{\nu_r}^{\epsilon_r}$ is deleted in freely reducing (17). Then $M_r x_{\nu_r}^{-\epsilon_r}$ is an initial segment of the right hand factor in (17). Since $M_r x_{\nu_r}^{-\epsilon_r}$ is not a representative (otherwise $M_r x_{\nu_r}^{-\epsilon_r} = K_r$), it cannot be an initial segment of K_{r+1}. Similarly,

$K_{r+1}x_{\nu_{r+1}}^{\epsilon_{r+1}}$ cannot be an initial segment of M_r. Therefore $M_r x_{\nu_r}^{-\epsilon_r} = K_{r+1}x_{\nu_{r+1}}^{\epsilon_{r+1}}$, since neither can be a proper initial segment of the other. But then taking representatives, $K_r = M_{r+1}$. Hence,

$$K_{r+1}x_{\nu_{r+1}}^{\epsilon_{r+1}}M_{r+1}^{-1} = (K_r x_{\nu_r}^{\epsilon_r}M_r^{-1})^{-1},$$

so that (17) is not freely reduced in the Schreier generators. The same type of argument shows $x_{\nu_{r+1}}^{\epsilon_{r+1}}$ cannot be deleted. Thus any Schreier generating system satisfies condition (i). Hence every minimal Schreier generating system is Nielsen reduced.

Conversely, suppose $\{W_i(x_\nu)\}$ is a Nielsen reduced set of generators for H. We may assume the right halves of each W_j of even length is isolated (otherwise we replace W_j by W_j^{-1}). Let

$$(18) \qquad W_i(x_\nu) = U_i x_{\nu_i}^{\epsilon_i} V_i^{-1}, \qquad \epsilon_i = \pm 1,$$

where V_i^{-1} is the minor terminal segment of W_i. Therefore, $L_x(V_i) < \frac{1}{2}L_x(W_i)$ and $L_x(U_i) \leq \frac{1}{2}L_x(W_i)$. We first show that $U_i(x_\nu)$, $V_i(x_\nu)$ are words of minimum length in their respective cosets mod H. Suppose U_i, R determine the same coset of H and $L_x(R) < L_x(U_i)$. Then $U_iR^{-1} = Q(W_{t_j})$, where Q is freely reduced in the W-symbols. Since $L_x(U_iR^{-1}) < L_x(W_i)$, neither W_i nor W_i^{-1} occurs in Q. Therefore, $L_x(W_i^{-1}Q) = L_x(V_i x_{\nu_i}^{-\epsilon_i}R^{-1}) < L_x(W_i)$, contrary to (ii). In the same way, we show that $V_i(x_\nu)$ has minimum length in its coset. It follows immediately that the initial segments of U_i and V_i are of minimum length in their own cosets. For example, if $U_i = U_i'U_i''$ and $HU_i' = HP$ where $L_x(P) < L_x(U_i')$, then PU_i'' determines the same coset as U_i and yet has shorter length.

Next, we show that if S_i is an initial segment of U_i or V_i, and S_j is an initial segment of U_j or V_j, and $S_i \neq S_j$, then S_i, S_j determine different cosets of H; for suppose $S_iS_j^{-1} \in H$ and

$$S_iS_j^{-1} = W_{t_1}^{\eta_1} \ldots W_{t_r}^{\eta_r} = Q(W_{t_q}), \qquad \eta_\rho = \pm 1.$$

Since S_i, S_j are of minimum length in their respective cosets, $L_x(S_i) = L_x(S_j)$. By (ii), $L_x(S_iS_j^{-1}) \geq L_x(W_{t_1})$, $L_x(W_{t_r})$, so that $L_x(S_i) \geq \frac{1}{2}L_x(W_{t_1})$, $\frac{1}{2}L_x(W_{t_r})$. By (iii), the freely reduced form of $S_iS_j^{-1}$ begins with an isolated initial segment of $W_{t_1}^{\eta_1}$ and ends with an isolated terminal segment of $W_{t_r}^{\eta_r}$, which do not overlap. We may assume S_i begins with an isolated initial segment of $W_{t_1}^{\eta_1}$ (otherwise S_j^{-1} ends with an isolated terminal segment of $W_{t_r}^{\eta_r}$ and we then consider $S_jS_i^{-1}$). Therefore, $W_{t_1}^{\eta_1}$ equals W_i or W_i^{-1}. But $L_x(S_i) \leq \frac{1}{2}L_x(W_i)$, so that $L_x(S_i) = \frac{1}{2}L_x(W_i)$, whence $S_i = U_i$. It then follows that S_j^{-1} ends with an isolated terminal segment of $W_{t_r}^{\eta_r}$, so that $S_j = U_j$. Therefore, by (i), $Q(W_{t_q}) = W_iW_j^{-1} = U_iU_j^{-1}$, contrary to the isolation of the right half of W_i.

We can now show the existence of a minimal Schreier representative system for F mod H in which each U_i and V_i above is a representative; for in the set of all initial segments of all U_i and V_i, different words define different cosets of H. Hence, by Problem 2.3.23, this set can be extended to a minimal Schreier system for F mod H.

Finally, we prove that either W_i or W_i^{-1} will be a Schreier generator arising from the representative system chosen above; for if W_i is as in (18) with $\epsilon_i = 1$, then $W_i = U_i x_{\nu_i} V_i^{-1}$ and by construction V_i is a representative, so that V_i is the representative of $U_i x_{\nu_i}$. On the other hand, if $\epsilon_i = -1$, then $W_i^{-1} = V_i x_{\nu_i} U_i^{-1}$, so that U_i is the representative of $V_i x_{\nu_i}$. This completes the proof. ◀

Thus Nielsen and Schreier, using entirely different approaches, were led to the same system of generators for a subgroup of the free group.

COROLLARY 3.4. *Let $\{x_\nu\}$ freely generate F and let H be a subgroup of F. Then any word of shortest x-length l^* in H is a primitive element of H. Moreover, if $\{W_i\}$ is a Nielsen reduced set of generators for H and V $(\neq 1)$ is a word of H, then $L_x(V) - L_W(V) \geq l^* - 2$. If l^* is odd, the inequality can be improved to $L_x(V) - L_W(V) \geq l^* - 1$.*

(For the definition of "primitive element" see Section 3.1.)

PROOF. Let

$$V(W_i) = W_{i_1}^{\epsilon_1} \dots W_{i_r}^{\epsilon_r}, \qquad \epsilon_i = \pm 1,$$

be freely reduced in the W-symbols. By (iii) and (i)

$$L_x(V) \geq \tfrac{1}{2}L_x(W_{i_1}) + \tfrac{1}{2}L_x(W_{i_2}) + r - 2 \geq l^* + L_W(V) - 2.$$

Therefore, $L_x(V) - L_W(V) \geq l^* - 2$. Now, if l^* is odd, by (iii) again we have that $W_{i_1}^{\epsilon_1}$ and $W_{i_r}^{\epsilon_r}$ each contribute at least $\frac{1}{2}(l^* + 1)$ x-symbols, so that $L_x(V) - L_W(V) \geq l^* - 1$. Finally, if V has length l^*, by the first inequality, $L_W(V) \leq 2$. Hence, V is a primitive element of H unless V is the square of some W-symbol, say, W_1. But if $V = W_1^2$ and $W_1 = STS^{-1}$ where $T \neq 1$ and is cyclically reduced in the x-symbols then ST^2S^{-1} is the freely reduced form of $V(x_\nu)$. Clearly, then,

$$l^* = L_x(V) = L_x(ST^2S^{-1}) > L_x(STS^{-1}) = L_x(W_1),$$

contrary to the definition of l^*. ◀

Problems for Section 3.2

1. Show that $\{a^2, b^2, ab\}$ satisfies condition (ii) but not condition (i).

2. Let F be the free group on $x_1 = a$, $x_2 = b$; and let $W = (ba^{-1}, ba^2b, b^2, a^3)$. Show that one elementary Nielsen transformation will not decrease the total x-length of W, but a succession of two will.

3. Show, using Lemma 3.1, that in the free group F on a,b the set $\{b^n a b^{-n}\}$, $n = 0, \pm 1, \pm 2, \ldots$, is Nielsen reduced and therefore freely generates its subgroup.

4. Show that the subgroup in Problem 3 is the smallest normal subgroup H containing a. [*Hint:* A word is in H if and only if its b-exponent sum is zero. Let $W = b^{\beta_1}a^{\alpha_1}b^{\beta_2}a^{\alpha_2} \ldots b^{\beta_r}a^{\alpha_r} \in H$. Then W is freely equal to the product of factors

$$b^{\beta_1 + \ldots + \beta_k}a^{\alpha_k}b^{-\beta_1 \ldots -\beta_k} \approx (b^{\beta_1 + \ldots + \beta_k}ab^{-\beta_1 - \ldots -\beta_k})^{\alpha_k}.$$

Hence, $\{b^n a b^{-n}\}$ freely generates H.]

5. Use Problem 3 and the remark following Corollary 3.1 to show that a subgroup of a countably generated free group is free.

6. Let F be the free group on $x_1 = a$, $x_2 = b$, and let $W_1 = a^2$, $W_2 = ba^{-1}$. Show that $W_1{}^2 = a^4$ and $W_2 W_1{}^{-1} = ba^{-3}$ form a Nielsen reduced set in the W-symbols but not in the x-symbols; on the other hand, $W_1{}^2 = a^4$, $W_1 W_2 = a^2ba^{-1}$, and $W_2 W_1 = ba$ form a Nielsen reduced set in the x-symbols but not in the W-symbols.

7. Show by using Theorem 3.1 that the free group on $x_1, \ldots, x_n$, n finite, cannot have fewer than n-generators.

8. Show that all elementary Nielsen transformations of rank n, n finite, are generated by those corresponding to the free substitutions $\bar{\eta}_i$, $\bar{\beta}_{ij}$, $i \neq j$, where

$$\bar{\eta}_i : x_i \to x_i^{-1},\ x_k \to x_k, \qquad k \neq i$$

$$\bar{\beta}_{ij} : x_i \to x_i x_j,\ x_k \to x_k, \qquad k \neq i \neq j$$

[*Hint:* Note that $\bar{\beta}^{\epsilon}_{rs}$ corresponds to (2). Also, the elementary Nielsen transformation given by (3) is

$$(N_{\eta_s} \cdot N_{\eta_r} \cdot N_{\beta_{rs}} \cdot N_{\eta_r} \cdot N_{\eta_s})^{\epsilon} = N^{\epsilon}_{\gamma_{rs}}.$$

Moreover, the "transposition" $\bar{\tau}_{ij}$, $i \neq j$,

$$\bar{\tau}_{ij} : x_i \to x_j,\ x_j \to x_i,\ x_k \to x_k, \qquad k \neq i, j$$

induces the Nielsen transformation $N_{\eta_r} \cdot N_{\beta_{rs}} \cdot N_{\eta_s} \cdot N_{\beta_{sr}} \cdot N_{\eta_i} \cdot N_{\beta_{rs}}$.]

9. Show that if $W \neq 1$ is freely reduced, then neither its major initial segment nor its left half can be an initial segment of W^{-1}, and neither can the major terminal segment nor the right half of W be a terminal segment of W^{-1}.

10. Let F be the free group of infinite rank on $x_1, x_2, \ldots, x_n, \ldots$. Show that $x_1, x_1x_2, \ldots, x_1x_2 \ldots x_n, \ldots$, are also a set of free generators for F. Show that no finite sequence of elementary Nielsen transformations can transform the latter set of generators into $x_1, x_2, \ldots, x_n, \ldots$.

11. Give a procedure for determining in a finite number of steps whether two given finite sets of words of the free group F generate the same subgroup.

[*Hint:* Nielsen reduce both sets of words. Then use the solution of the generalized word problem to check whether each of the reduced sets is in the subgroup generated by the other.]

12. Let F be the free group on $\{x_\nu\}$ and let $\bar{F}$ be the free group on $\{y_\mu\}$. Given $W_1(x_\nu), \ldots, W_r(x_\nu)$ and $V_1(y_\mu), \ldots, V_r(y_\mu)$, r finite, find a procedure for deciding in a finite number of steps whether the mapping $W_i(x_\nu) \to V_i(Y_\mu)$ determines an isomorphism of H, the subgroup of F generated by $W_1, \ldots, W_r$, onto $\bar{H}$, the subgroup of $\bar{F}$ generated by $V_1, \ldots, V_r$. [*Hint:* Apply a Nielsen transformation N_α of rank r to $(W_1(x_\nu), \ldots, W_r(x_\nu))$ to obtain $(W_1', \ldots, W_t', 1, \ldots, 1)$ where $(W_1', \ldots, W_t')$ is Nielsen reduced. Next apply N_α to $(V_1, \ldots, V_r)$ obtaining $(V_1', \ldots, V_r')$. If $W_i \to V_i$ induces an isomorphism then $V'_{t+1} = \ldots = V'_r = 1$, and $V_1', \ldots, V_t'$ must be free. To test this we Nielsen reduce $V_1', \ldots, V_t'$. If now $V_1', \ldots, V_t'$ is free, then $W_j' \to V_j'$ induces an isomorphism of H onto $\bar{H}$. Since $(W_1', \ldots, W_t')$ is Nielsen reduced we may rewrite $W_i(x_\nu)$ as $Q_i(W_j'(x_\nu))$. If then $V_i(y_\mu) = Q_i(V_j'(y_\mu))$, the mapping $W_i \to V_i$ does induce an isomorphism of H onto $\bar{H}$.]

13. Let F be a free group on $x_1, \ldots, x_n$, n finite, and let H be the subgroup generated by $W_1, \ldots, W_m$, m finite. Give a constructive procedure for determining whether H is of finite index. [*Hint:* We may Nielsen reduce $W_1, \ldots, W_m$ to $\tilde{W}_1, \ldots, \tilde{W}_t$. Now if H has finite index j then by the Schreier rank formula, $t = j(n - 1) + 1$. If $n = 1$ then the procedure is trivial. Otherwise, $j = (t - 1)/(n - 1)$. Hence, there exists a Schreier representative system for F mod H, in which each representative has length $\leq j$. We may enumerate all the words of length $\leq j + 1$. Using the solution of the generalized word problem for H on the Nielsen reduced set of generators $\tilde{W}_1, \ldots, \tilde{W}_t$, we can divide the words of length $\leq j + 1$ into classes according to the coset of H they determine. If then we have exactly j different classes, each containing a word of length $\leq j$, it follows that H is of finite index j. For, since each word of length $j + 1$ is in the coset of a word of length $\leq j$, every word is in the coset of a word of length $\leq j$. But there are exactly j of these.]

14. Let F be the free group on $x_1, x_2, \ldots, x_n$, n finite, and let H be generated by $W_1, \ldots, W_m$, m finite. Give a constructive procedure for determining the normalizer of H. [*Hint:* Now N_H, the normalizer of H, is a free group and has H as a normal subgroup. Since H is finitely generated, so is N_H, and H is of finite index in N_H. Let $U_1, \ldots, U_r$ be a Nielsen reduced set of generators for N_H. Now, if H is contained in a subgroup generated by a proper subset of the U_i, H cannot be of finite index in N_H. Hence each U_i must occur in expressing some W_j in terms of the U_i. Since the U_i are Nielsen reduced, $L_x(U_i) \leq L_x(W_j)$ for some j. Hence, each U_i has x length $\leq \max L_x(W_j)$, $j = 1, \ldots, m$. There are only finitely many such words V. We may conjugate each W_j by V and using the solution to the generalized word problem for H, determine those V for which $VW_jV^{-1} \in H$. Clearly, then, these V generate N_H.]

15. A *retract* H of a group G is a subgroup H of G such that G can be

mapped homomorphically onto H in such a way that each element of H goes into itself.

(a) Show that a subgroup H of G is a retract of G if and only if each homomorphism of H into a group K can be extended to a homomorphism of G into K.

(b) Show that H is a retract of G if and only if G has a normal subgroup N such that $G = HN$ and $H \cap N = 1$.

(c) Show that if F is a free group of rank n, n finite, then all retracts of F are obtained as follows: choose a set of free generators $a_1, \ldots, a_n$ for F, and let

$$U_1 = a_1C_1, \ldots, U_r = a_rC_r, \qquad r \leq n,$$

where C_i is in the normal subgroup of F generated by $a_{r+1}, \ldots, a_n$; then H is generated by $U_1, \ldots, U_r$. [*Hint:* For (a), use the identity map of H onto H. For (b), map G/N onto H by mapping $hN \to h$, where h is in H, and show this is an isomorphism. For (c), show first that $a_\rho \to a_\rho$, $a_\sigma \to 1$ for $\rho \leq r$ and $\sigma > r$ is a retraction of F onto H. Conversely, suppose α is a retraction of F onto H. If $b_1, \ldots, b_n$ are free generators for F, then $\alpha(b_1), \ldots, \alpha(b_n)$ may be Nielsen transformed to $(U_1, \ldots, U_r, 1, 1, \ldots, 1)$, where U_ρ freely generate H. Applying the same Nielsen transformation to $b_1, \ldots, b_n$ yields a set of free generators $a_1, \ldots, a_n$ for F such that $\alpha(a_\rho) = U_\rho$, $\rho \leq r$ and $\alpha(a_\sigma) = 1$, $\sigma > r$. But then as in the proof of Theorem 3.3, the kernel of α is the normal subgroup N of F generated by a_σ, $\sigma > r$. Since α is the identity on H, $\alpha(U_\rho) = U_\rho = \alpha(a_\rho)$, $\rho \leq r$. Hence, $U_\rho = a_\rho C_\rho$, where C_ρ is in the normal subgroup generated by $a_{r+1}, \ldots, a_n$.]

3.3. The Commutator Quotient Group

Let F be a finitely generated free group and let H be a finitely generated subgroup. We have seen how to obtain a Nielsen-reduced set of generators for H from a given set of generators by using Nielsen transformations. In this section, we use Nielsen transformations to obtain another set of generators for H, which is convenient for studying the commutator subgroup and commutator quotient group of $F/\overline{H}$ where $\overline{H}$ is the normal subgroup generated by H.

THEOREM 3.5. *Let F be the free group on $x_1, \ldots, x_n$ and let H be the subgroup generated by $W_1(x_\nu), \ldots, W_m(x_\nu)$, $m \geq n$. Then we can construct a set of free generators $y_1, \ldots, y_n$ for F and a set of generators $v_1, \ldots, v_m$ for H such that*

$$v_i = y_i^{d_i} \cdot Q_i(y_\nu), \qquad i = 1, \ldots, n \tag{1}$$

$$v_j = Q_j(y_\nu), \qquad n < j \leq m, \tag{2}$$

where $0 \leq d_i$ *divides* d_{i+1} *and* Q_i, Q_j *have zero exponent sum on each* y_ν. *Moreover, H determines the* d_i *uniquely.*

PROOF. Note first that the condition $m \geq n$ is no restriction since otherwise we may adjoin 1's to the given list of generators for H.

The conditions (1) and (2) may be expressed in terms of exponent sums. Let σ_{y_i} be the exponent sum with respect to y_i. Then

$$\sigma_{y_i}(v_k) = d_i \delta_{ik}$$

where $\delta_{ik} = 1$ if $i = k$, and $\delta_{ik} = 0$ otherwise. This suggests considering an "exponent sum matrix" for a set of generators of H relative to a set of free generators for F. Specifically, it will be convenient to introduce a *titled exponent sum matrix*. This is a matrix of integers with its rows labeled by a system of free generators for F, its columns labeled by a set of words of F, such that the (i, j)th entry in the matrix is the exponent sum of the jth word (column label) relative to the ith generator (row label). For example, if F is the free group on a, b, then

$$\begin{array}{c} \\ a \\ b \end{array} \begin{array}{c} \begin{matrix} a^2b^{-3} & (ab)^2 & b^{-1}a^3 \end{matrix} \\ \begin{pmatrix} 2 & 2 & 3 \\ -3 & 2 & -1 \end{pmatrix} \end{array} \tag{3}$$

is a titled exponent sum matrix.

Since F and H are free groups, the y's and v's of Theorem 3.5 must be obtainable from the x's and W's by Nielsen transformations. This suggests starting with the titled exponent sum matrix

$$\begin{array}{c} \\ x_1 \\ \cdot \\ \cdot \\ \cdot \\ x_n \end{array} \begin{array}{c} W_1 \ldots W_m \\ \left(\begin{array}{c} \\ \\ \sigma_{x_i}(W_j) \\ \\ \\ \end{array} \right) \end{array}$$

and transforming it into the titled exponent sum matrix

$$\begin{array}{c} \\ y_1 \\ \cdot \\ \cdot \\ \cdot \\ y_n \end{array} \begin{array}{c} v_1 \ldots v_m \\ \left(\begin{array}{c} \\ \\ d_i \delta_{ij} \\ \\ \\ \end{array} \right) \end{array}$$

using Nielsen transformations applied to both row and column labels with the appropriate changes in the entries.

If we apply an elementary Nielsen transformation to the row labels or column labels, we must perform "corresponding" transformations on the matrix in order that the result be a titled exponent sum matrix.

We consider first the matrix transformation corresponding to an elementary Nielsen transformation of the column labels. If the Nielsen transformation calls for permuting (and taking inverses of some of) the column labels, the corresponding matrix transformation consists in permuting (and changing the sign of) the associated columns. If u, v are two column labels and u is to be replaced by uv^ϵ (or $v^\epsilon u$), $\epsilon = \pm 1$, the uth column C_u is to be replaced by $C_u + \epsilon C_v$. [For, $\sigma_x(uv^\epsilon) = \sigma_x(u) + \epsilon\sigma_x(v)$.]

In the case of a Nielsen transformation which permutes (and takes inverses of some of) the row labels, we permute (and change the signs of) the associated rows. However, if a, b are two row labels, and a is replaced by ab^ϵ (or $b^\epsilon a$), we replace the bth row R_b by $R_b - \epsilon R_a$. [For let $W = u(\ldots, a, \ldots, b, \ldots)$ and let $c = ab^\epsilon$. Then $a = cb^{-\epsilon}$ and W when written as a word in $\ldots, c, \ldots, b, \ldots$ equals $u(\ldots, cb^{-\epsilon}, \ldots, b, \ldots)$. Therefore, $\sigma_c(W) = \sigma_a(W)$ and $\sigma_b(W) = \sigma_b(u) - \epsilon\sigma_a(u)$.]

(This asymmetry between the matrix column and the matrix row transformation corresponding to a given type of Nielsen transformation of the labels will prevent us from extending our results to the case of an F of infinite rank n, although we shall extend them to the case where m is infinite.)

If we examine the types of matrix transformations (either row or column) induced by a Nielsen transformation, we find they are the so-called elementary row and column operations on integer-valued matrices:

(4) permuting rows (columns),

(5) changing the sign of a row (column), and

(6) replacing a row (column) by its sum or difference with an integral multiple of another row (column).

Conversely, to each elementary matrix row (column) operation we may associate a Nielsen transformation of the row (column) labels, so that if both are performed, a titled exponent sum matrix is transformed into another titled exponent sum matrix. Thus, to elementary row (column) matrix operations of type (4) or (5), we associate the obvious Nielsen transformations of the row (column) labels. If a matrix column operation of type (6) replaces C_u by $C_u + \epsilon C_v$, we replace the column label u by uv^ϵ. If a matrix row operation of type (6) replaces R_a by $R_a + \epsilon R_b$, we replace the row label b by $ba^{-\epsilon}$.

To any succession of row (column) matrix operations there corresponds a succession of row (column) Nielsen transformations under which a titled exponent sum matrix is transformed into a titled exponent sum matrix.

These preliminary remarks allow us to reduce the proof of the existence part of Theorem 3.5 to a problem in matrix theory. Given an n-by-n matrix of integers, transform it by means of elementary row and column operations to a diagonal matrix of the form

$$\begin{pmatrix} d_1 & & & & & 0 & \cdots & 0 \\ & \ddots & & 0 & & \vdots & & \vdots \\ & & d_i & & & \vdots & & \vdots \\ & 0 & & \ddots & & \vdots & & \vdots \\ & & & & d_n & 0 & \cdots & 0 \end{pmatrix} \tag{7}$$

where $d_i \geq 0$ and d_i divides d_{i+1}. A constructive solution to this problem is known. The following is an outline of the process:

Permuting rows, columns and changing signs if necessary, we arrive at a matrix (a_{ij}) such that $0 < a_{11} \leq |a_{ij}|$ for $a_{ij} \neq 0$. Suppose

$$a_{21} = k_2 a_{11} + r_{21},\ 0 \leq r_{21} < a_{11}.$$

Then if we add $(-k_2)R^{(1)}$ to $R^{(2)}$ (where $R^{(i)}$ denotes the ith row) our new $a_{21} = r_{21}$. If $r_{21} \neq 0$, it is smaller than a_{11} and we interchange the new $R^{(2)}$ and $R^{(1)}$. We repeat the above procedure until we obtain an $a_{21} = 0$. Applying the procedure to the third row, we obtain an $a_{31} = 0$. Continuing in this way we have a new matrix (a_{ij}) in which $a_{21} = a_{31} = \ldots = a_{n1} = 0$. Next apply the procedure to the first row using column operations with the proviso that each time the first column is replaced by another column, we apply row operations so that as before the new first column has $a_{21} = a_{31} = \ldots = a_{n1} = 0$. Continuing in this way, we obtain a new matrix (a_{ij}) in which $a_{i1} = a_{1j} = 0$, $i, j \neq 1$. If now there is some a_{ij} such that

$$a_{ij} = k_{ij} a_{11} + r_{ij}, \qquad 0 < r_{ij} < a_{11}.$$

add the jth column to the first column, making the new $a_{i1} = a_{ij}$ and a_{11} is unchanged. Proceeding as before, we obtain a new $a_{11} = r_{ij}$. Note that each of the above operations which changes a_{11} decreases it. Hence, after finitely many replacements (no more than the original a_{11} steps), we arrive at a matrix (a_{ij}) in which

$$a_{i1} = a_{1j} = 0, \qquad i, j \neq 1,$$

and a_{11} divides a_{ij}. Starting over again with the submatrix obtained by ignoring the first row and column, we arrive at a matrix (a_{ij}) in which

$$a_{i1} = a_{1j} = 0, \quad i, j \neq 1; \qquad a_{i2} = a_{2j} = 0, \quad i, j \neq 2,$$

and a_{11} divides a_{22} divides a_{ij}, $i, j > 1$. Continuing in this way, we finally arrive at a matrix in which a_{ii} divides a_{jj}, $i < j \leq n$ and $a_{ij} = 0$, $i \neq j$. This proves the existence part of Theorem 3.5. We prove uniqueness later in this section. ◀

As an illustration of the above process, let F be the free group on a, b and let H be the subgroup generated by a^2, b^2, $a^{-4}b$. Form the titled exponent sum matrix,

$$\begin{array}{c} \\ a \\ b \end{array}\!\!\begin{array}{c} \begin{matrix} a^2 & b^2 & a^{-4}b \end{matrix} \\ \begin{pmatrix} 2 & 0 & -4 \\ 0 & 2 & 1 \end{pmatrix} \end{array}.$$

We have then successively,

$$\begin{array}{c} \\ a \\ b \end{array}\!\!\begin{array}{c} \begin{matrix} a^{-4}b & b^2 & a^2 \end{matrix} \\ \begin{pmatrix} -4 & 0 & 2 \\ 1 & 2 & 0 \end{pmatrix} \end{array},$$

$$\begin{array}{c} \\ b \\ a \end{array}\!\!\begin{array}{c} \begin{matrix} a^{-4}b & b^2 & a^2 \end{matrix} \\ \begin{pmatrix} 1 & 2 & 0 \\ -4 & 0 & 2 \end{pmatrix} \end{array},$$

$$\begin{array}{c} \\ ba^{-4} \\ a \end{array}\!\!\begin{array}{c} \begin{matrix} a^{-4}b & b^2 & a^2 \end{matrix} \\ \begin{pmatrix} 1 & 2 & 0 \\ 0 & 8 & 2 \end{pmatrix} \end{array},$$

$$\begin{array}{c} \\ ba^{-4} \\ a \end{array}\!\!\begin{array}{c} \begin{matrix} a^{-4}b & b^2(a^{-4}b)^{-2} & a^2 \end{matrix} \\ \begin{pmatrix} 1 & 0 & 0 \\ 0 & 8 & 2 \end{pmatrix} \end{array},$$

$$\begin{array}{c} \\ ba^{-4} \\ a \end{array}\!\!\begin{array}{c} \begin{matrix} a^{-4}b & a^2 & b^2(a^{-4}b)^{-2} \end{matrix} \\ \begin{pmatrix} 1 & 0 & 0 \\ 0 & 2 & 8 \end{pmatrix} \end{array},$$

$$\begin{array}{c} \\ ba^{-4} \\ a \end{array}\!\!\begin{array}{c} \begin{matrix} a^{-4}b & a^2 & b^2(a^{-4}b)^{-2}a^{-8} \end{matrix} \\ \begin{pmatrix} 1 & 0 & 0 \\ 0 & 2 & 0 \end{pmatrix} \end{array}.$$

To write the column labels as words in $c = ba^{-4}$, $d = a$, we first express a, b in terms of c, d. In general, this can be done by applying in reverse order the inverses of the Nielsen transformations leading from a, b to c, d. Thus one has successively,

$$c, d; \quad c, d; \quad c, d; \quad c, d; \quad cd^4, d; \quad d, cd^4; \quad d, cd^4;$$

so that $a = d$, $b = cd^4$, which is of course obvious in this simple case.

Even if H is infinitely generated there exists a set of generators of the form (1) and (2). A slight modification of the above proof shows this. Instead of the column operation (6), we allow infinitely many columns to be replaced simultaneously, i.e., allow the additional column operation:

(8) Replace C_{u_i} with $C_{u_i} + \alpha_i C_v$, and leave C_v unchanged, where α_i is an integer and u_i, v are column labels.

The associated transformation of the column labels is:

(9) Replace u_i with $u_i v^{\alpha_i}$, and leave v unchanged.

Although this transformation is not a Nielsen transformation, it transforms one set of generators for H into another.

The existence of a set of generators of the form asserted in Theorem 3.5 has application to finitely generated Abelian groups.

COROLLARY 3.5.1. *Let A_n be the free Abelian group on $x_1, \ldots, x_n$, n finite. Then every Nielsen transformation on $x_1, \ldots, x_n$ induces an automorphism of A_n and, conversely, every automorphism of A_n is induced by a Nielsen transformation on $x_1, \ldots, x_n$.*

PROOF. Any automorphism of a group (here, the free group on free generators x_ν) induces an automorphism in any quotient group (here, A_n) of a characteristic subgroup (here, the commutator subgroup). This proves the first statement. To show the converse, let α be an automorphism of A_n. It must be shown that $\alpha(x_1), \ldots, \alpha(x_n)$ can be obtained from $x_1, \ldots, x_n$ by a Nielsen transformation. By Theorem 3.5 there exist Nielsen transformations N, M such that

$$M[\alpha(x_i)] = [N(x_i)]^{d_i} \cdot Q_i,$$

where d_i is a non-negative integer and Q_i is in the commutator subgroup of the free group on $x_1, \ldots, x_n$. Therefore, in A_n,

$$M[\alpha(x_i)] = [N(x_i)]^{d_i}.$$

Since $M[\alpha(x_i)]$ are free Abelian generators ($y_1, \ldots, y_n$ are *free Abelian generators* if every element can be expressed uniquely as $y_1^{r_1} \ldots y_n^{r_n}$, where r_ν is an integer),

$$1 = \sigma_{M[\alpha(x_i)]}[N(x_i)]^{d_i} = d_i \cdot \sigma_{M[\alpha(x_i)]}[N(x_i)],$$

so that $d_i = 1$. Hence, in A_n,

$$\alpha(x_i) = M^{-1}[N(x_i)]. \blacktriangleleft$$

COROLLARY 3.5.2. *Let A_n be the free Abelian group of rank n, and let H be a subgroup of A_n. Then H is a free Abelian group of rank $\leq n$. Moreover, there exist free Abelian generators $x_1, \ldots, x_n$ of A_n such that $x_1^{d_1}, \ldots, x_m^{d_m}$ are free Abelian generators of H, where $d_i > 0$ and d_i divides d_{i+1}.*

PROOF. Let $y_1, \ldots, y_n$ be free Abelian generators for A_n and let $W_1, W_2, \ldots,$ be generators for H. Then in the free group F_n on $y_1, \ldots, y_n$ there exists a set of generators $x_1, \ldots, x_n$ such that the subgroup generated by $W_1, W_2, \ldots$ has a set of generators of the form

$$x_i^{d_i}Q_i, \; Q_j, \qquad i = 1, \ldots, n, \qquad j > n$$

where d_i, Q_i, Q_j are as described in Theorem 3.5. Thus, in A_n, H is generated by $x_i^{d_i}$. Since the x_i are free Abelian generators for A_n, it is easily seen that if $d_i \neq 0$ then $x_i^{d_i}$ are free Abelian generators for H. ◀

THEOREM 3.6. *A finitely generated Abelian group G of rank n is the direct product of ρ infinite cyclic groups and $n - \rho$ finite cyclic groups of orders $\tau_1, \ldots, \tau_{n-\rho}$ where τ_i divides τ_{i+1}. Moreover, ρ and the τ_i are uniquely determined by G.*

PROOF. To show that G is a direct product, note that $G \simeq A_n/H$ where A_n is the free Abelian group of rank n. By Corollary 3.5.2, A_n is generated by elements $x_1, x_2, \ldots, x_n$ such that H is generated by $x_1^{d_1}, \ldots, x_n^{d_n}$ where $d_i \geq 0$ and d_i divides d_{i+1}. Clearly then, A_n/H is isomorphic to the direct product of cyclic groups Z_i where Z_i is infinite cyclic if $d_i = 0$ and is cyclic of order d_i otherwise. Since G has rank n, no $d_i = 1$. Thus G is the direct product of cyclic groups of orders $\tau_1, \ldots, \tau_{n-\rho}$ and ρ infinite cyclic groups, where $\tau_i > 1$ and τ_i divides τ_{i+1}. The number ρ is called the *Betti number* of G and the numbers $\tau_1, \ldots, \tau_{n-\rho}$ are called the *torsion numbers* of G.

To show the uniqueness of the cyclic decomposition, it is convenient to introduce the *relation matrix* of a presentation. The *relation matrix* of the presentation

$$\langle x_1, \ldots, x_n; \; R_1(x_i), \ldots, R_m(x_i)\rangle \tag{10}$$

is the n-by-m matrix of integers whose (i, j) entry is the exponent sum of R_j relative to x_i. An Abelian group is completely determined by any one of its relation matrices. From the existence of a cyclic decomposition, it follows that a finitely generated Abelian group G of rank n has a relation matrix of the form

$$(11) \qquad \begin{pmatrix} d_1 & & & & \\ & d_2 & & 0 & \\ & & \cdot & & \\ & & & \cdot & \\ & 0 & & & \cdot \\ & & & & d_n \end{pmatrix}$$

where $d_i \geq 0$ and d_i divides d_{i+1}. Although different presentations of a group have different relation matrices, the "invariant factors" (different from 0, 1) of these matrices will be the same. The kth *invariant factor* $\epsilon_k(M)$ of a matrix M is the greatest common divisor of all k-by-k subdeterminants of M. For convenience we take $\epsilon_q(M) = 0$ if q exceeds the number of rows or columns of M. Clearly ϵ_k divides ϵ_{k+1} since expansion by minors of a determinant of order $k + 1$ yields a linear combination of kth order determinants. To determine the effect of a change of presentation on a relation matrix, we may restrict ourselves (by Corollary 1.5) to elementary Tietze transformations.

If a relator is adjoined which is derivable from (and so is a product of conjugates of) the original relators, then the relation matrix M' of the new presentation is obtained from M by adjoining to it a column which is an integral linear combination of the original columns of M. For example, if

$$\langle x, y; \quad x^2y^3, y^{-1}x^3\rangle$$

is changed to

$$\langle x, y; \quad x^2y^3, y^{-1}x^3, x^5y^2\rangle,$$

then the relation matrix

$$(12) \qquad \begin{pmatrix} 2 & 3 \\ 3 & -1 \end{pmatrix}$$

becomes

$$(13) \qquad \begin{pmatrix} 2 & 3 & 5 \\ 3 & -1 & 2 \end{pmatrix}$$

and the last column of (13) is the sum of the first two columns of (12). It is easily verified that the sequences of invariant factors of (12) and (13) are identical.

In general, M and M' have the same sequence of invariant factors. For each k-by-k subdeterminant of M is a k-by-k subdeterminant of M'. On the other hand, the determinant of a k-by-k submatrix of M' containing entries from the last column is seen to be an integral linear combination of

k-by-k submatrices of M (since the determinant, viewed as a function of its columns, is multilinear and is zero when two columns coincide). Thus $\epsilon_k(M')$, which is the gcd of all k-by-k subdeterminants of M, equals $\epsilon_k(M)$.

Consider next the effect of adjoining to a presentation a new generator together with a defining relator which specifies this new generator in terms of the original generators. The new relation matrix M' is obtained from M by adjoining a row of the form

$$(0, 0, \ldots, 0, 1)$$

which has one more column than M, and entering the appropriate integers above the 1 to fill out the last column. For example, if

$$\langle x, y; \quad x^2y^3, y^{-1}x^3\rangle$$

becomes

$$\langle x, y, z; \quad x^2y^3, y^{-1}x^3, zx^{-3}y^2\rangle,$$

then

$$\begin{pmatrix} 2 & 3 \\ 3 & -1 \end{pmatrix}$$

is changed to

$$\begin{pmatrix} 2 & 3 & -3 \\ 3 & -1 & 2 \\ 0 & 0 & 1 \end{pmatrix}.$$

In the case of this type of Tietze transformation we show that $\epsilon_{k+1}(M') = \epsilon_k(M)$ for $k \geq 1$ and $\epsilon_1(M') = 1$, so that the sequence of invariant factors of M' is obtained by placing a 1 in front of the sequence of invariant factors of M. For, a k-by-k submatrix of M has the same determinant as the $(k + 1)$-by-$(k + 1)$ submatrix of M' obtained from it by adjoining a part of the last row and last column of M' in such a way that 1 is the $(k + 1, k + 1)$ entry. On the other hand, expansion of a $(k + 1)$-by-$(k + 1)$ subdeterminant of M' about its last column shows that it is an integral linear combination of k-by-k subdeterminants of M. Clearly then, $\epsilon_{k+1}(M') = \epsilon_k(M)$, $k \geq 1$. Since M' has the entry 1, $\epsilon_1(M') = 1$.

Thus both types of elementary Tietze transformations when applied to a presentation leave its sequence of invariant factors (different from 1) unaltered. On the other hand, the number of invariant factors equal to 1 may increase by one. When this occurs the number of generators of the presentation also increases by one. Therefore, the difference between the number of generators and the number of invariant factors equal to 1 is left unaltered. Since the number of invariant factors different from 1 or 0 is left unchanged, it follows that the number of generators minus the number of non-zero invariant factors is left unaltered.

The uniqueness of the cyclic decomposition of a finitely generated Abelian group G asserted in Theorem 3.6 now follows immediately. For G has a presentation whose relation matrix is given by (11), where $d_i = \tau_i > 1$, $i = 1, \ldots, n - \rho$, and $d_i = 0$, $i > n - \rho$. Since $\epsilon_k(M) = d_1 \ldots d_k$, $1 \leq k \leq n$, $\epsilon_k(M) \neq 0$, for $1 \leq k \leq n - \rho$ and $\epsilon_k(M) = 0$ for $k > n - \rho$. Thus ρ may be characterized as the number of generators minus the number of non-zero invariant factors. On the other hand, $\tau_i = \epsilon_i(M)/\epsilon_{i-1}(M)$, $2 \leq i \leq n - \rho$ is the ratio of consecutive invariant factors different from 0 or 1. Finally, $\tau_1 = \epsilon_1(M) \neq 0, 1$. Consequently, ρ and $\tau_1, \ldots, \tau_{n-\rho}$, i.e., the Betti and torsion numbers, are isomorphic invariants of G. ◀

Having completed the proof of Theorem 3.6, we can now establish the uniqueness of the d_i asserted in Theorem 3.5. Using the notation of Theorem 3.5, let H be generated by

$$y_i^{d_i}Q_i, \; Q_j, \qquad i = 1, \ldots, n, \qquad j > n,$$

and H^* be the subgroup of F generated by H and the commutator subgroup of F. Then H^* is normal and

$$F/H^* \simeq Z_1 \times Z_2 \times \ldots \times Z_n,$$

where Z_i is the cyclic group of order d_i for $d_i \neq 0$ and Z_i is the infinite cyclic group for $d_i = 0$ (see Corollary 2.1). Clearly the number of d_i equal to 0 is just the Betti number of F/H^* while those $d_i \neq 1, 0$ are the torsion numbers of F/H^*. The number of d_i equal to 1 is n (the rank of F) minus the rank of F/H^*. This proves the uniqueness of the d_i for a given subgroup H of F. ◀

It follows from Theorem 3.5, that given a finite presentation of a group G, we can find generators $y_1, \ldots, y_n$ and a set of defining relators $V_1, \ldots, V_m$ in these generators having the form

$$V_i = y_i^{d_i}Q_i, \qquad i = 1, \ldots, n$$
$$V_j = Q_j, \qquad n < j < m$$

where $0 \leq d_i$ divides d_{i+1} and Q_i, Q_j have zero exponent sum on each y_i. Such a presentation of G will be called a *pre-Abelian* presentation of G. The direct product decomposition of Theorem 3.6 for G/G' (the commutator quotient group of G) is immediately read from a pre-Abelian presentation for G. Moreover, the words

$$y_1^{\alpha_1}y_2^{\alpha_2} \ldots y_n^{\alpha_n},$$

where $0 \leq \alpha_i < d_i$ for $d_i \neq 0$ and α_i an arbitrary integer otherwise, form a Schreier representative system for G mod G'. Hence, we can construct the Reidemeister-Schreier presentation for G' by using this representative system for G mod G'.

This leads to a test for the isomorphism of two finitely presented groups. For, from a finite presentation for a group G, we may obtain a pre-Abelian presentation for G, from which we may read off the canonical decomposition of G/G' given by Theorem 3.6. If groups G_1 and G_2 are isomorphic, then necessarily $G_1/G_1' \simeq G_2/G_2'$. Hence, if these Abelian factor groups are not isomorphic, we are assured $G_1 \not\simeq G_2$. On the other hand, if the factor groups are isomorphic, we must test further. Suppose the quotient groups are finite. Then G_1', G_2' have finite presentations which may be obtained as indicated in the preceding paragraph. To test the isomorphism of G_1', G_2' we repeat the above process using G_1', G_2' in place of G_1, G_2 and consider the Abelian factor groups G_1'/G_1'', G_2', G_2''. If these are not isomorphic, then $G_1' \not\simeq G_2'$ and so $G_1 \not\simeq G_2$. If the Abelian factor groups are isomorphic and finite, we may continue the above process. A breaking-off point is reached when we arrive at isomorphic Abelian factor groups which are infinite. Even in this case we can go one step further to test the isomorphism of G_1 and G_2. In the next section, we consider the simplest case, namely, where the Abelian factor groups are infinite cyclic.

Problems for Section 3.3

1. Let F be the free group on a, b, and let H be the subgroup generated by a^2, b^2, ab^{-1}. Find a set of generators for F and for H satisfying Theorem 3.5.

2. Let F be the free group on a, b, c, and let H be the subgroup generated by $abc^{-1}b^{-1}$, $bca^{-1}c^{-1}$, $cab^{-1}a^{-1}$. Find a set of generators for F and for H satisfying Theorem 3.5.

3. Obtain a pre-Abelian presentation for the fundamental group of the trefoil knot given by

$$T = \langle a, b; a^2b^{-3} \rangle.$$

4. Compute the Betti and torsion numbers of

$$\langle a, b, c; a^2b^{-3}c^4, (abc)^2 \rangle.$$

5. Show that if $G = \langle a, b; a^kb^{-l} \rangle$ with k, l relatively prime, then G/G' is infinite cyclic. [*Hint:* Compute the Betti and torsion numbers of G from its relation matrix.]

6. Show that $\langle a, b; a^2b^{-3}, (ab^{-1})^2\rangle$ and $\langle a, b; (ab^{-1})^2b^{-1}, a^2ba^{-1}b\rangle$ are not isomorphic.

7. Find a presentation for G' and the Betti and torsion numbers of G/G' and G'/G''; where $G = \langle a, b; (ab)^2, bab^{-2}a^{-1}\rangle$.

8. Show that $\langle a, b; a^2, b^3ab^{-2}a\rangle$ and $\langle a, b; a^3ba^{-1}b^{-1}, b^2a^2b^{-1}a^{-2}\rangle$ are not isomorphic. [*Hint:* Consider G'/G''.]

9. Show that $G' = G''$ where $G = \langle a, b; a^2b^2ab^{-1}, a^3b^4a^{-2}b^{-3}\rangle$.

10. Show that a cyclic group of order rs with r, s relatively prime is the direct product of a cyclic group of order r and one of order s, and conversely. Show that any finite cyclic group is the direct product of cyclic groups of prime power orders.

11. Show that any finite Abelian group A is the direct product of cyclic groups of prime power order and that, moreover, the decomposition is unique. [*Hint:* To show uniqueness note that for any prime p, the set of elements of A whose order is a power of p form a subgroup $A(p)$ which is the direct product of those cyclic factors of A of order p to a power. Since powers of p can always be arranged in a chain, each dividing the succeeding, the orders of these cyclic factors are just the torsion numbers of $A(p)$.]

12. Show how to compute the torsion numbers of the direct product of cyclic groups of prime power order without forming the relation matrix and compute the torsion numbers of $Z_3 \times Z_4 \times Z_4 \times Z_9 \times Z_9$ and of $Z_6 \times Z_{10} \times Z_{15}$ where Z_n is the cyclic group of order n. [*Hint:* Since τ_i divides τ_{i+1}, the exponent of p occurring in τ_{i+1} is no less than that occurring in τ_i. Hence the last torsion number is the product of the highest powers of each prime occurring.]

13. Show that a cyclic group of order p^n has p^m elements of exponent p^m (i.e., solutions of $x^{p^m} = 1$) for $m \leq n$ and has $p^m - p^{m-1}$ elements of order p^m. How many elements of exponent 3, 9, 27 are there in $Z_3 \times Z_3 \times Z_9 \times Z_{27}$ and of exponent 6 in $Z_3 \times Z_3 \times Z_4 \times Z_5$?

14. Let k_i, $i = 1, \ldots, n$, be positive integers, $k_i \leq k_{i+1}$, and let $e_i = p^{k_i}$ where p is a prime number. Show that if A is the direct product of the cyclic groups Z_{e_i}, of order e_i, then the number of elements of exponent p^k is $r_k = p^{k_1+k_2+\ldots+k_s+(n-s)k}$ where s is the largest integer such that $k_s \leq k$. Show that if A has q cyclic factors of order $\geq p^k$, then $p^q = r_k/r_{k-1}$. From this, prove that the number of elements of each order in A determines the cyclic decomposition of A.

15. Show that two finitely generated Abelian groups which have the same Betti number and the same number of elements of each prime power order are isomorphic.

16. Let G be a finitely generated Abelian group generated by $x_1, \ldots, x_n$ and let H be a subgroup of G. Show that $x_1, \ldots, x_n$ can be Nielsen transformed to $y_1, \ldots, y_n$ so that H is finitely generated by powers of the y_i. Show that any chain of subgroups $H_1 \subset H_2 \subset \ldots$ of G is finite.

17. Let $G = \langle x_1, \ldots, x_n; \; R_1, \ldots, R_m\rangle$ be a finite presentation. Show that the Betti number ρ of G satisfies $\rho \geq n - m$. [*Hint:* The relation matrix for G/G' is n-by-m. When this matrix is put into "diagonal" form,

the Betti number of G is n − (the numbers of 1's) − (the number of torsion numbers). Since the 1's and torsion numbers occur on the main diagonal, there are no more than m of them.]

18. Show that a finite group must have at least as many relators as generators in any of its presentations.

19. Let F be the free group on a, b, and let W_1, W_2 be words which generate F/F'. Show that there exist generators $c = C(a, b)$, $d = D(a, b)$, and words $\tilde{W}_1(c, d)$, $\tilde{W}_2(c, d)$ such that $[W_1, W_2] = [\tilde{W}_1, \tilde{W}_2]$ and $\sigma_c(\tilde{W}_1) = 1$, $\sigma_c(\tilde{W}_2) = 0$, $\sigma_d(\tilde{W}_1) = 0$, $\sigma_d(\tilde{W}_2) = 1$, where $[x, y]$ denotes $xyx^{-1}y^{-1}$. [*Hint:* In reducing a two-by-two titled exponent matrix to the form (7), the operations on the column titles can be restricted to replacing x, y by x, yx^k or xy^k, y. Now, $[x, y] = [xy^k, y] = [x, yx^k]$.]

20. A group G is residually finite if every element $g \neq 1$ is excluded from some normal subgroup of finite index (see Problem 2.4.24).

(a) Show that if A and B are residually finite then $A \times B$ is also residually finite.

(b) Show that a finitely generated free Abelian group A_n is residually finite.

(c) Show that any finitely generated Abelian group is residually finite.

(d) Show that a finitely generated Abelian group is Hopfian, i.e., cannot be isomorphic to a proper factor group of itself.

[*Hint:* For (a), if (c, d) is in $A \times B$ and H, K are a normal subgroup of A and B excluding c or d, respectively, then $H \times B$ or $A \times K$ excludes (c, d). For (b), use (a). For (c), use (a), (b), and Theorem 3.6. For (d), use Problem 2.4.24(d).]

21. A set of elements $b_1, \ldots, b_r$ of an Abelian group $(A, +)$ are called (integrally) independent if $\lambda_1 b_1 + \cdots + \lambda_r b_r = 0$ implies $\lambda_i = 0$, λ_i an integer.

(a) Show that under an elementary Nielsen transformation of rank r, a set of independent elements $b_1, \ldots, b_r$ go into a set of independent elements $c_1, \ldots, c_r$.

(b) Show that the free Abelian group A on n generators $a_1, \ldots, a_n$ has n independent elements, but any subset of $n + 1$ elements is dependent.

(c) Show that the Betti number of a finitely generated Abelian group A is the maximum number of independent elements in A.

(d) Show that if H is a subgroup of a finitely generated Abelian group A, and H has an element of infinite order, then the Betti number of A is greater than the Betti number of A/H.

(e) Show that a finitely generated Abelian group is Hopfian.

[*Hint:* For (b), form a titled exponent sum matrix using $a_1, \ldots, a_n$ as row labels and $b_1, \ldots, b_{n+1}$ as column labels. Then by elementary Nielsen transformations applied to $a_1, \ldots, a_n$ and $b_1, \ldots, b_{n+1}$ (and the corresponding matrix operations), we obtain a matrix whose last column has all zeros. Hence, $b_1, \ldots, b_{n+1}$ goes into $c_1, \ldots, c_n$, 0 which is not an independent set; use (a).

For (c), suppose that $A = A_n \times B$, where A_n is a free Abelian group and B is finite of order m. Then $b_1, \ldots, b_r$ is an independent set if and only if $mb_1, \ldots, mb_r$ is; but mb_i lies in A_n, and (b) can be used.

For (d), if $b_1H, \ldots, b_rH$ is an independent set in A/H then $mb_1H, \ldots, mb_rH$ is an independent set in $A_nH/H \simeq A_n/A_n \cap H$. But suppose H has an element h of infinite order; then $mh \in A_n$ and $mh \neq 0$, $A_n \cap H \neq 0$ and hence, using Corollary 3.5.2, the Betti number of $A_n/A_n \cap H$ is less than n.

For (e), use (d) and also if $H \subset B$, then $(A_n \times B)/H \simeq A_n \times (B/H)$. Since B is the subgroup of A of all elements of finite order and B/H is the subgroup of A/H of all elements of finite order, if $A \simeq A/H$, then $B \simeq B/H$.]

3.4. A Test for Isomorphism

In this section, we discuss a test for the isomorphism of finitely presented groups with infinite cyclic commutator quotient groups. All three-dimensional knot groups have this property (see Reidemeister, 1932b, p. 44, or Crowell and Fox, 1963, p. 110).

Let K be a finitely presented group which is such that K/K' is infinite cyclic. For ease of notation, we assume K has three generators. Then K has a pre-Abelian presentation of the form

$$(1) \qquad \langle a, b, c; bB(a, b, c), cC(a, b, c), \ldots, D_\nu(a, b, c), \ldots \rangle,$$

where B, C, D_ν have zero exponent sum on a, b, c. We may choose $\{a^n\}$, $n = 0, \pm 1, \pm 2, \ldots$, as a Schreier representative system for K mod K' and $\overline{W(a, b, c)} = a^\sigma$ where $\sigma = \sigma_a(W)$ is the exponent sum of a in W. Using the method of Section 2.3, K' is generated by

$$(2) \qquad b_n = a^n b a^{-n}, \qquad c_n = a^n c a^{-n}, \qquad (n = 0, \pm 1, \pm 2, \ldots).$$

Since B, C, D_ν are words in K', we may rewrite them as words in $\ldots, b_n, c_n, \ldots$, obtaining B_0, C_0, $D_{0,\nu}$. Then a set of defining relators for K' on the generators (2) is

$$(3) \qquad b_\lambda B_0(\ldots, b_{n+\lambda}, c_{n+\lambda}, \ldots), c_\lambda C_0(\ldots, b_{n+\lambda}, c_{n+\lambda}, \ldots),$$

and

$$(4) \qquad D_{0,\nu}(\ldots, b_{n+\lambda}, c_{n+\lambda}, \ldots), \qquad \lambda = 0, \pm 1, \pm 2, \ldots.$$

As a group then, K' is not finitely presented. However, using K' we may associate to K a *finitely presented module* as follows:

An Abelian group G is called an A-module if with G as "vectors" and A as "scalars" the postulates for a vector space are satisfied, except that A need only be a ring and not necessarily a field. If $g \in G$ and $\alpha \in A$, we denote "scalar multiplication" of g with α by g^α. Elements $x_1, \ldots, x_n$

of G *generate* G *over* A if each $g \in G$ can be written in the form

$$x_1^{\alpha_1} x_2^{\alpha_2} \ldots x_n^{\alpha_n}, \qquad \alpha_i \in A, \tag{5}$$

i.e., if the x_i together with their images under A generate the Abelian group G. G is a *free* A-*module* and $x_1, \ldots, x_n$ are *free generators* if the word (5) defines the identity of G only if all $\alpha_i = 0$. A set of words $\hat{R}_j(x_i)$ in the x_i with exponents in A is a *set of defining relators for an* A-*module* G (generated by $x_1, \ldots, x_n$) if $\hat{W}(x_i)$ defines the identity in G implies that in the free A-module on $x_1, \ldots, x_n$,

$$\hat{W} = \prod_j \hat{R}_j^{\beta_j}, \qquad \beta_j \in A.$$

In this case, we write

$$G = \langle x_1, \ldots, x_n; \hat{R}_j(x_i)/A \rangle. \tag{6}$$

If the number of generators and defining relators $\hat{R}_j$ in (6) is finite, then (6) is a *finite presentation for* G *over* A.

Returning to the group K above with infinite cyclic commutator quotient group, we construct a finitely presented module as follows: As "vectors," use the Abelian group K'/K''; as "scalars," use L-*polynomials*, i.e., polynomials in θ and θ^{-1} with integral coefficients, or in other words, finite sums of the form

$$L(\theta) = \sum_n j_n \theta^n,$$

where j_n and n are integers. Addition and multiplication of L-polynomials are defined in the obvious manner. Scalar multiplication by θ will correspond to conjugation by a. This if $q \in K'/K''$, we define,

$$q^\theta = aqa^{-1}, \qquad q^{\theta^n} = a^n q a^{-n}, \qquad q^{j_n \theta^n} = a^n q^{j_n} a^{-n}, \tag{7}$$
$$q^{L(\theta)} = \prod_n q^{j_n \theta^n} = \prod_n a^n q^{j_n} a^{-n}.$$

With this definition of scalar multiplication, K'/K'' is an L-polynomial module.

Lemma 3.4 states that the L-polynomial module K'/K'' is an invariant of K.

LEMMA 3.4. *Let ϕ be an isomorphism of the group K onto the group K_1 where K, K_1 have infinite cyclic commutator quotient groups generated by a, a_1, respectively. Then the L-polynomial modules K'/K'' and K_1'/K_1'' are isomorphic either under the mapping*

$$kK'' \to \phi(k)K_1'', \qquad L(\theta) \to L(\theta) \tag{8}$$

or under the mapping

(9) $$kK'' \to \phi(k)K_1'', \qquad L(\theta) \to L(\theta^{-1}),$$

where $k \in K'$, *and* $\phi(k) \in K_1'$.

PROOF. The mapping of $K'/K'' \to K_1'/K_1''$ given by $kK'' \to \phi(k)K_1''$ is clearly a group isomorphism. To show that

$$\phi[(kK'')^{L(\theta)}] = [\phi(k)K_1'']^{L(\theta^\epsilon)}.$$

where $\epsilon = 1$ or -1 (independent of k), it suffices to show that

$$\phi[(kK'')^\theta] = [\phi(k)K_1'']^{\theta\epsilon}.$$

Now,

$$\phi[(kK'')^\theta] = \phi[(aka^{-1})K''] = \phi(a)\phi(k)\phi(a)^{-1}K_1''.$$

On the other hand,

$$[\phi(k)K_1'']^{\theta\epsilon} = a_1^\epsilon\phi(k)a_1^{-\epsilon}K_1''.$$

Since the mapping

$$gK' \to \phi(g)K_1', \qquad (g \in K)$$

is an isomorphism of K/K' onto K_1/K_1', the generator aK' goes into $a_1^\epsilon K_1'$ where ϵ is either 1 or -1. Thus $\phi(a) = a_1^\epsilon k_1$ where $k_1 \in K_1'$. Hence,

$$\begin{aligned}\phi(a)\phi(k)\phi(a)^{-1}K_1'' &= a_1^\epsilon k_1\phi(k)k_1^{-1}a_1^{-\epsilon}K_1'' \\ &= (a_1^\epsilon k_1 a_1^{-\epsilon})K_1'' \cdot (a_1^\epsilon\phi(k)a_1^{-\epsilon})K_1'' \cdot (a_1^\epsilon k_1^{-1}a_1^{-\epsilon})K_1'' \\ &= a_1^\epsilon\phi(k)a_1^{-\epsilon}K_1'',\end{aligned}$$

since K_1'/K_1'' is Abelian. ◀

Thus there is associated to each group K with infinite cyclic commutator quotient group, the L-polynomial module K'/K'' which (up to changing the sign of the exponents in all L-polynomials) is an isomorphic invariant of K. Any invariants of the module are invariants of K. It remains, then, to get invariants of the module. This can be done constructively if K is finitely presented. In this case the module K'/K'' is finitely presented over the L-polynomials. For, a group presentation of K'/K'' is given by the generators b_n, c_n of (2) and the defining relators (3), (4), together with the words

(10) $$b_nb_mb_n^{-1}b_m^{-1}, \qquad c_nc_mc_n^{-1}c_m^{-1}, \qquad b_nc_mb_n^{-1}c_m^{-1}.$$

Since each of the generators b_n, c_n is simply b^{θ^n} or c^{θ^n}, it follows that bK'', cK'' generate the module K'/K''. Moreover, (3) and (4) can be written as

(11) $$[bB_0(\ldots, b^{\theta^n}, c^{\theta^n}, \ldots)]^{\theta^\lambda}, \qquad [cC_0(\ldots, b^{\theta^n}, c^{\theta^n}, \ldots)]^{\theta^\lambda}$$

and

(12) $$[D_{0,\nu}(\ldots, b^{\theta^n}, c^{\theta^n}, \ldots)]^{\theta\lambda}, \qquad \lambda = 0, \pm 1, \pm 2, \ldots.$$

Since K'/K'' has the relators (10), (3) and (4) can be simplified by collecting the b's and c's in B_0, C_0, $D_{0,\nu}$, thereby obtaining

(13) $$[b\hat{B}_0(b, c)]^{\theta\lambda}, \qquad [c\hat{C}_0(b, c)]^{\theta\lambda}, \qquad [\hat{D}_{0,\nu}(b, c)]^{\theta\lambda},$$

where $\hat{B}_0$, $\hat{C}_0$, $\hat{D}_{0,\nu}$ are words in b, c, with L-polynomial exponents. A module is necessarily Abelian and hence the relators (10) can be omitted. Thus we can present K'/K'' by writing

(14) $$K'/K'' = \langle b, c; b\hat{B}_0, c\hat{C}_0, \ldots, \hat{D}_{0,\nu}, \ldots/\mathscr{L}\rangle$$

where $\mathscr{L}$ is the ring of L-polynomials. Consequently, a finite presentation for the group K yields a finite presentation for the $\mathscr{L}$-module K'/K''.

To illustrate the procedure for obtaining a presentation of the $\mathscr{L}$-module K'/K'' from a presentation of K, let

(15) $$K = \langle a, b, c; \quad ba^2b^3a^{-1}cb^{-3}a^{-1}c^{-1}, ca^{-1}cbac^{-1}b^{-1}, abca^{-1}b^{-1}c^{-1}\rangle.$$

Then by rewriting the relators directly in terms of the L-polynomials, we obtain

$$K'/K'' = \langle b, c; b^{3\theta^2-3\theta+1}c^{\theta-1}, b^{\theta^{-1}-1}c^{\theta^{-1}}, b^{\theta-1}c^{\theta-1}/\mathscr{L}\rangle$$

As in the case of an Abelian group (which is a module over the ring of integers), we associate a relation matrix to the $\mathscr{L}$-module K'/K'' and consider its invariant factors. If

(16) $$K'/K'' = \langle x_1, \ldots, x_n; \hat{R}_1, \ldots, \hat{R}_m/\mathscr{L}\rangle$$

and $\hat{R}_j = \prod_i x_i^{L_{ij}(\theta)}$ then the matrix M, whose (i, j) entry is $L_{ij}(\theta)$, is called the *relation matrix of* (16).

Since $\mathscr{L}$ is a commutative ring, we may form the determinants of the square submatrices of M in the usual way. Moreover, it is easy to show the existence of greatest common divisors in $\mathscr{L}$. For, every L-polynomial $L(\theta) \neq 0$ can be written uniquely in the form

$$L(\theta) = \eta\theta^\lambda P(\theta)$$

where $\eta = \pm 1$, λ is an integer and $P(\theta)$ is an ordinary polynomial with integral coefficients and positive constant term. $P(\theta)$ is called the *stem of* $L(\theta)$. Clearly, $\pm\theta^\lambda$, $\lambda = 0, \pm 1, \pm 2, \ldots$ are the units in $\mathscr{L}$ and so $L_1(\theta)$ divides $L_2(\theta)$ if and only if the stem of L_1 divides the stem of L_2. Thus we have for the greatest common divisor (gcd):

$$\gcd(L_1, L_2) = \gcd(P_1, P_2).$$

Since P_1, P_2 are ordinary polynomials with integral coefficients, we can constructively compute their gcd (see Problems 6 and 7). Hence, we may define the rth *invariant factor* $\epsilon_r(M)$ as the gcd of the r-by-r subdeterminants of M.

In addition to considering the gcd of all r-by-r subdeterminants of M, we may consider a stronger invariant, namely, all linear combinations of these subdeterminants using L-polynomial coefficients, i.e., the ideal of $\mathscr{L}$ generated by them. The rth *Alexander ideal* $\alpha_r(M)$ of M is the ideal of $\mathscr{L}$ generated by the r-by-r subdeterminants of M. (In general, $\alpha_r(M)$ will not contain all multiples of $\epsilon_r(M)$, e.g., see Problem 9.) Expanding an $(r+1)$-by-$(r+1)$ subdeterminant of M about a column shows that $\alpha_r \supset \alpha_{r+1}$. For convenience, we take $\alpha_r = 0$ if r exceeds n or m of (16).

THEOREM 3.7. *Let K, K_1 be isomorphic groups with infinite cyclic commutator quotient group. Then the relation matrices for the $\mathscr{L}$-modules K'/K'', K_1'/K_1'' have the same (except for the possible replacement of θ by θ^{-1}) sequences of invariant factors $\neq 1$ and sequence of Alexander ideals $\neq \mathscr{L}$.*

PROOF. By Lemma 3.4, the $\mathscr{L}$-module K'/K'' is isomorphic to the $\mathscr{L}$-module K_1'/K_1'' in such a way that $\theta \to \theta$ or $\theta \to \theta^{-1}$. Assume first the case where $\theta \to \theta$.

Since invariant factors and Alexander ideals were defined by means of presentations, we require some connection between the presentations of isomorphic $\mathscr{L}$-modules. This is given by the analogue for module presentations of the Tietze Theorem 1.5 for group presentations. Specifically, if the modules

$$\langle x_1, \ldots, x_n; \hat{R}_1, \ldots, \hat{R}_m/\mathscr{L}\rangle,$$
$$\langle y_1, \ldots, y_p; \hat{S}_1, \ldots, \hat{S}_p/\mathscr{L}\rangle$$

are isomorphic (in such a way that $\mathscr{L}$ is left fixed), then both presentations can be transformed to a common presentation by

(a) adding or deleting relators which are products of powers (with L-polynomial exponents) of the other relators,

(b) adding a new generator and a new relator defining this generator in terms of the old generators.

The proof of this analogue is similar to that of Theorem 1.5 and is left as an exercise.

If we examine the relation matrix of a module presentation, it can be shown, as in the Abelian group case, that operations (a) and (b) above on a presentation correspond to simple operations on its relation matrix M. These are:

(a′) adjoining or deleting a column which is a linear combination (with L-polynomial coefficients) of the other columns, (b′) adjoining a new row $(0, \ldots, 0, 1)$ which has one more column than M and filling the last column above 1 with the appropriate L-polynomials. The operation (a′) will leave the sequence of invariant factors and of Alexander ideals unchanged. The operation (b′) places a 1 in front of the sequence of invariant factors of M and places an $\mathscr{L}$ in front of the sequence of Alexander ideals.

There remains the case in which the $\mathscr{L}$-module $K'/K'' \simeq K_1'/K_1''$ in such a way that $\theta \to \theta^{-1}$. In this case, replace θ by θ^{-1} in the presentation (16) of K'/K'' obtaining a new $\mathscr{L}$-module. Then the invariant factors and Alexander ideals of this new module can be obtained from those of (16) by replacing θ by θ^{-1}. Moreover, this new module is isomorphic to K'/K'' in such a way that $\theta \to \theta^{-1}$ and so is isomorphic to K_1'/K_1'' with $\theta \to \theta$. Consequently, combining this with the preceding case completes the proof of Theorem 3.7. ◀

Next we specialize our considerations to the group K for which the D_ν in (1) are absent, i.e., groups with infinite cyclic commutator quotient group having some presentation on $n + 1$ generators with n defining relators. In this case, the $\mathscr{L}$-module K'/K'' has n generators with n defining relators and, hence, a square n-by-n relation matrix M. The determinant $\Delta(\theta)$ of M is then an invariant factor. The stem of $\Delta(\theta)$ is called the *Alexander polynomial* of K and, in the case that K is a knot group, the *knot polynomial* of K.

THEOREM 3.8. *If K is a group on $n + 1$ generators with n defining relators having infinite cyclic commutator quotient group, then the Alexander polynomial of K is an annihilator ($\neq 0$) of the $\mathscr{L}$-module K'/K'', i.e., scalar multiplication by it always yields* 1.

PROOF. For convenience of notation, we take $n = 2$. Then K has a presentation

$$K = \langle a, b, c;\, bB, cC \rangle.$$

The $\mathscr{L}$-module K'/K'' has the presentation

$$K'/K'' = \langle b, c;\, b\hat{B}_0, c\hat{C}_0 / \mathscr{L} \rangle.$$

and the relation matrix

$$M = \begin{pmatrix} 1 + L_{11}(\theta) & L_{12}(\theta) \\ L_{21}(\theta) & 1 + L_{22}(\theta) \end{pmatrix},$$

where $L_{11}(\theta)$, $L_{12}(\theta)$ are the b-exponent sums of $\hat{B}_0$, $\hat{C}_0$ and $L_{21}(\theta)$, $L_{22}(\theta)$ are the c-exponent sums of $\hat{B}_0$, C_0. In the construction of $\hat{B}_0$ from B, note that any occurrence b^k in B is replaced by $b^{k\theta^\lambda}$ in $\hat{B}_0$ where λ is the sum of the a-exponents preceding b^k in B (see Problem 3). Hence, $L_{11}(1)$ is the b-exponent sum of B which is zero. Similarly, $L_{21}(1) = L_{12}(1) = L_{22}(1) = 0$. Hence, $\Delta(1) = 1$, and so $\Delta(\theta)$ is not identically zero. Since in K'/K'',

$$b^{1+L_{11}(\theta)}c^{L_{21}(\theta)} = 1,$$

$$b^{L_{12}(\theta)}c^{1+L_{22}(\theta)} = 1,$$

we have by Cramer's rule for solving linear equations,

$$b^{\Delta(\theta)} = 1, \qquad c^{\Delta(\theta)} = 1.$$

But b, c generate K'/K''. Consequently, $q^{\Delta(\theta)} = 1$ for each $q \in K'/K''$. ◀

As a simple example of the calculation of a knot polynomial, we use the fundamental group of the trefoil knot (Kleeblattschlinge, see M. Dehn, 1919) which can be presented as

$$T = \langle a, b; ba^2ba^{-1}b^{-1}a^{-1}\rangle.$$

The module T'/T'' has the presentation

$$T'/T'' = \langle b; b^{\theta^2-\theta+1}/\mathscr{L}\rangle,$$

and so the knot polynomial of T is $\theta^2 - \theta + 1$.

It should be noted that whereas the invariant factors and Betti number of a finitely presented Abelian group are complete, i.e., determine the group, the invariant factors and the difference between the number of generators and the number of invariant factors equal to 1 do not determine a module completely.

References and Remarks. The polynomial of a knot was introduced by Alexander, 1928. For its calculation from the fundamental group of the knot, see also K. Reidemeister, 1932b, or Crowell and Fox, 1963.

For a group K which is the fundamental group of a knot, the knot polynomial always satisfies $P(\theta) = \theta^r P(\theta^{-1})$.

Theorem 3.7 can be generalized in various ways. If we study groups K for which K/K' is not cyclic but a fixed finitely generated Abelian group, instead of the ring of scalars being polynomials in a single variable and its inverse, we would use residue class rings of polynomials in several variables with respect to certain types of ideals. Examples of this kind of investigation can be found in Magnus, 1934c.

Also, we can introduce "higher" commutator subgroups and their quotient groups. For results obtained here, see Chapter 5.

Problems for Section 3.4

1. Let G be a group generated by $x_n, y_n, n = 0, \pm 1, \ldots$ with defining relators $x_n x_m x_n^{-1} x_m^{-1}$, $y_n y_m y_n^{-1} y_m^{-1}$, $x_n y_m x_n^{-1} y_m^{-1}$ and $R_{\nu,\lambda}(\ldots, x_n, y_n, \ldots)$ such that $R_{\nu,\lambda}(\ldots, x_n, y_n, \ldots)$ can be obtained from $R_{\nu,0}$ by replacing each x_n, y_n by $x_{n+\lambda}, y_{n+\lambda}, n = 0, \pm 1, \ldots, \lambda = 0, \pm 1, \ldots, \nu = 1, \ldots, k$. Prove that G is a finitely presented $\mathscr{L}$-module. [*Hint:* Define $x_0^{\theta^\lambda} = x_\lambda, y_0^{\theta^\lambda} = y_\lambda$.]

2. Show that if

$$A = \langle a, b, c; \hat{R}_i/\mathscr{L}\rangle, \qquad i = 1, \ldots, k,$$

and each $\hat{R}_i$ has only integer exponents, then A is the direct product of infinitely many isomorphic Abelian groups. [*Hint:* Define $a_\lambda = a^{\theta^\lambda}$, etc. Then the group A is the direct product of

$$A_\lambda = \langle a_\lambda, b_\lambda, c_\lambda, \ldots; R_i(a_\lambda, b_\lambda, c_\lambda), \ldots, a_\lambda b_\lambda a_\lambda^{-1} b_\lambda^{-1}, \ldots\rangle.]$$

3. Let $K = \langle a, b, c; bB, cC, \ldots, D_\nu, \ldots\rangle$ where B, C, D_ν have zero exponent-sum on a, b, c. For any word $W(a, b, c)$ define $\hat{W}_0(b, c)$ to be the word obtained from W as follows:

Replace each occurrence of b^k by $b^{k\theta^\lambda}$ where λ is the a-exponent sum preceding b^k in W. Then the $\mathscr{L}$-module K'/K'' has the presentation

$$K'/K'' = \langle b, c; b\hat{B}_0, c\hat{C}_0, \ldots, \hat{D}_{0,\nu}, \ldots/\mathscr{L}\rangle.$$

[*Hint:* If W is rewritten, using the Reidemeister-Schreier process, and after that, $b_\lambda = a^\lambda b a^{-\lambda}$, $c_\lambda = a^\lambda c a^{-\lambda}$ are replaced by b^{θ^λ}, c^{θ^λ} respectively, we obtain $\hat{W}_0$.]

4. A presentation for the trefoil knot group T is $\langle a, b;\ a^2b^{-3}\rangle$. Derive the knot polynomial from this presentation.

5. A primitive polynomial over the integers is one whose coefficients have gcd $= 1$. Show that the product of two primitive polynomials is primitive. [*Hint:* Consider the natural homomorphism of polynomials over the integers Z into polynomials with coefficients over Z_p the integers mod p, where p is a prime. If p is a common factor of the coefficients of $P_1(\theta) \cdot P_2(\theta)$, then over Z_p, $P_1(\theta) \cdot P_2(\theta) = 0$, which implies $P_1(\theta) = 0$ or $P_2(\theta) = 0$ over Z_p. Hence, p is a common factor of the coefficients of $P_1(\theta)$ or of $P_2(\theta)$.]

6. Let $R(\theta)$ be a polynomial with rational coefficients. Then clearly $R(\theta) = (a/b)\bar{R}(\theta)$ where a, b are relatively prime integers and $\bar{R}(\theta)$ is a primitive polynomial. Prove that if $R(\theta)$ divides (over the rationals) a primitive polynomial $P(\theta)$, then $\bar{R}(\theta)$ divides (over the integers) $P(\theta)$. [*Hint:* If $P(\theta) = Q(\theta)R(\theta) = (c/d)\bar{Q}(\theta)(a/b)\bar{R}(\theta)$, then $bdP(\theta) = ac\bar{Q}(\theta)\bar{R}(\theta)$ and $\bar{Q}(\theta)\bar{R}(\theta)$ is primitive. Thus $bd = \pm ac$ and so $P(\theta) = \pm\bar{Q}(\theta)\bar{R}(\theta)$.]

7. Let $P_1(\theta)$, $P_2(\theta)$ be polynomials over the integers. Show how to constructively obtain their gcd. [*Hint:* Using the Euclidean algorithm, we can compute the gcd $D(\theta)$ (over the rationals) of $\bar{P}_1(\theta)$ and $\bar{P}_2(\theta)$. Then $\bar{D}(\theta)$ is the gcd (over the integers) of $\bar{P}_1(\theta)$, $\bar{P}_2(\theta)$.]

8. Let $K = \langle a, b; b^2a^{-1}b^{-1}a, a^3ba^{-1}b^{-1}a^{-1}b^2a^{-1}b^{-2}\rangle$ and

$$K_1 = \langle a, b; b^3a^{-1}b^{-2}a, a^3ba^{-1}b^{-1}a^{-1}b^4a^{-1}b^{-4}\rangle.$$

Show that the first invariant factor ϵ_1 of K'/K'' and K_1'/K_1'' are both 1 and that the Alexander ideal $\alpha_1(K'/K'')$ is generated by $2\theta - 1$ and $(\theta - 1)(\theta^2 + 2)$ while $\alpha_1(K_1'/K_1'')$ is generated by $3\theta - 2$ and $(\theta - 1)(\theta^2 + 4)$. Prove these ideals are different and hence, that K is not isomorphic to K_1. [*Hint:* Substituting -1 for θ, we see that neither $3\theta - 2$ nor $3\theta^{-1} - 2$ can be linear combinations of $2\theta - 1$ and $(\theta - 1)(\theta^2 + 2)$.]

9. Show that the first invariant factor of the $\mathscr{L}$-module K'/K'' where $K = \langle a, b; b^2aba^{-1}, a^2ba^{-2}b^{-1}\rangle$ is 1 and the first Alexander ideal is generated by 3 and $\theta - 2$ and is $\neq \mathscr{L}$. [*Hint:* To show $\alpha_1 \neq \mathscr{L}$, substitute -1 for θ.]

10. Show that if the Alexander polynomial of K is 1, then $K' = K''$. [*Hint:* Use Theorem 3.8.]

11. (E. Rapaport) Let K be a group on $n + 1$ generators with n relators and infinite cyclic commutator quotient group. Show that if the Alexander polynomial $A(\theta)$ of K has highest coefficient 1 or -1 and constant term 1, then K'/K'' is a finitely presented Abelian group. [*Hint:* For convenience of notation take $n = 2$. As an Abelian group, K'/K'' is generated by b^{θ^λ}, c^{θ^λ}, $\lambda = 0, \pm 1, \pm 2, \ldots$. Since $A(\theta)$ is an annihilator and has highest and lowest coefficients 1 or -1, we may express the effect of θ^λ on K'/K'' by a polynomial of degree one less than that of $A(\theta)$. For if $A(\theta) = \pm\theta^k + B(\theta) + 1$ where $B(\theta)$ involves only the $\theta, \theta^2, \ldots, \theta^{k-1}$ terms of $A(\theta)$, then since $b^{\theta^k} = b^{\mp[B(\theta)+1]}$, all b^{θ^λ}, $\lambda > 0$, can be expressed in terms of $b, b^\theta, \ldots, b^{\theta^{k-1}}$. Similarly for c^{θ^λ}. Hence, $b, c, b^\theta, c^\theta, \ldots, b^{\theta^{k-1}}, c^{\theta^{k-1}}$ generate K'/K'' as an Abelian group. Moreover, the defining relators of K'/K'' as an Abelian group are those of K'/K'' as an $\mathscr{L}$-module together with their $\theta, \theta^2, \ldots, \theta^{k-1}$ powers expressed as words in $b, c, \ldots, b^{\theta^{k-1}}, c^{\theta^{k-1}}$. Thus K'/K'' is a finitely presented Abelian group.]

12. Show that the knot polynomial of Listing's knot whose group is $K = \langle a, b;\ b^3ab^{-1}a^{-2}b^{-1}a\rangle$ is $\theta^2 - 3\theta + 1$. Prove that K is not isomorphic to T of Problem 4 and that K'/K'', T'/T'' are both free Abelian groups on two generators.

13. Find the Betti and torsion numbers of K'/K'' where $K = \langle a, b, c; b^3aca^{-1}b^{-2}c^{-1}, c^2a^2c^2a^{-1}bc^{-3}a^{-1}b^{-1}\rangle$. [*Hint:* First present the $\mathscr{L}$-module K'/K'' and then use the method of the hint to Problem 11 to present the Abelian group K'/K''.]

14. Let $K = \langle a, b; b^2ab^{-1}a^{-1}\rangle$. Show that K has Alexander polynomial $2 - \theta$ and the group K'/K'' is infinitely generated. [*Hint:* The mapping $b^{\theta^\lambda} \to e^{(\pi i)2^{\lambda+1}}$ determines a homomorphism of the group K'/K'' onto all 2^mth roots of unity, $m = 1, 2, 3, \ldots$.]

15. Show that $K = \langle a, b; b\rangle$ and $K_1 = \langle a, b; aba^{-1}b^2ab^{-1}a^{-1}b^{-1}\rangle$ have infinite cyclic commutator quotient group and $K'/K'' = K_1'/K_1'' = 1$ but that $K \not\approx K_1$. [*Hint:* Use Problem 10 to show that $K_1'/K_1'' = 1$. To show that K_1 is not infinite cyclic, use Theorem 4.10 to show that b does not define the identity in K_1; if K_1 is Abelian, its defining relator would reduce to b.]

16. Let $G = \langle a_1, \ldots, a_n; \hat{R}_1, \ldots, \hat{R}_n/\mathscr{L}\rangle$ be an $\mathscr{L}$-module, and let α_{n-1} and α_n be the $(n-1)$st and nth Alexander ideals of G. Prove that the annihilators of G are precisely those polynomials $A(\theta)$ such that $A(\theta)\cdot\alpha_{n-1} \subset \alpha_n$. [*Hint:* If $a_i^{A(\theta)} = 1$, then $a_i^{A(\theta)} = \prod_j \hat{R}_j^{E_{ji}(\theta)}$. Hence if $\hat{R}_j = \prod_k a_k^{L_{kj}(\theta)}$, then $\sum_j L_{kj}E_{ji} = \delta_{ki}\cdot A(\theta)$. By Cramer's rule $\Delta(\theta)\cdot E_{ji}(\theta) = A(\theta)(-1)^{i+j}\Delta_{ij}(\theta)$ where $\Delta(\theta)$ is the determinant of the relation matrix for G and $\Delta_{ij}(\theta)$ is the subdeterminant obtained by deleting the ith row and ith column. Hence, $\Delta(\theta)$ divides $A(\theta)\cdot\Delta_{ij}(\theta)$.]

3.5. The Automorphism Groups Φ_n of Free Groups

Let F_n be the free group of rank n and let x_ν $(\nu = 1, \ldots, n)$ be free generators of F_n. Let N be a Nielsen transformation which carries x_ν into y_ν, where y_ν is a word in the x_ν. Then we know from Theorem 3.2 of Section 3.2 that N defines an automorphism of F_n which is given by the substitution

$$x_\nu \to y_\nu,$$

and we also know that *all* automorphisms of F_n can be obtained in this manner. We repeat here the remark that N acts on the given set of generators of F_n and that every element $W(x_\nu)$ of F_n is mapped by N on the word $W(y_\nu)$. Now let M be another Nielsen transformation which carries the generators x_ν into certain elements z_ν of F_n. M defines another automorphism of F_n. For the rest of this chapter, we shall use the following

Rule of Composition. The product MN of two Nielsen transformations acting as automorphisms of F_n on the x_ν denotes the Nielsen transformation obtained by carrying out first M and then N. In symbols: If $z_\mu = V_\mu(x_\nu)$, $(\nu, \mu = 1, \ldots, n)$ is the expression for the map of x_μ under M in terms of the x_ν, then MN is defined by

$$x_\mu \to V_\mu(y_\nu).$$

From now on, we shall call M and N automorphisms of F_n (instead of saying that they define automorphisms of F_n). Nielsen, 1924a, and

B. H. Neumann, 1932, found and studied presentations for the automorphism group Φ_n of F_n. We are now going to summarize their results. For this purpose, we shall use the following notations:

The subscripts i, k denote fixed, different integers of the sequence $1, 2, \ldots, n$. The subscript j is variable and denotes all integers $1, 2, \ldots, n$ which are different from both i and k. The symbol $\rightleftarrows$ means *commutes with*. If L, M, N are any elements of Φ_n, a relation

$$L \rightleftarrows M, N$$

means that

$$LM = ML, \qquad LN = NL.$$

We shall need the following automorphisms:

$$P_{i,k}: x_i \to x_k, x_k \to x_i, x_j \to x_j.$$

The $P_{i,k}$ generate the symmetric group Σ_n of permutations of the generators. Furthermore, we need the σ_i defined by

$$\sigma_i: \quad x_i \to x_i^{-1}, \qquad x_j \to x_j, \qquad x_k \to x_k.$$

The $P_{i,k}$ and the σ_i generate the *extended symmetric group* Ω_n of order $2^n n!$. The σ_i alone generate an Abelian group of order 2^n in which the square of every element is the identity. Any element of Ω_n will be called a *level transformation* since its application to any element $W(x_\nu)$ of F_n will not change the length of the freely reduced word W. The elements of Ω_n are obviously the only automorphisms with this property (since a free generator must be mapped into a word of length one). Next, we need the automorphisms

$$U_{i,k}: \quad x_i \to x_i x_k, \qquad x_k \to x_k, \qquad x_j \to x_j$$

$$V_{i,k}: \quad x_i \to x_k x_i, \qquad x_k \to x_k, \qquad x_j \to x_j.$$

The automorphisms $U_{i,k}$, $V_{i,k}$, together with the $P_{i,k}$ and the σ_i, are called *elementary automorphisms*. The symbol $W_{i,k}$ shall denote either $U_{i,k}$ or $V_{i,k}$; a relation in which a $W_{i,k}$ appears is therefore the equivalent of two relations. Nielsen, 1924a, proved the following result, the easy part of which (concerning the generators) is contained in Theorem 3.2.

THEOREM N1. *Φ_n is generated by the elementary automorphisms ($P_{i,k}$, σ_i, $W_{i,k}$). The following is a set of defining relations for Φ_n* (with the understanding that all subscripts run from 1 to n and that different letters

i, k, etc., denote different numbers):

$$
\begin{aligned}
&P_{i,k} = P_{k,i}, \qquad P_{i,k}^2 = 1 \\
&P_{i,k} \rightleftarrows P_{r,s} \\
&P_{i,k}P_{k,r} = P_{i,r}P_{i,k} = P_{k,r}P_{i,r} \\
&\sigma_i^{\,2} = 1, \quad \sigma_i \rightleftarrows \sigma_k, \quad \sigma_j \rightleftarrows P_{i,k} \\
&P_{i,k}\sigma_i P_{i,k} = \sigma_k \\
&W_{i,k} \rightleftarrows P_{r,s}, \sigma_r \\
&P_{i,k}W_{i,k} = W_{k,i}P_{i,k}, \qquad P_{i,j}W_{i,k} = W_{j,k}P_{i,j} \\
&P_{i,j}W_{k,i} = W_{k,j}P_{i,j} \\
&\sigma_k W_{i,k} = W_{i,k}^{-1}\sigma_k, \qquad \sigma_i U_{i,k} = V_{i,k}^{-1}\sigma_i \\
&U_{i,k}^{-1}U_{k,i}V_{i,k}^{-1} = \sigma_i P_{i,k}, \qquad U_{i,k}V_{k,i}^{-}V_{i,k} = \sigma_k P_{i,k} \\
&W_{i,k} \rightleftarrows W_{l,m}, W_{l,k} \\
&U_{i,k} \rightleftarrows V_{i,k}, V_{i,l};
\end{aligned}
$$

and, with $\epsilon = \pm 1$,

$$
\begin{aligned}
&U_{i,k}U_{k,l}^{\epsilon}U_{i,k}^{-1}U_{k,l}^{-\epsilon} = U_{i,l}^{\epsilon} = U_{k,l}^{-\epsilon}U_{i,k}U_{k,l}^{\epsilon}U_{i,k}^{-1} \\
&V_{i,k}V_{k,l}^{\epsilon}V_{i,k}^{-1}V_{k,l}^{-\epsilon} = V_{i,l}^{\epsilon} = V_{k,l}^{-\epsilon}V_{i,k}V_{k,l}^{\epsilon}V_{i,k}^{-1} \\
&U_{i,k}^{-1}V_{k,l}^{\epsilon}U_{i,k}V_{k,l}^{-\epsilon} = U_{i,l}^{-\epsilon} = V_{k,l}^{-\epsilon}U_{i,k}^{-1}V_{k,l}^{\epsilon}U_{i,k} \\
&V_{i,k}^{-1}U_{k,l}^{\epsilon}V_{i,k}U_{k,l}^{-\epsilon} = V_{i,l}^{-\epsilon} = U_{k,l}^{-\epsilon}V_{i,k}^{-1}U_{k,l}^{\epsilon}V_{i,k}.
\end{aligned}
$$

It is obvious that the number of generators used for the presentation of Φ_n in Theorem N1 is too large. Nielsen, 1924a, needed a rather large set of generators satisfying a system of rather simple relations for the purpose of proving that this particular system can be used to put every element of Φ_n into a standard form. Nielsen's proof is a very difficult one and it has not been simplified since. However, Nielsen also showed that Φ_n can be generated by four automorphisms. We state his results in the form of

Corollary N1. *Let the automorphisms P, Q, σ, U of Φ_n be given by the following table, in which the row labeled P, Q, . . . contains the image of the generator x_ν listed at the top of each column resulting from an application of P, Q, ...*

	x_1	x_2	x_3	. . .	x_{n-1}	x_n
P	x_2	x_1	x_3	. . .	x_{n-1}	x_n
Q	x_2	x_3	x_4		x_n	x_1
σ	x_1^{-1}	x_2	x_3		x_{n-1}	x_n
U	x_1x_2	x_2	x_3		x_{n-1}	x_n

Then P, Q, σ, U generate Φ_n and the following is a set of relations between them which defines Φ_n:

$$
\begin{aligned}
&P^2 = 1, \qquad (QP)^{n-1} = Q^n = 1 \\
&P \rightleftarrows Q^{-i}PQ^i, \qquad (i = 2, 3, \ldots, [n/2]) \\
&\sigma^2 = 1, \qquad \sigma \rightleftarrows Q^{-1}PQ,\ QP,\ Q^{-1}\sigma Q, \\
&(P\sigma PU)^2 = 1, \qquad U^{-1}PUP\sigma U\sigma P\sigma = 1, \\
&(PQ^{-1}UQ)^2UQ^{-1}U^{-1}QU^{-1} = 1, \\
&U \rightleftarrows Q^{-2}PQ^2,\ QPQ^{-1}PQ, \\
&\qquad Q^{-2}\sigma Q^2,\ Q^{-2}UQ^2, \\
&\qquad \sigma U\sigma,\ PQ^{-1}\sigma U\sigma QP, \\
&\qquad PQ^{-1}PQPUPQ^{-1}PQP.
\end{aligned}
$$

B. H. Neumann, 1932, introduced the automorphisms T, S, R, defined by the following table:

	x_1,	x_2,	$x_3, \ldots,$	x_{n-1},	x_n
T	x_2,	x_1^{-1},	$x_3, \ldots,$	x_{n-1},	x_n
S	x_2^{-1},	x_3^{-1},	$x_4^{-1}, \ldots,$	x_n^{-1},	x_1^{-1}
R	x_2^{-1},	x_1,	$x_3, \ldots,$	$x_n x_{n-1}^{-1}$,	x_{n-1}^{-1}

and showed that Φ_n is generated by Q and R if $n = 4, 6, 8, \ldots$, and that S and R generate Φ_n for an odd $n \geq 5$. Φ_2 is generated by σ, P, U, and Φ_3 is generated by S, T, U. The corresponding systems of defining relations and related topics are also discussed in Neumann, 1932.

Nielsen, 1918, proved the following

THEOREM 3.9. *Let F_2 be the free group on two free generators x, y. Let ξ, η be the maps of x, y under any automorphism of F_2. Then*

$$\xi\eta\xi^{-1}\eta^{-1} = T \cdot (xyx^{-1}y^{-1})^{\pm 1} \cdot T^{-1}$$

where T is a word in x, y.

PROOF. The proof follows from the remark that the theorem is true for the generating automorphisms σ, P, U of Φ_2 (see also Problem 4.4.19). ◀

We shall say that the commutator $xyx^{-1}y^{-1}$ is invariant under automorphisms of F_2. No analogue of Theorem 3.9 exists for any F_n with $n > 2$. However, there exist weaker substitutes for Theorem 3.9 which have been studied by Wever, 1949, and by Burrow, 1958. It can be shown that there always exist elements in F_n which belong to a certain member of the lower central series of F_n and are invariant under automorphisms of F_n module the next member of the lower central series. For a precise statement and for details, see Section 5.8.

A word $W(x_\nu)$ in the x_ν is called a *primitive element* of F_n if it can be mapped onto x_1 by an automorphism of F_n. Similarly, a set of words

$$W_1, W_2, \ldots, W_k, \qquad 1 < k \leq n$$

is called a set of *associated primitive elements* if there exists an automorphism of F_n which maps W_i onto x_i for $i = 1, 2, \ldots, k$. If $k = n$, the reduction process described in Section 3.1 permits us to decide in a finite number of steps whether a set of words W_ν ($\nu = 1, \ldots, n$) is a set of associated primitive elements or not. However, for $n > 2$, it is a very difficult problem to decide the same question if $k < n$. (For $n = 2$, see the remarks after Corollary N4.). This problem has been solved by J. H. C. Whitehead, 1936a, b, and by Elvira S. Rapaport, 1958. Whitehead used topological methods to establish an algorithm which makes it possible to find, in a finite number of steps, the shortest set (or sets) of words into which a given set of words in F_n can be mapped by an automorphism of F_n. His results were derived algebraically by Rapaport. Her proof has been greatly simplified by P. J. Higgins and R. C. Lyndon, 1962.

In order to describe Whitehead's algorithm, we need the following:

Definition of a T-transformation. Let F_n be a free group on $n \geq 2$ free generators x_ν. Consider any decomposition of the set of the x_ν into five disjoint subsets (three of which may be empty) called A, B, C, D, Z such that D consists of exactly one element d (which is one of the x_ν) and A, B, C, Z consist respectively of the elements a_ρ ($\rho = 1, \ldots, i$), b_σ ($\sigma = 1, \ldots, j$), c_τ ($\tau = 1, \ldots, k$), z_μ ($\mu = 1, \ldots, m$) where

$$i + j + k + m + 1 = n$$

and where i, j, k, m will be zero if the corresponding subset is empty. Then the following automorphisms and their inverses are called T-transformations:

$$a_\rho \to a_\rho d, \quad b_\sigma \to d^{-1}b_\sigma, \quad c_\tau \to d^{-1}c_\tau d, \quad z_\mu \to z_\mu, \quad d \to d.$$

It should be noted that the T-transformations form a finite collection of automorphisms of F_n. Now the basic results found by J. H. C. Whitehead can be stated as follows:

THEOREM N2. *Let $W_\rho(x_\nu)$, $\rho = 1, \ldots, r$, be a set of freely reduced words in the x_ν. If there exists an automorphism of F_n which maps the set W_ρ onto a set $W_\rho{}'$ such that the sum of the lengths of the freely reduced words $W_\rho{}'$ is less than the sum of the lengths of the W_ρ, then there exists at least one*

T-transformation which (in combination with free reductions) maps the set W_ρ onto a set of words W_ρ'' with a smaller sum of lengths.

If the two sets W_ρ^* and W_ρ^{**} can be mapped onto each other by an automorphism of F_n and, if the sum of the lengths of the freely reduced words W_ρ^* as well as the sum of the lengths of the freely reduced words W_ρ^{**} cannot be shortened by any T-transformations, then it is possible to map the set W_ρ^* onto the set W_ρ^{**} by using a sequence of T-transformations, each of which (in combination with free reductions) does not change the sum of the lengths of the W_ρ^*, and by using automorphisms P_{ik} and σ_j of Theorem N1 (none of which changes the length of any word).

THEOREM N3. *Let G be a group on generators x_ν, $\nu = 1, 2, \ldots, n$ with a single defining relator $R(x_\nu)$. Then G will be isomorphic to a free group if and only if R is a primitive element in the x_ν.*

We cannot go into the proof of Theorem N2 and N3 here. Even in the algebraic versions due to E. S. Rapaport, 1958, and Higgins and Lyndon, 1962, the proof of Theorem N2 is a very elaborate one. Theorem N3 can be proved by using Grushko's Theorem (Section 4.1) and Theorem N5. Together with Theorem N2, Theorem N3 makes it possible to decide whether a group with a single defining relator is a free group or not. For the question of possible extensions of Theorem N3, the following remarks are relevant:

J. H. C. Whitehead, 1936, has exhibited two subgroups Γ and Γ^* of F_2, such that Γ and Γ^* are, respectively, freely generated by α, β, and α^*, β^*, where α, β, α^*, β^* are freely reduced words in the generators of F_2, such that

(i) the sum of lengths of α and β is less than the sum of lengths of α^* and β^*,

(ii) it is not possible to map the set α^*, β^* onto the set α, β by a sequence of T-transformations none of which increases the sum of lengths of α^* and β^*;

(iii) Γ can be mapped onto Γ^* by an automorphism of F_2.

This example shows that Theorem N3 cannot be generalized much further. An example given by E. S. Rapaport, 1958, shows that the set of T-transformations cannot be replaced by the set of generators of Φ_n as listed in Theorem N1.

It should be noted that Theorem N2 does not exclude the possibility that a word W in the x_ν is mapped onto itself by a T-transformation. An example of such a word can be found from Theorem 3.9 and others will be considered in Section 3.6.

Apart from its presentation in terms of generators and defining relations, not much is known about the group Φ_n except for its connection with the group Λ_n of n-dimensional *lattice transformations*. This is the group of affine transformations of Euclidean n-space which keep the origin fixed and map the lattice of points with integral coordinates onto itself. Λ_n can also be described as the multiplicative group of n-by-n matrices with determinant ± 1 and integers as entries, or as the group of automorphisms of the direct product A_n of n infinite cyclic groups (i.e., the free Abelian group on n free generators). It is obvious that there exists a homomorphic mapping of Φ_n into Λ_n. To show this, let us write A_n additively, its general element being

$$j_1a_1 + j_2a_2 + \cdots + j_na_n,$$

where the j_ν are integers and the a_ν are basis elements of A_n. Then the mapping $x_\nu \rightarrow a_\nu$ (with the understanding that a product in F_n is mapped onto the corresponding sum in A_n) defines a homomorphic mapping of F_n onto A_n under which a fully invariant subgroup of F_n (namely, the commutator subgroup of F_n) is mapped onto the identity of A_n. Therefore Φ_n is mapped into Λ_n; if the element Φ of Φ_n is given by

$$x_\mu \rightarrow W_\mu(x_\nu) \tag{1}$$

then the corresponding automorphism λ of A_n is given by the linear transformation

$$a_\mu \rightarrow \sum_{\nu=1}^{n} \sigma_\nu(W_\mu)a_\nu \tag{2}$$

where $\sigma_\nu(W_\mu)$ is the sum of exponents on x_ν in W_μ. It is not difficult to show that (2) maps the generators of Φ_n, as defined in Theorem N1 onto a set of generators of Λ_n (see Section 3.2). Therefore, (2) defines a homomorphic mapping of Φ_n onto Λ_n and we may consider every generator of Φ_n as a generator of Λ_n, knowing that any relation between these generators which is true in Φ_n is also true in Λ_n. Let

$$\Lambda_n \simeq \Phi_n/K$$

where the kernel K of the homomorphism of Φ_n onto Λ_n consists of those elements of Φ_n which leave the cosets of the commutator subgroup F_n' of F_n fixed. Then it has been shown for $n \leq 3$ by Nielsen, 1924b, and, for all n, by Magnus, 1934b, that the following result holds:

THEOREM N4. *A presentation of the group Λ_n of n-by-n matrices with integral entries and determinant ± 1 arises from the presentation for Φ_n as given in Corollary N1 by adding the relation*

$$(U\sigma)^2 = 1$$

to the set of defining relations for Φ_n. *The kernel* K *of the homomorphism* $\Phi_n \to \Lambda_n$ *is generated by the automorphisms*

$$K_{i,l}\colon\ a_i \to a_l a_i a_l^{-1},\quad a_j \to a_j \qquad (j \neq i)$$

$$K_{i,l,s}\colon\ a_i \to a_i a_l a_s a_l^{-1} a_s^{-1},\quad a_j \to a_j \qquad (j \neq i),$$

all of which are products of transforms of

$$K_{1,2} = (\sigma U)^2.$$

The main difficulty in proving Theorem N4 is deriving a set of relations for Λ_3 which were obtained by Nielsen, 1924b. For $n = 2$, the kernel K can be described by

COROLLARY N4. *The kernel of the homomorphic mapping* $\Phi_2 \to \Lambda_2$ *consists of the inner automorphisms of* F_2.

This result was proved by Nielsen, 1918. It shows that a primitive element p of F_2 is uniquely determined by the sum of exponents (in p) of the two generators, apart from an inner automorphism. This fact makes it possible to decide whether a given element of F_2 is a primitive element and to find all primitive elements associated with it.

The automorphisms of Λ_n have been investigated by Hua and Reiner, 1951, 1952. For other problems concerning Λ_n, see Reiner, 1955a.

Problems for Section 3.5

1. Show that a group on generators $P_{i,k}$ and σ_i $(i, k = 1, \ldots, n; i \neq k)$ is of order $2^n n!$ if the generators satisfy those relations for the P_{ik} and the σ_i stated in Theorem N1 which do not involve the $W_{i,k}$, $U_{i,k}$, $V_{i,k}$. (See Nielsen, 1924a.)

2. Show that Φ_2 can be presented by

$$\langle P, \sigma, U;\ P^2, \sigma^2, (\sigma P)^4, (P\sigma PU)^2, (UP\sigma)^3, [U, \sigma U\sigma]\rangle$$

where $[U, \sigma U \sigma]$ denotes the commutator $U^{-1}(\sigma U\sigma)^{-1}U\sigma U\sigma$. (See B. H. Neumann, 1932.)

3. Let p, q be two associated primitive elements of F_2. Show that the elements

$$p^\alpha q^\epsilon p^\beta \qquad (\alpha, \beta = 0, \pm 1, \pm 2, \ldots, \epsilon = \pm 1)$$

are the only primitive elements associated with p.

3.6. Free Automorphisms and Free Isomorphisms

Applying a Nielsen transformation to a set S of generators of any group G will always result in a new set S' of generators for G. In general, this mapping of one set of generators of G onto another one will not define an automorphism of G. The necessary and sufficient conditions for this to happen have been stated in Lemma 3.3.

If α is an automorphism of a group G on generators x_ν, $\nu = 1, \ldots, n$, such that there exists a Nielsen transformation N acting on the x_ν which maps x_ν (for all ν) onto the same element of G as α, then we shall say that α can be *presented as a free automorphism* on the x_ν. It should be noted that the term "free automorphism" makes sense only in connection with a preassigned set of generators x_ν of G and that, even if the x_ν are given, a free automorphism need not be written in the form of a Nielsen transformation acting on the x_ν since the words onto which the x_ν are mapped by the automorphism can be changed by using the defining relators of G. For example, every automorphism of the free Abelian group

$$\langle x_1, x_2;\, x_1x_2x_1^{-1}x_2^{-1}\rangle$$

can be presented as a free automorphism, but the automorphism

$$x_1 \rightarrow x_1^2x_2^3, \qquad x_2 \rightarrow x_1x_2^2$$

is not presented in the form of a Nielsen transformation since $x_1^2x_2^3$ is not a primitive element in x_1, x_2. For if $x_1^2x_2^3$ is a primitive element of the free group F on x_1, x_2 then

$$\langle x_1, x_2;\; x_1^2x_2^3\rangle$$

would be an infinite cyclic group, whereas it has the symmetric group of degree three as a homomorphic image. However, since $x_1x_2 = x_2x_1$,

$$x_1^2x_2^3 = (x_1x_2^2)^2x_2^{-1}$$

and the right-hand side of this equation is a primitive element associated with $x_1x_2^2$.

Let G and G' be two isomorphic groups and let x_ν and x_ν' ($\nu = 1, \ldots, n$), respectively, be sets of generators of G and G'. We shall say that the x_ν and the x_ν' are related by a *free isomorphism* of G onto G' if there exists a Nielsen transformation N of rank n such that

$$N(x_\nu) \rightarrow x_\nu'$$

defines an isomorphic mapping of G onto G'. Here $N(x_\nu)$ denotes the map of x_ν under an application of N. It should be noted that we do not exclude the case where one or several of the x_ν (or the x_ν') are equal to the identity of G (or G').

The concept of a free isomorphism can be used to state Grushko's Theorem (see Section 4.1), according to which a finitely generated group

G which is isomorphic to a free product can be mapped onto it by a free isomorphism. In this section, we shall consider mostly free automorphisms and we shall start by proving the following result due to Rapaport, 1959:

THEOREM 3.10. *Let G be a group on a finite set of n generators x_ν ($\nu = 1, \ldots, n$) and with defining relators R_i ($i = 1, 2, \ldots$) and let α be an automorphism of G. Then there exists a set y_μ ($\mu = 1, \ldots, 2n$) of $2n$ generators of G and a set S_j ($j = 1, 2, \ldots$) of defining relators such that α can be presented as a free automorphism on the y_μ.*

Before proving Theorem 3.10 it should be mentioned that not every automorphism of a given finitely generated group can be presented as a free automorphism on the given set of generators. Apart from finite cyclic groups, an example is provided by the group L (the fundamental group of Listing's knot; see Dehn, 1914, and Magnus, 1931) which has a presentation on two generators and a single relator:

$$L = \langle u, v; u^3vu^{-1}v^{-2}u^{-1}v\rangle.$$

It has been shown by Rapaport, 1959, that the automorphism

$$(1) \qquad u \to v^{-1}u^{-1}vu, \qquad v \to u^{-1}v^{-1}u^{-2}v^{-1}uv$$

cannot be effected by a Nielsen transformation acting on u and v. This example is interesting for two reasons: L has only one defining relator and therefore this example is of the simplest type conceivable since for free generators of a free group every automorphism is necessarily presented as a free automorphism. Also, it has been shown by Rapaport, 1959, that every isomorphism of L with a group having a single defining relator can be presented as a free isomorphism. It seems reasonable to conjecture that two groups G and G', both of which are presented in terms of a single defining relator, will be isomorphic only if there exists a free isomorphism mapping G onto G'.

Now we shall prove Theorem 3.10. Suppose that α maps x_ν onto $W_\nu(x_\mu)$, where W_ν is one of the words in the x_μ which represents the map of x_ν under α. Vice versa, let $V_\nu(x_\mu)$ be a word in the x_μ ($\mu = 1, \ldots, n$) representing the element of G which is mapped on x_ν by α. Now we see from Theorem 1.5 that by using a sequence of elementary Tietze transformations G can also be presented in terms of $2n$ generators x_ν, z_ν as

$$G = \langle x_\nu, z_\nu; R_i(x_\nu), z_\nu^{-1}W_\nu(x_\mu)\rangle.$$

For this presentation of G, α can now be defined by the mapping

$$x_\nu \to z_\nu, z_\nu \to x_\nu V_\nu^{-1}(z_\mu)W_\nu(z_\mu), \qquad (\nu = 1, 2, \ldots, n)$$

which is obviously a Nielsen transformation. This proves Theorem 3.10, if we identify the x_ν, z_ν (in some order) with the y_μ ($\mu = 1, \ldots, 2n$) of the theorem. We also see that the number of defining relators for the new presentation of G has been increased by n, no matter how many defining relators were used in the original presentation. ◀

For free automorphisms acting on the generators of a group with a single defining relator, a generalization of Theorem 3.9 holds. We have (see Magnus, 1930):

THEOREM N5. *Let G be a group on generators x_ν $(\nu = 1, 2, \ldots, n)$ with a single defining relator $R(x_\nu)$. If there exists a set of words $W_\nu(x_\mu)$ such that the mapping*

$$x_\nu \to W_\nu(x_\mu) \qquad (\nu = 1, 2, \ldots, n)$$

is a Nielsen transformation acting on the x_ν which defines an automorphism of G, then $R(W_\nu)$ is freely equal (as a word in the x_ν) to a transform

$$T(x_\nu) \cdot R(x_\nu)^{\pm 1} \cdot T^{-1}(x_\nu)$$

of $R^{\pm 1}$. (See Theorem 4.11.)

In Section 3.7 we shall discuss some important groups with a single defining relator for which all automorphisms are free automorphisms. Theorem N5 shows that for these groups the relator is an invariant in the sense defined after Theorem 3.9 for a free group F_2.

Problems for Section 3.6

1. Show that Theorem 3.9 can be derived from Theorem N5 without using any explicit formulas for the automorphisms of a free group on two free generators.

2. Find a presentation of the group L of Listing's knot (as defined after Theorem 3.10) for which the automorphism (1) becomes a free automorphism. Show that such a presentation of L exists which involves only three generators. [*Hint:* Start with the presentation of L as given in Dehn, 1914, or Magnus, 1931, and use the defining relations to eliminate some of the generators.]

3.7. Braid Groups and Mapping Class Groups

The groups Φ_n of automorphisms of a free group F_n on n free generators contain subgroups which arise from topological problems. We shall discuss first the braid groups B_n $(n = 2, 3, 4, \ldots)$ which were introduced by Artin, 1925. The definition of B_n, as given by Artin, arises from a topological construction. Artin then showed that B_n is isomorphic to a subgroup of Φ_n, and he derived a set of generators and defining relators for B_n from his topological construction. Later, it was shown by Magnus, 1934a, Bohnenblust, 1947, and Markoff, 1945, how generators and defining relators for B_n can be derived algebraically from the group-theoretical definition of B_n as a certain subgroup of Φ_n. We shall summarize this result by stating:

THEOREM N6. *Let F_n be the free group on $n > 2$ free generators x_ν, $\nu = 1, 2, \ldots, n$. Let B_n be the group of automorphisms of F_n which map x_ν into a transform*

$$T_\nu x_\mu T_\nu^{-1}$$

of one of the x_μ, $\mu = 1, 2, \ldots, n$ and which map

$$x_1 x_2 \ldots x_n$$

onto itself. Then B_n is generated by the automorphisms σ_ν, $\nu = 1, 2, \ldots, n - 1$ defined by

$$x_\nu \to x_{\nu+1}, \qquad x_{\nu+1} \to x_{\nu+1}^{-1} x_\nu x_{\nu+1}$$

$$x_\mu \to x_\mu, \qquad \mu \neq \nu, \nu + 1.$$

A set of defining relations for B_n is given by

$$\sigma_\nu \sigma_{\nu+1} \sigma_\nu = \sigma_{\nu+1} \sigma_\nu \sigma_{\nu+1}, \qquad \sigma_\nu \sigma_\mu = \sigma_\mu \sigma_\nu, \qquad |\nu - \mu| > 1.$$

The proof of Theorem N6 and of related results requires a certain knowledge of subgroups and quotient groups of B_n. The following result has been proved by Artin as an obvious corollary of his topological construction of B_n. It can be verified algebraically without much trouble. We have:

THEOREM N7. *Let Σ_n be the symmetric group of permutations of n symbols $1, 2, \ldots, n$. The mapping of σ_ν onto the transposition $(\nu, \nu + 1)$ of the two symbols ν and $\nu + 1$ defines a homomorphic mapping of B_n onto Σ_n such that the kernel K_n of the mapping is the smallest normal divisor of B_n which contains σ_1^2.*

The structure of K_n and of certain subgroups of B_n containing K_n has been investigated by Burau, 1936, Magnus, 1934, Markoff, 1945, Bohnenblust, 1947, Artin, 1947a, and Chou, 1948. We shall state the most important result in the formulation of Artin, 1947a, since his paper also contains a topological interpretation of the generators of K_n. For equivalent formulations, see Markoff, 1945, and Chou, 1948. We have:

THEOREM N8. *Let $A_{i,k}$ $(i < k, i, k = 1, \ldots, n)$ denote the elements of B_n which are given by the automorphisms of F_n:*

$$x_r \to x_r, \qquad \text{if } r < i, \qquad \text{or} \qquad r > k$$

$$x_i \to x_i C_{i,k}, \qquad x_k \to C_{i,k}^{-1} x_k$$

$$x_r \to C_{i,k}^{-1} x_r C_{i,k}, \qquad \text{if } i < r < k$$

where

$$C_{i,k} = x_i^{-1}x_k^{-1}x_ix_k.$$

Then the kernel K_n *of the homomorphic mapping*

$$B_n \to \Sigma_n, \qquad K_n \to 1, \qquad (\sigma_1^2 \to 1)$$

is generated by the $A_{i,k}$.

Let U_j, $j = 1, \ldots, n-1$, *be the subgroup of* K_n *generated by the* $A_{j,k}$, *for which* $j < k$. *Then* U_j *is a free group, the* $A_{j,k}$ *are free generators of* U_j, *and every element of* K_n *has a unique expression of the form*

$$u_1u_2 \ldots u_{n-1}, \qquad u_j \in U_j, \qquad j = 1, \ldots, n-1.$$

A set of defining relations for K_n *is given by the following formulas where different subscripts denote different integers of the set* $1, 2, \ldots, n$ *and where the first subscript of an* A *is always less than the second one:*

$$A_{r,s}A_{i,k}A_{r,s}^{-1} = A_{i,k} \qquad \text{if } s < i \text{ or } k < r$$

$$A_{k,s}A_{i,k}A_{k,s}^{-1} = A_{i,s}^{-1}A_{i,k}A_{i,s}, \qquad i < k < s$$

$$A_{r,k}A_{i,k}A_{r,k}^{-1} = A_{i,k}^{-1}A_{i,r}^{-1}A_{i,k}A_{i,r}A_{i,k}, \qquad i < r < k$$

$$A_{r,s}A_{i,k}A_{r,s}^{-1} = A_{i,s}^{-1}A_{i,r}^{-1}A_{i,s}A_{i,r}A_{i,k}A_{i,r}^{-1}A_{i,s}^{-1}A_{i,r}A_{i,s} \qquad (i < r < k < s)$$

The definition of B_n used in Theorem N6 makes the word problem for B_n trivial since we can decide at once whether a given automorphism of a free group is the identical mapping or not. But the transformation problem for B_n is still unsolved for $n > 4$, in spite of the fact that it is of interest in topology because of the following result due to Artin, 1925. The groups of knots and linkages (see Reidemeister, 1932b, Crowell and Fox, 1963) can be obtained as follows: Take any element β of B_n and let $\beta(x_\nu)$ be the map of $x_\nu \in F_n$ under the automorphism of F_n defined by β. Now consider all groups $L_n(\beta)$ on generators x_ν, $\nu = 1, \ldots, n$, with the defining relations

$$\beta(x_\nu) = x_\nu, \qquad (\nu = 1, 2, \ldots, n).$$

If we let n and β vary, the groups $L_n(\beta)$ give us all possible groups of knots and linkages. A solution of the transformation problem for B_n would be relevant for the classification of the groups $L_n(\beta)$ since $L_n(\beta) \simeq L_n(\beta')$ if β and β' are conjugate elements of B_n. In this connection, the investigation of the automorphisms of B_n carried out by Artin, 1947c, is also relevant. For $n = 3$, the transformation problem can be solved since B_3 can be presented as $\langle a, b; a^2 = b^3 \rangle$. For $n = 4$, the transformation problem for elements of K_4 has been solved by Froehlich, 1936.

For the problem of representing B_n in terms of finite matrices, see Burau, 1933, B. J. Gassner, 1956, and Lipschutz, 1961.

The braid group B_n is closely related to the mapping class group of a two-dimensional sphere from which n boundary points have been removed. For the concept of a mapping class group, see, for example, Nielsen, 1924c, and 1927 ("Gruppe der Abbildungsklassen"). We shall be concerned here with the mapping class group $M(n, g)$, $n, g = 0, 1, 2, \ldots$ of a two-dimensional surface which arises from a closed, two-dimensional orientable manifold of genus g by removing n boundary points. It has been proved by Fricke, 1897, and Magnus, 1934a, in the cases where $g = 0$ or $g = 1$, and by Nielsen, 1927, in the cases where $n = 0, 1$ and $g \geq 2$, that $M(n, g)$ can be described algebraically as follows:

Let $F(n, g)$ be the group on $n + 2g$ generators $u_1, \ldots, u_n, a_i, b_i$ $(i = 1, \ldots, g)$ with the single defining relator

$$u_1u_2 \ldots u_na_1b_1a_1^{-1}b_1^{-1} \ldots a_gb_ga_g^{-1}b_g^{-1}.$$

Let $A(n, g)$ be the group of all those automorphisms of $F(n, g)$ which map each u_ν, $(\nu = 1, \ldots, n)$, into a conjugate of the same or of another u_ν and let $I(n, g)$ be the group of inner automorphisms of $F(n, g)$. Obviously, $I(n, g)$ is contained in $A(n, g)$ as a normal divisor. Then

$$M(n, g) \simeq A(n, g)/I(n, g).$$

This we shall take as the algebraic definition of $M(n, g)$ even in those cases where the equivalence between algebraic and topological definition has not yet been proven. Incidentally, the proof of this equivalence is an algebraic problem. We have to find generators for $A(n, g)$. Once this has been done, it is, in all known cases, easy to see that these generators are automorphisms of the fundamental group $F(n, g)$ of the surface which actually can be induced by a suitably chosen topological mapping of the surface onto itself. For $g = 0$, we have

THEOREM N9. *The mapping class group $M(n, 0)$ is isomorphic to a quotient group B_n/T_n of the braid group B_n. The normal divisor T_n is generated by the n elements τ_ν, $\nu = 1, \ldots, n - 1$, and τ defined by*

$$\tau_1 = (\sigma_2\sigma_3 \ldots \sigma_{n-1})^{1-n}$$

$$\tau_2 = \sigma_1^2(\sigma_3 \ldots \sigma_{n-1})^{2-n},$$

$$\ldots\ldots\ldots\ldots\ldots\ldots$$

$$\tau_{l+1} = (\sigma_1 \ldots \sigma_l)^{l+1}(\sigma_{l+2} \ldots \sigma_{n-1})^{l+1-n},$$

$$\ldots\ldots\ldots\ldots\ldots\ldots\ldots\ldots\ldots\ldots$$

$$\tau_{n-1} = (\sigma_1\sigma_2 \ldots \sigma_{n-2})^{n-1}$$

$$\tau = (\sigma_1\sigma_2 \ldots \sigma_{n-1})^n.$$

T_n is the direct product of the free group freely generated by $\tau_1, \ldots, \tau_{n-1}$ and the infinite cyclic group generated by τ.

For a proof of Theorem N9 and a geometrical interpretation of the relationship between B_n and $M(n, 0)$, see Magnus, 1934a.

The groups $M(1, 1)$ and $M(2, 1)$ have been investigated by Fricke, 1897. For generators and a method to determine defining relators for $M(n, 1)$, see Magnus, 1934a.

The real difficulties of the investigation of the groups $M(n, g)$ start when $g \geq 2$. Nielsen, 1927, using geometrical methods, proved

THEOREM N10. *All automorphisms α of the group*

$$F(0, g) = \langle a_i, b_i; \prod_i a_i b_i a_i^{-1} b_i^{-1} \rangle, \qquad i = 1, \ldots, g$$

can be represented as free isomorphisms acting on the generators a_i, b_i. If α maps a_i onto a_i' and b_i onto b_i', then

$$\prod_i a_i' b_i' a_i'^{-1} b_i'^{-1}$$

is freely equal to a word

$$T(\prod_i a_i b_i a_i^{-1} b_i^{-1})^{\epsilon} T^{-1}$$

where T is a word in the a_i, b_i and $\epsilon = \pm 1$.

The second statement of Theorem N10 could be derived from the first by using Theorem N5. The automorphisms α for which $\epsilon = +1$ are called proper automorphisms of $F(0, g)$. They correspond to topological mappings of the closed, orientable surface of genus g onto itself which preserve the orientation of the surface. Dehn, 1939, proved

THEOREM N11. *The group of proper automorphisms of the group $F(0, g)$ defined in Theorem N10 can be generated by at most $2g^2$ generators if $g \geq 2$. Of these, $2g$ generators may be chosen as inner automorphisms.*

Explicit formulas for these generators can be derived from Dehn's topological description of the corresponding mappings of an orientable closed surface of genus g onto itself.

Dehn, 1938, gives five homeomorphisms which generate the mapping class group $M(0, 2)$ in the case of a closed orientable two-dimensional surface of genus 2. According to P. Gold, 1960, these five homeomorphisms induce automorphisms $\Delta_1, \Delta_2, \ldots, \Delta_5$ of $F(0, 2)$. These are listed in Table 3.1 which should be read as follows: The first row contains the generators of $F(0, 2)$. The following rows contain the maps of these generators under the automorphisms listed in the first column.

If we use the automorphisms α_1 and α_2 of $F(0, 2)$ which are defined in Theorem N12 below, we find that

$$\Delta_1 = \alpha_1^{-1}\alpha_2^{-1}\alpha_1, \qquad \Delta_2 = \alpha_2,$$
$$\Delta_3 = (\alpha_1^2\alpha_2\alpha_1\alpha_2\alpha_1^{-1}\alpha_2\alpha_1)\alpha_2^{-1}(\alpha_1^2\alpha_2\alpha_1\alpha_2\alpha_1^{-1}\alpha_2\alpha_1)^{-1}, \qquad \Delta_4 = \alpha_1^3\alpha_2\alpha_1^{-3},$$
$$\Delta_4 = (\alpha_1^4\alpha_2^{-1}\alpha_1^4\alpha_2\alpha_1^4\alpha_2)\Delta_1(\alpha_1^4\alpha_2^{-1}\alpha_1^4\alpha_2\alpha_1^4\alpha_2)^{-1}.$$

Table 3.1

	a_1	a_2	b_1	b_2
Δ_1	a_1	a_2	b_1a_1	b_2
Δ_2	a_1b_1	a_2	b_1	b_2
Δ_3	a_1	a_2	b_1	b_2a_2
Δ_4	a_1	a_2b_2	b_1	b_2
Δ_5	$x^{-1}a_1x$	a_2	b_1x	$x^{-1}b_2$

$(x = a_2^{-1}b_1a_1b_1^{-1})$

Therefore, a proof of Theorem N12 can be derived from the results of Dehn, 1938. Other investigations of $M(0, 2)$ may be found in Baer, 1927, 1929, and in Nielsen, 1924c, who states without proof:

THEOREM N12. *The group of proper automorphisms of $F(0, 2)$, as defined in Theorem N10, is generated by the inner automorphisms and by the automorphisms*

$$\alpha_1: \quad a_1 \to a_2^{-1}b_1, \quad b_1 \to a_1^{-1}$$
$$\qquad a_2 \to a_1^{-1}b_2, \quad b_2 \to a_2^{-1}$$

and

$$\alpha_2: \quad a_1 \to a_1b_1, \quad b_1 \to b_1, \quad a_2 \to a_2, \quad b_2 \to b_2$$

Every automorphism α of $F(0, g) = F$ induces a unimodular linear substitution σ of $2g$ variables with integral coefficients which describes the automorphism induced by α in the quotient group F/F' of F with respect to its commutator subgroup. (Obviously, F/F' is the direct product of $2g$ infinite cyclic groups.) However, these linear substitutions cannot be arbitrary; it can be shown (see Section 5.8, or Magnus, 1935) that the defining relator of $F(0, g)$ enforces a restriction on σ. Let the generators of $F(0, g)$ and, correspondingly, the variables on which σ acts, be ordered as follows:

$$a_1, a_2, \ldots, a_g, b_1, b_2, \ldots, b_g.$$

Then the linear substitution σ will be described by a $2g$-by-$2g$ matrix S which has the property

$$SJS' = \pm J$$

where S' is the transpose of S and where J is the matrix described by the scheme

$$\begin{pmatrix} 0 & I \\ -I & 0 \end{pmatrix}$$

in which I denotes the g-by-g unit matrix and 0 denotes the g-by-g matrix, all the entries of which are zero. S is called a *symplectic* matrix. Now we have:

THEOREM N13. *The automorphisms of $F(0, g) = F$ (as defined in Theorem N10) induce in the quotient group F/F', of F with respect to its commutator subgroup, a group of automorphisms which is isomorphic to the multiplicative group Γ of all $2g$-by-$2g$ symplectic matrices with integral elements.*

A purely algebraic proof of Theorem N13 can be obtained as follows. Hua and Reiner, 1949, have found a set of four generators for the *proper symplectic group*, i.e., for the group of integral matrices S_+ for which

$$S_+JS_+' = +J.$$

These four generators correspond to the four matrices arising from the automorphisms $\eta_1, \ldots, \eta_4$ of $F(0, g)$, as listed in Table 3.2 below. The fifth automorphism η_5 corresponds to a symplectic matrix S_- for which

$$S_-JS_-' = -J.$$

Therefore, $\eta_1, \ldots, \eta_5$ induce the full symplectic group as automorphisms of F/F'.

Table 3.2

	a_1	a_2	$\ldots$	a_g	b_1	b_2	$\ldots$	b_g
η_1	$a_1b_1a_1^{-1}$	a_2	$\ldots$	a_g	a_1^{-1}	b_2	$\ldots$	b_g
η_2	a_1b_1	a_2	$\ldots$	a_g	b_1	b_2	$\ldots$	b_g
η_3	$T_ga_gT_g^{-1}$	$T_{g-1}a_{g-1}T_{g-1}^{-1}$	$\ldots$	$T_1a_1T_1^{-1}$	$T_gb_gT_g^{-1}$	$T_{g-1}b_{g-1}T_{g-1}$	$\ldots$	$T_1b_1T_1^{-1}$
η_4	a_2a_1	$b_1a_2b_1^{-1}$	$\ldots$	a_g	b_1	$a_2b_2a_2^{-1}b_1^{-1}$	$\ldots$	b_g
η_5	b_g	b_{g-1}	$\ldots$	b_1	a_g	a_{g-1}	$\ldots$	a_1

$$T_g = C_1C_2 \ldots C_{g-1},\ T_{g-1} = C_1 \ldots C_{g-2}, \quad T_2 = C_1, \quad T_1 = 1$$
$$C_i = a_ib_ia_i^{-1}b_i^{-1}, \qquad (i = 1, 2, \ldots, g).$$

Generators for the group of matrices S_+ were found earlier by Clebsch and Gordan, 1866, and it has been shown by H. Burkhardt, 1890, that these can be induced by certain topological mappings of the underlying surface [the fundamental group of which is $F(0, g)$] onto itself.

It should be noted that the determinant of a symplectic matrix is always $+1$ if $g \geq 2$. For a simple algebraic proof of this fact, see Siegel, 1943.

The symplectic matrices with integral entries form an important group. For its connection with algebraic functions of a complex variable see, for example, Burkhardt, 1890. For the full theory of the symplectic group and its relation to number theory and the theory of functions of several complex variables, see Siegel, 1943.

The automorphisms of the group of integral symplectic matrices have been studied by Reiner, 1955b. See also Reiner, 1954.

The significance of the mapping class groups for Riemann's problem of the moduli of Riemann surfaces was exhibited in full first by R. Fricke in Fricke and Klein, 1897. Fricke calls these groups "*automorphic modular groups*."

Chapter 4

Free Products and Free Products with Amalgamations

4.1. Free Products

In this chapter, we shall study the concepts of the free product of groups and of the free product with amalgamated subgroups. We shall begin with the free product of groups, which we may introduce by comparing it to the familiar concept of the direct product $A \times B$ of two groups A and B. The free product defines a composition of groups which has many properties in common with the composition defined by the direct product. For details see the end of this section and the beginning of Section 6.4. The free product may even be considered as the source of all other compositions (with certain natural restrictions) of groups; for a report on this fact, see Section 6.4. In some respects, the properties of the free product are simpler than those of the direct product; this is true especially for the structure of the subgroups. (See Section 4.3.) Finally, we may mention that both the direct and the free product occur naturally in topology. The fundamental group F of the Cartesian product of two spaces with fundamental groups A and B is the direct product $A \times B$. If the two spaces are joined merely in the reference points of their fundamental groups, the fundamental group of the resulting space is the free product of the groups A and B. We define:

The *free product* $A * B$ of the groups

(1) $$A = \langle a_1, \ldots, a_n; R_1(a_\nu), \ldots, R_p(a_\nu)\rangle$$

and

(2) $$B = \langle b_1, \ldots, b_m; S_1(b_\mu), \ldots, S_q(b_\mu)\rangle$$

is the group

(3) $$A * B = \langle a_1, \ldots, a_n, b_1, \ldots, b_m; R_1(a_\nu), \ldots, R_p(a_\nu), S_1(b_\mu), \ldots, S_q(b_\mu)\rangle.$$

A and B are called the *free factors* of $A * B$.

Although $A * B$ has been defined by using particular presentations for A and B, it is nevertheless independent of these presentations.

LEMMA 4.1. *The free product $A * B$ is uniquely determined by the groups A and B. Moreover, $A * B$ is generated by two subgroups $\bar{A}$, $\bar{B}$* which are isomorphic to A, B respectively, and such that $\bar{A} \cap \bar{B} = 1$.

PROOF. For suppose A and B are given as in (1) and (2) and also

$$A = \langle a_1', \ldots, a_s'; R_1'(a_\sigma'), \ldots \rangle \tag{4}$$

$$B = \langle b_1', \ldots, b_t'; S_1'(b_\tau'), \ldots \rangle. \tag{5}$$

Then $A * B$ is given by (3) and also by

$$A * B = \langle a_1', \ldots, a_s', b_1', \ldots, b_t'; R_1'(a_\sigma'), \ldots, S_1'(b_\tau'), \ldots \rangle. \tag{6}$$

The isomorphism of (1) onto (4) and of (2) onto (5) naturally defines a homomorphism of (3) onto (6) since defining relators in (3) are mapped into relators in (6). This homomorphism can be inverted by using the inverses of the isomorphism of (1) onto (4) and of (2) onto (5). Thus $A * B$ is uniquely determined (up to isomorphism) by A and B.

Let $\bar{A}$ be the subgroup of (3) generated by $a_1, \ldots, a_n$ and let $\bar{B}$ be the subgroup of (3) generated by $b_1, \ldots, b_m$. Clearly, $A * B$ is generated by $\bar{A}$ and $\bar{B}$. To show that $A \simeq \bar{A}$, map (1) into (3) by sending a_ν into a_ν. This is a homomorphism of A onto $\bar{A}$. Moreover, the homomorphism of $A * B$ into A given by sending a_ν into a_ν and b_μ into 1, maps $\bar{A}$ onto A and is the inverse of the homomorphism of A onto $\bar{A}$. Thus $A \simeq \bar{A}$ and similarly $B \simeq \bar{B}$.

Moreover, if $g \in \bar{A} \cap \bar{B}$, then the homomorphism of $A * B$ into A given by sending a_ν into a_ν and b_μ into 1, maps g into 1 since $g \in \bar{B}$. But this homomorphism is one-one on $\bar{A}$ and so $g = 1$. Thus $\bar{A} \cap \bar{B} = 1$. ◀

Since $\bar{A}$, $\bar{B}$ are isomorphic to A, B respectively, it is customary to identify $\bar{A}$ with A and $\bar{B}$ with B, and so to consider A and B subgroups of $A * B$.

Definition. If $g \neq 1$ is in $A * B$ and g is in A or B then *the free factor of $A * B$ containing g is denoted by $F(g)$.* Thus $F(g)$ is A or B. Moreover, if W is a word in the a_ν alone then $F(W) = A$. Similarly, if W is a word in the b_μ alone then $F(W) = B$. This notation is valid even if W is a relator, provided only that W is not the empty word.

In Theorem 1.2, we established a canonical form for elements of a free group. We wish to do the same for the free product of two groups.

As an analogue of a freely reduced word in a free group we use a "reduced sequence" of elements of $A * B$. *A sequence of elements $g_1, g_2, \ldots, g_n$ from $A * B$ is called reduced if $g_i \neq 1$, g_i is in A or B, and g_i, g_{i+1} are not in the same free factor.*

For example, if $A * B = \langle a, b; a^5, b^3 \rangle$, then 1, a, a^2, a^3, a^4 are the elements of A and 1, b, b^2 are the elements of B; the sequence a^4, b, a, b^2, a^3 is reduced; the sequence a^2, 1, a is not reduced since the second term is 1; the sequence a^2, b, b^2 is not reduced since the consecutive terms b and b^2 are in the same free factor.

(The condition that consecutive terms of a reduced sequence do not come from the same free factor implies that the terms must be $\neq 1$, unless the sequence has only one term.)

Moreover, in addition to the "letter" length $L(W)$ of a word W which counts the number of generators and their inverses in the word, it is useful to introduce the "syllable" length of a word.

Definition. Let $W(a_\nu, b_\mu)$ be a word in $a_1, a_2, \ldots, a_n, b_1, b_2, \ldots, b_m$. If

$$W(a_\nu, b_\mu) = W_1 W_2 \ldots W_r,$$

where each W_i is a word in the a_ν alone or the b_μ alone, no W_i is the empty word (though W_i may define 1 in $A * B$), W_i and W_{i+1} are not both in the same free factor of $A * B$, then the *syllable length* $\lambda(W)$ *of* W *in* A *and* B *is* r, $W_1, W_2, \ldots, W_r$ *are called the syllables of* W.

For example, if $A * B = \langle a_1, a_2, b_1, b_2, b_3;\ a_1{}^2 a_2{}^2,\ b_1{}^3 b_2{}^3 b_3{}^3 \rangle$ and $W(a_\nu, b_\mu) = a_1 a_2{}^{-2} a_1{}^{-1} b_1 b_2{}^{-1} a_1{}^2 a_2{}^2$, then $L(W) = 10$, but $\lambda(W) = 3$.

The syllable length of an element g *in* $A * B$ *is the minimum syllable length of a word defining* g. Thus the element 1 of $A * B$ has syllable length zero; an element g in A or B, $g \neq 1$ has syllable length one; an element g not in A or B has syllable length ≥ 2.

THEOREM 4.1. *Each element* g *of* $A * B$ *can be uniquely expressed as a product*

$$g = g_1 g_2 \cdots g_r \tag{7}$$

where $g_1, g_2, \ldots, g_r$ *is a reduced sequence.*

PROOF. Since $A * B$ is generated by A and B, any element g can be expressed as a product $h_1 h_2 \ldots h_n$ where each h_i is in A or B.

Let $g_1 g_2 \cdots g_r$ be such a product expressing g with the smallest number r of terms. Then $g_1, g_2, \ldots, g_r$ is a reduced sequence; for if $g_i = 1$, or g_i and g_{i+1} come from the same free factor, then

$$g_1 \cdots g_{i-1} g_{i+1} \cdots g_r \quad \text{or} \quad g_1 \cdots g_{i-1}(g_i g_{i+1}) g_{i+2} \cdots g_r$$

is a product expressing g with fewer than r terms. Hence, every g can be expressed in the form (7).

It remains to show the uniqueness. To do this we introduce a specific process ρ similar to the one used in the proof of Theorem 1.2, to "reduce" any word $W(a_\nu, b_\mu)$ to a reduced sequence.

Before defining ρ formally, we illustrate the process for

$$W(a, b) = a^2b^5b^{-1}a^{-3}b^3a^{-2}$$

in the group $A * B = \langle a, b; a^5, b^3 \rangle$. To compute ρ of $W(a, b)$ we first break $W(a, b)$ into its syllables W_1, W_2, W_3, W_4, W_5

$$W(a, b) = a^2 \cdot b^5b^{-1} \cdot a^{-3} \cdot b^3 \cdot a^{-2}.$$

Next, we replace each syllable by the element of A or B it defines, thus obtaining the sequence of elements $a^2, b, a^2, 1, a^3$. We then compute successively:

$$\begin{aligned}
&\rho(W_1) = \rho(a^2) = a^2 \\
&\rho(W_1W_2) = \rho(a^2, b) = a^2, b \\
&\rho(W_1W_2W_3) = \rho(a^2, b, a^2) = a^2, b, a^2 \\
&\rho(W_1W_2W_3W_4) = \rho(a^2, b, a^2, 1) = a^2, b, a^2 \\
&\rho(W_1W_2W_3W_4W_5) = \rho(a^2, b, a^2, 1, a^3) = a^2, b
\end{aligned}$$

Thus ρ applies both to a word in a_ν and b_μ, and to a sequence of elements of $A * B$, each term of which is from the free factor A or B. Formally, ρ is defined as follows:

Definition. The reduced form $\rho(g_1, g_2, \ldots, g_n)$ of a sequence of elements $g_1, g_2, \ldots, g_n$ where g_i is in A or in B (but consecutive terms may be from the same factor) is defined inductively as follows:

ρ(empty sequence) = empty sequence

$$\rho(g_1) = \begin{cases} \text{empty sequence} & \text{if } g_1 = 1 \\ g_1 & \text{if } g_1 \neq 1. \end{cases}$$

Moreover, if $\rho(g_1, \ldots, g_n) = h_1, \ldots, h_r$, then

$$\rho(g_1, \ldots, g_n, g_{n+1}) = \begin{cases} h_1, \ldots, h_r & \text{if } g_{n+1} = 1 \\ h_1, \ldots, h_{r-1} & \text{if } g_{n+1} = h_r^{-1} \\ h_1, \ldots, h_r \cdot g_{n+1} & \text{if } g_{n+1} \neq h_r^{-1} \text{ but } F(g_{n+1}) = F(h_r) \\ h_1, \ldots, h_r, g_{n+1} & \text{if } F(g_{n+1}) \neq F(h_r). \end{cases}$$

Finally, ρ is defined for words $W(a_\nu, b_\mu)$ in a_ν and b_μ as follows:

If $W_1, W_2, \ldots, W_n$ are the syllables of W and g_i is the element of A or B defined by W_i, then

$$\rho(W_1W_2 \ldots W_n) = \rho(g_1, g_2, \ldots, g_n).$$

The theorem will follow easily, once we establish the following properties of ρ:

(a) $\rho(g_1, \ldots, g_n)$ is a reduced sequence, of length at most n.
(b) If $\rho(g_1, \ldots, g_n) = h_1, \ldots, h_r$, then $g_1 g_2 \ldots g_n = h_1 h_2 \ldots h_r$.
(c) If $g_1, \ldots, g_n$ is a reduced sequence then $\rho(g_1, \ldots, g_n) = g_1, \ldots, g_n$.
(d) $\rho(g_1, \ldots, g_k, g_{k+1}, \ldots, g_n) = \rho(\rho(g_1, \ldots, g_k), g_{k+1}, \ldots, g_n)$.
(e) $\rho(g_1, \ldots, g_n, 1) = \rho(g_1, \ldots, g_n)$.
(f) $\rho(g_1, \ldots, g_i, 1, g_{i+1}, \ldots, g_n) = \rho(g_1, \ldots, g_i, g_{i+1}, \ldots, g_n)$.
(g) If $F(g_n) = F(g_{n+1})$, then $\rho(g_1, \ldots, g_n, g_{n+1}) = \rho(g_1, \ldots, g_n \cdot g_{n+1})$
(h) If $F(g_i) = F(g_{i+1})$, then $\rho(g_1, \ldots, g_i, 1, g_{i+1}, \ldots, g_n)$
$= \rho(g_1, \ldots, g_i \cdot g_{i+1}, \ldots, g_n)$.

The properties (a), (b), and (c) follow immediately from the definition of ρ by using induction on n; property (d) follows by induction on $n - k$, from the definition of ρ and from (c); property (e) follows immediately from the definition of ρ; property (f) follows from (d) and (e). Property (g) requires a longer argument, using induction on n, which we now give.

Clearly, from the definition of ρ, $\rho(g_1, g_2) = \rho(g_1 \cdot g_2)$. If $\rho(g_1, \ldots, g_{n-1}) = h_1, \ldots, h_r$, then it suffices by (d) to show that

$$\rho(h_1, \ldots, h_r, g_n, g_{n+1}) = \rho(h_1, \ldots, h_r, g_n \cdot g_{n+1}).$$

If $g_{n+1} = 1$, then $\rho(h_1, \ldots, h_r, g_n, 1) = \rho(h_1, \ldots, h_r, g_n)$ by (e), which is $\rho(h_1, \ldots, h_r, g_n \cdot 1)$, and so the result holds. If $g_n = 1$, then

$$\rho(h_1, \ldots, h_r, 1, g_{n+1}) = \rho(h_1, \ldots, h_r, g_{n+1})$$

by (f), which is $\rho(h_1, \ldots, h_r, 1 \cdot g_{n+1})$ and so again we have our result.

We may assume then that g_n and g_{n+1} are $\neq 1$. Two cases arise, $F(h_r) \neq F(g_n)$ and $F(h_r) = F(g_n)$.

If $F(h_r) \neq F(g_n)$, then

$$\rho(h_1, \ldots, h_r, g_n) = h_1, \ldots, h_r, g_n$$

and

$$\rho(h_1, \ldots, h_r, g_n, g_{n+1}) = \begin{cases} h_1, \ldots, h_r & \text{if } g_n = g_{n+1}^{-1} \\ h_1, \ldots, h_r, g_n \cdot g_{n+1} & \text{if } g_n \neq g_{n+1}^{-1} \end{cases}$$

Similarly,

$$\rho(h_1, \ldots, h_r, g_n \cdot g_{n+1}) = \begin{cases} h_1, \ldots, h_r & \text{if } g_n = g_{n+1}^{-1} \\ h_1, \ldots, h_r, g_n \cdot g_{n+1} & \text{if } g_n \neq g_{n+1}^{-1}, \end{cases}$$

and so again our result holds.

Suppose then that $F(h_r) = F(g_n)$. Now,

$$\rho(h_1, \ldots, h_r, g_n, g_{n+1}) = \rho(\rho(h_1, \ldots, h_r, g_n), g_{n+1})$$

by (d). Since $r \leq n - 1$ by (a), the inductive hypothesis yields successively:

$$\rho(\rho(h_1, \ldots, h_r, g_n), g_{n+1}) = \rho(\rho(h_1, \ldots, h_r \cdot g_n), g_{n+1}) =$$
$$\rho(h_1, \ldots, h_r \cdot g_n, g_{n+1}) = \rho(h_1, \ldots, h_r \cdot g_n \cdot g_{n+1}) = \rho(h_1, \ldots, h_r, g_n \cdot g_{n+1}),$$

where (d) has been used.

Thus property (g) holds. Property (h) follows from (d), (f), and (g).

We may now proceed to prove our theorem. Suppose that $g_1, g_2, \ldots, g_r$ and $h_1, h_2, \ldots, h_s$ are reduced sequences and that

$$g_1 \cdot g_2 \cdot \ldots \cdot g_r = h_1 \cdot h_2 \cdot \ldots \cdot h_s.$$

Now g_i and h_j are defined by words of one syllable U_i and V_j, respectively. Then $U = U_1 U_2 \ldots U_r$ and $V = V_1 V_2 \ldots V_s$ are words in a_ν and b_μ which define the same element of $A * B$. Thus we can proceed from U to V by means of insertions and deletions of the defining relators $R(a_\nu), \ldots, S(b_\mu), \ldots$ or the trivial relators $a_\nu^{\epsilon} a_\nu^{-\epsilon}$ or $b_\mu^{\epsilon} b_\mu^{-\epsilon}$, $\epsilon = \pm 1$. We shall show that under the insertion or deletion of a relator of one syllable, ρ of a word remains unchanged. It suffices to consider the case of the insertion of a relator P of one syllable, since if P is deleted from X to get Y, P can be inserted in Y to get X.

Let $X = X_1 X_2 \ldots X_n$ be a word in a_ν, b_μ where $X_1, X_2, \ldots, X_n$ are the syllables of X, and let k_i be the element of $A * B$ defined by X_i. Then if Y is obtained from X by the insertion of the one-syllable relator P, P is inserted either at the beginning or end of X, in between consecutive syllables X_i and X_{i+1}, or in between symbols in some syllable X_i.

If P has a syllable of X on its left or right which is in the same free factor as P or if P is inserted into such a syllable of X, then the sequence of elements defined by the syllables of Y are the same as the sequence of elements defined by the syllables of X. Hence, $\rho(Y) = \rho(k_1, \ldots, k_n) = \rho(X)$.

Otherwise, P is inserted in front of X_1 and $F(X_1) \neq F(P)$, or P is inserted after X_n and $F(X_n) \neq F(P)$, or $X_i = X_i' X_i''$, $F(X_i) \neq F(P)$ and P is inserted in front of X_i'', We consider each of these cases:

$\rho(Y) = \rho(P X_1 X_2 \ldots X_n) = \rho(1, k_1, k_2, \ldots, k_n) = \rho(k_1, \ldots, k_n) = \rho(X)$ by property (f).

$\rho(Y) = \rho(X_1 X_2 \ldots X_n P) = \rho(k_1, k_2, \ldots, k_n, 1) = \rho(k_1, \ldots, k_n) = \rho(X)$ by property (f).

$\rho(Y) = \rho(X_1 \ldots X_1' P X_1'' \ldots X_n) = \rho(k_1, \ldots, k_i', 1\, k_i'', \ldots, k_n) = \rho(k_1, \ldots, k_i' k_i'', \ldots, k_n) = \rho(k_1, \ldots, k_i, \ldots, k_n) = \rho(X)$ by property (h).

Returning now to U and V, we must have $\rho(U) = \rho(V)$; for there is a sequence $U, \ldots, X, Y, \ldots, V$ such that consecutive terms differ by the insertion of a one-syllable relator and so $\rho(U) = \ldots = \rho(X) = \rho(Y) = \ldots = \rho(V)$. But $\rho(U_1 \ldots U_r) = \rho(g_1, \ldots, g_r) = (g_1, \ldots, g_r)$ and $\rho(V_1 \ldots V_s) = \rho(h_1, \ldots, h_s) = (h_1, \ldots, h_s)$ by property (c). Hence, $(g_1, \ldots, g_r) = (h, \ldots, h_s)$ and the theorem is proved. ◀

Corollary 4.1.1. *Let A, B be subgroups of a group G such that $A \cap B = 1$, and suppose that each element of G can be written uniquely as a product $g = g_1 \ldots g_r$ where $(g_1, \ldots, g_r)$ is a reduced sequence. Then G is the free product of A and B.*

Proof. Let $A = \langle a_1, \ldots, a_n;\ R(a_\nu), \ldots \rangle$ and $B = \langle b_1, \ldots, b_m;\ S(b_\mu), \ldots \rangle$. Then $a_1, \ldots, a_n, b_1, \ldots, b_m$ will generate G and $R(a_\nu), \ldots, S(b_\mu), \ldots,$ will be relators. To show that $G = A * B$ we must show that every relator $W(a_\nu, b_\mu)$ of G can be derived from $R(a_\nu), \ldots, S(b_\mu), \ldots$. Let $W = W_1 W_2 \ldots W_r$, where the W_i are the syllables of W. We use induction on the syllable length of W. If $r = 0$, then W is derivable from $R(A_\nu), \ldots, S(b_\mu), \ldots$.

Suppose that any relator with fewer than r syllables is derivable from $R(a_\nu), \ldots, S(b_\mu), \ldots$ and consider a relator $W_1 W_2 \ldots W_r$ of G. If g_i is the element of G defined by W_i and $g_i \neq 1$ for all i, then $(g_1, g_2, \ldots, g_r)$ is a reduced sequence with $r \geq 1$ and so cannot define 1. Thus some g_i is 1. Hence, W_i is a relator in A or B and so is derivable from $R(a_\nu), \ldots$ or $S(b_\mu), \ldots$. But then $W_1 \ldots W_{i-1} W_{i+1} \ldots W_r$ is a relator and hence, by inductive hypothesis, is derivable from $R(a_\nu), \ldots, S(b_\mu), \ldots$. Thus $W(a_\nu, b_\mu)$ is derivable from $R(a_\nu), \ldots, S(b_\mu), \ldots$. ◀

Corollary 4.1.2. *Let $G = A * B$ and suppose C is a subgroup of A, and D is a subgroup of B. If H is the subgroup of G generated by C and D then $H = C * D$.*

Proof. Since C and D generate H, each element of H is a product $g_1 g_2 \ldots g_r$ where each g_i is in C or D and $(g_1, g_2, \ldots, g_r)$ is a reduced sequence in $C * D$. But a reduced sequence in $C * D$ is a reduced sequence in $A * B$. Hence, the representation of an element of H as the product of a reduced sequence in $C * D$ is unique. Since $C \cap D \subset A \cap B = 1$, it follows from the previous corollary that $H = C * D$. ◀

Corollary 4.1.3. *If $(g_1, g_2, \ldots g_r)$ is a reduced sequence in $A * B$, then $\lambda(g_1 g_2 \ldots g_r) = r$.*

Proof. For if $W(a_\nu, b_\mu)$ defines $g_1 g_2 \ldots g_r$, then the sequence $\rho(W(a_\nu, b_\mu))$ has no more terms than the number of syllables in $W(a_\nu, b_\mu)$, from (a). But if U_i is a word of one syllable defining g_i, then

$$\rho(U_1 \ldots U_r) = \rho(g_1, \ldots, g_r) = (g_1, \ldots, g_r).$$

Since $U_1 \ldots U_r$ and W define the same element of $A * B$, $\rho(W) = \rho(U_1 \ldots U_r)$, and so $\lambda(W) \geq r$. Thus r is the minimal syllable length of any word defining $g_1 g_2 \ldots g_r$. ◀

Corollary 4.1.4. *An element of finite order in $A * B$ is a conjugate of an element of finite order in A or in B.*

Proof. Let g be an element of finite order in $A * B$. We use induction on the syllable length of g. If the syllable length of g is zero or one, the assertion clearly holds. If $g = g_1 g_2 \ldots g_r$ is the reduced form of g in $A * B$ and g_1, g_r come from different free factors, then for $k > 0$,

$$g^k = g_1 g_2 \ldots g_r g_1 g_2 \ldots g_r \ldots g_1 g_2 \ldots g_r$$

is the reduced form of g^k in $A * B$. Hence in this case, g cannot have finite order. Thus g_1, g_r are in the same free factor, and

$$g_1^{-1} g g_1 = g_2 \ldots g_{r-1}(g_r g_1)$$

has shorter syllable length than g, and is of finite order. By the inductive hypothesis some conjugate of $g_1^{-1} g g_1$ lies in a free factor. Hence, the same holds for g. ◀

Corollary 4.1.5. *If g is in $A * B$ and both a and gag^{-1} are in A, $a \neq 1$, then g is in A. In particular, if g is not in A, then*

$$gAg^{-1} \cap A = 1.$$

Proof. We use induction on the syllable length of g in $A * B$. If $\lambda(g) = 0$, then the assertion holds. Suppose that $g = g_1 g_2 \ldots g_r$, $r \geq 1$, is the reduced form of g in $A * B$ and that $g_r \notin A$. Then

$$gag^{-1} = g_1 g_2 \ldots g_r a g_r^{-1} \ldots g_2^{-1} g_1^{-1}$$

is the reduced form of gag^{-1} and has syllable length > 1. Hence, $gag^{-1} \notin A$, contrary to hypothesis. Thus $g_r \in A$, and $g_r a g_r^{-1} \in A$ and is $\neq 1$. Since $(g_1 g_2 \ldots g_{r-1})(g_r a g_r^{-1})(g_1 g_2 \ldots g_{r-1})^{-1}$ is in A, by inductive hypothesis $g_1 g_2 \ldots g_{r-1}$ is in A and so $g = (g_1 g_2 \ldots g_{r-1}) g_r$ is in A. ◀

Corollary 4.1.6. *Let U, V be elements of $A * B$ such that $UV = VU$. Then either both U and V are in the same conjugate of a free factor, or U and V are both powers of the same element W.*

Proof. If U or V is 1 then they are both powers of the same element. Hence, we may assume that $U, V \neq 1$. If now V is in the conjugate of some free factor, say, gAg^{-1}, then $g^{-1}Vg \in A$ and is different from 1. Since $U^{-1}VU = V$,

$$(g^{-1}Ug)^{-1} \cdot g^{-1}Vg \cdot g^{-1}Ug = g^{-1}Vg$$

is in A. Hence, by the preceding corollary, $g^{-1}Ug$ is in A, and so U, V are in gAg^{-1}.

Thus we must show that if two elements of $A * B$ commute and neither is in the conjugate of a free factor then they are powers of the same element. Suppose this assertion false and choose U, V so that V has the smallest syllable length of any element for which there is a U falsifying the above assertion; moreover choose U of smallest syllable length for any V of such length. Clearly, $\lambda(V) \leq \lambda(U)$. Let $U = g_1g_2 \ldots g_r$, $V = h_1h_2 \ldots h_s$ be the reduced forms of U and V in $A * B$; then $r \geq s$.

If h_1 and h_s are in the same free factor, then $h_1^{-1}Vh_1 = h_2 \ldots h_{s-1}(h_sh_1)$ has smaller syllable length than V and commutes with $h_1^{-1}Uh_1$; moreover, neither $h_1^{-1}Vh_1$ nor $h_1^{-1}Uh_1$ is in the conjugate of a free factor since neither V nor U is. Hence, by the minimality of the syllable length of V, $h_1^{-1}Vh_1 = W^k$ and $h_1^{-1}Uh_1 = W^p$; but then $V = (h_1Wh_1^{-1})^k$ and $U = (h_1Wh_1^{-1})^p$ are powers of the same element. Thus h_1 and h_s are not in the same free factor of $A * B$.

If h_1 is not in the same free factor of $A * B$ as g_r, then

$$UV = g_1g_2 \ldots g_rh_1h_2 \ldots h_s$$

is the reduced form of UV in $A * B$. Hence, $\lambda(UV) = r + s$; since

$$UV = VU, \qquad \lambda(VU) = r + s$$

and so

$$VU = h_1h_2 \ldots h_sg_1g_2 \ldots g_r$$

is the reduced form of VU in $A * B$. By uniqueness of the reduced form of $UV = VU$ in $A * B$, we must have $r > s$ and $UV^{-1} = g_1g_2 \ldots g_{r-s}$; for otherwise, $U = V$. Since U commutes with V, UV^{-1} also commutes with V. If UV^{-1} were in the conjugate of a free factor of $A * B$, then by the first part, V would be also. Thus, UV^{-1} and V commute, are neither in the conjugate of a free factor of $A * B$, and $\lambda(UV^{-1}) < \lambda(U)$. Hence, by the minimality of $\lambda(U)$ for a V with minimal $\lambda(V)$, it follows that $V = W^k$ and $UV^{-1} = W^p$. But then $U = UV^{-1}V = W^{p+k}$ and $V = W^k$. Hence, h_1 and g_r must be in the same free factor of $A * B$.

But then h_s^{-1} and g_r are not in the same free factor of $A * B$. Hence, the above paragraph applied to U and V^{-1} in place of U and V shows that U and V^{-1}, and so U and V, are powers of the same element. ◀

Just as for the free group, the transformation problem for the free product can be solved by using the notion of cyclic reduction.

A reduced sequence of elements $g_1, g_2, \ldots, g_r$ of $A * B$ *is called cyclically reduced if* g_1 *and* g_r *are from different free factors of* $A * B$.

THEOREM 4.2. *Each element* g *of* $A * B$ *is conjugate to an element* $g_1g_2 \ldots g_r$ *where the sequence* $g_1, g_2, \ldots, g_r$ *is cyclically reduced. Moreover,* *if* $g_1, g_2, \ldots, g_r$ *and* $h_1, h_2, \ldots, h_s$ *are cyclically reduced sequences such that* $g_1g_2 \ldots g_r$ *and* $h_1h_2 \ldots h_s$ *are conjugate in* $A * B$ *and* $r \neq 1$ *then the sequence* $g_1, g_2, \ldots, g_r$ *and* $h_1, h_2, \ldots, h_s$ *are cyclic permutations of one another; if* $r = 1$, *then* $s = 1$ *and* g_1 *and* h_1 *are conjugate in some free factor.*

PROOF. To show that g is conjugate to $g_1g_2 \ldots g_r$ where $g_1, g_2, \ldots, g_r$ is cyclically reduced, suppose that r is the smallest syllable length of any conjugate of g. Then some conjugate UgU^{-1} has the reduced form $g_1g_2 \ldots g_r$. If g_1 and g_r are from the same free factor, then

$$g_rUgU^{-1}g_r^{-1} = (g_rg_1)g_2 \ldots g_{r-1}$$

is a conjugate of g with smaller syllable length than UgU^{-1}. Hence, $g_1, g_2, \ldots, g_r$ is cyclically reduced.

To show the uniqueness up to cyclic permutation, of the cyclically reduced form of conjugates of an element not in the conjugate of a free factor, we introduce a specific process σ for cyclically reducing an element of $A * B$. Specifically, σ is first defined inductively for reduced sequences $g_1, g_2, \ldots, g_r$ of $A * B$ as follows:

$$\sigma(\text{empty sequence}) = \text{empty sequence};$$

$$\sigma(g_1) = g_1;$$

if $F(g_1) \neq F(g_r)$, then $\sigma(g_1, g_2, \ldots, g_r) = g_1, g_2, \ldots, g_r$;

if $F(g_1) = F(g_r)$, then $\sigma(g_1, g_2, \ldots, g_r) = \sigma(\rho(g_2, \ldots, g_r \cdot g_1))$.

Finally,

if g is any element of $A * B$, then $\sigma(g) = \sigma(\rho(g))$.

It is easily shown by induction on the syllable length of g that $\sigma(g)$ defines an element of $A * B$ which is conjugate to g in $A * B$, and that $\sigma(g)$ is a cyclically reduced sequence in $A * B$. We proceed now with the uniqueness proof.

First, if $\sigma(g)$ is the empty sequence, then g is conjugate to 1 in $A * B$ and so $g = 1$. Hence if h is conjugate to g, then $h = 1$ and $\sigma(h) = \sigma(\rho(h)) = \sigma(\text{empty sequence}) = \text{empty sequence}$. Hence, $\sigma(h)$ and $\sigma(g)$ are cyclic permutations of one another.

Next, suppose that $\sigma(g) = g_1$; then $g = Ug_1U^{-1}$, since $\sigma(g)$ defines a conjugate of g. We wish to prove that $\sigma(VgV^{-1}) = h_1$ where h_1 and g_1 are conjugate in $F(g_1)$. Since $\sigma(VgV^{-1}) = \sigma(VUg_1U^{-1}V^{-1})$, it suffices to prove that $\sigma(Wg_1W^{-1})$ is a conjugate of g_1 in $F(g_1)$. This we do by induction on the syllable length of W. If $\lambda(W) = 0$, the assertion clearly holds. Suppose then that $W = p_1 \ldots p_t$ is the reduced form of W in $A * B$. Then $Wg_1W^{-1} = p_1 \ldots p_tg_1p_t^{-1} \ldots p_1^{-1}$. If $F(p_t) = F(g_1)$, then

$\lambda(p_t g_1 p_t^{-1}) = 1$, $\lambda(p_1 \ldots p_{t-1}) < \lambda(W)$ and so by inductive hypothesis

$$\sigma(p_1 \ldots p_{t-1}(p_t g_1 p_{t-1}^{-1}) p_{t-1}^{-1} \ldots p_1^{-1})$$

is a conjugate of $p_t g_1 p_t^{-1}$ in $F(p_t g_1 p_t^{-1}) = F(g_1)$. Since $p_t g_1 p_t^{-1}$ is a conjugate of g_1 in $F(g_1)$, $\sigma(W g_1 W^{-1})$ is a conjugate of g_1 in $F(g_1)$.

On the other hand, if $F(p_t) \neq F(g_1)$, then

$$W g_1 W^{-1} = p_1 \ldots p_t g_1 p_t^{-1} \ldots p_1^{-1}$$

is the reduced form of $W g_1 W^{-1}$ in $A * B$. Hence, by definition of σ,

$$\begin{aligned} \sigma(W g_1 W^{-1}) &= \sigma(\rho(p_2, \ldots, p_t, g_1, p_t^{-1}, \ldots, p_2^{-1}, p_1^{-1} \cdot p_1)) \\ &= \sigma(p_2, \ldots, p_t, g_1, p_t^{-1}, \ldots, p_2^{-1}) \\ &= \sigma(p_2 \ldots p_t g_1 p_t^{-1} \ldots p_2^{-1}); \end{aligned}$$

hence, by inductive hypothesis $\sigma(W g_1 W^{-1})$ is a conjugate of g_1 in $F(g_1)$. Thus, in particular, we have shown that $g \neq 1$ lies in a conjugate of a free factor if and only if $\lambda(\sigma(g)) = 1$.

Finally, suppose that $\lambda(\sigma(g)) > 1$.

We wish to show that $\sigma(UgU^{-1})$ is a cyclic permutation of $\sigma(g)$. It suffices to show that $\sigma(aga^{-1})$ is a cyclic permutation of $\sigma(g)$, where a is in a free factor of $A * B$; for then we can use induction on $\lambda(U)$.

Suppose that $\rho(g) = g_1, g_2, \ldots, g_r$, where $r > 1$.

Case 1. $F(g_1) \neq F(g_r)$, $F(a) \neq F(g_1)$, $F(a) \neq F(g_r)$.

In this case, $\rho(aga^{-1}) = a, g_1, g_2, \ldots, g_r, a^{-1}$ and by definition of σ, $\sigma(g) = g_1, g_2, \ldots, g_r$ and

$$\sigma(aga^{-1}) = \sigma(\rho(g_1, g_2, \ldots, g_r, a^{-1} \cdot a)) = \sigma(g_1, g_2, \ldots, g_r);$$

hence $\sigma(aga^{-1}) = \sigma(g)$.

Case 2. $F(g_1) \neq F(g_r)$, $F(a) = F(g_1)$, $a \cdot g_1 \neq 1$.

In this case $\rho(aga^{-1}) = a \cdot g_1, g_2, \ldots, g_r, a^{-1}$ and $\sigma(aga^{-1}) = \sigma(\rho(g_2, \ldots, g_r, a^{-1}ag_1)) = \sigma(g_2, \ldots, g_r, g_1) = g_2, \ldots, g_r, g_1$ which is a cyclic permutation of $\sigma(g)$.

Case 3. $F(g_1) \neq F(g_r)$, $a \cdot g_1 = 1$.

In this case, $\rho(aga^{-1}) = g_2, \ldots, g_r, a^{-1} = g_2, \ldots, g_r, g_1$ and so $\sigma(aga^{-1}) = g_2, \ldots, g_r, g_1$ is a cyclic permutation of $\sigma(g)$.

Case 4. $F(g_1) \neq F(g_r)$, $F(a) = F(g_r)$, $g_r \cdot a^{-1} \neq 1$.

Similar to Case 2.

Case 5. $F(g_1) \neq F(g_r)$, $g_r \cdot a^{-1} = 1$.

Similar to Case 3.

CASE 6. $F(g_1) = F(g_r)$, $F(a) \neq F(g_1)$.
In this case, $\rho(aga^{-1}) = a, g_1, g_2, \ldots, g_r, a^{-1}$ and so $\sigma(aga^{-1}) = \sigma(\rho(g_1, g_2, \ldots, g_r, a^{-1} \cdot a)) = \sigma(g_1, g_2, \ldots, g_r) = \sigma(g)$.

CASE 7. $F(g_1) = F(g_r) = F(a)$, $a \cdot g_1 \neq 1$, and $g_r \cdot a^{-1} \neq 1$.
In this case, $\rho(aga^{-1}) = a \cdot g_1, g_2, \ldots, g_r \cdot a^{-1}$ and so $\sigma(aga^{-1}) = \sigma(\rho(g_2, \ldots, g_r \cdot a^{-1} \cdot a \cdot g_1)) = \sigma(\rho(g_2, \ldots, g_r \cdot g_1)) = \sigma(g_1, g_2, \ldots, g_r) = \sigma(g)$.

CASE 8. $F(g_1) = F(g_r) = F(a)$, $a \cdot g_1 = 1$, but $g_r \cdot a^{-1} \neq 1$.
In this case, $\rho(aga^{-1}) = g_2, \ldots, g_r \cdot a^{-1} = g_2, \ldots, g_r \cdot g_1$ and so $\sigma(aga^{-1}) = g_2, \ldots, g_r \cdot g_1 = \sigma(g_1, g_2, \ldots, g_r) = \sigma(g)$.

CASE 9. $F(g_1) = F(g_r) = F(a)$, $a \cdot g_1 \neq 1$, but $g_r \cdot a_1^{-1} = 1$.
Similar to Case 8.

CASE 10. $F(g_1) = F(g_r) = F(a)$, $a \cdot g_1 = 1 = g_r \cdot a^{-1}$.
In this case, $\rho(aga^{-1}) = g_2, \ldots, g_{r-1}$ and so $\sigma(aga^{-1}) = \sigma(\rho(g_2, \ldots, g_{r-1})) = \sigma(\rho(g_2, \ldots, g_{r-1}, 1)) = \sigma(\rho(g_2, \ldots, g_{r-1}, g_r \cdot g_1)) = \sigma(g_1, g_2, \ldots, g_r) = \sigma(g)$.

Thus in all cases $\sigma(aga^{-1})$, and hence $\sigma(UgU^{-1})$, is a cyclic permutation of $\sigma(g)$.

Using that, if $g_1, g_2, \ldots, g_r$ is cyclically reduced, then

$$\sigma(g_1, g_2, \ldots, g_r) = g_1, g_2, \ldots, g_r,$$

the assertion of the theorem follows immediately from the preceding. ◀

COROLLARY 4.2 *If g is in $A * B$ then the smallest syllable length of any conjugate of g is $\lambda(\sigma(g))$.*

PROOF. It follows easily from the definition of σ that if h is in $A * B$, then $\lambda(\sigma(h)) \leq \lambda(h)$.

Let h be a conjugate of g with smallest syllable length s. Since $\sigma(h)$ defines a conjugate of h, and so of g, and $\lambda(\sigma(h)) \leq \lambda(h)$, we must have $\lambda(\sigma(h)) = s$. But either $\lambda(\sigma(g)) = 1$ and then $\lambda(\sigma(h)) = 1$, or $\sigma(h)$ is a cyclic permutation of $\sigma(g)$. In any case, $\lambda(\sigma(g)) = \lambda(\sigma(h))$ and so $\lambda(\sigma(g))$ is the minimal syllable length of a conjugate of g. ◀

The last theorem in this section gives a survey of the possible sets of generators for a finitely generated free product $A * B$.

GRUSHKO-NEUMANN THEOREM. *Let $G = A * B$ and let $g_1, g_2, \ldots, g_k$, k finite, be a set of generators for G. Then $g_1, g_2, \ldots, g_k$ can be obtained by a Nielsen transformation from a set of generators, part of which lie in A and the rest of which lie in B.* (These generators in A or B need not be minimal in number, and may include 1.)

The proof of this theorem involves complicated cancellation arguments and will not be given here. For a proof see B. H. Neumann, 1943a. ◀

Corollary. *The rank (i.e., minimum number of generators) of $A * B$ is the rank of A added to the rank of B.*

Proof. Since A and B are homomorphic images of $A * B$, the rank of $A * B$ is at least the maximum of the ranks of A and B. Moreover, since a set of generators for A and one for B when put together form a set of generators for $A * B$, rank $(A * B) \leq$ rank A + rank B.

If rank A or rank B is infinite, then rank A + rank $B \geq$ rank $(A * B)$ $\geq$ max (rank A, rank B) = rank A + rank B.

If rank $A = n$ and rank $B = m$, n, m finite, then $A * B$ is finitely generated. Let k be the rank of $A * B$ and $g_1, g_2, \ldots, g_k$ a set of generators for $A * B$. Then by Nielsen transformations $g_1, g_2, \ldots, g_k$ can be transformed to $h_1, h_2, \ldots, h_k$ where $h_1, \ldots h_r$ is in A and $h_{r+1}, \ldots, h_k$ is in B. By mapping A identically into itself and B into 1, we send $A * B$ onto A. Since $h_1, \ldots, h_k$ goes into $h_1, \ldots, h_r, 1, \ldots, 1$ we must have that $h_1, \ldots, h_r$ generate A. Similarly, $h_{r+1}, \ldots, h_k$ generate B. Thus $r \geq n$ and $k - r \geq m$, and so $n + m \geq$ rank $(A * B) = k \geq n + m$. Hence,

$$\text{rank } (A * B) = \text{rank } A + \text{rank } B. \quad ◀$$

Thus far we have restricted our definition of the free product of groups to the case of two factors. To define the free product of finitely many or even infinitely many factors, we merely choose a presentation for each of the factors, making sure to use distinct sets of generators, collect all the generators together as a new set of generators and collect all the defining relators together as a new set of defining relators. The group defined by this new presentation is called the free product of the original groups; as for two factors, the free product depends only on the factors and not the particular presentations used. For example, the free product of three cyclic groups of order α, β, γ, respectively is given by

$$\langle a, b, c; a^\alpha, b^\beta, c^\gamma \rangle.$$

If we look upon the free product $A * B$ of the groups A and B as the result of a binary composition of A and B, we see that this composition is commutative, i.e.,

$$A * B \simeq B * A.$$

This follows from the fact that the definition of $A * B$ is symmetric in A and B. From the solution of Problem 3 of this section, we see that our composition is also associative, i.e.,

$$(A * B) * C \simeq A * (B * C)$$

for any three groups A, B, C. Our definition of the free product of finitely many groups (which is independent of their arrangement) could therefore have been replaced by a recursive definition which would describe a free product of n factors as a free product of one factor and the free product of the remaining $n - 1$ factors.

In Section 6.4, six properties of the free product are stated. So far, we have proved the first, third, and fourth of these. The second one is dealt with in Problem 4 of this section. The fifth one is contained in the statement of Problem 5. The sixth one is the statement of Corollary 4.1.2.

The direct product $A \times B$ of two groups A and B is the homomorphic image of the free product $A * B$ under a mapping which maps the replica of A in $A * B$ onto the replica of A in $A \times B$ (and similarly for B). The kernel of this mapping is called the *Cartesian subgroup* of $A * B$ (and can, of course, be defined in the same manner for products of any number of groups). This Cartesian subgroup can be described very easily and explicitly in the case of a product $A * B$ of two groups. See Problem 13, 23, and 24.

References and Remarks. Free products appear implicitly in Fricke and Klein, 1897, in the theory of groups of fractional linear substitutions of a complex variable z which are discontinuous in a part of the z-plane. The process called composition of groups ("Gruppen Komposition") by Fricke and Klein is exactly the formation of the free product of certain discontinuous groups. The contents of their theorem express the fact that the free product of discontinuous groups will be discontinuous again if the fundamental regions of the free factors are related in a certain manner.

Explicitly, and as an object of a special investigation, free products were introduced by Schreier, 1927. The algebraic importance of the concept of the free product of groups was fully established by Kurosh.

Problems for Section 4.1

1. Let $G = A_1 * A_2 * \ldots * A_n$. Prove that every element $\neq 1$ of G can be represented uniquely by a product $g_1 \cdot g_2 \cdot \ldots \cdot g_r$, where $g_i \in A_{j_i}$, $g_i \neq 1$ and $j_i \neq j_{i+1}$. [*Hint:* Use induction on n.]

2. Let $G = A_1 * A_2 * \ldots * A_n$. Show that two elements in A_i are conjugate in G if and only if they are conjugate in A_i. Show that an element not in the conjugate of any A_i is itself conjugate to a "cyclically reduced" product $g_1 \cdot g_2 \cdot \ldots \cdot g_r$, unique up to cyclic permutation. [*Hint:* Use induction on n.]

3. Let $\alpha: A \to A'$, $\beta: B \to B'$, $\gamma: C \to C'$ be isomorphisms. Show that α and β can be extended to an isomorphism of $A * B$ onto $B' * A'$. Similarly show that α, β, and γ can be extended to an isomorphism of $(A * B) * C$ onto $A' * (B' * C')$.

4. Let $N(A)$ and $N(B)$ be the normal subgroups of $A * B$ generated by A and B, respectively. Show that $A \cap N(B) = B \cap N(A) = 1$. Show that no set of generators for $A * B$ can be contained entirely in $N(A)$ or $N(B)$, if A and B are $\neq 1$. [*Hint:* Map $A * B$ onto A and B by mapping the elements of B and of A, respectively, onto the identity element.]

5. Let H be a normal subgroup of A and K be a normal subgroup of B. If N is the normal subgroup of $A * B$ generated by H and K, show that

$$(A * B)/N \simeq (A/H) * (B/K)$$

[*Hint:* Use Theorem 2.1. Observe that $(A * B)/N$ can be defined by adding relators, each of which is in A or in B, to the relators of $A * B$.]

6. Show that $A * B$ is a free group if and only if A and B are free groups.

7. Show that if $A, B \neq 1$, then A has infinitely many cosets in $A * B$. Similarly, show that A has infinitely many conjugates in $A * B$. [*Hint:* Let $a \in A$, $b \in B$, $a, b \neq 1$. Then $A(ab)^n \neq A(ab)^k$ and $(ab)^n A(ab)^{-n} \neq (ab)^k A(ab)^{-k}$, if $k \neq n$.]

8. Show that an element $\neq 1$ in the conjugate of a free factor of $A * B$ has odd syllable length. Show that the conjugates of A and B do not exhaust $A * B$.

9. Let $g = g_1 g_2 \ldots g_r$ be a cyclically reduced element of $A * B$, $r > 1$. Show that if $x^n = g$ has a solution, it has a unique solution and n divides r, $n \neq r$. In particular, if an element of $A * B$ has infinitely many roots, show that it must be in a conjugate of A or B.

10. Show that if $g, h \in A * B$, $gA \neq hA$, then the subgroup S of $A * B$ generated by gAg^{-1} and hAh^{-1} is isomorphic to $A * A$ and that gAg^{-1} and hAh^{-1} are free factors of S.

11. Show that the subgroup S of $A * B$ generated by gAg^{-1} and hBh^{-1} is isomorphic to $A * B$, and that gAg^{-1} and hBh^{-1} are free factors of S. Show that S need not be $A * B$.

12. Show that a finite subgroup of $A * B$ is contained entirely in a conjugate of A or B. [*Hint:* If x and y are elements of finite order, they are each in a conjugate of some free factor, by Corollary 4.1.4. Then use Problems 10 and 11 to show that if x and y are in different conjugates, xy has infinite order.]

13. Show that if $[A, B]$ is the normal subgroup of $A * B$ generated by the mixed commutators $aba^{-1}b^{-1}$, for each $a \in A$, $b \in B$, then $A * B/[A, B] \simeq A \times B$.

14. Show that $A * B$ has a subgroup of finite index not containing A or B if and only if A and B have proper subgroups of finite index. [*Hint:* If $A * B$ has a subgroup H of finite index, consider $H \cap A$ and $H \cap B$. On the other hand, if A and B have proper subgroups of finite index, H, K respectively, then $H \times K$ is of finite index in $A \times B$. Consider the pre-image of $H \times K$ under the homomorphism of $A * B$ onto $A \times B$ given in Problem 13.]

15. Let $G = \langle a, b; a^r, b^s \rangle$, where $r, s \geq 0$. Show that if a and gb^tg^{-1} generate G then $g = a^k b^n$.

16. Show that the automorphisms of $\langle a, b; a^2, b^3\rangle$ are all given by $a \to gag^{-1}$, $b \to gb^tg^{-1}$ where $t = 1$ or 2.

17. Show that the automorphisms of

$$\langle a, b; a^r, b^s\rangle,$$

where $r \neq s$ and $r, s > 0$, are all given by

$$a \to ga^tg^{-1}, \qquad b \to gb^ug^{-1}$$

where t and r are coprime, and also u and s are coprime.

18. Show that the automorphisms of $\langle a, b;\ a^r, b^r\rangle$, where $r > 0$ are all given by

$$a \to ga^tg^{-1}, \qquad b \to gb^ua^{-1}$$

or by

$$a \to gb^tg^{-1}, \qquad b \to ga^ug^{-1},$$

where t and u are each coprime to r.

19. Show that $A * B$ has a free subgroup of rank two unless $A = 1$, or $B = 1$, or A and B are both of order two. [*Hint:* If A and B each have an element of infinite order then by Corollary 4.1.2, we have our result. If A or B has an element of order greater than two, then $A * B$ has a subgroup $\langle c, d; c^r, d^s\rangle$, where $r > 2$, $s \geq 0$, $s \neq 1$. Using Theorem 4.1 show that if $x = c(cd)c^{-1}$ and $y = dc(cd)c^{-1}d^{-1}$, then x and y generate a free group of rank two. Finally, if A and B only contain elements of order one or two, then $A * B$ contains a subgroup

$$\langle c, d, e; c^2, d^2, (cd)^2, e^2\rangle.$$

Show then by Theorem 4.1 that $cdec$ and $ecdece$ generate a free subgroup of rank two.]

20. *A group G is indecomposable (relative to free products) if $G = A * B$ implies $A = 1$ or $B = 1$.* Show that a finitely generated group is the free product of finitely many finitely generated indecomposable groups. [*Hint:* Use the corollary to the Grushko-Neumann theorem.]

21. Using the definition in Problem 20, show that each of the following groups are indecomposable:

(a) A finite group.
(b) A simple group.
(c) A group with a non-trivial center and, in particular, an Abelian group.
(d) A group in which each element has a square root.
(e) A group in which every element has an nth root for some n greater than one.
(f) A group in which any element $\neq 1$ with an nth root for n greater than one, has at least two nth roots.
(g) A group which has a finite normal subgroup.

[*Hint:* For (a), use Problem 12: for (b), use Problem 13; for (c), use Corollary 4.1.6. For (d) and (e), show that if $a \in A$, $b \in B$ and $a, b \neq 1$, then ab has no roots in $A * B$; for (f) use Problem 9; for (g) use Problem 12 and Corollary 4.1.5.]

22. Show that if C, $D \neq 1$, then $C \times D$ is indecomposable with respect to free products. [*Hint:* Let c, $d \neq 1$ be in C, D, respectively. Using Corollary 4.1.6, if c is in a conjugate of A or B, then so is D; but then so is C. This contradicts Problem 4. Hence, again by Corollary 4.1.6, $c = x^k$, $d = x^p$. If $x = c_1d_1$ where $c_1 \in C$, $d_1 \in D$, then $d = c_1{}^pd_1{}^p$. Hence, $c_1{}^p = 1$ and so by Corollary 4.1.4, c_1 is in a conjugate of A or B; but $c = c_1{}^kd_1{}^k = c_1{}^k$, and so c is in a conjugate of A or B.]

23. Show that if A, $B \neq 1$, then the subgroup $[A, B]$ generated by all mixed commutators $aba^{-1}b^{-1}$, $a \in A$, $b \in B$ is a normal subgroup of $A * B$. [*Hint:* Show that

$$Wa_1 \cdot (aba^{-1}b^{-1}) \cdot a_1{}^{-1}W^{-1} = W \cdot (a_1a)b(a_1a)^{-1}b^{-1} \cdot ba_1b^{-1}a_1{}^{-1} \cdot W^{-1},$$

and similarly that

$$Wb_1 \cdot aba^{-1}b^{-1} \cdot b_1{}^{-1}W^{-1} = W \cdot b_1ab_1{}^{-1}a^{-1} \cdot a(b_1b)a^{-1}(b_1b)^{-1} \cdot W^{-1}.]$$

24. Show that if A, $B \neq 1$, then the elements $aba^{-1}b^{-1}$, for any a, $b \neq 1$, $a \in A$, $b \in B$ freely generate $[A, B]$. [*Hint:* Let $x_{i_1}^{\epsilon_1} \ldots x_{i_r}^{\epsilon_r}$, $\epsilon_j = \pm 1$, be a freely reduced word in $x_1, \ldots, x_q$. Suppose that distinct x_{i_j} are replaced by distinct commutators. Then show by induction on r that if x_{i_r} is replaced by $aba^{-1}b^{-1}$, the reduced form of the word in $A * B$ after replacement ends with $a^{-1}b^{-1}$ if $\epsilon_r = +1$, and ends in $b^{-1}a^{-1}$ if $\epsilon_r = -1$.]

25. Show that if A, $B \neq 1$, then $A * B$ has a proper normal free subgroup H_1. Show that $A * B$ has an infinite chain of free subgroups $H_1 \supset H_2 \supset \ldots$ $H_n \supset \ldots$ such that H_{i+1} is a proper normal subgroup of H_i. [*Hint:* Use Problems 23 and 24.]

26. Show that if G has a simple normal subgroup N so that G/N is simple then G is indecomposable. [*Hint:* Use Problem 25, and note that if $N \cap H_1 = 1$ and $NH_1 = G$, then G is $N \times H_1$.]

27. Show that if G has a finite chain of subgroups $G = N_0 \supset N_1 \supset N_2 \supset \ldots \supset N_n = 1$, where N_{i+1} is normal in N_i and N_i/N_{i+1} is simple, then G is indecomposable. [*Hint:* Use Problem 25 and the Jordan-Hölder theorem.]

28. Let $G = A * B$. Show that if $C(g)$ is the set of elements commuting with g, then g is in a conjugate of a factor or $C(g)$ is an infinite cyclic group. Show that if g is not in the conjugate of a factor and $h \neq 1$, $h \in C(g)$, then $C(h) = C(g)$. [*Hint:* Use Corollary 4.1.6 and Problem 9.]

29. Let $G = A * B$. Show that G is the union of subgroups of three types, the first type isomorphic to A, the second type isomorphic to B, and the third type infinite cyclic. Moreover, the subgroups are disjoint in pairs. [*Hint:* Use conjugates of A, conjugates of B, and the $C(g)$ discussed in Problem 28.]

30. Let $G = A * B$. Suppose that α: $A \to H$, β: $B \to H$ are homomorphisms of A and B into some group H. Show that α and β can be extended to a homorphism γ of $A * B$ into H. [*Hint:* Show that the defining relators for $A * B$ go into the identity under the mapping $a_\nu \to \alpha(a_\nu)$, $b_\mu \to \beta(b_\mu)$.]

31. Let G be a group generated by two of its subgroups A and B. Suppose that any homomorphisms $\alpha: A \to H$, $\beta: B \to H$ can be extended to a homomorphism $\gamma: G \to H$, where H is any group. Show that $G = A * B$. [*Hint:* Consider the extension to G of the identity mappings of $A \to A * B$, $B \to A * B$, and apply Corollary 4.1.1.]

32. Show that $A * B$ satisfies no identity $W(X_\nu) = 1$, unless $A = 1$, or $B = 1$, or A and B are cyclic of order two. Show that if A and B are cyclic of order two, then the relation $[[X_1, X_2], [X_3, X_4]] = 1$ is identically satisfied where $[U, V] = U^{-1}V^{-1}UV$. [*Hint:* For the first part use Problem 19 and the fact that a free group of rank two has free subgroups of any finite rank. For the second part, show that since the commutator subgroup of $A * B$ is generated by $a^m b^n a^{-m} b^{-n}$, if A and B are cyclic of order two, then the commutator subgroup of $A * B$ is generated by $aba^{-1}b^{-1}$, and so is cyclic.]

33. Let $A = \langle a_1, \ldots, a_n \rangle$ be a free group and let B be any group, and let $G = A * B$.

(a) Show that in G, $a_1 W_1(b_\mu), \ldots, a_n W_n(b_\mu)$ freely generate a free group F, where $W_\nu(b_\mu) \in B$.

(b) Show that F and B are free product factors for G.

[*Hint:* For (a), map G into A by sending a_ν into a_ν and b_μ into 1. Then F goes into A and so F is a free group. For (b), consider any homorphism φ of F into H and β of B into H, where H is any group. If $\varphi(a_\nu W_\nu(b_\mu)) = h_\nu$ and $\beta(b_\mu) = k_\mu$, then let α be the homomorphism of A into H given by sending a_ν into $h_\nu W_\nu(k_\mu)^{-1}$. Then if γ is the extension of α and β mapping G into H, it is also the extension of φ and β. Hence, by Problem 31, $G = F * B$.]

34. Show that if A and B are Abelian groups, the commutator subgroup of $G = A * B$ is the subgroup of mixed commutators $[A, B]$; conclude that the commutator subgroup of $A * B$ is free. [*Hint:* Clearly, $[A, B]$ is contained in the commutator subgroup of G. Since $G/[A, B] \simeq A \times B$ by Problem 13, $G/[A, B]$ is Abelian and so the commutator subgroup of G is contained in $[A, B]$. Now use Problem 24.]

4.2. Free Product with Amalgamated Subgroups

In this section we introduce an eminently applicable group-theoretical construction. The free product of groups with amalgamated subgroups reduces to the ordinary free product if the amalgamated subgroup is the identity. Like the ordinary free product, the more general concept which we shall now investigate arises in a natural manner from a topological construction. If S_1, S_2 are two arcwise connected spaces with fundamental groups F_1, F_2 then a space S arising from S_1, S_2 by identifying appropriate non-empty homeomorphic subspaces S_1', S_2' of S_1 and S_2, respectively, may have as fundamental group the free product of F_1 and F_2 with an amalgamated subgroup which is the fundamental group of both S_1' and S_2'.

The free product of groups with one amalgamated subgroup was introduced by O. Schreier, 1927. The applications to the theory of groups with a single defining relator, which are presented in Section 4.4, are based on his work. So are the numerous applications contained in the problems of this section and of Sections 4.3 and 4.4.

Hanna Neumann, 1948, 1949, generalized Schreier's original construction. A monograph on the whole theory with numerous applications, to embedding problems in particular, has been published by B. H. Neumann, 1954*a*. It should also be mentioned that the construction by Britton, 1958, of a finitely presented group with a recursively unsolvable word problem uses the free product with amalgamations. (For this problem, see Section 6.1.) Further references to the literature may be found at the end of Section 4.4.

The group presented by $\langle a, b; a^4, b^6\rangle$ is obviously the free product of the cyclic group $\langle a; a^4\rangle$ and the cyclic group $\langle b; b^6\rangle$. On the other hand, the group presented by

$$\langle a, b; a^4, b^6, a^2 = b^3\rangle \tag{1}$$

does not seem to be a free product. Indeed, a free product has a trivial center. On the other hand, the element a^2 in (1) commutes with a, and commutes with b since a^2 is a power of b; thus a^2 is in the center of (1). To show that a^2 is not the identity, note that (1) may be mapped homomorphically onto the cyclic group $\langle x; x^{12}\rangle$ by sending a into x^3 and b into x^2; under this mapping a^2 goes into x^6, which is not the identity. Thus (1) does not define a free product. Nevertheless, the form of the presentation given in (1) is particularly simple, and a group with such a presentation is given a special name.

Definition. If

$$\begin{aligned} (2)\quad G = \langle a_1, \ldots, a_n, b_1, \ldots, b_m; R(a_\nu), \ldots, \\ S(b_\mu), \ldots, U_1(a_\nu) = V_1(b_\mu), \ldots, U_q(a_\nu) = V_q(b_\mu)\rangle \end{aligned}$$

and A' is the subgroup of G generated by $a_1, \ldots, a_n$,

B' is the subgroup of G generated by $b_1, \ldots, b_m$,

H' is the subgroup of A' generated by $U_1(a_\nu), \ldots, U_q(a_\nu)$,

K' is the subgroup of B' generated by $V_1(b_\mu), \ldots, V_q(b_\mu)$,

then G is called the *free product of A' and B' with the subgroups H' and K' amalgamated under the mapping $U_i(a_\nu) \to V_i(b_\mu)$.*

For example, consider $G = \langle a, b; a^4, b^6, a^2 = b^3\rangle$; the homomorphism of G into $\langle x; x^{12}\rangle$ given by $a \to x^3$, $b \to x^2$ shows that a and b have orders four and six, respectively. Hence, $A' = \langle a; a^4\rangle$, $B' = \langle b; b^6\rangle$. Moreover, H' is the cyclic subgroup of A' with order two, K' is the cyclic subgroup of

B' with order two, and H' and K' are amalgamated under the mapping $a^2 \to b^3$.

Similarly, if

$$G = \langle c, d; c^8, d^{10}, c^2 = d^5 \rangle, \tag{3}$$

then G is the free product of the subgroup C' generated by c and the subgroup D' generated by d, with the subgroups L' and M', generated by c^2 and d^5, respectively, amalgamated under the mapping $c^2 \to d^5$. In this case however, $C' \neq \langle c; c^8 \rangle$. Indeed, since $c^2 = d^5$, $c^4 = d^{10} = 1$. Hence, c has order at most four. On the other hand, the homomorphism of (2) into the group $\langle x; x^{20} \rangle$ given by $c \to x^5$, $d \to x^2$, shows that c and d have orders four and ten, respectively. Thus (2) is the free product of the cyclic group C' of order four and the cyclic group D' of order ten with the cyclic subgroups L' and M', each of order two, amalgamated under the mapping $c^2 \to d^5$.

The question naturally arises as to when in the group G in (2) the subgroups A' and B' will have the "natural" presentations

$$\langle a_1, \ldots, a_n; R(a_\nu), \ldots \rangle \qquad \text{and} \qquad \langle b_1, \ldots, b_m; S(b_\mu), \ldots \rangle,$$

respectively. The answer is given in the following theorem.

THEOREM 4.3. *Let*

$$A = \langle a_1, \ldots, a_n; R(a_\nu), \ldots \rangle,$$
$$B = \langle b_1, \ldots, b_m; S(b_\mu), \ldots \rangle, \textit{ and}$$
$$G = \langle a_1, \ldots, a_n b_1, \ldots, b_m; R(a_\nu), \ldots, S(b_\mu), \ldots,$$
$$U_1(a_\nu) = V_1(b_\mu), \ldots, U_q(a_\nu) = V_q(b_\mu) \rangle.$$

If A' is the subgroup of G generated by $a_1, \ldots, a_n$ and B' is the subgroup of G generated by $b_1, \ldots, b_m$, then $A \simeq A'$ under $a_\nu \to a_\nu$ and $B \simeq B'$ under $b_\mu \to b_\mu$ if and only if the mapping $U_i(a_\nu) \to V_i(b_\mu)$ induces an isomorphism φ between the subgroup H of A generated by the $U_i(a_\nu)$ and the subgroup K of B generated by the $V_i(b_\mu)$.

PROOF. For, suppose that $A \simeq A'$ under $a_\nu \to a_\nu$, and that $B \simeq B'$ under $b_\mu \to b_\mu$. If H' and K' are the subgroups of G generated by the $U_i(a_\nu)$ and the $V_i(b_\mu)$, respectively, then $H \simeq H'$ under $U_i(a_\nu) \to U_i(a_\nu)$, and $K \simeq K'$ under $V_i(b_\mu) \to V_i(b_\mu)$. Since $U_i(a_\nu) \to V_i(b_\mu)$ is the identity isomorphism of H' onto K', it follows that $U_i(a_\nu) \to V_i(b_\mu)$ induces an isomorphism φ of H onto K.

Conversely, suppose that the mapping $U_i(a_\nu) \to V_i(b_\mu)$ induces an isomorphism φ of H onto K. Clearly, the mapping $a_\nu \to a_\nu$ and $b_\mu \to b_\mu$ induce homomorphisms of A onto A' and B onto B', respectively. To show that these homomorphisms are indeed isomorphisms is no easy matter. In fact, to do this, we shall have to solve the word problem for G. Hence, we postpone the completion of the proof of this theorem until we have solved the word problem for G under the hypothesis that

$$U_i(a_\nu) \to V_i(b_\mu)$$

induces an isomorphism of H onto K. ◀

In order to gain insight into the solution of the word problem, we consider some examples.

If

$$G = \langle a, b; a^{12}, b^{15}, a^4 = b^5\rangle \tag{4}$$
$$A = \langle a; a^{12}\rangle, \qquad B = \langle b; b^{15}\rangle,$$

then H is a cyclic group of order three, K is a cyclic group of order three, and $a^4 \to b^5$ induces an isomorphism of H onto K. As predicted by Theorem 4.3, the homomorphism of G into $\langle x; x^{60}\rangle$ given by $a \to x^5$, $b \to x^4$ shows that $A' \simeq A$ and $B' \simeq B$. To get an idea of the solution of the word problem for G let us try to reduce some specific word, say,

$$a^{15}b^{-21}a^{32}b^{42}a^{-19}. \tag{5}$$

We can immediately replace the exponents on a and b so that they range from 0 to 11 and 0 to 14, respectively. Thus (5) defines the same element as

$$a^3b^9a^8b^{12}a^5. \tag{6}$$

Moreover, since $a^4 = b^5$, we can simplify (6) further as follows:

$$\begin{aligned}
a^3b^9a^8b^{12} \cdot a^5 &= a^3b^9a^8b^{12}(a^4 \cdot a)\\
&= a^3b^9a^8b^{12}(b^5 \cdot a) = a^3b^9a^8 \cdot b^{17} \cdot a\\
&= a^3b^9a^8(b^{15} \cdot b^2)a = a^3b^9a^8(1 \cdot b^2)a\\
&= a^3b^9 \cdot a^8 \cdot b^2a = a^3b^9 \cdot b^{10} \cdot b^2a\\
&= a^3 \cdot b^{21} \cdot a = a^3 \cdot b^6 \cdot a = a^3(b^5 \cdot b)a\\
&= a^3(a^4 \cdot b)a = a^7 \cdot ba = (a^4 \cdot a^3)ba = a^4 \cdot a^3ba.
\end{aligned}$$

In a similar manner it can be shown that any word $W(a, b)$ can be reduced to a word of the form

$$a^{4k}a^{\alpha_1}b^{\beta_1} \ldots a^{\alpha_r}b^{\beta_r} \tag{7}$$

where $k = 0, 1, 2$; $0 \le \alpha_i \le 3$; $0 \le \beta_i \le 4$; and $\alpha_i \ne 0$ if $i \ne 1$, and $\beta_j \ne 0$ if $j \ne r$.

To show that no two different words of (7) can define the same element of G, it suffices to use the homomorphism of G into $\langle a, b; a^4, b^5\rangle$

given by $a \to a$, $b \to b$, and the homomorphism of G into $\langle x; x^{60}\rangle$ given by $a \to x^5$, $b \to x^4$.

What is the proper generalization of this solution to the group of Theorem 4.3? The factor a^{4k} in (7) defines the general element of H, in (4). Moreover, a^{α_i} can be 1, a, a^2 or a^3; these are simply the coset representatives of A mod H. Similarly, b^{β_i} can be 1, b, b^2, b^3, b^4; these are simply the coset representatives of B mod K. Thus every element of G can be expressed uniquely as a product of an element H and coset representatives alternately from A mod H and B mod K. This suggests the following generalization:

THEOREM 4.4. *Let*

$$A = \langle a_1, \ldots, a_n; R(a_\nu), \ldots\rangle,$$

$$B = \langle b_1, \ldots, b_m; S(b_\mu), \ldots\rangle,$$

(where $A \cap B = 1$) and let

$$G = \langle a_1, \ldots, a_n, b_1, \ldots, b_m; R(a_\nu), \ldots, S(b_\mu), \ldots,$$
$$U_1(a_\nu) = V_1(b_\mu), \ldots, U_q(a_\nu) = V_q(b_\mu)\rangle.$$

Suppose the mapping $U_i(a_\nu) \to V_i(b_\mu)$ induces an isomorphism φ between the subgroups H and K of A and B generated by the $U_i(a_\nu)$ and the $V_i(b_\mu)$, respectively; and further suppose specific right coset representative systems for A mod H and B mod K have been selected. Then to each element g of G we can associate a unique sequence

(8) $$(h, c_1, c_2, \ldots, c_r)$$

such that

(i) *h is an element, possibly 1, of H;*
(ii) *c_i is a coset representative of A mod H or of B mod K;*
(iii) *$c_i \neq 1$*
(iv) *c_i and c_{i+1} are not both in A, and are not both in B;*
(v) *if h', c_i' are the elements of G that correspond to h, c_i under the homomorphisms of A into G and B into G given by $a_\nu \to a_\nu$ and $b_\mu \to b_\mu$, respectively, then $g = h'c_1'c_2' \ldots c_r'$.*

PROOF. In order to prove this theorem, we introduce a specific process ρ for reducing a word in a_ν and b_μ to a reduced sequence of the type (8). It is convenient to first define ρ on a sequence of elements from A and B.

We define ρ inductively as follows:

$$\rho(\text{empty word}) = 1, \text{ the identity element of } H.$$

$$\rho(g_1) = \begin{cases} g_1, \text{ if } g_1 \in H; \\ \varphi^{-1}(g_1), \text{ if } g_1 \in K; \\ (h, \overline{g_1}) \text{ where } h = g_1\overline{g_1}^{-1}, \text{ if } g_1 \in A,\ g_1 \notin H; \\ (h, \overline{g_1}) \text{ where } h = \varphi^{-1}(g_1\overline{g_1}^{-1}), \text{ if } g_1 \in B,\ g_1 \notin K; \end{cases}$$

here $\bar{d}$ is the coset representative of d in A mod H or B mod K. Suppose that $\rho(g_2, \ldots, g_r) = (p, c_2, c_3, \ldots, c_s)$; then

$$\rho(g_1, g_2, \ldots, g_r) = \begin{cases} (g_1 \cdot p, c_2, c_3, \ldots, c_s), \text{ if } g_1 \in H; \\ (\varphi^{-1}(g_1) \cdot p, c_2, c_3, \ldots, c_s), \text{ if } g_1 \in K; \\ (h, \overline{g_1 p}, c_2, c_3, \ldots, c_s) \text{ where } h = (g_1 p) \cdot (\overline{g_1 p})^{-1}, \\ \qquad \text{if } g_1 \in A,\ g_1 \notin H,\ c_2 \notin A; \\ (h, \overline{g_1\varphi(p)}, c_2, c_3, \ldots, c_s) \text{ where} \\ \qquad h = \varphi^{-1}[g_1\varphi(p) \cdot \overline{g_1\varphi(p)}^{-1}], \\ \qquad \text{if } g_1 \in B,\ g_1 \notin K,\ c_2 \notin B; \\ (g_1 p c_2, c_3, \ldots, c_s), \\ \qquad \text{if } g_1 \in A,\ g_1 \notin H,\ c_2 \in A,\ g_1 p c_2 \in H; \\ (\varphi^{-1}(g_1\varphi(p)c_2), c_3, \ldots, c_s), \\ \qquad \text{if } g_1 \in B,\ g_1 \notin K,\ c_2 \in B,\ g_1\varphi(p)c_2 \in K; \\ (h, \overline{g_1 p c_2}, c_3, \ldots, c_s) \text{ where } h = (g_1 p c_2) \cdot (\overline{g_1 p c_2})^{-1}, \\ \qquad \text{if } g_1 \in A,\ g_1 \notin H,\ c_2 \in A,\ g_1 p c_2 \notin H; \\ (h, \overline{g_1\varphi(p)c_2}, c_3, \ldots, c_s) \text{ where} \\ \qquad h = \varphi^{-1}[(g_1\varphi(p)c_2) \cdot (\overline{g_1\varphi(p)c_2})^{-1}], \\ \qquad \text{if } g_1 \in B,\ g_1 \notin K,\ c_2 \in B,\ g_1 p c_2 \notin K. \end{cases}$$

Finally, if $W(a_\nu, b_\mu) = W_1 \cdot W_2 \cdot \ldots \cdot W_r$ where $W_1, \ldots, W_r$ are the syllables of $W(a_\nu, b_\mu)$ and g_i is the element of A or B defined by W_i, we define $\rho(W(a_\nu, b_\mu)) = \rho(W_1 W_2 \ldots W_r) = \rho(g_1, g_2, \ldots, g_r)$.

We note that the complicated form of the definition of ρ is due, in part, to the fact that we have not yet established the preceding theorem, i.e., we cannot assume that A and B are subgroups of G; hence, we cannot multiply elements of A and K, or B and H together, but must use φ^{-1} or φ to bring the elements involved into the appropriate subgroup, H or K.

Using the definition of ρ, we can show that $\rho(g_1, g_2, \ldots, g_r)$ is a reduced sequence with properties (i) through (iv); moreover, we can show that ρ satisfies

(vi) $\rho(1, g_1, \ldots, g_r) = \rho(g_1, g_2, \ldots, g_r)$ and more generally, using induction on s,

$$\rho(g_1, \ldots, g_s, 1, g_{s+1}, \ldots, g_r) = \rho(g_1, \ldots, g_s, g_{s+1}, \ldots, g_r);$$

(vii) if $c_1, c_2, \ldots, c_r$ are alternating coset representatives for A mod H and B mod K, $c_i \neq 1$, then

$$\rho(c_1, c_2, \ldots, c_r) = (1, c_1, c_2, \ldots, c_r);$$

(viii) $\rho(\rho(g_1, g_2, \ldots, g_r)) = \rho(g_1, g_2, \ldots, g_r)$;

(ix) $\rho(g_1, \ldots, g_s, g_{s+1}, \ldots, g_r) = \rho(g_1, \ldots, g_s, \rho(g_{s+1}, \ldots, g_r))$, using induction on s;

(x) if $\rho(g_1, g_2, \ldots, g_r) = (h, c_1, c_2, \ldots, c_s)$, then $g_1'g_2' \ldots g_r' = h'c_1'c_2' \ldots c_s'$ in G;

(xi) $\rho(g_1, \ldots, g_s, k, g_{s+1}, \ldots, g_r) = \rho(g_1, \ldots, g_s, \varphi^{-1}(k), g_{s+1}, \ldots, g_r)$ where $k \in K$ and so

$$\rho(g_1, \ldots, g_s, h, g_{s+1}, \ldots, g_r) = \rho(g_1, \ldots, g_s, \varphi(h), g_{s+1}, \ldots, g_r)$$

where $h \in H$;

(xii) if g_s and g_{s+1} both are in A, or both are in B, then

$$\rho(g_1, \ldots, g_s, g_{s+1}, \ldots, g_r) = \rho(g_1, \ldots, g_s \cdot g_{s+1}, \ldots, g_r).$$

As in the case of the free product, this last property is by far the most difficult to establish. The proof of this property is left for the problems (see Problems 13 and 14).

We may now show that if $W(a_\nu, b_\mu)$ and $T(a_\nu, b_\mu)$ define the same element of G, then $\rho(W) = \rho(T)$. For, since we may transform W into T by inserting and deleting the one-syllable relators $a_\nu{}^\epsilon a_\nu{}^{-\epsilon}$, $b_\mu{}^\epsilon b_\mu{}^{-\epsilon}$ $(\epsilon = \pm 1)$, $R(a_\nu), \ldots, S(b_\mu), \ldots$, or the two-syllable relators $U_i(a_\nu)V_i(b_\mu)^{-1}$, it suffices to consider the case where T can be obtained from W by the insertion of one of these relators.

$$\text{If } \rho(W) = \rho(g_1, g_2, \ldots, g_r)$$

and a one-syllable relator is inserted in W to obtain T, the situation is the same as in the case of the free product; namely $\rho(R)$ is either

$$\rho(g_1, \ldots, g_r) = \rho(W)$$

or

$$\rho(T) = \rho(g_1, \ldots, g_s, 1, g_{s+1}, \ldots, g_r) = \rho(g_1, \ldots, g_s, g_{s+1}, \ldots, g_r) = \rho(W)$$

or

$$\rho(T) = \rho(g_1, \ldots, g_s{}^*, 1, g_s{}^{**}, g_{s+1}, \ldots, g_r)$$

where $g_s = g_s^* \cdot g_s^{**}$ and so

$$\rho(T) = \rho(g_1, \ldots, g_s^*, g_s^{**}, g_{s+1}, \ldots, g_r)$$
$$= \rho(g_1, \ldots, g_s^* \cdot g_s^{**}, g_{s+1}, \ldots, g_r) = \rho(W).$$

If a two-syllable relator $U_iV_i^{-1}$ is inserted in W to obtain T, let u and v be the elements of H and K defined by U_i and V_i, respectively; then $\varphi(u) = v$. Moreover, $\rho(T)$, is either

$$\begin{aligned}\rho(g_1, &\ldots, g_s, u, v^{-1}, g_{s+1}, \ldots, g_r)\\ &= \rho(g_1, \ldots, g_s, \varphi(u), v^{-1}, g_{s+1}, \ldots, g_r)\\ &= \rho(g_1, \ldots, g_s, \varphi(u)v^{-1}, g_{s+1}, \ldots, g_r)\\ &= \rho(g_1, \ldots g_s, 1, g_{s+1}, \ldots, g_r) = \rho(W);\end{aligned}$$

or

$$\begin{aligned}\rho(T) &= \rho(g_1, \ldots, g_s \cdot u, v^{-1}, g_{s+1}, \ldots, g_r)\\ &= \rho(g_1, \ldots, g_s, u, v^{-1}, g_{s+1}, \ldots, g_r) = \rho(W);\end{aligned}$$

or

$$\begin{aligned}\rho(T) &= \rho(g_1, \ldots, g_s \cdot u, v^{-1} \cdot g_{s+1}, \ldots, g_r)\\ &= \rho(g_1, \ldots, g_s, u, v^{-1}, g_{s+1}, \ldots, g_r) = \rho(W);\end{aligned}$$

or

$$\rho(T) = \rho(g_1, \ldots, g_s^* \cdot u, v^{-1}, g_s^{**}, g_{s+1}, \ldots, g_r)$$

where $g_s = g_s^* \cdot g_s^{**}$, so

$$\begin{aligned}\rho(T) &= \rho(g_1, \ldots, g_s^*, u, v^{-1}, g_s^{**}, g_{s+1}, \ldots, g_r)\\ &= \rho(g_1, \ldots, g_s^*, 1, g_s^{**}, g_{s+1}, \ldots, g_r) = \rho(W);\end{aligned}$$

or, finally,

$$\begin{aligned}\rho(T) &= \rho(g_1, \ldots, g_s^*, u, v^{-1} \cdot g_s^{**}, g_{s+1}, \ldots, g_r)\\ &= \rho(g_1, \ldots, g_s^*, u, v^{-1}, g_s^{**}, g_{s+1}, \ldots, g_r) = \rho(W).\end{aligned}$$

Thus, if W and T define the same element of G, $\rho(W) = \rho(T)$. Consequently, we may apply ρ to an element g of G, by applying ρ to any word in a_ν and b_μ that defines g. We show now $\rho(g)$ is a reduced sequence of the type (8) in Theorem 4.4 satisfying (v). For, let g be defined by $W(a_\nu, b_\mu) = W_1 \ldots W_r$, where the W_i are the syllables of $W(a_\nu, b_\mu)$. If g_i is the element of A or B defined by W_i, then $g = g_1'g_2' \ldots g_r'$ and

$$\rho(g) = \rho(W) = \rho(g_1, g_2, \ldots, g_r) = (h, c_1, c_2, \ldots, c_s).$$

But we have established that $g_1'g_2' \ldots g_r' = h'c_1'c_2' \ldots c_s'$ and so $\rho(g)$ satisfies properties (i) through (v). To show that no other sequence can satisfy properties (i) through (v), suppose that $(p, d_1, \ldots, d_t)$ does. Then, if P, D_i are words in a_ν or b_μ defining p, d_i in A or B, respectively,

$Q = PD_1 \ldots D_t$ is a word defining g in G. Hence, $\rho(g) = \rho(Q) = \rho(PD_1 \ldots D_t)$. Now the syllables of Q are either $P, D_1, \ldots, D_t$ or $PD_1, D_2, \ldots, D_t$; thus $\rho(Q) = \rho(p, d_1, \ldots, d_t)$ or $\rho(pd_1, d_2, \ldots, d_t) = \rho(p, d_1, \ldots, d_t)$. Since $(p, d_1, \ldots, d_t)$ is a reduced sequence, $\rho(g) = \rho(Q) = (p, d_1, \ldots, d_t)$. Thus $\rho(g)$ is the unique sequence satisfying properties (i) through (v). ◀

We may now proceed to complete the proof of Theorem 4.3. We must show that the homomorphisms of A and B into G, given by $a_\nu \to a_\nu$, $b_\mu \to b_\mu$ are one-one. If g_1 is an element of A or B, then we note that g_1' is its image in G. Moreover, if W is a word in a_ν or b_μ defining g_1', then $\rho(g_1') = \rho(W) = \rho(g_1)$. Thus, if g_1 and g_2 are elements both in A or both in B and $g_1' = g_2'$, then $\rho(g_1) = \rho(g_1') = \rho(g_2') = \rho(g_2)$. We shall show that g_1 must be g_2.

Suppose $g_1 \in A$; then

$$\rho(g_1) = \begin{cases} g_1, & \text{if } g_1 \in H \\ (h, \overline{g_1}) \text{ where } h = g_1 \cdot \overline{g_1}^{-1}, & \text{if } g_1 \in A, g_1 \notin H. \end{cases}$$

In any event, the product of the terms in $\rho(g_1)$ is g_1. Hence, if $\rho(g_1) = \rho(g_2)$, we have $g_1 = g_2$.

On the other hand, if $g_1 \in B$, then

$$\rho(g_1) = \begin{cases} \varphi^{-1}(g_1), & \text{if } g_1 \in K \\ (h, \overline{g_1}) \text{ where } h = \varphi^{-1}(g_1\overline{g_1}^{-1}), & \text{if } g_1 \in B, g_1 \notin K. \end{cases}$$

In any event, g_1 is the product of the terms of $\rho(g_1)$, if the term from H is replaced by its image under φ. Hence, again, if $\rho(g_1) = \rho(g_2)$, we have $g_1 = g_2$. Thus the homomorphisms of A and B into G are isomorphisms. ◀

COROLLARY 4.4.1. *Let G be as in Theorem 4.4; select a right coset representative system for A' mod H' and B' mod K'. Then any element g in G can be represented uniquely as a product*

$$h'c_1'c_2' \ldots c_q'$$

where $h' \in H'$, $c_i' \neq 1$, c_i' is a representative for A' mod *H' or B'* mod *K', and c_i', c'_{i+1} are not both in A' or both in B'.*

PROOF. If h, c_i are the elements of H, A, or B that go into h', c_i', then

$$\rho(h'c_1'c_2', \ldots, c_q') = (h, c_1, c_2, \ldots, c_q)$$

and we have our result by the last theorem. ◀

COROLLARY 4.4.2. *Let G be as in Theorem 4.4. If an element g in G is a product $g = g_1'g_2' \ldots g_r'$ where $g_i' \notin H$, $g_i' \in A'$ or B', g_i' and g'_{i+1} are not* *both in A' or both in B', and if $\rho(g) = (h, c_1, \ldots, c_q)$, then $q = r$ and g_i' and c_i' are both in A' or both in B'. Moreover, if $g_k', g'_{k+1}, \ldots, g_r'$ are representatives, then*

$$c_k' = g_k', c'_{k+1} = g'_{k+1}, \ldots, c_r' = g_r'.$$

PROOF. This result is proved by induction on r. If $r = 1$ and $g_1' \in A'$, then $\rho(g_1') = \rho(g_1) = (h, \overline{g_1})$ where $h = g_1 \cdot \overline{g_1}^{-1}$; similarly, if $g_1' \in B'$, then $\rho(g_1') = \rho(g_1) = (h, \overline{g_1})$ where $h = \varphi^{-1}(g_1\overline{g_1}^{-1})$. Hence, if g_1' is a representative, $c_1' = g_1'$. In any event, we have our result. Suppose the result for r and consider $\rho(g_1'g_2' \ldots g_r'g'_{r+1}) = \rho(g_1, g_2, \ldots, g_{r+1})$. To compute this we must first compute $\rho(g_2, \ldots, g_{r+1})$ which by inductive hypothesis is $(h, c_2, \ldots, c_{r+1})$ where c_i and g_i are both in A or both in B. Hence, g_1 and c_2 are not both in A or both in B. If $g_1 \in A$, then

$$\rho(g_1, g_2, \ldots, g_{r+1}) = (p, c_1, c_2, \ldots, c_{r+1})$$

where $c_1 = \overline{g_1h}$ and $p = g_1h \cdot \overline{g_1h}^{-1}$; if $g_1 \in B$, then

$$\rho(g_1, g_2, \ldots, g_{r+1}) = (p, c_1, c_2, \ldots, c_{r+1})$$

where $c_1 = \overline{g_1\varphi(h)}$ and $p = \varphi^{-1}(g_1\varphi(h) \cdot \overline{g_1\varphi(h)}^{-1})$.

If now $g_k', g'_{k+1}, \ldots, g'_{r+1}$ are representatives and $k \geq 2$, we have our result by inductive hypothesis. If $g_1', g_2', \ldots, g'_{r+1}$ are all representatives, then by inductive hypothesis $g_2' = c_2', \ldots, g'_{r+1} = c'_{r+1}$. Since $\rho(g_2' \ldots g'_{r+1}) = (h, c_2, \ldots, c_{r+1})$, $c_2' \ldots c'_{r+1} = g_2' \ldots g'_{r+1} = h'c_2' \ldots c'_{r+1}$ and $h' = 1$; hence, $c_1 = \overline{g_1h} = \overline{g_1} = g_1$ and so $c_1' = g_1'$. Thus we have our results. ◀

Definition. If $g = h'c_1' \ldots c_q'$ is as in the preceding corollaries, then $h'c_1' \ldots c_q'$ is called the *reduced form* of g in G. Moreover, q is called the *representative length* of g.

(That q is independent of the particular representative systems used for A' mod H' and B' mod K' follows from the preceding corollary.)

Corollary 4.4.3. *If G is as in Theorem 4.4, then $A' \cap B' = H'$.*

PROOF. Suppose $g \in A' \cap B'$. Then g can be defined by two words $W(a_\nu)$ and $T(b_\mu)$. Hence, $\rho(g) = \rho(W(a_\nu)) = \rho(T(b_\mu))$. If now $W(a_\nu) \notin H$, then $\rho(W(a_\nu)) = (h, c)$, where c is the representative of $W(a_\nu)$ in A mod H. Similarly, if $T(b_\mu) \notin K$, then $\rho(T(b_\mu)) = (h_1, c_1)$, where c_1 is the representative of $T(b_\mu)$ in B mod K. Since $\rho(W(a_\nu)) = \rho(T(b_\mu))$, $c = c_1$; this is impossible because $A \cap B = 1$. Hence, $W(a_\nu) \in H$ or $T(b_\mu) \in K$. In either event, $g \in H' = K'$. Thus $A' \cap B' \subset H'$. Since $H' = K'$, $H' \subset A' \cap B'$; and so $A' \cap B' = H'$. ◀

It should be noted that in Theorem 4.4 we assumed that $A \cap B = 1$; on the other hand, $A' \cap B' = H'$. Thus we must be careful in identifying the isomorphic groups A, A' and B, B'. Nevertheless, it is customary to identify A, A' and B, B', and we shall continue this custom whenever no confusion is caused thereby. In keeping with this custom we introduce the following:

Definition. If A, B, H, K, φ, and G are as in Theorem 4.4, then we call *G the free product of A and B with the subgroups H and K amalgamated under φ*, and denote G by $*(A, B, H, K, \varphi)$; for brevity G is often called *the free product of A and B with an amalgamated subgroup H, and A and B* are called *the factors of the amalgamation.*

Although $G = *(A, B, H, K, \varphi)$ was defined in terms of the presentation (2) which seems to depend upon particular presentations for A and B and particular generators U_i for H (the generators V_i of K are determined by U_i and φ), the group G depends in fact only upon A, B, H, and φ. This is stated more precisely in the next corollary.

COROLLARY 4.4.4. *Suppose that A has the presentations*

(9) $$\langle a_1, \ldots, a_n; R(a_\nu), \ldots\rangle$$

and

(10) $$\langle c_1, \ldots, c_k; P(c_\kappa), \ldots\rangle$$

and also B has the presentations

(11) $$\langle b_1, \ldots, b_m; S(b_\mu), \ldots\rangle$$

and

(12) $$\langle d_1, \ldots, d_l; Q(d_\lambda), \ldots\rangle.$$

Moreover, suppose that the subgroup H of A is generated by $U_1(a_\nu), \ldots$ and also by $T_1(c_\kappa), \ldots$ and that the subgroup K of B is generated by $\varphi(U_1(a_\nu)) = V_1(b_\mu), \ldots$ and also by $\varphi(T_1(c_\kappa)) = W_1(d_\lambda), \ldots$, where φ is an isomorphism of H onto K. Then the isomorphisms of (9) *onto* (10) *and* (11) *onto* (12) *can be extended to an isomorphism between*

(13)
$$\langle a_1, \ldots, a_n, b_1, \ldots, b_m; R(a_\nu), \ldots, S(b_\mu), \ldots, U_1(a_\nu) = V_1(b_\mu), \ldots\rangle$$

and

(14)
$$\langle c_1, \ldots, c_k, d_1, \ldots, d_l; P(c_\kappa), \ldots, Q(d_\lambda), \ldots, T_1(c_\kappa) = W_1(d_\lambda), \ldots\rangle$$

PROOF. The proof is fairly straight forward; one shows that the isomorphisms of (9) onto (10) and (11) onto (12) carry defining relators in (13) into relators in (14); similarly, the inverses of the isomorphisms of

(9) onto (10) and of (11) onto (12) carry defining relators in (14) into relators in (13). We leave the details to the reader. ◄

It should be noted that the isomorphism φ plays an essential role, i.e., if A, B, H, K are kept fixed but φ is changed then $*(A, B, H, K, \varphi)$ may change. To illustrate this let A and B be the symmetries of the square,

$$A = \langle a_1, a_2; a_1^4, a^2{}_2, a_1a_2 = a_2a_1^{-1}\rangle \tag{15}$$

$$B = \langle b_1, b_2; b_1^4, b_2^2, b_1b_2 = b_2b_1^{-1}\rangle \tag{16}$$

The permutations

$$I, (xyzw), (xz)(yw), (xwzy), (xz), (yw), (xy)(zw), (xw)(yz) \tag{17}$$

also represent the symmetries of the square; in fact, $a_1 \to (xyzw)$, $a_2 \to (xz)$, and $b_1 \to (xyzw)$, $b_2 \to (xz)$ give isomorphisms between (15) and (17), and between (16) and (17). Let H be the subgroup of A generated by $(xz)(yw)$ and (xz), i.e., by a_1^2 and a_2; similarly, let K be the subgroup of B generated by $(xz)(yw)$ and (xz), i.e., by b_1^2 and b_2. Then H and K are each the Klein four-group; moreover, the mappings

$$\varphi\colon a_1^2 \to b_1^2, \quad a_2 \to b_2$$

and

$$\psi\colon a_1^2 \to b_2, \quad a_2 \to b_1^2$$

both induce isomorphisms of H onto K. We show that $G_1 = *(A, B, H, K, \varphi)$ and $G_2 = *(A, B, H, K, \psi)$ are not isomorphic. For, G_1 has a non-trivial center. Indeed, a_1^2 is in the center of A and b_1^2 is in the center of B and since $a_1^2 = b_1^2$, a_1^2 is in the center of G_1; moreover, since $A \subset G_1$, and $a_1^2 \neq 1$ in A, $a_1^2 \neq 1$ in G_1. On the other hand, as we shall presently show, the center of a free product with an amalgamated subgroup must lie in that subgroup. Thus the only candidates for non-trivial center elements in G_2 are a_1^2, a_2 and $a_1^2a_2$. Since only a_1^2 is in the center of A and $A \subset G$, only a_1^2 can be in the center of G. But $a_1^2 = b_2$ in G_2 and b_2 is not in the center of $B \subset G_2$. Thus G_2 has only a trivial center and so

$$*(A, B, H, K, \varphi) \not\approx *(A, B, H, K, \psi).$$

As another corollary to Theorem 4.4 we have:

Corollary 4.4.5. *If* $G = *(A, B, H, K, \varphi,)$ *then any element in* G *of finite order is in a conjugate of* A *or* B.

Proof. Let g be an element of finite order not in a conjugate of A or B. If g has the reduced form $g = hc_1c_2 \ldots c_r$, $r \geq 2$, where c_1 and c_r are from different amalgamation factors, then

$$\begin{aligned} g^k &= hc_1c_2 \ldots c_r \cdot hc_1c_2 \ldots c_r \cdot \ldots \cdot hc_1c_2 \ldots c_r \\ &= (hc_1) \cdot c_2 \cdot \ldots \cdot c_r \cdot (hc_1) \cdot c_2 \cdot \ldots \cdot c_r \cdot \ldots \cdot (hc_1) \cdot c_2 \cdot \ldots \cdot c_r \end{aligned}$$

has representative length kr by a previous corollary, and so cannot be 1. On the other hand, if c_1 and c_r are from the same amalgamation factor let g_1 be a conjugate of g with smallest representative length. Then g_1 must have finite order and yet have the reduced form $pd_1d_2 \ldots d_q$ with $q \geq 2$, $p \in H$, and d_1, d_q in different amalgamation factors. For, g is not in a conjugate of A or B and so $q \geq 2$; if d_1, d_q are in the same amalgamation factor, then $d_1g_1d_q{}^{-1} = (d_qpd_1) \cdot d_2 \cdot \ldots \cdot d_{q-1}$ has a shorter representative length than g_1. Thus we cannot have g_1 of finite order, and our assertion holds. ◀

The free product of groups is a generalization of a free group; for, a free group is the free product of infinite cyclic groups. Similarly, the free product of groups with an amalgamated subgroup is a generalization of the free product; for, if the subgroup amalgamated is 1, then the free product results. Hence, although we would hope that statements similar to those made for free products would hold also for free products with amalgamated subgroup, the statements will usually be more complicated. As an illustration of this growing complexity, consider the problem of commutativity, i.e., given two elements x and y which commute, what can be said about them? In the case of the free group, x and y must be powers of some element z. In the case of a free product, x and y are both in the same conjugate of a factor, or x and y are powers of some element z. In the case of a free product with an amalgamated subgroup, three possibilities arise as is indicated in the following theorem.

THEOREM 4.5. *Let $G = *(A, B, H, K, \varphi)$, and suppose $x, y \in G$ and $xy = yx$. Then*

(i) *x or y may be in a conjugate of H;*
(ii) *if neither x nor y is in a conjugate of H, but x is in the conjugate of a factor, then y is in that same conjugate of a factor;*
(iii) *if neither x nor y is in a conjugate of a factor, then $x = ghg^{-1} \cdot W^j$ and $y = gh'g^{-1} \cdot W^k$, where $g, W \in G$, $h, h' \in H$, and ghg^{-1}, $gh'g^{-1}$, and W commute in pairs.*

PROOF. If x or y is in a conjugate of H, there is nothing to prove. Assume then that neither x nor y is in a conjugate of H. If x is in a conjugate of one of the factors, say, A, then $txt^{-1} = a \in A$, $a \notin H$. Then tyt^{-1} commutes with a. Since $tyt^{-1} \notin H$, the reduced form of tyt^{-1} is $pc_1 \ldots c_r$, $r \geq 1$, where $p \in H$. Now

$$pc_1 \cdot \ldots \cdot c_{r-1} \cdot c_r \cdot a \cdot c_r^{-1} \cdot c_{r-1}^{-1} \cdot \ldots \cdot c_1^{-1}p^{-1} \tag{18}$$

is a; if $c_r \notin A$, then (18) has representative length $2r + 1$. Thus, $c_r \in A$.

But if $r > 1$, then $c_r a c_r^{-1} \notin H$ (since x is not in a conjugate of H) and so (18) has representative length $2(r - 1) + 1$; thus, $r = 1$ and $tyt^{-1} \in A$. Hence, x and y are both in $t^{-1}At$.

To complete the proof of the theorem, we shall suppose the assertion false and obtain a contradiction. Let x be an element of smallest representative length for which there exists some y falsifying the assertion; let y be an element of smallest representative length falsifying the assertion for such an x of minimal representative length. By the first part of our proof, neither x nor y can lie in the conjugate of a factor. Suppose, then, that $x = pc_1c_2 \ldots c_r$, $r > 1$, and $y = p'd_1d_2 \ldots d_s$, $s > 1$, are the reduced forms of x and y; clearly $r \leq s$. Now, c_1 and c_r are in different amalgamation factors; for otherwise $c_r x c_r^{-1} = (c_r p c_1)c_2 \ldots c_{r-1}$ and $c_r y c_r^{-1}$ falsify the assertion of the theorem, although the representative length of $c_r x c_r^{-1}$ is less than that of x. Hence, not both c_1 and c_r are in the same amalgamation factor as d_s; suppose c_1 is not in the same factor as d_s.

Then $yx = p'd_1 \cdot \ldots \cdot d_s p \cdot c_1 \cdot \ldots \cdot c_r$ has representative length $r + s$ and so $yx = xy = pc_1 \cdot \ldots \cdot c_r p' \cdot d_1 \cdot \ldots \cdot d_s$ must also have representative length $r + s$; hence, c_r and d_1 are not in the same amalgamation factor. Since $c_1, \ldots, c_r$ are representatives, by a previous corollary, $c_1, \ldots, c_r$ will occur as the last r representatives when yx is in reduced form; similarly, when xy is in reduced form, $d_1, \ldots, d_s$ will occur as the last s representatives. But $r \leq s$ and so

$$c_r = d_s, \qquad c_{r-1} = d_{s-1}, \ldots, c_1 = d_{s-r+1}.$$

Hence, $yx^{-1} = p'd_1 \cdot \ldots \cdot d_{s-r}p^{-1}$. Now, yx^{-1} commutes with x and yet yx^{-1} has smaller representative length than y. Hence, yx^{-1} and x satisfy the assertion of the theorem. If yx^{-1} is in a conjugate of H, then $yx^{-1} = ghg^{-1}$; hence, $x = g1g^{-1} \cdot x$ and $y = yx^{-1} \cdot x = ghg^{-1} \cdot x$, and the assertion of the theorem would hold for x and y contrary to assumption. If yx^{-1} is in a conjugate of a factor, then x is also, and so is $y = yx^{-1} \cdot x$; this contradicts our assumption on x and y. Thus, it must be that $x = ghg^{-1} \cdot W^j$, $yx^{-1} = gh'g^{-1} \cdot W^k$, where ghg^{-1}, $gh'g^{-1}$ and W commute in pairs. Hence, $y = yx^{-1} \cdot x = gh'g^{-1} \cdot W^k \cdot ghg^{-1} \cdot W^j = gh'hg^{-1} \cdot W^{k+j}$, and again the assertions of the theorem would hold for x and y. Thus, assuming that c_1 and d_s are not in the same factor leads to a contradiction.

On the other hand, if c_1 and d_s are in the same factor, d_s and c_r are not. But then $x^{-1} = c_r^{-1} \cdot c_{r-1}^{-1} \cdot \ldots \cdot c_1^{-1}p^{-1}$ can be used in place of x in the preceding paragraph; for x^{-1} has the same representative length as x, and x and y falsify the assertion of the theorem if and only if x^{-1} and y falsify the assertion. Thus we again are led to a contradiction as in the preceding paragraph.

Hence, our theorem holds. ◀

COROLLARY 4.5. *The center of* $G = *(A, B, H, K, \varphi)$, *if* $A \neq H$ *and* $B \neq K$, *is* $H \cap C(A) \cap C(B)$, *where* $C(A)$, $C(B)$ *are the centers of* A *and* B, *respectively.*

PROOF. Since A and B generate G, $H \cap C(A) \cap C(B)$ is contained in the center of G. Moreover, if x is in the center of G, then x is in H; for, if $x = hc_1 \ldots c_r$ is the reduced form of x and $r \geq 1$, then choose a representative $c \neq 1$ from the factor A or B not containing c_r. Then $xc = hc_1 \cdots c_r \cdot c$ is a reduced form. Hence, $cx = chc_1 \ldots c_r$ must have representative length $r + 1$; thus c and c_1 are from different factors, and so by Corollary 4.4.2, cx ends in c_r. This contradicts $xc = cx$. Hence, x is in H, and therefore is in the center of A and the center of B, since it is in the center of G. ◀

Thus, for example, if A and B are Abelian groups, the center of $G = *(A, B, H, K, \varphi)$ is $H \cap C(A) \cap C(B) = H \cap A \cap B = H$. If A or B has a trivial center, then G has a trivial center. Even if A and B each have non-trivial centers, unless φ sends elements of $H \cap C(A)$ into $K \cap C(B)$, G will have no center. Thus, if A and B are given as in (15) and (16), and ψ is given by $a_1^2 \to b_2$, $a_2 \to b_1^2$, then A and B have non-trivial centers $\{1, a_1^2\}$ and $\{1, b_1^2\}$, respectively; yet $G_2 = *(A, B, H, K, \psi)$ has only a trivial center; for $H \cap C(A) = \{1, a_1^2\}$, but $a_1^2 = b_2 \notin C(B)$.

The growing difficulty of a given problem when posed for the free group, the free product, and then the free product with amalgamation is also illustrated by the word problem.

Thus, given a free group F on free generators $x_1, x_2, \ldots, x_n$, we can directly solve the word problem. On the other hand, solution of the word problem for $A * B$ requires a knowledge of the word problem for A and B. The word problem for $*(A, B, H, K, \varphi)$ requires more for its solution. Indeed, we must be able to solve the generalized word problem for A mod H and for B mod K; in addition, we must have an effective method for computing $\varphi(h)$ and $\varphi^{-1}(k)$ for $h \in H$ and $k \in K$ (for example, a method for writing h and k as words in the $U_i(a_\nu)$ and $V_i(b_\mu)$, respectively, if φ is given by $U_i(a_\nu) \to V_i(b_\mu)$).

The transformation problem affords yet another illustration of the increasing difficulty. For the free group F on free generators $x_1, x_2, \ldots, x_n$, it can be solved directly; two elements are conjugate if and only if, when cyclically reduced, one of the elements is a cyclic permutation of the other. On the other hand, to solve the transformation problem for $A * B$ we must be able to solve it for A and B; indeed, two elements of one of the factors are conjugate in $A * B$ if and only if they are conjugate in that factor. Moreover, if the transformation problem for A and B can be solved, then it can be solved for $A * B$ (see Theorem 4.2). The situation for $*(A, B, H, K, \varphi)$ is much more complicated. Even if the transformation problem can be solved in A and B and the word problem can be solved in $*(A, B, H, K, \varphi)$, the solution of the transformation problem for $*(A, B, H, K, \varphi)$ is unknown in general. However, in special cases, the transformation problem can be solved using the following definition and theorem.

An element g of $G = *(A, B, H, K, \varphi)$ is called *cyclically reduced* if g has a reduced form $g = hg_1 \ldots g_r$, where g_1 and g_r are not in the same factor, unless $r = 1$.

It is clear that if $g = p_1 \ldots p_r$ where $r \geq 2$ and p_i, p_{i+1} are in distinct factors, then g is cyclically reduced if and only if p_1 and p_r are in distinct factors; for, if g is written in reduced form as $hg_1 \ldots g_r$, then g_i is in the same factor as p_i by Corollary 4.4.2.

THEOREM 4.6. *Let $G = *(A, B, H, K, \varphi)$. Then every element of G is conjugate to a cyclically reduced element of G. Moreover, suppose that g is a cyclically reduced element of G. Then:*

(i) *If g is conjugate to an element h in H, then g is in some factor and there is a sequence $h, h_1, h_2, \ldots, h_t, g$ where h_i is in H and consecutive terms of the sequence are conjugate in a factor.*
(ii) *If g is conjugate to an element g' in some factor, but not in a conjugate of H, then g and g' are in the same factor and are conjugate in that factor.*
(iii) *If g is conjugate to an element $p_1 \ldots p_r$, where $r \geq 2$, and p_i, p_{i+1} as well as p_1, p_r are in distinct factors, then g can be obtained by cyclically permuting $p_1 \ldots p_r$ and then conjugating by an element of H.*

PROOF. To show that every element of G is conjugate to a cyclically reduced element, let g' be an element of smallest representative length conjugate to g. If $g' = hg_1' \ldots g_s'$ is the reduced form of g' and $s > 1$, then g_1' and g_s' cannot be in the same factor; for then $g_s'g'g_s'^{-1} = (g_s'hg_1') \cdot g_2' \cdot \ldots \cdot g_{s-1}'$ is conjugate to g and has smaller representative length than g'. Thus g' is cyclically reduced.

To prove (i), suppose that g is a cyclically reduced element conjugate to an element h in H, and that $g = WhW^{-1}$, where $W = kc_1 \ldots c_s$ is the reduced form of W. If $s \leq 1$, the sequence h, g is of the required type since $g = (kc_1)h(kc_1)^{-1}$. Let $s > 1$. Suppose there is a largest integer q such that $c_q \ldots c_s h c_s^{-1} \ldots c_q^{-1}$ is not in H but $c_{q+1} \ldots c_s h c_s^{-1} \ldots c_{q+1}^{-1} = h'$ is in H. Then $c_j \ldots c_s h c_s^{-1} \ldots c_j^{-1}$ is in H, if $j > q$; for, h is in H and the existence of $j > q$ with $c_j \ldots c_s h c_s^{-1} \ldots c_j^{-1}$ not in H would contradict the maximality of q. If now $q > 1$, then

$$g = kc_1 \cdot \ldots \cdot (c_q c_{q+1} \ldots c_s h c_s^{-1} \ldots c_{q+1}^{-1} c_q^{-1}) \cdot \ldots \cdot c_1^{-1} k^{-1}$$

cannot be a cyclically reduced element. Hence, $q = 1$ or there is no q, i.e., $c_j \ldots c_s h c_s^{-1} \ldots c_j^{-1}$ is in H for each j. In any event, the sequence

$$h, \quad c_s h c_s^{-1}, \quad \ldots, \quad c_2 \ldots c_s h c_s^{-1} \ldots c_2^{-1}, \quad g$$

is of the required type.

To prove (ii) suppose that g is a cyclically reduced element conjugate to an element g' in some factor but not in a conjugate of H, and that $g = Wg'W^{-1}$, where $W = kc_1 \ldots c_s$ is the reduced form of W. If $s = 0$, then g' and $g = kg'k^{-1}$ are in the same factor and are conjugate in that factor. Suppose then $s \geq 1$. Then g is

$$kc_1 \cdot \ldots \cdot c_s \cdot g' \cdot c_s^{-1} \cdot \ldots \cdot c_1^{-1}k^{-1}.$$

If g' and c_s are in different factors then

$$kc_1 \cdot \ldots \cdot c_s \cdot g' \cdot c_s^{-1} \cdot \ldots \cdot c_1^{-1}k^{-1}$$

cannot be a cyclically reduced element. Hence, c_s and g' are in the same factor. Since g' is not in a conjugate of H, $c_s g' c_s^{-1}$ is not in H. Thus, if $s > 1$,

$$kc_1 \cdot \ldots \cdot (c_s g' c_s^{-1}) \cdot \ldots \cdot c_1^{-1}k^{-1}$$

cannot be a cyclically reduced element. Hence, $s = 1$, and both g' and $g = kc_1 g' c_1^{-1} k^{-1}$ are in the same factor and are conjugate in that factor.

To prove (iii), let $g' = p_1 \ldots p_r$ be an element, where $r \geq 2$ and p_i, p_{i+1} are in different factors, and also p_1, p_r are in different factors.

Suppose $W = kc_1 \ldots c_s$ is in reduced form and $g = Wp_1 \ldots p_r W^{-1}$. We use induction on s to prove our result.

If $s = 0$, then g can be obtained directly from $p_1 \ldots p_r$ by conjugating with an element of H.

Suppose that $s > 0$ and consider g which is

$$kc_1 \ldots c_s p_1 \ldots p_r c_s^{-1} \ldots c_1^{-1} k^{-1}.$$

Suppose c_s is not in the same factor as p_1; then c_s is in the same factor as p_r. If $p_r c_s^{-1}$ is not in H, then

$$kc_1 \cdot \ldots \cdot c_s \cdot p_1 \cdot \ldots \cdot p_r c_s^{-1} \cdot \ldots \cdot c_1^{-1} k^{-1}$$

cannot be a cyclically reduced element. Thus $c_s = hp_r$ and g is

$$kc_1 \cdot \ldots \cdot c_{s-1} \cdot hp_r \cdot p_1 \cdot \ldots \cdot p_{r-1} \cdot h^{-1} \cdot c_{s-1}^{-1} \cdot \ldots \cdot c_1^{-1} k^{-1}$$
$$= (kc_1 \cdots c_{s-1}) \cdot h(p_r p_1 \cdots p_{r-1}) h^{-1} \cdot (kc_1 \ldots c_{s-1})^{-1}.$$

Now,

$$hp_r p_1 \ldots p_{r-1} h^{-1} = hp_r h^{-1} \cdot hp_1 h^{-1} \cdot \ldots \cdot hp_{r-1} h^{-1} \tag{19}$$

satisfies the same conditions as $p_1 \ldots p_r$. Hence, using the inductive hypothesis, g, the conjugate of (19) by $kc_1 \cdots c_{s-1}$, may be obtained by cyclically permuting $hp_rh^{-1} \cdot hp_1h^{-1} \cdot \ldots \cdot hp_{r-1}h^{-1}$ and then conjugating by another element of H, say, h'. The final result will be the conjugate by $h'h$ of a cyclic permutation of $p_1 \ldots p_r$, and we have our result. If c_s is in the same factor as p_1, c_s is not in the same factor as p_r, and an argument completely analogous to the above gives the desired conclusion. ◀

We apply Theorem 4.6 to a special case in the following corollary.

COROLLARY 4.6. *Let $G = *(A, B, H, K, \varphi)$, where H and K are in the centers of A and B, respectively. Then two elements of G are conjugate if and only if both elements when cyclically reduced are in the same factor and are conjugate in that factor, or one of the elements when cyclically reduced has the reduced form $hg_1 \ldots g_r$ and the other element when cyclically reduced has the reduced form $hg_s \ldots g_rg_1 \cdots g_{s-1}$.*

PROOF. Since H and K are in the centers of A and B, respectively, H is in the center of G. Hence, any element conjugate to an element h in H must be h. Suppose now that g and g' are cyclically reduced elements conjugate in G. If g' is in a conjugate of H, i.e., H, then $g = g'$, and so g and g' are in the same factor and are conjugate in that factor. Suppose g' is in a factor but that g' is not in a conjugate of H, then case (ii) in Theorem 4.5 gives us our result. Finally, suppose that g' has the reduced form $hg_1 \ldots g_r$, with $r \geq 2$. Then by applying case (iii) of Theorem 4.5 with $p = hg_1, p_2 = g_2, \ldots, p_r = g_r$ we have

$$g = k(g_s \cdot \ldots \cdot g_r \cdot hg_1 \cdot \ldots \cdot g_{s-1})k^{-1},$$

where $k \in H$. Since H is in the center of G,

$$g = hg_s \cdots g_rg_1 \cdots g_{s-1}. \quad ◀$$

Other special cases are considered in the problems at the end of this section. In particular, if H is finite, then the transformation problem for $G = *(A, B, H, K, \varphi)$ can be solved if the solution to the transformation problems in A and B and the solution to the word problem in G are known (see Problem 41). However, H need not be finite nor in the center of G (nor be one of the factors) and yet the transformation problem may be solvable in G (see Problem 42).

Although we have so far considered only an amalgamated product for two factors A, B, the construction can be easily generalized to include finitely many, or even arbitrarily many, such factors. Moreover, Theorem 4.3 and 4.4 carry over for these cases. For example, the group

$$G = \langle a, b, c; a^4, b^6, c^{10}, a^2 = b^3, a^2 = c^5 \rangle$$

is the free product of the cyclic groups $A = \langle a; a^4 \rangle$ $B = \langle b; b^6 \rangle$ and $C = \langle c; c^{10} \rangle$ with the subgroups H generated by a^2, K generated by b^3, and L generated by c^5 amalgamated under the isomorphisms induced by $a^2 \to b^3$, $a^2 \to c^5$. Every element of G can be written uniquely as the product of an element of H and coset representatives for A mod H, B mod K, and C mod L. Thus, for example, $a^2bc^3ac^2b^2$ cannot be 1.

Before concluding this section we give some applications of the free product with an amalgamated subgroup construction.

A. G. Howson, 1954, proved that the intersection of two finitely generated subgroups of a free froup is finitely generated. An example of a group in which two finitely generated free subgroups intersect in an infinitely generated subgroup is given by

$$G = \langle a_1, a_2, b_1, b_2; \ldots, [a_1{}^k, a_2{}^l] = [b_1{}^k, b_2{}^l] \ldots \rangle$$

where $[u, v] = u^{-1}v^{-1}uv$; for, the infinitely generated commutator subgroups of the free groups $A = \langle a_1, a_2 \rangle$ and $B = \langle b_1, b_2 \rangle$ are isomorphic under the mapping $[a_1{}^k, a_2{}^l] \to [b_1{}^k, b_2{}^l]$. Hence, G contains the finitely generated free groups A and B whose intersection is the infinitely generated commutator subgroup of A.

As a second application of the amalgamated subgroup construction, we show that given an element g of a group G, G can be embedded in a group S in which g has at least one nth root for each positive integer n. For, consider the group of non-negative real numbers less than one under addition modulo 1, i.e., the factor group R/Z of the reals under addition by the subgroup of integers. If g has order r, then amalgamate g in G with $1/r$ in R/Z; if g has infinite order, then amalgamate g in G with $\sqrt{2}/2$ in R/Z. In either case, we obtain S, a free product of G and R/Z with cyclic subgroup amalgamated; moreover, g is in the subgroup R/Z. Since every element of R/Z has an nth root for each positive integer n, g has an nth root in S for each positive integer n. Since G is imbedded in S, we have our result.

As yet another application, we show that the group

$$G = \langle a_1, a_2, b_1, b_2, c_1, c_2, c_3; a_1{}^2 = b_1{}^3, b_1b_2{}^5 = c_1c_2{}^5 \rangle$$

has no element $\neq 1$ of finite order. First consider the group $D = \langle a_1, a_2, b_1, b_2; a_1{}^2 = b_1{}^3 \rangle$. This is the free product of the free groups $A = \langle a_1, a_2 \rangle$ and $B = \langle b_1, b_2 \rangle$ with amalgamated subgroups. Hence, A and B are subgroups of D. Now, $b_1b_2{}^5 \to c_1c_2{}^5$ clearly induces an isomorphism between the cyclic groups generated by $b_1b_2{}^5$ in the free group B and $c_1c_2{}^5$ in the free group $C = \langle c_1, c_2, c_3 \rangle$. Thus $b_1b_2{}^5 \to c_1c_2{}^5$ induces an isomorphism between a subgroup of D and a subgroup of C. Hence, G is the free product of D and C with amalgamated subgroups. By Corollary 4.4.5, since A and B have no elements $\neq 1$ of finite order, neither does D. Again, by Corollary 4.4.5, G can have no element $\neq 1$ of finite order, since neither D nor C does.

We shall give more significant applications of the amalgamated subgroup construction in later sections and especially in Section 4.4. Applications to other constructions are given in the problems (see Problems 22 through 30).

Problems for Section 4.2

1. Find a reduced form for each of the words a^3b^{-4}, $a^{-3}b^5$, $b^5a^2ba^3b^3a$ in the group $\langle a, b; a^4, b^6, a^2 = b^3\rangle$. (This is a presentation of the two-by-two matrices with integer entries and determinant one.)

2. Show that the group $B = \langle x, y; xyx = yxy\rangle$ is the free product, with amalgamated subgroup, of the infinite cyclic groups generated by xy and xyx; hence, conclude that the group has no elements $\neq 1$ of finite order. (This is a presentation of Artin's braid group on three strings.) Show that the center of B is a cyclic group. [*Hint:* Let $a = xy$, $b = xyx$.]

3. Find cyclic groups and their appropriate subgroups such that each of the following groups is a free product of the cyclic factors with the subgroups amalgamated:

(a) $\langle x, y; y^6, x^3 = y^3\rangle$.
(b) $\langle x, y; x^4, y^6, x^3 = y^3\rangle$.
(c) $\langle x, y; x^5, y^{10}, x^3 = y^2\rangle$.
(d) $\langle x, y; x^5, y^6, x^3 = y^3\rangle$.
(e) $\langle x, y; x^{30}, y^{70}, x^3 = y^5\rangle$.

4. Show that the group

$$G = \langle x, y, z; x^2 = y^3, y^5 = z^7\rangle$$

can be obtained by first taking the free product A of the infinite cyclic groups generated by x and y amalgamating the subgroups generated by x^2 and y^3, and then taking the free product of A with the infinite cyclic group generated by z, amalgamating the subgroups generated by y^5 and z^7.

5. Find the center of each of the following groups:

(a) $\langle x, y, z; x^4 = y^3, y^2 = z^5\rangle$.
(b) $\langle x, y, z; x^4 = y^3, y^6 = z^5\rangle$.
(c) $\langle x, y, z; x^4 = y^{10}, y^{15} = z^5\rangle$.
(d) $\langle x, y, z, w; xy = yx, x^6 = z^3, x^4 = w^5\rangle$.
(e) $\langle x, y, z, w; xy = yx, zw = wz, x^3y = z^3w\rangle$.
(f) $\langle x, y, z, w; xy = yx, zw = wz, x^3 = z^3, y^3 = w^3\rangle$.
(g) $\langle x, y, z, w; x^4, y^2, xy = yx^3, z^4, w^4, zw = wz, x^2 = z^2w^2\rangle$.
(h) $\langle x, y, z, w; x^4, y^2, xy = yx^3, z^4, w^4, zw = wz, x^2y = z^2w^2\rangle$.

6. Show that $G = *(A, B, H, K, \varphi)$ is Abelian if and only if A is Abelian, B is Abelian, and $A = H$ or $B = K$. [*Hint:* Use the corollary to Theorem 4.5.]

7. Show that $G = *(A, B, H, K, \varphi)$ is finite if and only if A is finite, B is finite, and $A = H$ or $B = K$. [*Hint:* Show that if $a \in A$, $a \notin H$ and $b \in B$, $b \notin K$, then ab has infinite order in G.]

8. Let A be the free group on a, b, c, d, e, and let X be the free group on x, y, z, w, t. Show that $a, b, c, ede^{-1}, e^2de^{-2}, e^3de^{-3}$ freely generate a free group, and similarly for $x, y, z, twt^{-1}, t^2wt^{-2}, t^3wt^{-3}$. Show that the group

$$G = \langle a, b, c, d, e, x, y, z, w, t; a = twt^{-1}, b = t^2wt^{-2},$$
$$c = t^3wt^{-3}, x = ede^{-1}, y = e^2de^{-2}, z = e^3de^{-3}\rangle$$

is the free product of two free groups each of rank five with free subgroups of rank six amalgamated. Show that G is the free group of rank four. Conclude that the rank of a free product with amalgamated subgroups may be less than the rank of either factor.

9. Show how to obtain a free group of rank four or more as a free product of two infinitely generated free groups with amalgamated subgroups. [*Hint:* If a free group of rank r is desired, let $x_1, x_2, \ldots$ and $y_1, y_2, \ldots$ be the free generators of two free groups, and amalgamate $y_3, y_4, \ldots$ with the commutator subgroup of $x_1, x_2, \ldots, x_{r-2}$ and amalgamate $x_{r-1}, x_r, \ldots$ with the commutator subgroup of y_1, y_2.]

10. Show that if H does not contain the verbal subgroup $A(X^2)$ and K does not contain the verbal subgroup $B(X^2)$, then $G = *(A, B, H, K, \varphi)$ contains a free subgroup of rank two. Conclude that in this case G does not satisfy any identical relations. [*Hint:* Suppose $a^2 \notin H$ and $b^2 \notin K$. Then show that $(aba)^n$ starts and ends with a coset representative from A, and that $(bab)^n$ starts and ends with a coset representative from B, where $n \neq 0$. Hence, show that aba and bab freely generate a free subgroup of G. Then use that a free subgroup of rank two contains free subgroups of any finite rank.]

12. Show that if $A = \langle a; a^4\rangle$, $B = \langle b; b^4\rangle$ and H and K are generated by a^2 and b^2, respectively, then $G = \langle a, b; a^4, b^4, a^2 = b^2\rangle$ satisfies the identical relation $((X_1, X_2), (X_3, X_4)) = 1$, where $(u, v) = u^{-1}v^{-1}uv$. [*Hint:* Show that the commutator subgroup of G is generated by $aba^{-1}b^{-1}$ and so is cyclic.]

13. Show that if g_1, g_2 are both in A or both in B and ρ is the reduction process for $*(A, B, H, K, \varphi)$ given in Theorem 4.4, then

$$\rho(g_1, g_2, g_3, \ldots, g_n) = \rho(g_1 \cdot g_2, g_3, \ldots, g_n).$$

[*Hint:* Suppose $\rho(g_3, \ldots, g_n) = (p, c_1, \ldots, c_r)$ and then compute $\rho(g_1, g_2, p, c_1, \ldots, c_r)$ and $\rho(g_1 \cdot g_2, p, c_1, \ldots, c_r)$ from the definition of ρ.]

14. Show that if g_s, g_{s+1} are both in A or both in B and ρ is the reduction process in Theorem 4.4, then

$$\rho(g_1, \ldots, g_s, g_{s+1}, \ldots, g_n) = \rho(g_1, \ldots, g_s \cdot g_{s+1}, \ldots, g_n).$$

[*Hint:* Use property (ix) of ρ and Problem 13.]

15. Show: If $A = \langle a, b, c; R(a, b, c)\rangle$, then the subgroup of

$$G = \langle x, b, c; R(x^s, b, c)\rangle,$$

(where $s \geq 1$ is an integer) which is generated by x^s, b, c, is isomorphic to A under the mapping $a \to x^s$, $b \to b$, $c \to c$. [*Hint:* Suppose a has order n in A. (If a is of infinite order, we put $n = 0$.) Let X be the group generated by x and defined by $x^{ns} = 1$. Then the subgroup H of A generated by a is isomorphic to the subgroup K of X generated by x^s under the mapping $a \to x^s$. Therefore, the group

$$G^* = \langle a, b, c, x; R(a, b, c), ax^{-s}\rangle$$

contains an isomorphic replica of A which is generated by a, b, c, since G^* is the free product of A and X with H and K amalgamated. A Tietze transformation shows that $G^* \simeq G$.]

16. Show that every non-cyclic finitely generated group G with a single defining relator can be imbedded in a group with the same number of generators and a single defining relator which has zero exponent sum on one of the generators of the supergroup, and in which each of the other generators occur no more often than in the original relator. [*Hint:* Since the group is non-cyclic, it has at least two generators. Suppose, for example, the group is generated by a, b, c and has the single defining relator $R(a, b, c)$; thus

$$G = \langle a, b, c; R(a, b, c)\rangle.$$

If R has zero exponent on a, b, or c, we take G itself as the supergroup. Otherwise, let r and s be the exponent sum of $R(a, b, c)$ on a and b, respectively. Then, by Problem 15, G is imbedded in the group

$$G' = \langle x, b, c; R(x^s, b, c)\rangle.$$

We apply a Tietze transformation to this group to introduce a new generator y so that $b = yx^{-r}$, i.e., $y = bx^r$. Hence, G' can be presented

$$\langle x, b, c, y; R(x^s, b, c), y = bx^r\rangle;$$

applying a Tietze transformation to eliminate b, we obtain

$$G' = \langle x, c, y; R(x^s, yx^{-r}, c)\rangle.$$

Now the exponent sum on x of $R(x^s, yx^{-r}, c)$ is zero; moreover, G' has the same number of occurrences of the generators other than x in its relator as originally (y replaces b), the same number of generators as G, and a single defining relator.]

17. Let $R(a, b)$ have non-zero exponent sum on a and b. Show that the subgroup of

$$G_n = \langle a, b_1, b_2, \ldots, b_n; R(a, b_1), R(a, b_2), \ldots, R(a, b_n)\rangle$$

generated by a and $b_1, b_2, \ldots, b_k (k \leq n)$ is

$$G_k = \langle a, b_1, b_2, \ldots, b_k; R(a, b_1), R(a, b_2), \ldots, R(a, b_k)\rangle.$$

[*Hint:* Show that a has infinite order in $\langle a, b; R(a, b)\rangle$. Form the amalgamated product of

$$G_1 = \langle a, b_1; R(a, b_1)\rangle \quad \text{and} \quad \langle a_2, b_2; R(a_2, b_2)\rangle$$

amalgamating a and a_2; G_2 is obtained. Then form the amalgamated product of G_2 with $\langle a_3, b_3; R(a_3, b_3)\rangle$ amalgamating a and a_3; G_3 is obtained. Continue in this way.]

18. Let $R(a, b)$ have non-zero exponent sum on a and b. Show that if

$$G = \langle x_1, x_2, \ldots, x_n; R(x_1, x_2), R(x_2, x_3), \ldots, R(x_{n-1}, x_n)\rangle$$

then the subgroup of G generated by $x_1, x_2, \ldots, x_k$ $(k \leq n)$ is isomorphic to

$$\langle x_1, x_2, \ldots, x_k; R(x_1, x_2), R(x_2, x_3), \ldots, R(x_{k-1}, x_k)\rangle$$

[*Hint:* Show that x_2, x_2', and x_3 in the groups $\langle x_1, x_2; R(x_1, x_2)\rangle$ and $\langle x_2', x_3; R(x_2', x_3)\rangle$ have infinite order. Form the free product amalgamating x_2 and x_2'. Next, take the free product of the resulting group with $\langle x_3', x_4; R(x_3', x_4)\rangle$, amalgamating x_3 and x_3'. Continue in this way until G is obtained.]

19. Let $R(a, b)$ have non-zero exponent sum on a and b. Show that the subgroup of

$$G = \langle x_1, x_2, \ldots, x_n, \ldots; R(x_1, x_2), R(x_2, x_3), \ldots, R(x_n, x_{n+1}), \ldots\rangle$$

generated by x_1, x_2 is isomorphic to $\langle a, b; R(a, b)\rangle$. [*Hint:* Let G_n be the group

$$\langle x_1, x_2, \ldots, x_n; R(x_1, x_2), \ldots, R(x_{n-1}, x_n)\rangle.$$

Show that the subgroup of G_n generated by $x_1, x_2, \ldots, x_{n-1}$ is just G_{n-1}, by Problem 18. Then show that G is the union of $G_1, G_2, \ldots, G_n, \ldots$ by means of Problem 18 in Section 1.3.]

20. Let $R(a, b)$ have non-zero exponent sum on a and b. Show that the subgroup of

$$G = \langle \ldots, x_{-2}, x_{-1}, x_0, x_1, x_2, \ldots; \ldots, R(x_{-2}, x_{-1}),$$
$$R(x_{-1}, x_0), R(x_0, x_1), R(x_1, x_2), \ldots\rangle$$

generated by $x_{-k}, x_{-k+1}, \ldots, x_{-1}, x_0, x_1, \ldots, x_{k-1}, x_k$ is the group

$$G_k = \langle x_{-k}, x_{-k+1}, \ldots, x_{-1}, x_0, x_1, \ldots, x_{k-1}, x_k;$$
$$R(x_{-k}, x_{-k+1}), \ldots, R(x_{-1}, x_0), R(x_0, x_1), \ldots, R(x_{k-1}, x_k)\rangle$$

[*Hint:* First take the amalgamated product of

$$\langle x_0, x_1; R(x_0, x_1)\rangle \quad \text{and} \quad \langle x_{-1}, x_0'; R(x_{-1}, x_0')\rangle$$

amalgamating x_0 and x_0'; G_1 is thus obtained. Next take the amalgamated product of G_1 and $\langle x_1', x_2; R(x_1', x_2)\rangle$; the free product of the resulting group and the group $\langle x_{-2}, x_{-1}'; R(x_{-2}, x_{-1}')\rangle$, amalgamating x_{-1}' and x_{-1} yields G_2. Continuing in this way, we obtain G_{n-1} is a subgroup of G_n. Finally, G is the union of $G_0, G_1, G_2, \ldots, G_n, \ldots$ by Problem 1.3.18.]

21. Show that the elements of finite order in

$$G = \langle a, b; a^{10}, b^{15}, a^2b^5 = b^5a^2\rangle$$

are conjugates of a^r, b^s, or $a^{2p}b^{5q}$ where $0 \leq r \leq 9$, $1 \leq s \leq 14$, $0 \leq p \leq 4$, $0 \leq q \leq 2$. [*Hint:* Let $C = \langle x, y; x^5, y^3, xy = yx\rangle$, i.e., C is what will play the role of the subgroup of G generated by a^2 and b^5. Let $A = \langle a; a^{10}\rangle$, $B = \langle b; b^{15}\rangle$. First form D, the amalgamated product of A and C amalgamating a^2 with x; thus $D = \langle a, x, y; a^{10}, x^5, y^3, xy = yx, x = a^2\rangle$. The subgroup generated by y is, of course, cyclic of order three. Applying a Tietze transformation to eliminate x, we have $D = \langle a, y; a^{10}, y^3, a^2y = ya^2\rangle$ and that the subgroup generated by y is still cyclic of order three. Hence, we may form the amalgamated product of D with B, amalgamating y and b^5, obtaining $\langle a, b, y; a^{10}, b^{15}, y^3, a^2y = ya^2, y = b^5\rangle$; upon applying a Tietze transformation to eliminate y we have $\langle a, b; a^{10}, b^{15}, a^2b^5 = b^5a^2\rangle$ which is G. Since G is obtained by taking the amalgamated product of D and B, by Corollary 4.4.5, its elements of finite order are conjugates of the elements of finite order in D and in B. Since D is itself an amalgamated product of A and G, the elements of finite order in D are conjugates of the elements of finite order in A and in C. Hence, we have the result.]

22. If in the group

$$G = \langle a_1, \ldots, a_n, b_1, \ldots, b_m; R_1(a_\nu), \ldots, R_r(a_\nu), S_1(b_\mu), \ldots, S_s(b_\mu),$$
$$U_i(a_\nu)V_j(b_\mu) = V_j(b_\mu)U_i(a_\nu)\rangle, \ (i = 1, \ldots, p; j = 1, \ldots, q)$$

the subgroup generated by $a_1, \ldots, a_n$ is A, the subgroup generated by $b_1, \ldots, b_m$ is B, the subgroup generated by $U_1(a_\nu), \ldots, U_p(a_\nu)$ is H, and the subgroup generated by $V_1(b_\mu), \ldots, V_q(b_\mu)$ is K, then G is called the *free product of A and B with commuting subgroups H and K*. Show that A has the presentation $\langle a_1, \ldots, a_n; R_1(a_\nu), \ldots, R_r(a_\nu)\rangle$, that B has the presentation $\langle b_1, \ldots, b_m; S_1(b_\mu), \ldots, S_s(b_\mu)\rangle$ and that $A \cap B = 1$. [*Hint:* Let

$$A' = \langle a_1', \ldots, a_n'; R_1(a_\nu'), \ldots, R_r(a_\nu')\rangle,$$
$$B' = \langle b_1', \ldots, b_m'; S_1(b_\mu'), \ldots, S_s(b_\mu')\rangle,$$

and map A' into G by sending a_ν' into a_ν. Then φ maps A' onto A. Let ψ map G into $A' \times B'$ by sending $a_\nu \to a_\nu'$ and $b_\mu \to b_\mu'$. Then $\psi\varphi$ is the identity of A' onto A', hence, φ is one-to-one. Thus A' is isomorphic to A. Similarly, B' is isomorphic to B. Since ψ is one-to-one on A and B in G, if $A \cap B \neq 1$ in G then $A' \cap B' \neq 1$ in $A' \times B'$; thus $A \cap B = 1$.]

23. Let G, A, B, H, K be as in Problem 22. Show that the elements of finite order in G are conjugates of elements of finite order in A, elements of finite order in B, or products of an element of finite order in H and an element of finite order in K. [*Hint:* Let $C = H \times K$ where

$$H = \langle u_1, \ldots, u_p; X_1(u_i), \ldots, X_x(u_i)\rangle$$

and

$$K = \langle v_1, \ldots, v_q; Y_1(v_j), \ldots, Y_y(v_j)\rangle.$$

Forming the amalgamated product D of A and C, amalgamating $U_i(a_\nu)$ and u_i, we obtain

$$\begin{aligned} D = \langle a_1, \ldots, a_n, u_1, \ldots, u_p, v_1, \ldots, v_q; R_1(a_\nu), \ldots, R_r(a_\nu), \\ X_1(u_i), \ldots, X_x(u_i), Y_1(v_j), \ldots, Y_y(v_j), \\ u_iv_j = v_ju_i, u_i = U_i(a_\nu)\rangle, \qquad (i = 1, \ldots, p; j = 1, \ldots, q). \end{aligned}$$

Applying a Tietze transformation to eliminate each u_i, and realizing that $X_k(U_i(a_\nu))$ is derivable from $R_1(a_\nu), \ldots, R_r(a_\nu)$, we have

$$\begin{aligned} D = \langle a_1, \ldots, a_n, v_1, \ldots, v_q; R_1(a_\nu), \ldots, R_r(a_\nu), Y_1(v_j), \ldots, Y_y(v_j), \\ U_i(a_\nu)v_j = v_jU_i(a_\nu)\rangle, \qquad (i = 1, \ldots, p; j = 1, \ldots, q). \end{aligned}$$

Now forming the amalgamated product of D and B, amalgamating v_j and $V_j(b_\mu)$, and then performing the necessary Tietze transformations we obtain G. Hence, the elements of finite order in G are conjugates of the elements of finite order in A or in B or in $H \times K$; our result then follows.]

24. Let G, A, B, H, K be as in Problem 22, and suppose further that $A \neq H$ and $B \neq K$. Show that G has only a trivial center. [*Hint:* Use Corollary 4.5 and the construction of G as a repeated amalgamated product given in the hint to Problem 23, and also use Problem 22.]

25. Let G, A, B, H, K be as in Problem 22 and suppose further that $A = H$ but $B \neq K$. Show that the center of G is $K \cap$ center (B). [*Hint:* See the hint of Problem 24.]

26. Show that the elements of finite order in $\langle a, b; a^{10}, b^{15}, ab^5 = b^5a, a^2b = ba^2\rangle$ are conjugates of a^rb^{5q} or $a^{2p}b^s$, where $0 \leq r \leq 9$, $0 \leq q \leq 2$, $0 \leq p \leq 4$, $0 \leq s \leq 14$. [*Hint:* Let $C = \langle a, x; a^{10}, x^3, ax = xa\rangle$ and let $D = \langle y, b; y^5, b^{15}, yb = by\rangle$. The subgroup of C generated by a^2 and x is isomorphic to the subgroup of D generated by y and b^5. Hence, we may form the amalgamated product of C and D, amalgamating a^2 with y, and x with b^5. The group G results. The elements of finite order in G are just conjugates of elements of finite order in C or in D; thus, our result follows.]

27. If in the group

$$\begin{aligned} G = \langle a_1, \ldots, a_n, b_1, \ldots, b_m; K_1(a_\nu), \ldots, K_r(a_\nu), S_1(b_\mu), \ldots, S_s(b_\mu), \\ a_\nu V_j(b_\mu) = V_j(b_\mu)a_\nu, U_i(a_\nu)b_\mu = b_\mu U_i(a_\nu)\rangle, \\ (\nu = 1, \ldots, n; \mu = 1, \ldots, m; i = 1, \ldots, p; j = 1, \ldots, q), \end{aligned}$$

the subgroup generated by $a_1, \ldots, a_n$ is A, the subgroup generated by $b_1, \ldots, b_m$ is B, the subgroup generated by $U_1(a_\nu), \ldots, U_p(a_\nu)$ is H, and the subgroup generated by $V_1(b_\mu), \ldots, V_q(b_\mu)$ is K, then G is called the *free product of A and B with centralized subgroups H and K*. Show that A has the presentation $\langle a_1, \ldots, a_n; R_1(a_\nu), \ldots, R_r(a_\nu)\rangle$, that B has the presentation

$\langle b_1, \ldots, b_m; S_1(b_\mu), \ldots, S_s(b_\mu)\rangle$ and that $A \cap B = 1$. [*Hint:* See the hint to Problem 22.]

28. Let G, A, B, H, K be as in Problem 27. Show that any element of finite order in G is the conjugate of either the product of an element of finite order in A and an element of finite order in K, or the product of an element of finite order in H and an element of finite order in B. [*Hint:* Let $C = A \times K$ and $D = H \times B$. Show that G is the amalgamated product of C and D amalgamating $H \times K$. Then use Corollary 4.4.5.]

29. Let G, A, B, H, K be as in Problem 27. Show that the center of G is the direct product of center $(A) \cap H$ and center $(B) \cap K$. [*Hint:* Use the construction of G as given in the hint to Problem 28, Corollary 4.5, and $(P \times Q) \cap (M \times N) = (P \cap M) \times (Q \cap N)$.]

30. Find the center of the group in Problem 26. [*Hint:* Use Problem 29.]

31. Let G be the free product of A and B with the subgroups H and K amalgamated. Suppose that M and N are normal subgroups of A and B, respectively, such that $H \cap M = 1 = K \cap N$. If P is the normal subgroup of G generated by M and N show that G/P is the free product of A/M and B/N with the isomorphic subgroups HM/M and KN/N amalgamated. [*Hint:* Use the presentation of G and add on as relators words that generate M and N; since $H \cap M = 1 = K \cap N$, $HM/M \simeq H$ and $KN/N \simeq K$. If $U_i(a_\nu)$ generates H in G, then $U_i(a_\nu)$ generates HM/M in G/P. Similarly, if $V_j(b_\mu)$ generates K in G, $V_j(b_\mu)$ generates KN/N in G/P. Moreover, if $U_i(a_\nu)$ into $V_i(b_\mu)$ induces an isomorphism between H and K, it also induces an isomorphism between HM/M and KN/N.]

32. Show that if $G = *(A, B, H, K, \varphi)$ and right coset representatives for $A \bmod H$ and $B \bmod K$ are chosen, then a right coset representative system for $G \bmod H$ is given by the reduced forms in G with $h = 1$. [*Hint:* For, if $g = hg_1 \ldots g_r$ is the reduced form of an element in G, then $g_1 \ldots g_r$ is in the same coset of H as g. Moreover, any element in the same coset of H as $g_1 \ldots g_r$ has a reduced form $hg_1 \ldots g_r$.]

33. Show that if G is the free product of A and B with centralized subgroups H and K (see Problem 27), then every element in G has a unique representation $hkg_1 \ldots g_r$ where $h \in H$, $k \in K$, $g_1, \ldots, g_r$ is each a member of a given right coset representative system for $A \bmod H$ or a given right coset representative system for $B \bmod K$, $g_i \neq 1$, and g_i, g_{i+1} are not both in A or both in B. [*Hint:* Use the construction of G as the amalgamated product of $A \times K$ and $H \times B$, with $H \times K$ amalgamated; also use that a representative system of $A \times K \bmod H \times K$ is given by a representative system for $A \bmod H$; and similarly for $H \times B \bmod H \times K$.]

34. Show that if H is a subgroup of A, and G is the free product of A and $H \times K$, with H amalgamated, then the right coset representative system for $G \bmod K$ is given by elements of the form

$$hg_1k_1 \ldots g_tk_t$$

where $h \in H$, $k_i \in K$, $g_1, \ldots, g_r$ is each a member of a given right coset representative system for $A \bmod H$, $g_i \neq 1$. [*Hint:* Since K is a representative system for $H \times K \bmod H$, every element of G can be written uniquely in the form

$$hkg_1k_1 \ldots g_tk_t$$

where $k_i \neq 1$ if $i \neq t$. Since H and K commute elementwise, this element is in the same coset of $G \bmod K$ as $hg_1k_1 \ldots g_tk_t$. Moreover, any element in the same coset of $G \bmod K$ as $hg_1k_1 \ldots g_tk_t$ has the form $khg_1k_1 \ldots g_tk_t = hkg_1k_1 \ldots g_tk_t$. Hence, we have our result.]

35. Show that if G is the free product of A and B with commuting subgroups H and K (see Problem 22) then every element of G can be written uniquely in the form

$$ke_1f_1e_2f_2 \ldots e_zf_z$$

where $k \in K$, e_i is an element of the form

$$hg_1k_1 \ldots g_tk_t$$

as given in Problem 34, and f_i is a member of a given right coset representative system for $B \bmod K$ and e_i and f_j are different from 1, if $i \neq 1$ or $j \neq z$. [*Hint:* Use the construction of G as given in the hint of Problem 23, and use Problem 34.]

36. Let us define two new products for groups A and B, whose centers are H and K respectively, as follows: $A \odot B$ is the free product of A and B with the commuting subgroups H and K; $A \otimes B$ is the free product of A and B with the centralized subgroups H and K.

(a) Show that $A \odot B = B \odot A$ and $A \otimes B = B \otimes A$.
(b) Show that if A and B are Abelian then $A \odot B$ and $A \otimes B$ are just the direct product $A \times B$.
(c) Show that if A and B have trivial centers then $A \odot B$ and $A \otimes B$ are just the free product $A * B$.
(d) Show that if A is Abelian but B has a trivial center, then $A \odot B$ is just $A * B$ and $A \otimes B$ is just $A \times B$.
(e) Show that if A and B are Abelian but C has a trivial center then $(A \odot B) \odot C = (A \times B) * C$ but $A \odot (B \odot C) = A * (B * C)$. Hence, show that if A is a cyclic group of order two, B a cyclic group of order five, and C the symmetric group on three objects, then $(A \odot B) \odot C$ has elements of order ten whereas $A \odot (B \odot C)$ has no such elements. Conclude that $\odot$ is not asociative.
(f) Show that the center of $A \otimes B$ is generated by the center of A and B.
(g) Use (f) to show that $(A \otimes B) \otimes C = A \otimes (B \otimes C)$.
(h) Show that if A and B are both symmetric groups on three objects then $A \otimes B$ is $A * B$ and has no elements of order six. On the other hand, if C is a subgroup of A generated by a three-cycle and D is a subgroup of B generated by a two-cycle then $C \otimes D$ is $C \times D$ and has elements of order six. Conclude that if C is a subgroup of A and D is a subgroup of B then $C \otimes D$ need not be a subgroup of $A \otimes B$.

37. Let

$$A = \langle a, c; a^3, c^2, (ac)^2 \rangle$$

and

$$B = \langle b, d; b^3, d^2, (bd)^2 \rangle$$

be the symmetric group on three objects; let G be the free product of A and B with the cyclic group generated by c and d amalgamated. Show that ab^2 and a^2b are cyclically reduced elements in G which are conjugate, though neither is a cyclic permutation of the other. [*Hint:* Conjugate ab^2 by c.]

38. (a) Show that two cyclically reduced conjugate elements of $G = *(A, B, H, K, \varphi)$ may have different representative lengths.

(b) Show that if a cyclically reduced element of G has representative length $r \geq 2$ then any cyclically reduced conjugate of it has representative length r.

[*Hint:* For (a) take h and ghg^{-1} where $h \in H$, $g \in A$ but $ghg^{-1} \notin H$. For (b) use case (iii) of Theorem 4.6.]

39. Find all cyclically reduced elements conjugate to the given element in the given group:

(a) $x^2y^2xyxy^2x$ in $\langle x, y; x^2 = y^3 \rangle$
(b) $xyz^3y^{-1}x^3$ in $\langle x, y, z; x^4, xy = yx, y^2 = z^2 \rangle$
(c) $xyxy^2$ in $\langle x, y; x^4, y^6, x^2 = y^3 \rangle$.

[*Hint:* Use Theorem 4.6.]

40. Let $G = *(A, B, H, K, \varphi)$.

(a) Show that the h_i in the sequence described in case (i) of Theorem 4.6 can be chosen so that no term is repeated in the sequence.

(b) Show that if H is finite, there are only finitely many sequences $h_1, \ldots, h_t$ of distinct elements of H such that consecutive terms are conjugate in some factor.

[*Hint:* For (a), if $h_j = h_s$, $j < s$ then $h, h_1, \ldots, h_j, h_{s+1}, \ldots, g$ is a shorter sequence of the required type. Hence, any sequence of the required type with fewest terms in it has distinct terms. For (b), there are only finitely many sequences of distinct elements of H.]

41. Let $G = *(A, B, H, K, \varphi)$ with H finite and suppose that the transformation problems for A and B can be constructively solved, and also that elements of G can be constructively put into reduced form. Show that the transformation problem for G can be solved constructively. [*Hint:* Given g and g', elements of G, put them in reduced form and then cyclically reduce them. Assume then that g and g' are cyclically reduced. If g' has representative length zero, i.e., $g' \in H$, g must have representative length zero or one to be conjugate to g'. Assume that g has representative length zero or one. Form all sequences $g', h_1, h_2, \ldots, h_t$ such that $h_i \in H$, consecutive terms are conjugate in some factor, and all terms are distinct. There are only finitely many since H is finite. For each such sequence, check whether h_t and g are conjugate in some factor. By Case (i) of Theorem 4.6 if g is conjugate

to g' there must be such an h_t; clearly if there is such an h_t then g is conjugate to g'. If g' has representative length one, then if g' is conjugate to an element of H, it must be conjugate in its own factor to some element of H, by Case (i) of Theorem 4.6. This can be checked. If g' is conjugate to some element in H replace g' by this element and repeat the preceding paragraph. If g' is not in a conjugate of H then g and g' must be in the same factor and conjugate in that factor, in order to be conjugate in G. Finally, if g' has reduced form $hg_1 \ldots g_r$, then cyclically permuting $hg_1, \ldots, g_r$ and then conjugating the results by the elements of H will yield finitely many cyclically reduced elements that can be conjugate to g'. Check g against these.

41. Let

$$A = \langle h, a_1, a_2, \ldots, a_n; h^n, ha_1 = a_2h, \ldots, ha_n = a_1h\rangle$$
$$B = \langle k, b_1, b_2, \ldots, b_n; k^n, kb_1 = b_2k, \ldots, kb_n = b_1k\rangle$$

and let G be the free product of A and B with h and k amalgamated. Find the cyclically reduced elements conjugate to a_1b_1. [*Hint:* First show that neither a_i nor b_i is in H, by adjoining h as a relator. Then use Case (iii) of Theorem 4.6.]

42. Let A be the group $\langle h, a; a^2, ah = h^{-1}a\rangle$, B be the group $\langle k, b; b^2, bk = k^{-1}b\rangle$, and let G be the free product of A and B amalgamating h and k. Solve the transformation problem for G as follows:

(a) Show that H is a normal subgroup of A and that every element of A can be written uniquely as $h^n a^\alpha$ where n is an integer and $\alpha = 0$ or 1.
(b) Show that every cyclically reduced element of G has the form h^n, $h^n a$, $h^n b$, $h^n(ab)^m$, or $h^n(ba)^m$ where n is an integer and m is a positive integer.
(c) Show that the conjugates in G of h^n are h^n and h^{-n}.
(d) Show that the cyclically reduced conjugates of $h^n a$ in G are just its conjugates in A, i.e., $h^{n+2s}a$ where s is any integer. Similarly for $h^n b$.
(e) Show that the cyclically reduced conjugates of $h^n(ab)^m$ in G are its cyclic permutations $h^n(ab)^m$ and $h^{-n}(ba)^m$. Similarly for $h^n(ba)^m$.

[*Hint:* Use Theorem 4.6.]

43. If

$$A = \langle a_1, a_2, a_3, a_4, a_5, a_6; a_1{}^2 = a_3a_2{}^2a_3{}^{-1}, a_4{}^2 = a_6a_5{}^2a_6{}^{-1}\rangle$$

and

$$B = \langle b_1, b_2, b_3, b_4; b_2{}^2 = b_1b_4{}^2b_1{}^{-1}\rangle,$$

then show the following:

(a) A is the free product of $\langle a_1, a_2, a_3; a_1{}^2 = a_3a_2{}^2a_3{}^{-1}\rangle$ and $\langle a_4, a_5, a_6; a_4{}^2 = a_6a_5{}^2a_6{}^{-1}\rangle$.
(b) The group $\langle x, y, z; x^2 = yz^2y^{-1}\rangle$ is the same as the group $\langle x, y, w; x^2 = w^2\rangle$ where $w = yzy^{-1}$. Hence, x and $z = y^{-1}wy$ freely generate a free group.

(c) The elements b_2 and b_4 freely generate a free group in B, and the elements a_2 and a_4 freely generate a free group of A.
(d) The elements a_1^2 and a_5^2 of A are not conjugate in A.
(e) In the group G obtained by taking the free product of A and B amalgamating a_2, a_4 with b_2, b_4 the elements a_1^2 and a_5^2 are conjugate.
(f) Show that the hypothesis g' is not in a conjugate of H is necessary for Case (ii) of Theorem 4.6.

44. Let $G = *(A, B, H, K, \varphi)$ and suppose $A \neq H$ and $B \neq K$. Then G is not the union of the conjugates of A and the conjugates of B. [*Hint:* Let $a \in A$, $a \notin H$, $b \in B$, $b \notin K$. Then ab is cyclically reduced and not in a conjugate of A or B by Theorem 4.6. For, by (i) and (ii) of Theorem 4.6, if a cyclically reduced element g is conjugate to an element in a factor, then g must have representative length zero or one.]

45. Let $G = *(A, B, H, K, \varphi)$. Then $A \cap WBW^{-1}$ is a subgroup of a conjugate of H. [*Hint:* For, if W is in B, then $A \cap WBW^{-1} = H$. Otherwise, $W = Ub$ where U ends in a syllable from A. Then $WBW^{-1} = UBU^{-1}$. If $c \in B$ but $c \notin H$ then UcU^{-1} will have representative length greater than one and so cannot be in A. Hence, the only elements of $WBW^{-1} = UBU^{-1}$ that can be in A are in UHU^{-1}. Thus $A \cap WBW^{-1} \subset UHU^{-1}$.]

46. Let $G = \langle a, b, h, k; hah^{-1} = a^2, kbk^{-1} = b^2, h^2 = k^2\rangle$. Suppose $A = \langle a, h; hah^{-1} = a^2\rangle$, $B = \langle b, k; kbk^{-1} = b^2\rangle$, H is the cyclic subgroup of A generated by h^2 and K is the cyclic subgroup of B generated by k^2. Then A is isomorphic to B under $a \to b$, $h \to k$, and so in particular H is isomorphic to K under $\varphi: h^2 \to k^2$. Thus $G = *(A, B, H, K, \varphi)$. Show that ab has infinitely many cyclically reduced elements of G which are conjugate to it. [*Hint:* Show that in A the element a has infinite order, e.g., by mapping a into the transformation $x \to x + 1$ of the rationals and mapping h into the transformation $x \to \frac{1}{2}x$ of the rationals. Moreover, this shows that the powers of a define different cosets of H. Thus the powers of a and of b may be used as part of a representative system for A mod H and B mod K, respectively. Then ab is conjugate to $h^{2n}abh^{-2n} = a^{4n} \cdot b^{4n}$.]

47. (a) Show that

$$G_n = \langle a, b, c, d; ab = cd, a^2b^2 = c^2d^2, \ldots, a^nb^n = c^nd^n\rangle$$

is a free product of the free groups

$$A = \langle a, b\rangle, \qquad C = \langle c, d\rangle$$

with the free subgroups H_n and K_n, freely generated by a^ib^i and c^id^i, $i = 1, 2, \ldots, n$, respectively, amalgamated.
(b) Show that $a^{n+1}b^{n+1} \neq c^{n+1}d^{n+1}$ in G_n.
(c) Show that

$$G = \langle a, b, c, d; ab = cd, \ldots, a^ib^i = c^id^i, \ldots\rangle, \qquad i = 1, 2, 3, \ldots,$$

cannot be finitely related.

[*Hint:* For (a), use Problem 1.4.12. For (b), use Problem 1.4.12 and Corollary 4.4.2. For (c), use Problem 1.1.14 and (b).]

48. Let $G = *(A, B, H, K, \varphi)$. Show that any homomorphisms α of A into a group L and β of B into a group L, such that $\alpha = \beta\varphi$ on H, can be extended to a homomorphism γ of G into L.

49. A group G is *strongly indecomposable* if $G = *(A, B, H, K, \varphi)$ implies $A = H$ or $B = K$. Show that the following groups are strongly indecomposable:

(a) a finite group;
(b) an Abelian group;
(c) a group in which all elements have finite order.

4.3. Subgroup Theorems for Free and Amalgamated Products

In this section we shall examine the structure of the subgroups of a free product. To determine this structure, we find it convenient to review the notion of a rewriting process and its use in presenting a subgroup.

Let G be a group and suppose that G has the presentation

$$G = \langle a_1, \ldots, a_n; R_1(a_\nu), \ldots, R_m(a_\nu)\rangle; \tag{1}$$

moreover, suppose that H is a subgroup of G generated by the elements defined in G by the words $J_i(a_\nu)$. To obtain a presentation for H on these generators, we first introduce the generating symbol s_i for $J_i(a_\nu)$. We next require a *rewriting process*, i.e., a mapping τ of each word $U(a_\nu)$ that defines an element of H into a word $V(s_i)$ so that $U(a_\nu)$ and $V(J_i(a_\nu))$ define the same element in G. (Intuitively, τ expresses elements of H as words in the $J_i(a_\nu)$.) Then as was shown in Section 2.3, H can be presented on the s_i with the following defining relations:

$$s_i = \tau(J_i(a_\nu)); \tag{2}$$

$$\tau(U(a_\nu)) = \tau(U'(a_\nu)) \tag{3}$$

where U and U' are freely equal words in a_ν which define elements of H;

$$\tau(U_1(a_\nu)U_2(a_\nu)) = \tau(U_1(a_\nu))\tau(U_2(a_\nu)) \tag{4}$$

where U_1 and U_2 define elements of H;

$$\tau(W(a_\nu)R_\mu(a_\nu)W(a_\nu)^{-1}) = 1 \tag{5}$$

where $W(a_\nu)$ is any word in a_ν and $R_\mu(a_\nu)$ is a defining relator in (1).

To simplify this bulky presentation we must choose τ in a special way.

One such method for choosing τ is to make use of a right coset representative function for G mod H, i.e., a mapping $W(a_\nu) \rightarrow \overline{W(a_\nu)}$ of words in a_ν into words in a_ν, carrying the empty word into itself, and such that

(i) $W(a_\nu)$ and $\overline{W(a_\nu)}$ define elements in the same coset of H, and

(ii) the elements defined by $\overline{W(a_\nu)}$ form a right coset representative system for G mod H. Given a right coset representative function for G mod H, the elements defined by

$$(6) \qquad Ka_\nu\overline{Ka_\nu}^{-1},$$

where K is a representative, generate H. Moreover, if s_{K,a_ν} is used as the generating symbol for (6), then τ may be defined on a word $U(a_\nu)$, which itself defines an element of H, as follows:

(iii) replace each symbol a_λ in the word $U(a_\nu)$ by the symbol $s_{K,a\lambda}$, where K is the representative of the initial segment of $U(a_\nu)$ preceding a_λ;

(iv) replace each symbol a_λ^{-1} in the word $U(a_\nu)$ by the symbol s_{K,a_λ}^{-1}, where K is the representative of the initial segment of $U(a_\nu)$ that ends with a_λ^{-1}.

A rewriting process constructed in this way from a representative function is called a Reidemeister rewriting process. Using such a process in conjunction with a Schreier representative system allows us to simplify the defining relators (2) through (5) for a presentation of H as follows:

(2) is replaced by $s_{K,a_\nu} = 1$, where Ka_ν is freely equal to $\overline{Ka_\nu}$;

(3) and (4) can be eliminated;

(5) is replaced by $\tau(KR_\mu K^{-1}) = 1$, where K is a representative. Hence, H can be presented

$$(7) \quad \langle \ldots, s_{K,a_\nu}, \ldots ; s_{L,a_\lambda}, \ldots \text{ (where } La_\lambda \approx \overline{La_\lambda}), \tau(KR_\mu K^{-1}), \ldots \rangle.$$

Now although this presentation is fairly simple, and is especially so if there are no defining relators R_μ in (1), i.e., G is a free group, it may be more complicated than necessary. Thus, for example, if $G = \langle a, b; a^3, b^2 \rangle$ and H is the commutator subgroup of G, then H is freely generated by $aba^{-1}b^{-1}$ and $a^2ba^{-2}b^{-1}$ (see Problems 4.1.24 and 4.1.34). On the other hand, the Reidemeister-Schreier process using the representatives $1, a, a^2, b, ab, a^2b$ presents H on the generators

$$1 \cdot a \cdot \overline{1a}^{-1} = aa^{-1} = 1, \qquad 1 \cdot b \cdot \overline{1b}^{-1} = bb^{-1} = 1,$$
$$a \cdot a \cdot \overline{aa}^{-1} = a^2a^{-2} = 1, \; a \cdot b \cdot \overline{ab}^{-1} = ab(ab)^{-1} = 1$$
$$a^2 \cdot a \cdot \overline{a^2a}^{-1} = a^3 = 1, \; a^2 \cdot b \cdot \overline{a^2b}^{-1} = a^2b(a^2b)^{-1} = 1$$
$$b \cdot a \cdot \overline{ba}^{-1} = (ba)(ab)^{-1} = bab^{-1}a^{-1}, \; b \cdot b \cdot \overline{bb}^{-1} = b^2 = 1$$
$$ab \cdot a \cdot \overline{aba}^{-1} = aba(a^2b)^{-1} = abab^{-1}a^{-2}, \; ab \cdot b \cdot \overline{ab^2}^{-1} = ab^2a^{-1} = 1$$
$$a^2b \cdot a \cdot \overline{a^2ba}^{-1} = a^2bab^{-1}, \; a^2b \cdot b \cdot \overline{a^2bb}^{-1} = a^2b^2a^{-2} = 1,$$

that is, on the generators $bab^{-1}a^{-1}$, $abab^{-1}a^{-2}$, and a^2bab^{-1}. These generators are not free generators, and indeed it is not obvious from them that H is a free group on two generators.

One difficulty with a Reidemeister-Schreier rewriting process is that it treats all generators for G in the same way, whereas there may be a natural way to partition the generators, e.g., if G is presented as a free product or as a free product with amalgamations. Specifically, the generators for the Reidemeister-Schreier presentation have the form $Ka\overline{Ka}^{-1}$, where the same representative system $\{K\}$ is used for each generator. We shall now modify the process to allow several representative systems.

Suppose G is presented; arbitrarily partition the generators of the presentation into disjoint classes. For definiteness, we assume in the definitions and theorems throughout this section three partitioning sets; thus the presentation of G takes the form

(8) $G = \langle a_1, \ldots, a_n, b_1, \ldots, b_m, c_1, \ldots, c_z; \ldots, R_i(a_\nu, b_\mu, c_\zeta), \ldots \rangle.$

Each a_ν is called an *α-generator*, each b_μ is called a *β-generator*, and each c_ζ is called a *γ-generator*. If H is any subgroup of G, we next choose any four (right) coset representative functions for G mod H, and refer to them as the α-, β-, γ-, and *neutral representative functions*, respectively. If W is any word in a_ν, b_μ, c_ζ then

$$^{\alpha}W, {}^{\beta}W, {}^{\gamma}W, {}^{*}W$$

denote the various representatives of W. [*Note:* $^{\alpha}W^{-1}$ will be used to denote $(^{\alpha}W)^{-1}$.]

To obtain a presentation for H we introduce the following generating symbols:

$s_{K,a}$ for $Ka \cdot {}^{\alpha}(Ka)^{-1}$, where K is any α-representative and a is any α-generator; $s_{L,b}$ for $Lb \cdot {}^{\beta}(Lb)^{-1}$, where L is any β-representative and b is any β-generator; $s_{M,c}$ for $Mc \cdot {}^{\gamma}(Mc)^{-1}$, where M is any γ-representative and c is any γ-generator; t_N for $N \cdot {}^{*}N^{-1}$, where N is any α-, β-, or γ-representative.

For example, let $G = \langle a, b \rangle$, and let a be the α-generator and b be the β-generator. Choose H to be the commutator subgroup of G. Let the α-representatives be $\{b^pa^q\}$, the β-representatives be $\{a^pb^q\}$ and let the neutral representatives be the same as the α-representatives. Then

$$s_{b^pa^q,a} \text{ corresponds to } b^pa^q \cdot a \cdot {}^{\alpha}(b^pa^{q+1})^{-1} = 1$$
$$a_{a^pb^q,b} \text{ corresponds to } a^pb^q \cdot b \cdot {}^{\beta}(a^pb^{q+1})^{-1} = 1$$
$$t_{b^pa^q} \text{ corresponds to } b^pa^q \cdot {}^{*}(b^pa^q)^{-1} = 1$$
$$t_{a^pb^q} \text{ corresponds to } a^pb^q \cdot {}^{*}(a^pb^q)^{-1} = a^pb^qa^{-p}b^{-q}.$$

Thus our new process gives us the "natural" generators for H.

We wish to use the s and t generating symbols for a rewriting process τ. Again, τ will be a "symbol by symbol replacement." Suppose then that G has the presentation (8) and H is a subgroup of G. Let U be a word in a_ν, b_μ, c_ζ which defines an element of H,

(9) $$U = x_1^{\epsilon_1} \ldots x_r^{\epsilon_r}$$

where x_i is an α-, β-, or γ-generator and ϵ_i is 1 or -1. Then to obtain $\tau(U)$, we proceed almost as in the Reidemeister-Schreier process. However, we can only use an s-symbol in which the representative and the generator are of the same type, and so to "neutralize" the effect of the type, we precede and follow the s-symbol by a t-symbol. Specifically, suppose x_i is a δ-generator (δ is α, β, or γ) and V is the initial segment of U preceding $x_i^{\epsilon_i}$; then to obtain $\tau(U)$, replace each $x_i^{\epsilon_i}$ with

(10) $$t_{\delta_V}^{-1} \cdot s_{\delta_V, x_i} \cdot t_{\delta_{(Vx_i)}}, \qquad \text{if } \epsilon_i = 1,$$

and with

(11) $$t_{\delta_V}^{-1} \cdot s_{\delta_{(Vx_i^{-1})}, x_i}^{-1} \cdot t_{\delta_{(Vx_i^{-1})}}, \qquad \text{if } \epsilon_i = -1.$$

As an illustration of this rewriting process, again suppose that $G = \langle a, b; a^3, b^2 \rangle$, H is the commutator subgroup of G, a is the α-generator, b is the β-generator, $\{b^p a^q\}$ are the α-representatives and neutral representatives, and $\{a^q b^p\}$ are β-representatives for $G \bmod H$. Then $U = aba^2b^{-1}$ defines an element of H and

$$\tau(U) = (t_1^{-1} \cdot s_{1,a} \cdot t_a) \cdot (t_a^{-1} \cdot s_{a,b} \cdot t_{ab}) \cdot (t_{ba}^{-1} \cdot s_{ba,a} \cdot t_{ba^2}) \cdot (t_{ba^2}^{-1} \cdot s_{ba^2,a} \cdot t_b) \cdot (t_b^{-1} \cdot s_{1,b}^{-1} \cdot t_1).$$

LEMMA 4.2. *Let G be given as in* (8) *and let H be any subgroup of G. Then H is generated by the elements of G defined by the words*

(12) $$Ka \cdot {}^{\alpha}(Ka)^{-1},\ Lb \cdot {}^{\beta}(Lb)^{-1},\ Mc \cdot {}^{\gamma}(Mc)^{-1},\ N \cdot {}^{*}N^{-1}$$

where K and a are α-type, L and b are β-type, M and c are γ-type representatives and generators respectively, and N is an α-, β-, or γ-representative. Moreover, the process τ given by (10) *and* (11) *is a rewriting process for $G \bmod H$ (τ is called a Kurosh rewriting process).*

PROOF. If the s- and t-symbols in (10) and (11) are replaced by those words in a_ν, b_μ, c_ζ for which they are the generating symbols, we obtain

(13) $$[({}^{\delta}V \cdot {}^{*}V^{-1})^{-1}] \cdot [{}^{\delta}V \cdot x_i \cdot {}^{\delta}(Vx_i)^{-1}] \cdot [{}^{\delta}(Vx_i) \cdot {}^{*}(Vx_i)^{-1}] \approx {}^{*}V \cdot x_i \cdot {}^{*}(Vx_i)^{-1}$$

and

$$(14)\quad [({}^{\delta}V \cdot {}^{*}V^{-1})^{-1}] \cdot [{}^{\delta}(Vx_i^{-1}) \cdot x_i \cdot ({}^{\delta}V)^{-1}]^{-1} \cdot [{}^{\delta}(Vx_i^{-1}) \cdot {}^{*}(Vx_i^{-1})^{-1}]$$
$$\approx {}^{*}V \cdot x_i^{-1} \cdot {}^{*}(Vx_i^{-1})^{-1} = [{}^{*}(Vx_i^{-1}) \cdot x_i \cdot {}^{*}V^{-1}]^{-1},$$

respectively. Now in applying the Reidemeister rewriting process for G mod H which employs the neutral representative system, the symbol $x_i^{\epsilon_i}$ of U is replaced by

$$(15)\qquad s_{{}^{*}V,x_i}, \qquad \text{if } \epsilon_i = 1$$

and

$$(16)\qquad s^{-1}_{{}^{*}(Vx_i^{-1}),x_i}, \qquad \text{if } \epsilon_i = -1.$$

These s-symbols are just the generating symbols for (13) and (14) respectively, in the neutral representative Reidemeister rewriting process for G mod H. Hence, when the s- and t-symbols in $\tau(U)$ are replaced by the appropriate words from (12), a word freely equal to U results. Thus τ is a rewriting process and the words in (12) define generators for H. ◀

If τ is a Kurosh rewriting process, it follows that a presentation for H on the s- and t- symbol generators has as defining relations the following correspondents of (2) through (5):

$$(17)\qquad s_{N,x} = \tau(Nx \cdot {}^{\delta}(Nx)^{-1}) \qquad \text{and} \qquad t_N = \tau(N \cdot {}^{*}N^{-1}),$$

where N is a representative and x is a generator, both of δ-type;

$$(18)\qquad \tau(U) = \tau(U')$$

where U and U' are freely equal and define an element of H;

$$(19)\qquad \tau(U_1U_2) = \tau(U_1)\tau(U_2)$$

where U_1 and U_2 define elements of H;

$$(20)\qquad \tau(WRW^{-1}) = 1$$

where W is any word in a_ν, b_μ, c_ζ and R is a defining relator in (8).

The relators in (20) can be simplified by means of the relators in (18) and (19). For, if in (19) we let $U_1 = U_2$ be the empty word, we have τ(empty word) $= \tau$(empty word) $\cdot\ \tau$(empty word). Hence, τ(empty word) $= 1$ is a consequence of (19).

Next if U defines an element of H then by (10) and (18), $\tau(U)\tau(U^{-1}) = \tau(UU^{-1}) = \tau$(empty word). Thus $\tau(U^{-1}) = \tau(U)^{-1}$ is a consequence of (18) and (19). Finally, if $\{N\}$ is a representative system for G mod H, any word W is of the form UN where U defines an element of H. Since R defines 1 in H, NRN^{-1} defines 1 in H, and so

$$\tau(WRW^{-1}) = \tau(U \cdot NRN^{-1} \cdot U^{-1}) = \tau(U)\tau(NRN^{-1})\tau(U)^{-1}$$

is a consequence of (18) and (19). Thus (20) is a consequence of (18), (19), and

$$(21) \qquad\qquad \tau(NRN^{-1}) = 1,$$

where N ranges over any representative system for G mod H.

As in the case of a Reidemeister rewriting process the defining relations for a presentation of H using a Kurosh rewriting process can be simplified further. Indeed, we shall show that the defining relations in (18) and (19) can be eliminated. To do this we use some of the following lemmas for a Kurosh rewriting process τ.

LEMMA 4.3.   *If W defines an element of H, then*

$$\tau(W^{-1}) = \tau(W)^{-1}$$

PROOF.   Let $W = Px^{\epsilon}Q$; then $W^{-1} = Q^{-1}x^{-\epsilon}P^{-1}$. Since W defines an element of H we note that Q^{-1} and Px^{ϵ} will have the same representative of any given type, and similarly for $Q^{-1}x^{-\epsilon}$ and P. Suppose $\epsilon = 1$ and x is of δ-type. Then in computing $\tau(W)$, x^{ϵ} is replaced by

$$(22) \qquad\qquad t_{\delta_P}^{-1} \cdot s_{\delta_{P,x}} \cdot t_{\delta_{(Px)}},$$

while in computing $\tau(W^{-1})$, $x^{-\epsilon}$ is replaced by

$$(23) \qquad t_{\delta_{(Q^{-1})}}^{-1} \cdot s_{\delta_{(Q^{-1}x^{-1}),x}}^{-1} \cdot t_{\delta_{(Q^{-1}x^{-1})}} = t_{\delta_{(Px)}}^{-1} \cdot s_{\delta_{P,x}}^{-1} \cdot t_{\delta_P},$$

which is clearly the inverse of (22).

Similarly, if $\epsilon = -1$ and x is of δ-type, then in computing $\tau(W)$, x^{ϵ} is replaced by

$$(24) \qquad\qquad t_{\delta_P}^{-1} \cdot s_{\delta_{(Px^{-1}),x}}^{-1} \cdot t_{\delta_{(Px^{-1})}},$$

while in computing $\tau(W^{-1})$, $x^{-\epsilon}$ is replaced by

$$(25) \qquad t_{\delta_{(Q^{-1})}}^{-1} \cdot s_{\delta_{(Q^{-1}),x}} \cdot t_{\delta_{(Q^{-1}x^{-1})}} = t_{\delta_{(Px^{-1})}}^{-1} \cdot s_{\delta_{(Px^{-1}),x}} \cdot t_{\delta_P},$$

which is clearly the inverse of (24). Hence, in computing $\tau(W)$ and $\tau(W^{-1})$, x^{ϵ} and $x^{-\epsilon}$ are replaced by inverse words in the s- and t-symbols. Since x^{ϵ} and $x^{-\epsilon}$ occur in symmetrical positions from the left and right of W and W^{-1}, respectively, (22) [or (24)] and (23) [or (25)] occur in symmetrical positions from the left and right of $\tau(W)$ and $\tau(W^{-1})$, respectively. Hence, $\tau(W)$ and $\tau(W^{-1})$ are inverse words in the s- and t-symbols.   ◀

LEMMA 4.4. *If W defines an element of H and $W = Px^{\epsilon}x^{-\epsilon}Q$, then in computing $\tau(W)$, the word in the s- and t-symbols that replaces x^{ϵ} is inverse to the word that replaces $x^{-\epsilon}$, where x is a generator in* (8) *and ϵ is* 1 *or* -1.

PROOF. First, assume $\epsilon = 1$. Then in computing $\tau(W)$, the s- and t-symbols replacing x are

$$t_{\delta_P}^{-1} \cdot s_{\delta_P, x} \cdot t_{\delta_{(Px)}} \tag{26}$$

and those that replace x^{-1} are

$$t_{\delta_{(Px)}}^{-1} \cdot s_{\delta_{(Pxx^{-1})}, x}^{-1} \cdot t_{\delta_{(Pxx^{-1})}} = t_{\delta_{(Px)}}^{-1} \cdot s_{\delta_P, x}^{-1} \cdot t_{\delta_P}. \tag{27}$$

Clearly, (26) and (27) are inverses. If now $\epsilon = -1$, then in computing $\tau(W)$, x^{-1} is replaced by

$$t_{\delta_P}^{-1} \cdot s_{\delta_{(Px^{-1})}, x}^{-1} \cdot t_{\delta_{(Px^{-1})}} \tag{28}$$

while x is replaced by

$$t_{\delta_{(Px^{-1})}}^{-1} \cdot s_{\delta_{(Px^{-1})}, x} \cdot t_{\delta_{(Px^{-1}x)}} = t_{\delta_{(Px^{-1})}}^{-1} \cdot s_{\delta_{(Px^{-1})}, x} \cdot t_{\delta_P}. \tag{29}$$

Clearly, (28) and (29) are inverses. ◀

From this last lemma, it follows that the defining relations (18) can be eliminated; for, if $U \approx U'$, we can get to U' from U by inserting or deleting words of the form $x^{\epsilon}x^{-\epsilon}$, where x is a generator in (8) and ϵ is 1 or -1. But each time $x^{\epsilon}x^{-\epsilon}$ is inserted in a word, τ of the new word is freely equal in the s- and t-symbols to τ of the original word. Hence, if $U \approx U'$, then $\tau(U) \approx T(U')$, and the defining relation $\tau(U) = \tau(U')$ is superfluous.

LEMMA 4.5. *If U_1 and U_2 define elements of H, then*

$$\tau(U_1U_2) = \tau(U_1)\tau(U_2).$$

PROOF. We first note that in rewriting a word W, the word in s- and t-symbols that replaces a given x^{ϵ} depends only upon x^{ϵ} and the representative of the initial segment of W preceding x^{ϵ}.

Thus the words in the s- and t-symbols that replace the symbols of U_1 in U_1 or in U_1U_2 will be the same. Moreover, since U_1 defines an element of H, if X is any word in a_{ν}, b_{μ}, c_{ζ}, then any type representative of U_1X is the same as that type representative of X alone. Hence, the s- and t-symbols that replace the symbols of U_2 in U_2 or in U_1U_2 will be the same. Thus $\tau(U_1U_2) = \tau(U_1)\tau(U_2)$. ◀

It follows from the previous lemmas that (18) and (19) can be deleted from the defining relations of a presentation for H using a Kurosh rewriting process. Moreover, the defining relations in (17) can be simplified somewhat; namely, if Nx is freely equal to ${}^{\delta}(Nx)$, (or N is freely equal to $*N$), then $s_{N,x}$ (or t_N) will be a relator and can replace the corresponding defining relation involving it; for, if $Nx\,{}^{\delta}(Nx)^{-1}$ is freely equal to 1, then $\tau(Nx\,{}^{\delta}(Nx)^{-1})$ is freely equal in the s- and t-symbols to $\tau(1) = 1$, by a previous lemma. Hence, the defining relation $s_{N,x} = \tau(Nx\,{}^{\delta}(Nx)^{-1})$ is equivalent to the defining relation $s_{N,x} = 1$, if $Nx \approx {}^{\delta}(Nx)$. Similarly, the defining relation $t_N = \tau(N\,{*N}^{-1})$ is equivalent to the defining relation $t_N = 1$, if $N \approx *N$.

Our results thus far can be summarized in the following theorem:

THEOREM 4.7. *Let G have the presentation* (8) *and let τ be a Kurosh rewriting process for G mod a subgroup H. If the generating symbols $s_{N,x}$ and t_N correspond to the elements $Nx \cdot {}^{\delta}(Nx)^{-1}$ and $N \cdot {*N}^{-1}$, respectively, where N is a representative and x is a generator in* (8), *each of δ-type, then H has a presentation using the generators $s_{N,x}$ and t_N with the following defining relations:*

$$s_{N,x} = 1, \qquad \text{if} \quad Nx \approx {}^{\delta}(Nx); \tag{30}$$

$$t_N = 1, \qquad \text{if } N \approx *N; \tag{31}$$

$$s_{N,x} = \tau(Nx \cdot {}^{\delta}(Nx)^{-1}), \qquad \text{if} \quad Nx \not\approx {}^{\delta}(Nx); \tag{32}$$

$$t_N = \tau(N \cdot {*N}^{-1}), \qquad \text{if } N \not\approx *N; \tag{33}$$

$$\tau(NRN^{-1}) = 1, \tag{34}$$

where R is a defining relator in (8) *and N runs over any representative system for G mod H.*

PROOF. This theorem summarizes results established before. ◀

To simplify the defining relations (32) and (33) further, we must restrict ourselves to special kinds of representative functions.

Definition. A collection of representative functions of α-, β-, γ- and neutral types is called an *extended Schreier system*, provided that any type representative N is such that if x^{ϵ} is the last symbol in N and $N = Mx^{\epsilon}$, then N and M are both representatives of the same type as the generator x.

If there is only one type of generator, then an extended Schreier system reduces to an ordinary Schreier system. Even if the generators are of different types, if each representative system is chosen to be the same ordinary Schreier system, then an extended Schreier system results. As a specific example of an extended Schreier system let $G = \langle a, b \rangle$, let H be the commutator subgroup of G, let a be the α-type generator, b the β-type generator and let the α- and neutral representatives be $\{b^p a^q\}$

while the β-representatives are $\{a^p b^q\}$. Now, if a representative ends in an α-symbol, it must be $b^p a^q$ where $q \neq 0$ or a^p where $p \neq 0$. In either case, both the representative and its initial segment obtained by deleting the last α-symbol will be α-representatives; similarly, for representatives ending in a β-symbol. Thus we have an extended Schreier system.

On the other hand, if we allow G and H to remain the same but make $\{a^p b^q\}$ the α-representatives, and $\{b^p a^q\}$ the β-representatives, and choose the neutral representatives to be the same as the α-type, we do not obtain an extended Schreier system. For example, ab ends in a β-type symbol, although its β-representative is not itself but is ba.

It is obvious that in an extended Schreier system the neutral representatives must be chosen from among the other type representatives; for, whatever type of symbol a neutral representative ends in, it must be that type of representative also. Conversely, if the α-, β- and γ-representatives satisfy the extended Schreier condition, and the neutral representative system is contained in the union of the other type representative systems, the whole system is an extended Schreier system.

Before investigating the structure of an extended Schreier system further, we shall show that its employment in a Kurosh rewriting process τ allows the elimination of the defining relations in (32) and (33).

LEMMA 4.6. *Let τ be a Kurosh rewriting process using an extended Schreier system, and let K and L be representatives. Then in computing $\tau(KWL^{-1})$, the s-symbols entering from K and L^{-1} are all of the kind in the defining relations* (30). *Moreover, the t-symbols entering from K and L^{-1} all cancel out or are in* (31), *except for the last in K, which is t_K, and the first in L^{-1}, which is t_L^{-1}.*

PROOF. By induction on the length of a representative in an extended Schreier system it is easily shown that an initial segment of a representative is a representative, though possibly of different type. Suppose now that the symbol x in K is being replaced in the process τ, and that M precedes x. If x is of δ-type, then Mx and M are also of δ-type; thus x will be replaced by

$$t_M^{-1} \cdot s_{M,x} \cdot t_{Mx}.$$

Since ${}^{\delta}(Mx) = Mx$, this s-symbol is of the kind in (30). Similarly, if the symbol x^{-1} in K is being replaced in the process τ and M precedes x^{-1}, then Mx^{-1} and M are δ-type representatives; thus x^{-1} will be replaced by

$$t_M^{-1} \cdot s^{-1}_{Mx^{-1},x} \cdot t_{Mx^{-1}}.$$

Since ${}^{\delta}(Mx^{-1} \cdot x) = M \approx (Mx^{-1} \cdot x)$, this s-symbol is of the kind in (30).

Moreover, in replacing either x or x^{-1}, the t-symbol $t_M{}^{-1}$ will be adjacent to the last t-symbol entering from the last symbol in M. This will be t_M, since M is a representative of the same type as the last symbol in M. Thus all t-symbols that enter from K will cancel except for the very first t-symbol $t_1{}^{-1}$ and the very last t-symbol. Since $1 = {}^*1$ by definition of a representative function, t_1 is in (31) and so $t_1{}^{-1}$ may be deleted. Suppose the last symbol in K is of δ-type; then K is a δ-representative, and so the last t-symbol inserted from K will be t_K. To show the same results for L^{-1}, note that by a previous lemma $\tau(KWL^{-1}) = [\tau(LW^{-1}K^{-1})]^{-1}$. Now, using the first part of this proof to describe the s- and t-symbols that enter $\tau(LW^{-1}K^{-1})$ from L, and then inverting we have our result. ◀

THEOREM 4.8. *If G has the presentation* (8), *and τ is a Kurosh rewriting process using an extended Schreier system for G mod H, then H has a presentation using the generating symbols $s_{N,x}$ and t_N (which correspond to $Nx \cdot {}^{\delta}(Nx)^{-1}$ and $N \cdot {}^*N^{-1}$) and the defining relations* (30), (31), *and* (34).

PROOF. We must show that the defining relations (32) and (33) can be deleted. Consider first $\tau(Nx \cdot {}^{\delta}(Nx)^{-1})$. By the preceding lemma, and in view of the defining relations (30) and (31), we can ignore the s-symbols that enter from N and from ${}^{\delta}(Nx)^{-1}$ in rewriting $Nx \cdot {}^{\delta}(Nx)^{-1}$; moreover, only the last t-symbol of N and the first t-symbol from ${}^{\delta}(Nx)^{-1}$ need be considered. Hence, $\tau(Nx \cdot {}^{\delta}(Nx)^{-1})$ is equivalent to the word

$$t_N \cdot t_N^{-1} \cdot s_{N,x} \cdot t_{\delta(Nx)} \cdot t^{-1}_{\delta(Nx)}$$

in the face of the defining relations (30) and (31). Thus the defining relation

$$s_{N,x} = \tau(Nx \cdot {}^{\delta}(Nx)^{-1})$$

is superfluous to (30) and (31).

Next, consider $\tau(N \cdot {}^*N^{-1})$. By the preceding lemma, and in view of the defining relations (30) and (31) in rewriting $N \cdot {}^*N^{-1}$ we need only consider the last t-symbol entering from N and the first t-symbol entering from ${}^*N^{-1}$. Thus $\tau(N \cdot {}^*N^{-1})$ is equivalent to the word

$$t_N \cdot t^{-1}_{*N},$$

and hence, in view of (31), it is also equivalent to t_N. Consequently, the defining relation

$$t_N = \tau(N \cdot {}^*N^{-1})$$

is superfluous to (30) and (31). ◀

In general, the defining relations (34) cannot be simplified. However, when G is exhibited as a free product of its subgroups generated by the generators of a given type, then Theorem 4.8 allows H to be exhibited as a free product, as the following corollary shows.

COROLLARY 4.8. *If G as in (8) is a free product of its subgroups generated by the generators of a given type, i.e., each defining relator R in (8) involves only one type of generator, then H is the free product of the subgroup generated by the t-symbols which is a free group, and each of the subgroups generated by the s-symbols $s_{N,x}$, where x is a given type of generator, and the same type representative N is in a fixed double-coset of G mod H and X, where X is the subgroup of G generated by the generators of the same type as x.*

PROOF. To prove our result, we must show that each of the defining relations (30), (31), and (34) involves only t-symbols, or only s-symbols $s_{N,x}$, where the x is a given type generator, and N is in a given double-coset of G mod H and X. Let us examine the defining relations (30) and (31). Since each involves only one s-symbol, or one t-symbol, each has the required form. In the defining relations (34), for each R choose N to be the same type of representative as the type of generators involved in R. Consider then $\tau(NRN^{-1})$; by the preceding Lemma 4.6, and in the face of (30) and (31), the only contributions from N and N^{-1} that need be considered are t_N and $t_N{}^{-1}$. Moreover, since all the generators in R are of the same type, all intermediate t-symbols introduced from the symbols of R will cancel. Also, since N is the same type of representative as the generators in R, say δ, the replacement of R will begin with $t_N{}^{-1}$ and end with

$$t_{\delta_{(NR)}} = t_{\delta_{(NRN^{-1}\cdot N)}} = t_{\delta_N} = t_N.$$

Thus, in computing $\tau(NRN^{-1})$, the last t-symbol from N and the first t-symbol from R cancel; moreover, the last t-symbol from R and the first t-symbol from N^{-1} cancel. Thus the word $\tau(NRN^{-1})$ is equivalent to a word involving only the s-symbols introduced from R. Such as s-symbol will have the form

$$s_{\delta_{(NT)},x} \quad \text{or} \quad s^{-1}_{\delta_{(NT)},x}$$

where T is an initial segment of R and x is a generator involved in R. If X is the subgroup generated by the δ-type generators of G, then T defines an element of X. Hence,

$$H \cdot {}^{\delta}(NT) \cdot X = H \cdot NT \cdot X = H \cdot N \cdot TX = H \cdot N \cdot X.$$

Thus the representative subscript of each s-symbol in $\tau(NRN^{-1})$ belongs to the H and X double-coset of G determined by N. Moreover, the generator subscript of each s-symbol in $\tau(NRN^{-1})$ is of δ-type.

Since the only defining relators involving the t-symbols are some of the t-symbol generators themselves, the t-symbols generate a free group. ◀

As an illustration of this last corollary, let $G = \langle a, b; a^2, b^2\rangle$, let H be the commutator subgroup of G, let a be the α-generator, b the β-generator, let $\{1, a, b, ba\}$ be the α- and neutral-type representatives and $\{1, a, b, ab\}$ be the β-type representatives. Then we have an extended Schreier system. The generating symbols for the presentation of H given by the Kurosh rewriting process using the extended Schreier system are

$$s_{1,a}, \quad s_{a,a}, \quad s_{b,a}, \quad s_{ba,a},$$
$$s_{1,b}, \quad s_{a,b}, \quad s_{b,b}, \quad s_{ab,b},$$
$$t_1, \quad t_a, \quad t_b, \quad t_{ba}, \quad t_{ab}.$$

Moreover, using the defining relations (30) and (31), we can eliminate all but

(35) $$s_{a,a}, \quad s_{ba,a}, \quad s_{b,b}, \quad s_{ab,b}, \quad t_{ab}.$$

Next we must consider the defining relations (34). Using the preceding lemma and the defining relations (30) and (31) we shall simplify as we rewrite. (We do only the first one in great detail.)

$$\tau(1 \cdot a^2 \cdot 1^{-1}) = \tau(a^2) = \tau(a \cdot a) = t_a \cdot t_a^{-1} \cdot s_{a,a} \cdot t_1 = s_{a,a}$$

[Here, since a is a representative, only the last t-symbol it introduces need be considered. Moreover, the last two equal signs are not word equality but rather word equivalence in view of free equality or of the defining relations (30) and (31).]

$\tau(a \cdot a^2 \cdot a^{-1}) = \tau(a^2)$, which was done before;

$\tau(b \cdot a^2 \cdot b^{-1}) = \tau(ba \cdot a \cdot b^{-1}) = t_{ba} \cdot t_{ba}^{-1} \cdot s_{ba,a} \cdot t_b \cdot t_b^{-1} = s_{ba,a}$;

$\tau(ba \cdot a^2 \cdot (ba)^{-1}) = \tau(bab^{-1})$, which was done before;

$\tau(1 \cdot b^2 \cdot 1^{-1}) = \tau(b \cdot b) = t_b \cdot t_b^{-1} \cdot s_{b,b} \cdot t_1 = s_{b,b}$;

$\tau(a \cdot b^2 \cdot a^{-1}) = \tau(ab \cdot b \cdot a^{-1}) = t_{ab} \cdot t_{ab}^{-1} \cdot s_{ab,b} \cdot t_a \cdot t_a^{-1} = s_{ab,b}$;

$\tau(b \cdot b^2 \cdot b^{-1}) = \tau(b^2)$, which was done before;

$\tau(ab \cdot b^2 \cdot (ab)^{-1}) = \tau(ab^2a^{-1})$, which was done before.

Thus, from (34), we obtain the defining relators

$$s_{a,a}, \quad s_{ba,a}, \quad s_{b,b}, \quad \text{and} \quad s_{ab,b}.$$

Hence, the only generator in (35) not also a defining relator is t_{ab}. Thus H is the infinite cyclic group generated by $(ab) \cdot {}^*(ab)^{-1} = ab(ba)^{-1} = aba^{-1}b^{-1}$, a familiar result.

To gain more insight into the structure of the free product factors of H generated by the s-symbols, we must restrict the extended Schreier system so that it is directly connected to double-cosets.

Definition. An extended Schreier system is called *regular* if, when the δ-generators are deleted completely from the ends of the δ-representatives, the resulting words form a double-coset representative system for G mod H and X, where X is the subgroup of G generated by the δ-generators.

The double-coset representative system, which arises from a regular extended Schreier system by completely deleting the δ-symbols from the ends of the δ-representatives, is called the *δ-double-coset representative system* of the extended Schreier system.

An an example of a regular extended Schreier system, let $G = \langle a, b; a^3, b^2 \rangle$; let H be the commutator subgroup of G; let a be the α-generator; let b be the β-generator; let $\{1, a, a^2, b, ba, ba^2\}$ be the α- and neutral representatives; and let $\{1, a, a^2, b, ab, a^2b\}$ be the β-representatives. Then if a representative ends in an α- or β-symbol, both it and its initial segment obtained by deleting the final symbol is an α- or β-representative, respectively, as required for an extended Schreier system, Moreover, if the α-symbols are completely deleted from the ends of the α-representatives, then $\{1, b\}$ results; similarly, if the β-symbols are completely deleted from the ends of the β-representatives, then $\{1, a, a^2\}$ results. Since $\{1, b\}$ forms a double-coset representative system for G mod H and A, and $\{1, a, a^2\}$ forms a double-coset representative system for G mod H and B, we have a regular extended Schreier system.

On the other hand, if we leave everything as before except to make $\{1, a, a^2, b, ba, ba^2\}$ the α-, β-, and neutral representative system for G mod H, then we have an extended Schreier system which is not regular; for, upon deleting all β-symbols from the ends of the β-representatives, $\{1, a, a^2, ba, ba^2\}$ results. Since, for example, a and ba determine the same double-coset of G mod H and B, $\{1, a, a^2, ba, ba^2\}$ cannot be a double-coset representative system for G mod H and B.

To make use of the condition of regularity on an extended Schreier system we must first examine the relation of ordinary and double-coset representatives.

LEMMA 4.7. *Let $\{D_r\}$ be a double-coset representative system for a group G mod two of its subgroups H and X. Then a collection of words $\{D_rP_{rs}\}$ with P_{rs} in X is a coset representative system for G mod H if and only if $\{P_{is}\}$ is a coset representative system for X mod $D_i^{-1}HD_i \cap X$. Moreover, if $\{D_rP_{rs}\}$ is a coset representative system for G mod H, and P is an element of X, then the representative of D_iP in the system $\{D_rP_{rs}\}$ for G mod H is D_iP_{ij}, where P_{ij} is the representative of P in the system $\{P_{is}\}$ for X mod $D_i^{-1}HD_i \cap X$.*

PROOF. Suppose first, that $\{D_rP_{rs}\}$ is a coset representative system for G mod H. We must show that $\{P_{is}\}$ is a representative system for X

mod $D_i^{-1}HD_i \cap X$, i.e., given P in X, then there exists a unique P_{ij} such that $P_{ij}P^{-1}$ is in $D_i^{-1}HD_i \cap X$. Since $P_{ij}P^{-1}$ is automatically in X for any P_{ij}, we require a unique P_{ij} such that $P_{ij}P^{-1}$ is in $D_i^{-1}HD_i$; this is equivalent to requiring a unique P_{ij} such that D_iP_{ij} is in HD_iP. But since $\{D_rP_{rs}\}$ is a representative system for G mod H, and since the double-coset representative of D_iP mod H and X is D_i, it follows that there is a unique P_{ij} such that D_iP_{ij} is in HD_iP. Thus we have that $\{P_{is}\}$ is a representative system for X mod $D_i^{-1}HD_i \cap X$.

Moreover, if P_{ij} is the representative of P in the system $\{P_{is}\}$ for X mod $D_i^{-1}HD_i \cap X$, then $P_{ij}P^{-1}$ is in $D_i^{-1}HD_i$; this means that D_iP_{ij} is in HD_iP. Hence, D_iP_{ij} is the representative of D_iP in the system $\{D_rP_{rs}\}$ for G mod H.

Next, we show that if $\{P_{is}\}$ is a coset representative system for X mod $D_i^{-1}HD_i \cap X$, then $\{D_rP_{rs}\}$ is a coset representative system for G mod H. For let W be any element of G. We must show that there is a unique D_iP_{ij} such that $HW = HD_iP_{ij}$. Since $\{D_r\}$ is a double-coset representative system for G mod H and X, there is a unique D_i such that $HWX = HD_iX$. Hence, $W = hD_iP$ where P is in X and h is in H. Thus $HW = HhD_iP = HD_iP$. But if P_{ij} is the representative of P in the system $\{P_{is}\}$ for X mod $D_i^{-1}HD_i \cap X$, then $HD_iP_{ij} = HD_iP$. Thus D_iP_{ij} is in the coset HW. Suppose in addition that D_tP_{tu} is in the coset HW. Then $HD_iP_{ij} = HD_tP_{tu}$, and so $HD_iP_{ij}X = HD_iX = HD_tP_{tu}X = HD_tX$. Since $\{D_r\}$ is a double-coset representative system for G mod H and X, $t = i$. Moreover, $(D_iP_{ij})(D_iP_{iu})^{-1} = D_i(P_{ij}P_{iu}^{-1})D_i^{-1}$ is in H. Hence, $P_{ij}P_{iu}^{-1}$ is in $D_i^{-1}HD_i \cap X$, since $P_{ij}P_{iu}^{-1}$ is automatically in X. Since $\{P_{is}\}$ is a coset representative system for X mod $D_i^{-1}HD_i \cap X$, $u = j$. Thus $\{D_rP_{rs}\}$ is a coset representative system for G mod H. ◀

LEMMA 4.8. *Let τ be a Kurosh rewriting process using a regular extended Schreier system for a subgroup H of a group G. Then the subgroup of H generated by the symbols $s_{N,x}$, where N and x are of δ-type and N is in a fixed double-coset of H and X, and X is the subgroup of G generated by all δ-generators, is $H \cap D_iXD_i^{-1}$, where D_i is the δ-double-coset representative of HNX.*

PROOF. For, the δ-double-coset representatives are obtained by deleting the δ-symbols from the ends of the δ-representatives. Thus the δ-representative system has the form $\{D_rP_{rs}\}$ where $\{D_r\}$ is a double-coset representative system for G mod H and X, and P_{rs} is in X. Hence, by the lemma, $\{P_{is}\}$ is a coset representative system for X mod $D_i^{-1}HD_i \cap X$. Suppose $N = D_iP_{ij}$. Then $s_{N,x}$ defines the element $Nx \cdot {}^{\delta}(Nx)^{-1} = D_iP_{ij}x \cdot {}^{\delta}(D_iP_{ij}x)^{-1}$. Again by the lemma, if $\overline{P_{ij}x}$ is the representative of

$P_{ij}x$ in the system $\{P_{is}\}$ of X mod $D_i^{-1}HD_i \cap X$, then $^\delta(D_iP_{ij}x) = D_i\overline{P_{ij}x}$. Hence,

$$\begin{aligned} Nx \cdot {}^\delta(Nx)^{-1} &= D_iP_{ij}x \cdot (D_i\overline{P_{ij}x})^{-1} \\ &= D_iP_{ij}x \cdot \overline{P_{ij}x}^{-1}D_i^{-1} \\ &= D_i(P_{ij}x \cdot \overline{P_{ij}x}^{-1})D_i^{-1}. \end{aligned} \tag{36}$$

But as P_{ij} ranges over $\{P_{is}\}$ and x ranges over the generators of X, the elements $P_{ij}x \cdot \overline{P_{ij}x}^{-1}$ generate the subgroup $D_i^{-1}HD_i \cap X$ of X. Thus the elements in (36) generate

$$D_i(D_i^{-1}HD_i \cap X)D_i^{-1} = H \cap D_iXD_i^{-1}. \quad \blacktriangleleft$$

To apply this lemma to the corollary of Theorem 4.8 on the structure of a subgroup of a free product, we must establish the existence of a regular extended Schreier system. We do this in the next theorem.

THEOREM 4.9. *Given a group G presented as in* (8) *and any subgroup H of G, there exists a regular extended Schreier system for G mod H.*

PROOF. We first note that any double-coset of H and X is the union of ordinary cosets of H, and moreover, every ordinary coset of H is in some double-coset of H and X.

Second, the neutral representative system can be constructed from the union of the other type representatives.

Now, if we define the syllable length of a double-coset as the minimal syllable length (using the α-, β- and γ-syllables) of a word in the double-coset, then we shall define our δ-double coset representatives, and δ-representatives inductively using the syllable length of the δ-double-cosets of H (δ is α, β, or γ).

For the α-double-coset of length zero, i.e., HA, we define 1 as the α-double coset representative. To complete this to a representative system for all the H-cosets in HA, choose a Schreier representative system $\{P_{1s}\}$ for A mod $H \cap A$, and let these be the α-representatives for all H-cosets in HA. Similarly, define 1 as the β- and γ-double-coset representative of HB and HC, respectively. We then find a Schreier representative system $\{Q_{1t}\}$ for B mod $H \cap B$ and $\{S_{1u}\}$ for C mod $H \cap C$, and let these be the β-representatives and γ-representatives for all H-cosets in HB and HC, respectively.

Suppose, then, as inductive hypothesis that we have constructed δ-double-coset representatives for all double-cosets of H and X of syllable length less than r (where δ is α, β, or γ, and X is A, B, or C, respectively);

moreover, suppose that δ-representatives have been defined for all H-cosets in a double-coset of H and X of syllable length less than r, and that these δ-representatives have the following properties:

(i) The syllable length of a double-coset representative is the same as that of the double-coset it represents.
(ii) If D is a double-coset representative and D ends in an ϵ-symbol (ϵ is α, β, or γ), then D is an ϵ-representative.
(iii) If D is a δ-double-coset representative, then the δ-representatives of the H-cosets in HDX have the form $\{DP_s\}$, where P_s is a Schreier representative system for X mod $D^{-1}HD \cap X$.

We now show how to define the double-coset representative of a δ-double coset of syllable length r, and how to define δ-representatives for each H-coset in it so that (i), (ii), and (iii) are satisfied. Suppose that for definiteness we take δ to be α. Consider then a double-coset HWA of syllable length r, where W itself has syllable length r. If W ended in an α-symbol, then HWA would have smaller syllable length than r; hence, W ends in some type of symbol other than α, say, β. Thus the double-coset HWB has syllable length less than r; thus the β-double-coset representative of HWB has already been defined, and also the β-representative of the H-coset HW, which is in this double-coset HWB, has been defined. Let D be ${}^{\beta}W$, and define D to be the α-double-coset representative of HWA. Finally, let the α-representatives of the H-cosets in HWA have the form $\{DP_s\}$ where $\{P_s\}$ is a Schreier representative system for A mod $D^{-1}HD \cap A$.

We show that the new α-double-coset representative and new α-representatives when added to those obtained already still satisfy (i), (ii), and (iii). Condition (iii) follows immediately from the construction of the new α-representatives. To show condition (i) and (ii) we observe that HWB must have syllable length $r - 1$; for, if E is the β-double-coset representative of HWB (E has been defined since HWB has syllable length less than r), then, by condition (iii) applied to HWB, the β-representative of W has the form EQ where Q is in B. Thus,

$$HWA = HEQA$$

and if λ represents syllable length, then

$$\lambda(E) = \lambda(HWB) \leq r - 1 = \lambda(HWA) - 1 \leq \lambda(EQ) - 1 \leq \lambda(E) + 1 - 1 = \lambda(E).$$

Hence, equality must hold all through and $\lambda(E) = r - 1$. But then $\lambda(EQ) \leq \lambda(E) + 1 = r = \lambda(HWA) = \lambda(HEQA) \leq \lambda(EQ)$, and again equality must hold all through; thus $\lambda(EQ) = r$. Since $EQ = {}^{\beta}W = D$, conditions (i) and (ii) easily follow. For D has the same syllable length as HWA, the double-coset it represents. Moreover, $\lambda(Q) = 1$, so that $D = EQ = {}^{\beta}W$ ends in a β-symbol and is a β-representative.

Thus we have established our inductive hypothesis for double-cosets of syllable length r.

We wish now to show that we have constructed a regular extended Schreier system for $G \bmod H$. By condition (i), a δ-double-coset representative cannot end in a δ-symbol. Hence, condition (iii) guarantees that if the δ-symbols are completely deleted from the ends of the δ-representatives, a δ-double-coset representative system results. Thus regularity is shown, once we have established that the system is an extended Schreier system. To show this, consider any representative. By condition (iii) every representative has the form DP where D is a δ-double-coset representative and P is a Schreier representative system for $X \bmod D^{-1}HD \cap X$. If $P \neq 1$, then DP ends in a δ-symbol x^{ζ}, $P = P'x^{\zeta}$. Since P' is also a Schreier representative for $X \bmod D^{-1}HD \cap X$, we have DP' and DP are both δ-representatives. On the other hand, if $P = 1$, i.e., our representative is a double-coset representative D, and D ends in an ϵ-symbol, then, by condition (ii), D is an ϵ-representative. Thus, by condition (iii), $D = EQ$ and E is a double-coset representative for H and Y (where Y is the subgroup of G generated by the ϵ-generators) and Q is a Schreier representative for $Y \bmod E^{-1}HE \cap Y$. Moreover, by condition (i), $\lambda(E) = \lambda(HDY) < \lambda(D)$, and so $Q \neq 1$. Thus, if $Q = Q'y^{\zeta}$, EQ and EQ' are both ϵ-representatives, as required for an extended Schreier system.

Hence, we have shown the existence of a regular extended Schreier system for $G \bmod H$. ◀

COROLLARY 4.9.1 (*Kurosh Subgroup Theorem*). *Every subgroup H of a free product G, of groups A, B, and C is itself a free product of a free group F and the intersection of H with certain conjugates of A, B, and C.*

PROOF. Present H by using a Kurosh rewriting process that employs a regular extended Schreier system. Then it suffices to combine Corollary 4.8 and Lemma 4.8, immediately preceding Theorem 4.9, to prove our result. In fact, we can be quite specific; namely, given a regular extended Schreier system for G mod H, the free group F is generated by the elements $N \cdot {}^{*}N^{-1}$, where N is a δ-representative not freely equal to ${}^{*}N$; moreover, if $\{D_i\}$ are the δ-double-coset representatives then $H \cap D_iXD_i^{-1}$, are the remaining factors of H as a free product. ◀

COROLLARY 4.9.2. *Let G be a free product of A, B, C with amalgamations from the factor A, i.e., all defining relations either involve one type of* *generator, or have the form* $U(a_\nu) = V(b_\mu)$ *or* $U(a_\nu) = W(C_\zeta)$. *Then any subgroup* H *of* G, *whose intersection with the conjugates of* A, B, *and* C *is* 1, *must be a free group.*

PROOF. Present H by using a Kurosh rewriting process that employs a regular extended Schreier system. Moreover, choose the neutral representatives to be the α-representatives, and conjugate the defining relators of the form $U(a_\nu)V(b_\mu)^{-1}$ or $U(a_\nu)W(c_\zeta)^{-1}$ in the presentation for G with α-representatives; defining relators only involving δ-generators are conjugated by δ-representatives.

Since the generators of H given by $s_{N,x}$ define elements of H intersect a conjugate of A, B, or C, they must define 1. Hence, they all may be thrown on as defining relators in the presentation for H, and in computing τ of a word we need only list the t-symbols introduced. Now, if R is a defining relator for G involving only δ-generators and N is a δ-representative, then, as in the proof of Corollary 4.8, it is easily shown that only the s-symbols that enter from $\tau(NRN^{-1})$ need be considered; but the s-symbols have all been added as defining relators to our presentation for H. Hence, $\tau(NRN^{-1})$ adds no contribution to the presentation of H.

Now let K be an α-representative and consider

$$\tau(K \cdot U(a_\nu) \cdot V(b_\mu)^{-1} \cdot K^{-1}).$$

For reasons discussed previously, we need only consider those t-symbols that enter from the ends of K, $U(\alpha_\nu)$ and $V(b_\mu)^{-1}$, and from the beginnings of $U(a_\nu)$, $V(b_\mu)^{-1}$, and K^{-1}. Thus,

$$\tau(K \cdot U(a_\nu) \cdot V(b_\mu)^{-1} \cdot K^{-1})$$

may be replaced by

$$t_K \cdot t_{\alpha_K}^{-1} \cdot t_{\alpha(KU)} \cdot t_{\beta(KU)}^{-1} \cdot t_{\beta(KUV^{-1})} \cdot t_K^{-1}.$$

Since K is an α-representative and the α-representatives are also the neutral representatives, it follows that t_K and $t_{\alpha(KU)}$ are already defining relators. Thus the above is equivalent to using the defining relator

$$t_{\beta(KU)}^{-1} \cdot t_{\beta(KUV^{-1})}$$

which yields the defining relation

$$t_{\beta(KU)} = t_{\beta K},$$

since $KUV^{-1}K^{-1}$ defines 1.

Thus the defining relations in our presentation for H have the form $s_{N,x} = 1$, $t_N = 1$ or $t_N = t_M$. By applying Tietze transformations we can eliminate each of the defining relations along with one of the corresponding generators. Hence, H will be a free group. ◀

As an application of the first corollary we show that *in a free product any subgroup* $\neq 1$ *which is indecomposable (with respect to free product) must be in a conjugate of a factor or intersect such conjugates trivially and be infinite cyclic*; for, if H is a subgroup of a free product G, then H is the free product of a free group and the intersections of H with various conjugates of the factors of G. Since H is indecomposable, only one of these factors of H can be different from 1. Hence, if H is not contained in a conjugate of factor of G, then H must be a free group. Since H is indecomposable, H must be infinite cyclic. Moreover, if A is a factor of G and WAW^{-1} is a conjugate of this factor, we must have $H \cap WAW^{-1} = 1$; for if D is the α-double-coset representative for W in a regular extended Schreier system, then $W = hDa$, where $h \in H$, $a \in A$. Thus,

$$\begin{aligned} H \cap WAW^{-1} &= H \cap hDaAa^{-1}D^{-1}h^{-1} \\ &= H \cap hDAD^{-1}h^{-1} \\ &= h(H \cap DAD^{-1})h^{-1}. \end{aligned}$$

If $H \cap WAW^{-1} \neq 1$, then $H \cap DAD^{-1} \neq 1$ contrary to the assumption that H is not contained in a conjugate of a factor of G. Thus any indecomposable subgroup of a free product is contained in a conjugate of some factor or has intersection 1 with all such conjugates and is infinite cyclic.

This last result implies that *if*

$$G = A_1 * A_2 * \ldots * A_n = B_1 * B_2 \ldots * B_m$$

where A_i *and* B_j *are indecomposable groups* $\neq 1$, *then* $n = m$ *and* $B_1, \ldots, B_n$ *can be rearranged so as to yield* $B_{j_1}, \ldots, B_{j_n}$ *where* $B_{j_i} \simeq A_i$. For suppose first that A_k is not infinite cyclic. Since A_k is indecomposable, A_k is contained in a conjugate of some B_j, say, B_{j_k}. Then B_{j_k} cannot be infinite cyclic and so B_{j_k} is contained in a conjugate of some A_i, say, A_{i_k}. Thus A_k is contained in a conjugate of A_{i_k}. But any factor intersects the conjugate of another factor in 1. Hence, $i_k = k$, and we have $A_k \subset TB_{j_k}T^{-1}$ and $B_{j_k} \subset WA_kW^{-1}$. Then $A_k \subset TWA_kW^{-1}T^{-1}$; but $A_k \cap TWA_kW^{-1}T^{-1}$ is 1 unless $TW \in A_k$. Thus $W = T^{-1}a_k$ and $B_{j_k} \subset WA_kW^{-1} = T^{-1}a_kA_ka_k^{-1}T = T^{-1}A_kT$. But we then have $A_k \subset TB_{j_k}T^{-1} \subset A_k$. Hence, $A_k = TB_{j_k}T^{-1}$. We have thus shown that the factors A_i which are not infinite cyclic can be paired with conjugate factors B_{j_i}. Moreover, the mapping is one-one and onto between the non-infinite-cyclic factors among the A_i and the non-infinite-cyclic factors among the B_j. Let $A_1, \ldots, A_r$ $(r \leq n)$ be the non-infinite-cyclic factors among the A_i, and let $B_1, \ldots, B_r$ be the non-infinite-cyclic factors among the B_j. If N is the normal subgroup generated by $A_1, \ldots, A_r$, then N is also the normal subgroup generated by $B_1, \ldots, B_r$. Since $A_{r+1} * \ldots * A_n \simeq G/N \simeq B_{r+1} * \ldots * B_m$,

G/N is a free group of rank $n - r$ and of rank $m - r$. Thus $n = m$, and the infinite cyclic factors $A_{r+1}, \ldots, A_n$ are clearly isomorphic to the infinite cyclic factors $B_{r+1}, \ldots, B_n$. Finally then, $A_1, \ldots, A_r, A_{r+1}, \ldots, A_n$ are isomorphic to $B_1, \ldots, B_r, B_{r+1}, \ldots, B_n$, respectively.

As another application of the first corollary, we show that *if A and B are indecomposable and $H \neq 1$, then $G = *(A, B, H, K, \varphi)$ is indecomposable.* For suppose that $G = C * D$, where C and D are different from 1. Since A and B are indecomposable subgroups of $C * D$, each is either in a conjugate of C or D, or intersects all such conjugates trivially and is infinite cyclic. Suppose then A or B is not infinite cyclic, say A. Then A is contained in a conjugate of C or D, say C. Since $B \cap A = H \neq 1$, B must intersect this conjugate of C and so A and B are both in that conjugate. But then G is in that conjugate of C, contrary to $D \neq 1$. Thus A and B must be infinite cyclic; but then H is in the center of G. Since $C * D$ has only a trivial center, G is indecomposable.

Further applications are given in the problems. For references see the remarks at the end of Section 4.4.

Problems for Section 4.3

1. Let G be $\langle a, b; a^2, b^2\rangle$ and let H be the commutator subgroup of G. If a is the α-generator and b is the β-generator, find a regular extended Schreier system for G mod H. Present H using this system.

2. Let G be $\langle a, b; a^2\rangle$ and let H be the X^2-verbal subgroup of G. Find a regular extended Schreier system for G mod H, if a is the α-generator and b is the β-generator. Use this system to obtain a presentation for H.

3. Let G be $\langle a_1, a_2, b_1, b_2; a_1b_1 = a_2b_2, b_1{}^2\rangle$ and let H be the normal subgroup generated by a_1 and a_2. If the α-generators are a_1, a_2 and the β-generators are b_1, b_2, find a regular extended Schreier system for G mod H. Use this system to present H.

4. Let G be $\langle x, y, z; x^2, yz = xy, zx = yz\rangle$ and let H be the commutator subgroup of G. Find a regular extended Schreier system for G mod H in each of the following cases:

(i) x, y, z are the α-, β-, γ-generators, respectively;
(ii) x, y are the α-generators and z is the β-generator;
(iii) x, z are the α-generators and y is the β-generator;
(iv) x is the α-generator and y, z are the β-generators.

Obtain a presentation for H, using each of these regular extended Schreier systems.

5. Let $G = \langle a, b; a^2, b^2\rangle$ and let H be the subgroup generated by a. Show that the words $(ba)^k$, $(ba)^kb$, $k = 0, 1, 2, \ldots$ form an α-representative system as well as a β-representative system in a regular extended Schreier

system for G mod H (where the α- and β-generators are a and b, respectively). Obtain a presentation for H using this regular extended system.

6. Show that the words in any extended Schreier system are freely reduced. [*Hint:* If $Uxx^{-1}V$ is a representative, then Uxx^{-1} is a representative. If x is a δ-generator, then Uxx^{-1}, Ux, and U are δ-representatives.]

7. Show that in any extended Schreier system for any finite subgroup of a free product $A * B * C$ the α-, β-, and γ-representatives systems must all be the same system of representatives. [*Hint:* For if ${}^{\alpha}K \not\approx {}^{*}K$, then $t_{\alpha K}$ has infinite order. Thus ${}^{\alpha}K \approx {}^{*}K$, and so by the preceding problem ${}^{\alpha}K = {}^{*}K$. Similarly, ${}^{\beta}K = {}^{*}K = {}^{\gamma}K$. Thus the α-, β- and γ-representative systems are just the neutral one.]

8. Let D be an α-double-coset representative in any regular extended Schreier system for a group G mod its subgroup H. If $\{U(a_\nu)\}$ is the set of α-words which follow D in the α-representatives for H cosets in the double-coset HDA, show that $\{U(a_\nu)\}$ is a Schreier representative system for A mod $A \cap D^{-1}HD$. [*Hint:* That $\{U(a_\nu)\}$ is a representative system follows from Lemma 4.7 preceding Theorem 4.9. That $\{U(a_\nu)\}$ is Schreier follows from the extended Schreier condition.]

9. Let D be an α-double-coset representative in a regular extended Schreier system for a group G mod its subgroup H, and let $\overline{W}$ be the representative of $W(a_\nu)$ in the Schreier system for A mod $A \cap D^{-1}HD$ which completes D to the α-representatives for the H cosets in HDA (see preceding problem).

(a) Show that $D\overline{W}a \approx {}^{\alpha}(D\overline{W}a)$ if and only if $\overline{W}a \approx \overline{\overline{W}a}$. Hence, the number of $D\overline{W}$ and a such that $D\overline{W}a \approx {}^{\alpha}(D\overline{W}a)$ is $j_D - 1$ where j_D is the index of $A \cap D^{-1}HD$ in A.

(b) Show that j_D is also the number of H cosets in the double coset HDA.

(c) Show that the number of s-symbols $s_{{}^{\alpha}K,a}$ such that ${}^{\alpha}Ka \approx {}^{\alpha}(Ka)$ is $j - j_A$ where j is the index of H in G and j_A is the double-coset index of H and A in G.

(d) Conclude that in any presentation for the subgroup H using a regular extended Schreier system with α-, β- and γ-representatives,

$$3j - j_A - j_B - j_C$$

of the s-symbols can immediately be deleted as generators.

[*Hint:* For (a), use Lemma 4.7 preceding Theorem 4.9 and the result that the number of $\overline{W}$ and a such that $\overline{W}a \approx \overline{\overline{W}a}$ in an ordinary Schreier representative system is one less than the index of the subgroup. For (b), use Lemma 4.7 again. For (c), sum $j_D - 1$ over all double-cosets of G mod H and A. For (d), use Theorem 4.8.]

10. Show that a finite subgroup of a free product $A * B$ must be in a conjugate of A or of B. [*Hint:* Use the Kurosh subgroup theorem and that a proper free product cannot be a finite group.]

11. Show that if $G = A * B$ and p, q are two elements of G that commute then p and q are in the same conjugate of A or B, or p and q are powers of

the same element. [*Hint:* Consider the subgroup generated by p and q and use the Kurosh subgroup theorem and that a proper free product cannot be Abelian.]

12. Let A and B be groups without elements of order two. Show that any infinite non-cyclic subgroup H of $A * B$ which satisfies an identical relation must be contained in a conjugate of A or B. [*Hint:* Use the Kurosh subgroup theorem on H and that a proper free product of two groups neither of which has order two cannot satisfy an identical relation.]

13. (a) Show that every subgroup of a free product of cyclic groups is also a free product of cyclic groups.
(b) Show that every subgroup of a free product of Abelian groups is also a free product of Abelian groups.

[*Hint:* Use the Kurosh subgroup theorem.]

14. Let P be a property of groups such that

(i) P is an isomorphic invariant, i.e., a group isomorphic to a group with property P must itself have property P;
(ii) P is hereditary, i.e., a subgroup of a group with property P has property P;
(iii) P is an integral property, i.e., P is possessed by the group of integers under addition.

We call P an *isomorphic hereditary integral property* and call groups with property P, *P-groups.*

(A) Which of the following properties of groups are ihi properties?

(1) being cyclic
(2) being free
(3) being finite
(4) being infinite
(5) having no elements ($\neq 1$) of finite order
(6) having no elements of order two
(7) being Abelian
(8) being nilpotent
(9) being solvable
(10) being finitely generated
(11) consisting of integers
(12) having a free subgroup of finite index
(13) having an Abelian subgroup of finite index
(14) having a cyclic subgroup of finite index
(15) having a unique nth root for any element that has some nth root.

(B) Show that if P is an ihi property then a subgroup of a free product of P-groups is also a free product of P-groups.

[*Hint:* Use the Kurosh subgroup theorem for (B).]

15. A subgroup A of a group G is called a *retract* of G if there is a homomorphism of G onto A which is the identity on A.

(A) Show that A is a retract of each of the following groups G:

(a) $G = A \times B$;
(b) $G = A * B$;
(c) G is generated by A and B, where B is a normal subgroup of G such that $A \cap B = 1$;
(d) $G = *(A, B, H, K, \varphi)$ where K is a retract of B;
(e) $G = *(A, A, H, H, \text{identity})$.

(B) (Takahasi) Show that if A is a retract of a group G and $G(W_\mu)$ is any verbal subgroup of G, then $G(W_\mu) \cap A = A(W_\mu)$.
[*Hint:* For (a), (b), and (c), map a_ν into a_ν and b_μ into 1, where the a_ν generate A and the b_μ generate B. For (d), map A onto A with the identity map and retract B onto K. For (e), map both the first factor A and the second factor A onto A by the identity map. To show that $G(W_\mu) \cap A = A(W_\mu)$, note first that $A(W_\mu)$ is clearly contained in $G(W_\mu) \cap A$. On the other hand, under the retractive mapping of G onto A, $G(W_\mu) \cap A$ is left fixed since it is in A, and is mapped into $A(W_\mu)$ since $G(W_\mu)$ is mapped into $A(W_\mu)$. Thus $G(W_\mu) \cap A$ is contained in $A(W_\mu)$, and so the equality holds.]

16. (Takahasi) Show that if A, B, and C satisfy the identical relations $W_\mu(X_\nu) = 1$ and $G = A * B * C$, then $G(W_\mu)$ is a free group. Show, in particular, that the commutator subgroup of a free product of Abelian groups is free. Show that if A, B, and C are finite groups with orders dividing n, then $G(X^n)$ is a free group. [*Hint:* Since $G(W_\mu)$ is a normal subgroup of G, the intersection of $G(W_\mu)$ with the conjugates of A, B, and C will be 1; for, $G(W_\mu) \cap A = A(W_\mu) = 1$, $G(W_\mu) \cap B = B(W_\mu) = 1$, and $G(W_\mu) \cap C = C(W_\mu) = 1$. Now use the Kurosh subgroup theorem.]

17. Show that if $A(W_\mu)$, $B(W_\mu)$, and $C(W_\mu)$ are free groups and $G = A * B * C$, then $G(W_\mu)$ is a free group. [*Hint:* Use the Kurosh subgroup theorem and Problem 15(B).]

18. Show that if $G = *(A, B, H, K, \varphi)$ has A and B as retracts where A and B satisfy identical relations $W_\mu(X_\nu) = 1$, then $G(W_\mu)$ is free. [*Hint:* Use Problem 15(B) and the second corollary to Theorem 4.9.]

19. Show that each of the following groups have a free commutator subgroup:

(a) $\langle x, y, z; x^3, y^5, z^6\rangle$;
(b) $\langle x, y, z, w; xy = yx, zw = wz\rangle$;
(c) $\langle x, y, z, w; xy = yx, zw = wz, x^2y^3 = z^2w^3\rangle$;
(d) $\langle r, s, t, x, y, z, w; rs = sr, tx = xt, r^2s^3 = t^2x^3, y^2 = z^2\rangle$.

[*Hint:* For (a) and (b), use Problem 16. For (c) use Problems 15(A)(e) and 18. For (d), use Problems 15(A)(e), 18, and 17.]

20. (a) Show that $A = \langle a; a^4\rangle$ and $B = \langle b; b^4\rangle$ are retracts of

$$G = \langle a, b; a^4, b^4, a^2 = b^2\rangle.$$

(b) Show that the commutator subgroup H of G is a free group.

(c) Use a regular extended Schreier system for $G \bmod H$ to obtain a presentation for H and find the rank of H (a is the α-generator and b is the β-generator).

[*Hint:* For (a), use Problem 15(A)(e). For (b), use Problem 18. For (c), use $\{a^r b^s\}$ as the β-representatives and $\{b^r a^s\}$ as the α- and neutral representatives, where $r = 0, 1$ and $s = 0, 1, 2, 3$.]

21. Let $G = \langle a, b; a^2, b^3, (ab)^6 \rangle$ and let H be the commutator subgroup of G. Find a regular extended Schreier system for $G \bmod H$ in which a is the α-generator, b is the β-generator and the neutral representatives are the same as the β-representatives. Use this system to find a presentation for H and show that H is a free Abelian group of rank two. [*Hint:* Note that $\tau(Ka^2K^{-1})$ and $\tau(Lb^3L^{-1})$ as defining relators imply that all s-symbols occur as relators. Thus in forming $\tau(L(ab)^6L^{-1})$ it is sufficient to consider only t-symbols. Moreover, since the neutral and β-representative system are the same, only the t-symbols from the α-generators need be considered.]

22. Show that if G is the free product of infinitely many indecomposable groups $A_1, A_2, \ldots$ (each $\neq 1$) and G is also the free product of indecomposable groups $B_1, B_2, \ldots$ (each $\neq 1$), then there is a one-one correspondence of the A_i onto the B_j such that corresponding groups are isomorphic. [*Hint:* Use an argument similar to the one given for the case of finitely many factors A_i.]

23. Show that each of the following pairs of groups are not isomorphic:

(a) $\langle x, y, z; x^2, y^3, z^4 \rangle$ and $\langle u, v, w; u^2, v^3, w^4, uv = vu \rangle$

(b) $\langle x, y, z; x^4, y^2, (xy)^2 \rangle$ and $\langle u, v, w; u^4, v^2, uv = vu \rangle$

(c) $\langle x, y, z, w; xy = yx, zw = wz \rangle$ and $\langle s, t, u, v; st^2 = t^2s, uv = vu \rangle$

[*Hint:* Use the uniqueness of a free product decomposition of a group into indecomposable groups; and for (a), that a finite group is indecomposable; for (b), that a finite or Abelian group is indecomposable; and for (c), that a group with non-trivial center is indecomposable.]

24. A group G is called *completely indecomposable* if every factor group of G is indecomposable (with respect to free products). Show that the following groups are completely indecomposable:

(a) a finite group,
(b) an Abelian group,
(c) a simple group,
(d) a group in which each element is of finite order,
(e) a nilpotent group.

Show, on the other hand, that although the following groups are indecomposable they need not be completely indecomposable:

(f) a group with a non-trivial center
(g) an amalgamated product $*(A, B, H, K, \varphi)$ with $H \neq 1$.

[*Hint:* For (e), note that the only free product satisfying an identical relation is $\langle a, b; a^2, b^2 \rangle$; but this group is not nilpotent since, e.g., $[\ldots[[a, b], b], \ldots, b]$, where there are n b's is $(ab)^{2n}$ if n is odd and $(ba)^{2n}$ if n is even, as may be proved by induction. For (f) and (g), the group $\langle a, b; a^2 = b^2 \rangle$ is indecomposable, but its factor group $\langle a, b; a^2, b^2 \rangle$ is obviously decomposable.]

25. Let A and B be indecomposable subgroups which generate a group G and suppose $A \cap B \neq 1$. Show that G is indecomposable. [*Hint:* Suppose $G = C * D$ where $C, D \neq 1$. If A or B is not infinite cyclic, say, A, then A must intersect a conjugate of C or D, say, KCK^{-1}. Since A is indecomposable, $A \subset KCK^{-1}$. But $A \cap B \neq 1$, and so B intersects KCK^{-1}. Since B is indecomposable, $B \subset KCK^{-1}$. Hence, $G \subset KCK^{-1}$ contrary to $D \neq 1$. Thus A and B are infinite cyclic. But then $A \cap B$ is in the center of G so that G has a non-trivial center and is indecomposable.]

26. Let A and B be completely indecomposable groups. Show $G = *(A, B, H, K, \varphi)$ is completely indecomposable if and only if A is the normal subgroup of A generated by H or B is the normal subgroup of B generated by K. [*Hint:* Suppose $H \subset L$ and $K \subset M$ where L and M are proper normal subgroups of A and B, respectively. If N is the normal subgroup of G generated by L and M, then $G/N \simeq A/L * B/M$ and so G is not completely indecomposable. On the other hand, suppose A is the normal subgroup of A generated by H. Consider any normal subgroup R of G; G/R is generated by AR/R and BR/R. Since A and B are completely indecomposable, their factor groups AR/R and BR/R are indecomposable. If $H \not\subset R$, then $AR/R \cap BR/R$ contains $HR/R \neq 1$. Thus, by Problem 25, G/R is indecomposable. If $H \subset R$, then the normal subgroup generated by H in A, i.e., A must be in R. Hence, $AR/R = 1$ and $G/R = BR/R$, which is indecomposable.]

27. Show that the group $G = *(A, B, H, K, \varphi)$ is completely indecomposable under the following conditions:

(a) A is a simple group, B is completely indecomposable, and $H \neq 1$.
(b) A is a finite symmetric group, B is completely indecomposable, and H contains an odd permutation.
(c) A is completely indecomposable, B is completely indecomposable, and H is a non-normal subgroup of prime index in A.

[*Hint:* Use Problem 26.]

28. Let

$$A = \langle a_1, \ldots, a_n; R(a_\nu), \ldots \rangle,$$
$$B = \langle b_1, \ldots, b_n; R(b_\nu), \ldots \rangle,$$

and let H, K be the subgroups of A and B generated by $a_p, a_{p+1}, \ldots, a_n$, and $b_p, b_{p+1}, \ldots, b_n$, respectively. Moreover, let the mapping φ be the isomorphism of H onto K given by

$$\varphi: a_i \to b_i, \; i = p, p+1, \ldots, n.$$

If $G = *(A, B, H, K, \varphi)$, then let J be the subgroup of G generated by

$$a_q, a_{q+1}, \ldots, a_r, b_q, b_{q+1}, \ldots, b_r,$$

and let N be the normal subgroup of G generated by

$$a_1b_1^{-1}, a_2b_2^{-1}, \ldots, a_nb_n^{-1}.$$

(i) Show that $G/N \simeq A$ under $a_\nu \to a_\nu$.
(ii) Show that N is a free group.
(iii) Show that $J \cap N$ is a free factor of N.

[*Hint:* For (ii) use that $N \cap A = N \cap B = 1$, and Corollary 4.9.2. For (iii), show that a regular extended Schreier system for G mod N can be obtained by choosing as α-representatives a Schreier system for A mod the identity subgroup of A, such that if the coset $NW(a_\nu)$ contains an element from J then its representative is a word in $a_q, a_{q+1}, \ldots, a_r$; moreover, β-representatives can be obtained by replacing each α-symbol a_ν in an α-representative by b_ν; neutral representatives are the same as α-representatives. Show that if the word W defines an element of N then in computing $\tau(W)$ only the t-symbols corresponding to β-representatives need be considered. Show that $J \cap N$ is generated by the set of all t-symbols t_U, where the β-representative U is a word in $b_q, b_{q+1}, \ldots, b_r$.]

4.4. Groups with One Defining Relator

In this section we shall apply the theory of free and amalgamated products to the study of groups which possess a presentation having a single defining relator. In particular we shall solve the word problem for such groups.

Fundamental to the results of this section is the *Freiheitssatz* or *independence theorem*:

THEOREM 4.10. *Let $R(a_1, a_2, \ldots, a_n)$ be a cyclically reduced word in $a_1, a_2, \ldots, a_n$ which involves a_n. Then the subgroup of*

$$G = \langle a_1, a_2, \ldots, a_n;\ R(a_1, a_2, \ldots, a_n)\rangle$$

generated by $a_1, a_2, \ldots, a_{n-1}$ is freely generated by them; in other words, every non-trivial relator of G must involve a_n. [*Note:* G may have infinitely many generators.]

PROOF. We shall use induction on the word length of the relator R. Clearly, if the length of R is 1, or if R involves only a_n, then G is the free product of the free group on $a_1, a_2, \ldots, a_{n-1}$ and the cyclic group generated by a_n. Thus our assertion holds.

Assume then that our assertion holds for all groups with one defining relator whose word length is less than r, and suppose that $R(a_1, a_2, \ldots, a_n)$ has word length r.

We may assume that R involves at least one generator other than a_n, say, a_1. It will be convenient at this point to relabel our generators so that the subscripts are not needed. (We shall require subscripts with different significance presently.) We represent a_n by t and $a_1, a_2, \ldots$ by $b, c, \ldots$.

Of course, R need not involve all the generators of G; in fact, if G is infinitely generated, then R can only involve finitely many of the generators of G. Those generators of G not involved in R freely generate a free subgroup K of G; and, in fact, G is the free product of K and the group whose generators are those of G which are involved in R, and whose defining relator is R. For example, if

$$G = \langle b, c, d, e, t;\ b^2c^2t^2\rangle$$

then

$$G = \langle b, c, t;\ b^2c^2t^2\rangle * \langle d, e\rangle.$$

Hence, to show that the generators of G other than t freely generate a free group, it suffices to show that the generators of G other than t which are involved in R freely generate a free group. (Here we use Corollary 4.1.2, namely, that if $C \subset A$ and $D \subset B$, then C and D generate their free product in $A * B$.) Thus we may assume that all the generators of G are involved in R.

CASE 1. R involves only b and t and $\sigma_t(R)$, the exponent sum of R on t, is different from zero. In this case, if

$$b^k = \prod_i S_i R^{\epsilon_i} S_i^{-1},$$

then

$$k = \sigma_b(b^k) = \sum_i \sigma_b(R^{\epsilon_i}) = \left(\sum_i \epsilon_i\right)\sigma_b(R);$$

$$0 = \sigma_t(b^k) = \sum_i \sigma_t(R^{\epsilon_i}) = \left(\sum_i \epsilon_i\right)\sigma_t(R).$$

Since $\sigma_t(R) \neq 0$, we must have $\sum_i \epsilon_i = 0$. Then $k = 0$, and so b has infinite order in $\langle b, t;\ R(b, t)\rangle$ and freely generates a free group.

CASE 2. R involves only b and t, and $\sigma_t(R) = 0$. In this case, we shall examine the behavior of b in a subgroup of $G = \langle b, t;\ R(b, t)\rangle$, namely, the normal subgroup N of G generated by b. To obtain a presentation of N we make use of a Reidemeister-Schreier rewriting process. As Schreier representatives for G mod N we choose t^k, where k ranges over all integers. A word $W(b, t)$ defines an element of N if and only if $\sigma_t(W) = 0$; the representative $\overline{W}$ of a word W mod N will be t^k where $k = \sigma_t(W)$. We find that N is generated by the elements b_k defined by

$$b_k = t^k b t^{-k} = t^k \cdot b \cdot \overline{t^k b}^{-1}$$

For every word W which defines an element of N, we obtain an expression $\tau(W)$ for W in terms of the b_k in the following manner: Every symbol b^ϵ ($\epsilon = \pm 1$) in W is replaced by $b_s{}^\epsilon$ where s is the sum of the exponents of the t-symbols preceding the particular b^ϵ in W. Thus, e.g.,

$$\tau(t^{-1}btb^2tb^{-3}t^{-1}) = b_{-1}b_0{}^2b_1{}^{-3}$$

We prove by a simple inspection: *If $R(b, t)$ is cyclically reduced as a word in b and t, then $\tau(R)$ is a cyclically reduced word in the b_k, $k = 0, \pm 1, \pm 2, \ldots$.* (See Problem 2.3.2.)

Since t occurs in R but is not replaced by any symbol b_k in writing $\tau(R)$, the word length of $\tau(R)$ in the b_k is less than the word length of R in b, t. Thus, we may apply our inductive hypothesis to $\tau(R)$ or any word of its length. Let

$$\tau(R) = P(\ldots, b_k, \ldots)$$

Then N is generated by the b_k and has as defining relators

$$P_m = \tau(t^m R t^{-m}) = P(\ldots, b_{k+m}, \ldots),$$

and so

(1) $$N = \langle \ldots, b_{-1}, b_0, b_1, \ldots; \ldots, P_{-1}, P_0, P_1, \ldots \rangle$$

In order to make the proof that b_0 has infinite order in N more comprehensible, we first consider a specific example. Let

$$G = \langle b, t; t^{-1}btb^2tb^{-3}t^{-1} \rangle.$$

Then

$$\tau(R) = b_{-1}b_0{}^2b_1{}^{-3}$$

and

$$N = \langle \ldots, b_{-2}, b_{-1}, b_0, b_1, b_2, \ldots; \ldots, b_{-2}b_{-1}^2b_0^{-3}, b_{-1}b_0{}^2b_1{}^{-3}, b_0b_1{}^2b_2{}^{-3}, \ldots \rangle.$$

Now the groups

$$N_m = \langle b_{-1+m}, b_m, b_{1+m}; b_{-1+m}b_m{}^2b_{1+m}^{-3} \rangle$$

may be used to construct N in a fairly simple fashion.

Since the relator in N_m has shorter word length than the relator in G we may assume the inductive hypothesis for N_m; hence, any two b-generators of N_m freely generate a free group. Thus

$$N_{0,1} = \langle b_{-1}, b_0, b_1, b_2; b_{-1}b_0{}^2b_1{}^{-3}, b_0b_1{}^2b_2{}^{-3} \rangle$$

is the free product of N_0 and N_1 with the free subgroup in each, freely generated by b_0, b_1, amalgamated under the identity mapping. Since $N_0 \subset N_{0,1}$, if follows that b_{-1}, b_0 freely generate a free group in $N_{0,1}$ and hence,

$$N_{-1,1} = \langle b_{-2}, b_{-1}, b_0, b_1, b_2; b_{-2}b_{-1}^2b_0{}^{-3}, b_{-1}b_0{}^2b_1{}^{-3}, b_0b_1{}^2b_2{}^{-3} \rangle$$

is the free product of N_{-1} and $N_{0,1}$ with the free subgroup in each, freely generated by b_{-1}, b_0 amalgamated under the identity mapping. Similarly, since $N_1 \subset N_{0,1} \subset N_{-1,1}$, and so b_1, b_2 freely generate a free subgroup of $N_{-1,1}$, it follows that

$$N_{-1,2} = \langle b_{-2}, b_{-1}, b_0, b_1, b_2, b_3; b_{-2}b_{-1}^2b_0^{-3}, b_{-1}b_0{}^2b_1{}^{-3}, b_0b_1{}^2b_2{}^{-3}, b_1b_2{}^2b_3{}^{-3}\rangle$$

is the free product of $N_{-1,1}$ and N_2 with the free subgroup in each, freely generated by b_1, b_2, amalgamated under the identity mapping. Continuing in this way, we obtain a chain of groups

$$N_0 \subset N_{0,1} \subset N_{-1,1} \subset N_{-1,2} \subset N_{-2,2} \subset \ldots \subset N_{-i+1,i} \subset N_{-i,i} \subset N_{-i,i+1} \subset \ldots$$

where $N_{-i,i}$ is the free product of N_{-i} and $N_{-i+1,i}$ with an amalgamated free subgroup, and $N_{-i,i+1}$ is the free product of $N_{-i,i}$ and N_{i+1} with an amalgamated free subgroup. Moreover, by Problem 1.3.18, N is the union of this chain of groups. Thus $N_0 \subset N$ and since b_0 generates a free group in N_0 by inductive hypothesis, b_0 generates a free group in N. Since b_0 defines the element b in $N \subset G$, it follows that b generates a free group in G.

We return now from our specific example to the situation where N is presented as in (1). Since $P(\ldots, b_k, \ldots) = \tau(R)$ and t is involved in R but with exponent sum zero, and also R is cyclically reduced, P must involve at least some b_k and b_q, $k \neq q$; for, if only a single power of b occurs in R, then $R = t^\alpha b^\beta t^\gamma$ where α or $\gamma \neq 0$ and $\alpha + \gamma = 0$, and so R is not cyclically reduced. Hence, at least two powers of b occur in R separated by a power of t, i.e., R contains a subword $b^\alpha t^\beta b^\gamma$ where $\alpha, \beta, \gamma \neq 0$. If the t-exponent sum of R preceding b^α is k, then this subword is replaced by $b_k{}^\alpha b_{k+\beta}^\gamma$. Since $\tau(R) = P$ is cyclically reduced and $k + \beta \neq k$, letting $q = k + \beta$, we find that b_k and b_q both occur in P. Let μ be the minimum subscript on b occurring in P and let M be the maximum subscript on b occuring in P; then $\mu < M$.

Define for each integer i:

$$(2) \qquad N_i = \langle b_{\mu+i}, b_{\mu+i+1}, \ldots, b_{M+i}; P_i\rangle$$

As in the example, we next construct $N_{0,1}, N_{-1,1}, \ldots$;

$$N_{0,1} = \langle b_\mu, b_{\mu+1}, \ldots, b_M, b_{M+1}; P_0, P_1\rangle$$

is the free product of N_0 and N_1, with the free subgroup in each, freely generated by $b_{\mu+1}, \ldots, b_M$ (which excludes b_μ in N_0 and b_{M+1} in N_1), amalgamated under the identity mapping. Similarly, since $N_0 \subset N_{0,1}$ and $b_\mu, \ldots, b_{M-1}$ freely generate a free group in N_0,

$$N_{-1,1} = \langle b_{\mu-1}, b_\mu, \ldots, b_M, b_{M+1}; P_{-1}, P_0, P_1\rangle$$

is the free product of N_{-1} and $N_{0,1}$, with the free subgroup in each freely generated by $b_\mu, \ldots, b_{M-1}$, amalgamated under the identity mapping. Continuing in this way, we construct a chain of groups

$$(3) \quad N_0 \subset N_{0,1} \subset N_{-1,1} \subset N_{-1,2} \subset \ldots \subset N_{-i+1,i} \subset N_{-i,i} \subset N_{-i,i+1} \subset \ldots,$$

where $N_{-i,i}$ is the free product of N_{-i} and $N_{-i+1,i}$ with an amalgamated free subgroup, and $N_{-i,i+1}$ is the free product of $N_{-i,i}$ and N_{i+1} with an amalgamated free subgroup. Problem 1.3.18 implies that N is the union of this chain of groups. Since each N_i in (2) is a subgroup of some group in (3), each N_i is a subgroup of N. Since $\mu < M$, any generator of N_i generates a free group. But b_0 is in some N_i and so generates a free group in N; hence, b generates a free group in G.

CASE 3. R contains at least three generators b, c, t and the exponent sum of R on some generator other than t is zero, say, $\sigma_b(R) = 0$.

Suppose that we could derive a relator $U(b, c, \ldots)$ for G which did not involve t. Since $R(b, c, \ldots, t)$ is the defining relator for G and $\sigma_b(R) = 0$, it follows that $\sigma_b(U) = 0$. It is therefore natural to examine the normal subgroup N of G generated by $c, \ldots, t$, i.e., by the generators of G excluding b. A word $W(b, c, \ldots, t)$ defines an element of N if and only if its b-exponent sum is zero. To obtain a presentation for N we use a Reidemeister-Schreier rewriting process τ which has b^k, k any integer, as its Schreier representative system. We denote the elements

$$b^k c b^{-k}, \ldots, b^k t b^{-k}, \qquad (k = 0, \pm 1, \pm 2, \ldots)$$

by

$$c_k, \ldots, t_k,$$

respectively. Since the representative of $W(b, c, \ldots, t)$ is b^k where $k = \sigma_b(W)$, each symbol $c^\epsilon, \ldots, t^\epsilon$ in a word $V(b, c, \ldots, t)$ defining an element of N, is replaced under τ by

$$c_k^{\ \epsilon}, \ldots, t_k^{\ \epsilon},$$

where k is the b-exponent sum of the initial segment of V preceding $c^\epsilon, \ldots, t^\epsilon$, respectively. For example,

$$\tau(bctb^{-1}d^2t) = c_1 t_1 d_0^{\ 2} t_0.$$

If $\tau(R) = P(\ldots, c_k, \ldots, t_k, \ldots)$, then

$$P_m = \tau(b^m R b^{-m}) = P(\ldots, c_{k+m}, \ldots, t_{k+m}, \ldots).$$

Thus,

$$(4) \qquad N = \langle \ldots, c_{-1}, c_0, c_1, \ldots, t_{-1}, t_0, t_1, \ldots; \ldots, P_{-1}, P_0, P_1, \ldots \rangle.$$

We shall show that those generators of N other than the t_k freely generate a free group. Again, as in Case 2, since R is cyclically reduced, $\tau(R) = P$ is cyclically reduced in the $c_k, \ldots, t_k$. Moreover, since R involves t, P involves some t_k. Let μ be the minimum subscript on t involved in P and let M be the maximum subscript on t involved in P. Finally, let N_i be generated by all the generators of (4) except those t_k with $k < \mu + i$ or $k > M + i$, and let

$$P(\ldots, c_{k+i}, \ldots, t_{k+i}, \ldots)$$

be the single defining relator of N_i. Thus,

$$N_i = \langle \ldots, c_{-1}, c_0, c_1, \ldots, t_{\mu+i}, t_{\mu+i+1}, \ldots, t_{M+i}; P_i \rangle \tag{5}$$

Since R involves b, $\tau(R) = P$ has smaller word length than R, and so we may apply our inductive hypothesis to N_i.

Thus the subgroup of N_i generated by all the generators of (5) but $t_{\mu+i}$, or by all but t_{M+i}, is freely generated by these. Hence,

$$N_{0,1} = \langle \ldots, c_{-1}, c_0, c_1, \ldots, t_\mu, t_{\mu+1}, \ldots, t_{M+1}; P_0, P_1 \rangle$$

(note that all generators of N occur except t_k with $k < \mu$ or $k > M + 1$) is the free product of N_0 and N_1 with the free subgroup in each, freely generated by $\ldots, c_{-1}, c_0, c_1, \ldots, t_{\mu+1}, \ldots, t_M$, amalgamated under the identity mapping. Since $N_0 \subset N_{0,1}$, the elements $\ldots, c_{-1}, c_0, c_1, \ldots, t_\mu, t_{\mu+1}, \ldots, t_{M-1}$ freely generate a free subgroup of $N_{0,1}$ and so

$$N_{-1,1} = \langle \ldots, c_1, c_0, c_1, \ldots, t_{\mu-1}, t_\mu, \ldots, t_M, t_{M+1}; P_{-1}, P_0, P_1 \rangle$$

(note that all generators of N occur except t_k with $k < \mu - 1$ or $k > M + 1$) is the free product of N_{-1} and $N_{0,1}$ with the free subgroup in each, freely generated by $\ldots, c_{-1}, c_0, c_1. \ldots, t_\mu, \ldots t_{M-1}$, amalgamated under the identity mapping. Continuing in this way we construct a chain of groups

$$N_0 \subset N_{0,1} \subset N_{-1,1} \subset N_{-1,2} \subset \ldots \subset N_{-i+1,i} \subset N_{-i,i} \subset N_{-i,i+1} \subset \ldots$$

where $N_{-i,i}$ is the free product of N_{-i} and $N_{-i+1,i}$ with an amalgamated free subgroup, and $N_{-i,i+1}$ is the free product of $N_{-i,i}$ and N_{i+1} with an amalgamated free subgroup. By Problem 1.3.18, N is the union of this chain. Since the generators of N_i in (5) other than the t_k freely generate a free group by inductive hypothesis, this is true for N_0 in particular; but $N_0 \subset N$ and so the generators of N in (4) other than the t_k freely generate a free group in N.

Returning now to the supposed relator in G, $U(b, c, \ldots)$, which does not involve t, we see that if $U(b, c, \ldots)$ is cyclically reduced, not the empty word, and defines the identity in G then $\tau(U)$ is cyclically reduced in the $c_k, \ldots, t_k$ and does not involve any t_k. Since $\sigma_b(U) = 0$, U must involve some generator of G other than b, and so $\tau(U)$ is not the empty word but defines the identity in N, contrary to the generators of N other than the t_k being free. Thus every non-trivial relator in G involves t, and the generators of G other than t are free.

CASE 4. R involves at least three generators b, c, t and the exponent sum on each generator in R, other than t, is different from zero.

We shall attempt to reduce this case to the previous one. To do this we change G (enlarging it as we shall show later; see Corollary 4.10.2). Suppose $\sigma_b(R) = \beta$, $\sigma_c(R) = \gamma$, where we know that β and γ are different from zero.

To introduce a zero exponent sum we first replace

$$G = \langle b, c, \ldots, t; R(b, c, \ldots, t)\rangle$$

by

$$E = \langle x, c, \ldots, t; R(x^\gamma, c, \ldots, t)\rangle$$

Clearly, G can be mapped homomorphically into E by mapping $b \to x^\gamma$, $c \to c, \ldots, t \to t$. Hence, if we can show that $x, c, \ldots$ (without t) freely generate a free subgroup of E, it follows immediately that $b, c, \ldots$ (without t) freely generate a free subgroup of G; for, if the homomorphic image of generators is a free set of generators, so is the original set of generators. By applying Tietze transformations to E we will now obtain a zero exponent sum. To do this, we replace c by $yx^{-\beta}$, i.e., let $y = cx^\beta$. Thus

$$E = \langle x, c, \ldots, t, y; R(x^\gamma, c, \ldots, t), y = cx^\beta\rangle,$$

$$E = \langle x, c, \ldots, t, y; R(x^\gamma, c, \ldots, t), c = yx^{-\beta}\rangle,$$

$$E = \langle x, y, \ldots, t; R(x^\gamma, yx^{-\beta}, \ldots, t)\rangle. \tag{6}$$

Again $x, c, \ldots$ will be a free set of generators if and only if $x, y, \ldots$ (excluding t) is a free set of generators (since $y = cx^\beta$, the second set of generators for E is obtained from the first by a Nielsen transformation). Now

$$\sigma_x(R(x^\gamma, yx^{-\beta}, \ldots, t)) = \beta\gamma - \gamma\beta = 0.$$

Moreover, when $R(x^\gamma, yx^{-\beta}, \ldots, t)$ is cyclically reduced in $x, y, \ldots, t$ it must still involve t. For, if $R(x^\gamma, yx^{-\beta}, \ldots, t)$ when cyclically reduced does not involve t, then $R(x^\gamma, (cx^\beta)x^{-\beta}, \ldots, t)$ when cyclically reduced will not involve t; hence $R(x^\gamma, c, \ldots, t)$ will not involve t, for it is cyclically reduced. But then $R(b, c, \ldots, t)$ would not involve t. Thus $R(x^\gamma, yx^{-\beta}, \ldots, t)$ when cyclically reduced in $x, y, \ldots, t$ still involves t.

Unfortunately, $R(x^\gamma, yx^{-\beta}, \ldots, t)$ may have greater word length in $x, y, \ldots, t$ than $R(b, c, \ldots, t)$ has in $b, c, \ldots, t$; however, this will then be due solely to extra x-symbols. In applying Case 3 to (6), we consider N

the normal subgroup generated by $y, \ldots, t$; thus the x-symbols in $R(x^\gamma, yx^{-\beta}, \ldots, t)$ will contribute no symbols to $\tau(R(x^\gamma, yx^{-\beta}, \ldots, t))$. Hence, $\tau(R)$ has smaller word length than R and we have our result as in Case 3.

Since Cases 1 through 4 take care of all possibilities, we have established our assertion for G. ◀

COROLLARY 4.10.1. *Every finitely generated group with one defining relator is a subgroup of some group with two generators and a single defining relator.*

PROOF. If H is a free group the result is clear. Suppose

$$H = \langle b_0, b_1, \ldots, b_n; P(\ldots, b_k, \ldots)\rangle.$$

where P is cyclically reduced, and all the generators of H are involved in P. Define $R(b, t) = P(\ldots, t^k b t^{-k}, \ldots)$. Then, as in Case 2 of the proof of Theorem 4.10, N, the normal subgroup of $G = \langle b, t; R(b, t)\rangle$ generated by b, has the presentation (1) and contains N_i, as given by (2) as a subgroup. When $i = 0$, $N_0 = H$, and so G has H as a subgroup.

If P does not involve all the generators of H, replace one generator involved in P by the product of some new generator and all generators not involved in P. Then H will be given by a single defining relator which involves all its generators. ◀

As an illustration of Corollary 4.10.1, let

$$H = \langle x, y, z, w; x^3yzyz^{-1}\rangle.$$

We first let $z = uw$. Then

$$H = \langle x, y, u, w; x^3yuwyw^{-1}u^{-1}\rangle.$$

If we let $x = b_0$, $y = b_1$, $u = b_2$, $w = b_3$, then

$$\begin{aligned} R(b, t) &= b^3 \cdot tbt^{-1} \cdot t^2bt^{-2} \cdot t^3bt^{-3} \cdot tbt^{-1} \cdot t^3b^{-1}t^{-3} \cdot t^2b^{-1}t^{-2} \\ &= b^3tbtbtbt^{-2}bt^2b^{-1}t^{-1}b^{-1}t^{-2}. \end{aligned}$$

Corollary 4.10.1 is a special case of the result that any denumerable group with r defining relators is a subgroup of a group with two generators and r defining relators (B. H. Neumann, 1954).

COROLLARY 4.10.2. *If*

$$E = \langle x, c, \ldots, t; R(x^\gamma, c, \ldots, t)\rangle, \qquad \gamma \neq 0,$$

then the subgroup G of E generated by x^γ, $c, \ldots, t$ has the presentation

$$G = \langle b, c, \ldots, t; R(b, c, \ldots, t)\rangle$$

where b corresponds to x^γ, c to $c, \ldots,$ and t to t.

PROOF. If $R(b, c, \ldots, t)$ involves only b, then the result follows easily from the theory of free products. If $R(b, c, \ldots, t)$ involves some generator other than b, then b has infinite order in $G = \langle b, c, \ldots, t; R(b, c, \ldots, t)\rangle$ by Theorem 4.10. Hence,

$$\langle x, b, c, \ldots, t; b = x^{\gamma}, R(b, c, \ldots, t)\rangle \tag{7}$$

is the free product of the infinite cyclic group on x and the group G with the infinite cyclic subgroups generated by x^{γ} and b, respectively, amalgamated under $x^{\gamma} \to b$. By a Tietze transformation we may delete b from the presentation (7) and obtain

$$\langle x, c, \ldots, t; R(x^{\gamma}, c, \ldots, t)\rangle$$

which is precisely E. Thus (7) also presents E, and the subgroup generated by $b = x^{\gamma}$ and $c, \ldots, t$ has the required presentation. (It should be mentioned that Corollary 4.10.2 can be proved directly and rather easily without using Theorem 4.10.) ◀

Before using the Freiheitssatz to help solve the word problem for any group with one defining relator, we shall apply the construction used in its proof to show that a certain group with two generators and one defining relator is non-Hopfian. (It can be shown that this group is the simplest example of a finitely generated non-Hopfian group.)

Consider the group

$$G = \langle b, t; tb^2t^{-1} = b^3\rangle; \tag{8}$$

G is non-Hopfian, i.e., isomorphic to a proper factor group. For when we suppose $b = (b^{-1}t^{-1}bt)^2$ is added to the presentation (8) for G, we obtain

$$H = \langle b, t; t^{-1}b^2t = b^3, b = (b^{-1}t^{-1}bt)^2\rangle. \tag{9}$$

Applying a Tietze transformation to introduce $c = b^{-1}t^{-1}bt$ as a generator, we obtain

$$H = \langle b, t, c; t^{-1}b^2t = b^3, b = (b^{-1}t^{-1}bt)^2, c = b^{-1}t^{-1}bt\rangle$$

and so

$$H = \langle b, t, c; t^{-1}b^2t = b^3, b = c^2, c = b^{-1}t^{-1}bt\rangle.$$

We can now eliminate b, replacing it by c^2, and obtain

$$H = \langle t, c; t^{-1}c^4t = c^6, c = c^{-2}t^{-1}c^2t\rangle.$$

Changing the form of the last defining relation we have

$$H = \langle t, c; t^{-1}c^4t = c^6, t^{-1}c^2t = c^3\rangle;$$

and, since the first defining relation is derivable from the second,

$$H = \langle c, t; t^{-1}c^2t = c^3\rangle \tag{10}$$

Therefore, H is isomorphic to G. We show now that the relation $b = (b^{-1}t^{-1}bt)^2$ is not valid in G and so H is a proper factor group of G. To do this we use relators, rather than relations. G has the single defining relator $t^{-1}b^2tb^{-3}$; we note that $b(b^{-1}t^{-1}bt)^{-2}$ defines an element in N, the normal subgroup of G generated by b. As in Case 2 of the proof of the Freiheitssatz, we obtain the presentation for N

$$N = \langle \ldots, b_{-1}, b_0, b_1, \ldots; \ldots, b_{-1}^2 b_0^{-3}, b_0^2 b_1^{-3}, \ldots \rangle$$

If $b(b^{-1}t^{-1}bt)^{-2}$ is rewritten in N, it becomes

$$b_0(b_0^{-1}b_{-1})^{-2} = b_0(b_0^{-1}b_{-1})^{-1} \cdot (b_0^{-1}b_{-1})^{-1} = b_0 b_{-1}^{-1} b_0 b_{-1}^{-1} b_0.$$

Now $N_0 = \langle b_{-1}, b_0; b_{-1}^2 b_0^{-3} \rangle$ is the subgroup of N generated by b_{-1}, b_0 as was demonstrated in the proof of Case 2 of Theorem 4.10; but

$$N_0 = \langle b_{-1}, b_0; b_{-1}^2 = b_0^3 \rangle$$

is the free product of the infinite cyclic groups $\langle b_{-1} \rangle$ and $\langle b_0 \rangle$ with the infinite cyclic subgroups generated by b_{-1}^2 and b_0^3 amalgamated. By the solution of the word problem for amalgamated free products, $b_0 b_1^{-1} b_0 b_1^{-1} b_0$ does not define 1 in N_0; for b_0, b_1^{-1}, b_0, b_1^{-1}, b_0 alternate from the factors $\langle b_0 \rangle$ and $\langle b_{-1} \rangle$, and do not define elements in the amalgamated subgroups. Thus G as given in (8) is isomorphic to its proper factor group H as given in (9) and (10), and so G is non-Hopfian.

As another application of Theorem 4.10 and its method of proof we prove the *Conjugacy Theorem for Groups with One Defining Relator.*

THEOREM 4.11. *Let* $G = \langle a_1, \ldots, a_n;\ R(a_1, \ldots, a_n) \rangle$ *and* $H = \langle a_1, \ldots, a_n; S(a_1, \ldots, a_n) \rangle$. *Then* G *is isomorphic to* H *under the mapping* $a_\nu \to a_\nu$ *if and only if* $R(a_1, \ldots, a_n)$ *and* $S^\epsilon(a_1, \ldots, a_n)$ *are conjugate in the free group on* $a_1, \ldots, a_n$, *for* $\epsilon = 1$ *or* -1.

PROOF. Before proceeding to the proof of our theorem, we note that the theorem is equivalent to proving that the two words $R(a_1, \ldots, a_n)$ and $S(a_1, \ldots, a_n)$ determine the same normal subgroup of the free group F on $a_1, \ldots, a_n$ if and only if R and S^ϵ are conjugate in F for $\epsilon = 1$ or -1. For the mapping $a_\nu \to a_\nu$ induces the identity mapping in F. If T and U are the normal subgroups of F generated by R and S, respectively, then $G \simeq F/T$ and $H \simeq F/U$; for the identity mapping of F to induce an isomorphism between F/T and F/U, T must be carried onto U and so $T = U$. Thus $a_\nu \to a_\nu$ induces an isomorphism between G and H if and only if R and S determine the same normal subgroup of F; and so the equivalence is established.

Clearly if R and S^ϵ, $\epsilon = 1$ or -1, are conjugate in F, then R and S determine the same normal subgroup of F; thus $a_\nu \to a_\nu$ will induce an isomorphism between G and H.

It is much more difficult to show that if $a_\nu \to a_\nu$ induces an isomorphism between G and H then R and S^ϵ are conjugate for $\epsilon = 1$ or -1. We can easily conclude that R is freely equal to a product of conjugates of S and S^{-1} and that S is freely equal to a product of conjugates of R and R^{-1}, since R and S determine the same normal subgroup of F.

Any attempt to show, by cancellation arguments in the free group F, that if R is the product of more than one conjugate of S or S^{-1} then S cannot be the product of conjugates of R and R^{-1}, presents great difficulty. Instead, we shall use the theory of groups with a single defining relator, and induction on the minimum of the lengths of R and S. Clearly, we may assume that R and S are cyclically reduced, and that R has length less than or equal to the length of S.

Suppose first that R has length zero, i.e., R is the empty word. Then G is the free group on $a_1, \ldots, a_n$. Since $G \simeq H$ under $a_\nu \to a_\nu$, and $S(a_1, \ldots, a_n)$ is a relator in H, S is the empty word. Thus we have our assertion.

Suppose that R or S involves only one of the generators, say R involves only a_n. Then $R = a_n^{\,k}$, $k \neq 0$, and so a_n has order k in G. Since $G \simeq H$ under $a_\nu \to a_\nu$, a_n has order k in H. If S involved any generator other than a_n, a_n would have infinite order by the Freiheitssatz (Theorem 4.10). Thus S involves only a_n, and since a_n has order k in H, $S = a_n^{\,\epsilon k}$, $\epsilon = 1$ or -1. Thus we again have our assertion, and so may assume that R and S involve at least two generators.

Next, we note that R and S must involve the same generators, for suppose that R involves, say , a_n, but S does not. Then, by Theorem 4.10, $a_1, \ldots, a_{n-1}$ freely generate a free group in G, although in H, $a_1, \ldots, a_{n-1}$ satisfy the non-trivial relation $S = 1$; this is contrary to $G \simeq H$ under $a_\nu \to a_\nu$. Moreover, if R has exponent sum zero on some generator, then S which is a product of conjugates of R and R^{-1} will also have exponent sum zero on this generator.

CASE 1. R has zero exponent sum on some generator, say, a_n, and involves some other generator, say, a_1.

Again we find it convenient to relabel our generators so as to free the subscripts $1, \ldots, n$ for other work; we shall represent a_n by t and $a_1, a_2, \ldots$ by $b, c, \ldots$. Thus R, and so S, has exponent sum zero on t.

If N and L are the normal subgroup of G and H, respectively, generated by $b, c, \ldots$, then we may use a Reidemeister-Schreier rewriting process τ to obtain a presentation for N and L using t^k, $k = 0, \pm 1, \pm 2, \ldots$ as representatives; abbreviating $t^k b t^{-k}, t^k c t^{-k}, \ldots$ by $b_k, c_k, \ldots$, respectively, if

$$\tau(t^i R t^{-i}) = P_i(\ldots, b_k, \ldots, c_k, \ldots)$$

and

$$\tau(t^i S t^{-i}) = Q_i(\ldots, b_k, \ldots, c_k, \ldots)$$

then

(11) $N = \langle \ldots, b_{-1}, b_0, b_1, \ldots, c_{-1}, c_0, c_1, \ldots; \ldots, P_{-1}, P_0, P_1, \ldots \rangle$

and

(12) $L = \langle \ldots, b_{-1}, b_0, b_1, \ldots, c_{-1}, c_0, c_1, \ldots; \ldots, Q_{-1}, Q_0, Q_1, \ldots \rangle$.

Since $t \to t$, $b \to b$, $c \to c, \ldots$ induces an isomorphism between G and H, and $b_k = t^k b t^{-k}$, $c_k = t^k c t^{-k}, \ldots$, it follows that $b_k \to b_k$, $c_k \to c_k, \ldots$ induces an isomorphism between N and L. Now we cannot show that $P_0(\ldots, b_k, \ldots, c_k, \ldots)$ and $Q_0^{\epsilon}(\ldots, b_k, \ldots, c_k, \ldots)$, $\epsilon = 1$ or -1, must be conjugate; for example, if $R = tbt^{-1}c$ and $S = t^{-1}ctb$, then $P_0 = \tau(R) = b_1 c_0$ while $Q_0 = \tau(S) = c_{-1} b_0$; however, P_0 is conjugate to $Q_1 = \tau(tSt^{-1}) = c_0 b_1$. Thus we shall try to show that P_0 is conjugate to some Q_j^{ϵ}, $\epsilon = 1$ or -1, j an integer.

Since R involves b, S must also involve b. Let μ be the minimum and M the maximum subscript k such that b_k is involved in P_0; let δ be the minimum and D the maximum subscript k such that b_k is involved in Q_0. Then $Q_{\mu-\delta}$ involves b_μ and no b_k with smaller subscript. We show that P_0 is a conjugate of $Q^{\epsilon}_{\mu-\delta}$, $\epsilon = 1$ or -1.

To do this, we first show that M is the largest subscript k such that b_k is involved in $Q_{\mu-\delta}$. Since $\mu - \delta + D$ is the largest subscript on b involved in $Q_{\mu-\delta}$, it suffices to prove that $\mu - \delta + D = M$, i.e., $M - \mu = D - \delta$. Suppose $M - \mu > D - \delta$ (intuitively, that P involves more of the b_k than Q). We show that $b_\delta, \ldots, b_D$ together with all the generators of L in (12) other than b_k, $k < \delta$ or $k > D$, freely generate a free group in L, contrary to the relation $Q_0 = 1$, which involves only $b_\delta, \ldots, b_D$ of the b_k generators of L. Since $N \simeq L$ under $b_k \to b_k$, $c_k \to c_k, \ldots$ it suffices to show that the generators of N in (11) other than b_k, $k < \delta$ or $k > D$, freely generate a free group in N. If

(13) $$N_i = \langle \ldots, c_k, \ldots, b_{\mu+i}, \ldots, b_{M+i}; P_i \rangle,$$

then

$$N_{\delta-\mu} = \langle \ldots, c_k, \ldots, b_\delta, \ldots, b_{\delta+(M-\mu)}; P_{\delta-\mu} \rangle.$$

Since $M - \mu > D - \delta$, $\delta + M - \mu > D$; by Theorem 4.10,

$$\ldots, c_k, \ldots, b_\delta, \ldots, b_D$$

freely generate a free group in $N_{\delta-\mu}$. Examining next

$$N_{\delta-\mu-1} = \langle \ldots, c_k, \ldots, b_{\delta-1}, \ldots, b_{\delta+M-\mu-1}; P_{\delta-\mu-1} \rangle,$$

we have by Theorem 4.10 that $\ldots, c_k \ldots, b_\delta, \ldots, b_{\delta+M-\mu-1}$ freely generate a free group in $N_{\delta-\mu-1}$. Hence,

$$N_{\delta-\mu-1,\delta-\mu} = \langle \ldots, c_k, \ldots, b_{\delta-1}, \ldots, b_{\delta+M-\mu}; P_{\delta-\mu-1}, P_{\delta-\mu} \rangle$$

is the free product of $N_{\delta-\mu-1}$ and $N_{\delta-\mu}$ with the free subgroup in each, freely generated by $\ldots, c_k, \ldots, b_\delta, \ldots, b_{\delta+M-\mu-1}$, amalgamated under the identity mapping. Since $N_{\delta-\mu-1}$ and $N_{\delta-\mu}$ are contained as subgroups of $N_{\delta-\mu-1,\delta-\mu}$ we may first construct $N_{\delta-\mu-1,\delta-\mu+1}$ by taking the free product of $N_{\delta-\mu-1,\delta-\mu}$ and $N_{\delta-\mu+1}$ amalgamating the free group in each, freely generated by $\ldots, c_k, \ldots, b_{\delta+1}, \ldots, b_{\delta+M-\mu}$, under the identity mapping; we may then construct $N_{\delta-\mu-2,\delta-\mu+1}$ by taking the free product of $N_{\delta-\mu-2}$ with $N_{\delta-\mu-1,\delta-\mu+1}$ amalgamating the free group in each, freely generated by $\ldots, c_k, \ldots, b_{\delta-1}, \ldots, b_{\delta+M-\mu-2}$ under the identity mapping. Continuing in this way, we obtain a chain of groups whose union is N. Thus $N_{\delta-\mu}$ is a subgroup of N. Since $\ldots, c_k, \ldots, b_\delta, \ldots, b_D$ freely generate a free group in $N_{\delta-\mu}$, they do so in N and also in L; this contradicts the possibility that Q_0 involves only $\ldots, c_k, \ldots, b_\delta, \ldots b_D$. Thus $M - \mu \not> D - \delta$. Repeating the above argument with the roles of P and Q interchanged we have that $D - \delta \not> M - \mu$. Hence, $M - \mu = D - \delta$. In a similar fashion we can show that if c_k is involved in P_0, then the difference between the maximum and minimum subscript occurring on c in P_0 is the same as that for Q_0; similarly for all of the generators $b, c, \ldots$ (other than t).

We now know that P_0 and $Q_{\mu-\delta}$ both involve $b_\mu, \ldots, b_M$. Moreover, if P_0 involves $c_\eta, \ldots, c_E$, then $Q_{\mu-\delta}$ must involve $c_{\eta+i}, \ldots, c_{E+i}$, since $E - \eta$ must be the difference between the maximum c-subscript and minimum c-subscript in $Q_{\mu-\delta}$. To show that $i = 0$, we show that if $i \neq 0$, then in L, $b_\mu, \ldots, b_M, c_\eta, \ldots, c_E$ and all of the other generators other than the b_k and c_k freely generate a free group. The proof of this is similar to the previous proof and is left as an exercise (see Problem 31). Since $N \simeq L$ under $b_k \to b_k$, $c_k \to c_k, \ldots$, it follows that the non-trivial relator P_0 in N cannot involve only $b_\mu, \ldots, b_M$ and $c_\eta, \ldots, c_E$ but must involve other b_k or c_k, contradicting the definition of μ, M, η, E. Thus P_0 and $Q_{\mu-\delta}$ contain the same maximum and minimum subscripts on all generators $b, c, \ldots$. But

$$\langle b_\mu, \ldots, b_M, c_\eta, \ldots, c_E, \ldots; P_0\rangle \tag{14}$$

(note that all other generators of N occur except b_k, $k < \mu$ or $k > M$, and c_k, $k < \eta$ or $k > E$) and

$$\langle b_\mu, \ldots, b_\mu, c_\eta, \ldots, c_E, \ldots; Q_{\mu-\delta}\rangle \tag{15}$$

[note that the same generators occur in (15) as in (16)] are the subgroups of N and L, respectively, generated by $b_\mu, \ldots, b_M, c_\eta, \ldots, c_E, \ldots$. Hence, (14) and (15) are isomorphic under the mapping $b_k \to b_k$, $c_k \to c_k$, $\ldots$, which induces an isomorphism between N and L. Since P_0 has smaller length than R, by inductive hypothesis P_0 is a conjugate of $Q^\epsilon_{\mu-\delta}$, $\epsilon = 1$ or -1, in the free group on $b_\mu, \ldots, b_M, c_\eta, \ldots, c_E, \ldots$. Thus

$P_0 \approx WQ^{\epsilon}_{\mu-\delta}W^{-1}$, where W is a word in $b_\mu, \ldots, b_M, c_\eta, \ldots, c_E, \ldots$. Now $P_0 = \tau(R)$, $W = \tau(U)$, and, $Q_{\mu-\delta} = \tau(t^{\mu-\delta}St^{\delta-\mu})$; hence, by familiar properties of any Reidemeister-Schreier rewriting process,

$$\tau(R) \approx \tau(Ut^{\mu-\delta}S^{\epsilon}t^{\delta-\mu}U^{-1}).$$

But if τ of two words are freely equal, the original words must have been freely equal (see Problem 2.3.2). Thus R is freely equal to a conjugate of S^{ϵ}, we have our assertion, and Case 1 is finished.

CASE 2. R involves at least two generators, say b, t, neither of which has exponent sum zero.

We shall reduce this case to the previous one. Suppose $\sigma_b(R) = \beta$, $\sigma_t(R) = \alpha$ and $\beta, \alpha \neq 0$. First we introduce a generator x so that $b = x^\alpha$. Next we introduce a generator y so that $t = yx^{-\beta}$. Thus G and H are replaced by

(16) $$\langle x, c, \ldots, y; R(x^\alpha, c, \ldots, yx^{-\beta})\rangle$$

and

(17) $$\langle x, c, \ldots, y; S(x^\alpha, c, \ldots, yx^{-\beta})\rangle,$$

respectively. Since $R(b, c, \ldots, t)$ and $S(b, c, \ldots, t)$ are each products of conjugates of the other or of the other's inverse in the free group on $b, c, \ldots t$, it follows that $R(x^\alpha, c, \ldots, yx^{-\beta})$ and $S(x^\alpha, c, \ldots, yx^{-\beta})$ define the same normal subgroup in the free group on $x, c, \ldots, y$. Thus (16) and (17) are isomorphic under the mapping $x \to x$, $c \to c, \ldots, y \to y$. Since $R(x^\alpha, c, \ldots, yx^{-\beta})$ has zero exponent sum on x, $S(x^\alpha, c, \ldots, yx^{-\beta})$ does also. Letting N and L be the normal subgroups of (16) and (17), respectively, generated by $c, \ldots, y$, we may obtain a presentation similar to (11) and (12). Since the x-symbols of $R(x^\alpha, c, \ldots, yx^{-\beta})$ are not replaced under τ, the length of $\tau(R(x^\alpha, c, \ldots, yx^{-\beta}))$ will be less than the length of R. Hence, we may make use of the inductive hypothesis as in Case 1 and obtain that $R(x^\alpha, c, \ldots, yx^{-\beta})$ is freely equal to a conjugate of $S^{\epsilon}(x^\alpha, c, \ldots, yx^{-\beta})$, $\epsilon = 1$ or -1, in the free group on $x, c, \ldots, y$. Letting $t = yx^{\beta}$, and using that $x \to x$, $c \to c, \ldots, t \to yx^{-\beta}$ is an isomorphism between the free group on $x, c, \ldots, t$ and the free group on $x, c, \ldots, y$, we have that $R(x^\alpha, c, \ldots, t) \approx$ a conjugate of $S^{\epsilon}(x^\alpha, c, \ldots, t)$ in the free group on $x, c, \ldots, t$. Finally, since $R(b, c, \ldots, t)$ and $S(b, c, \ldots, t)$ are cyclically reduced, $R(x^\alpha, c, \ldots, t)$ and $S^{\epsilon}(x^\alpha, c, \ldots, t)$ are cyclically reduced. Being conjugate in the free group on $x, c, \ldots, t$, $R(x^\alpha, c, \ldots, t)$ and $S^{\epsilon}(x^\alpha, c, \ldots, t)$ must be cyclic permutations of one another. Since $R(x^\alpha, c, \ldots, t)$ is a word in $x^\alpha, c, \ldots, t$, in cyclically permuting $S^{\epsilon}(x^\alpha, c, \ldots, t)$ to obtain $R(x^\alpha, c, \ldots, t)$ no x^α can be "split." Thus $R(b, c, \ldots, t)$ and $S^{\epsilon}(b, c, \ldots, t)$ are cyclic permutations of one another, and we have the assertion of the theorem.

This completes Case 2 and thereby the proof of Theorem 4.11. ◄

Corollary 4.11. *If* $G = \langle a_1, \ldots, a_n; V^k(a_1, \ldots, a_n)\rangle$, $k > 1$, V *non-empty, then* V *defines an element exactly of order* k *in* G.

Proof. Since cyclic reduction replaces V by a conjugate, we may assume that V is cyclically reduced. Clearly, the order d of V in G divides k. Since V^d is a relator in G, V^d is in the normal subgroup generated by V^k in the free group on $a_1, \ldots, a_n$. Thus, by Theorem 4.11, V^d and V^k, or V^d and V^{-k}, are conjugate in the free group on $a_1, \ldots, a_n$. Since V^d and V^k are cyclically reduced, they must have the same length, and so $d = k$. ◀

As another application of Theorem 4.11, we show that $G = \langle a, b; R(a, b)\rangle$ *is a free abelian group of rank two if and only if* $R(a, b)$ *is a conjugate of* $aba^{-1}b^{-1}$ *or its inverse.* For suppose that $\langle a, b; R(a, b)\rangle$ is a free Abelian group of rank two. Consider the homomorphism of $A = \langle a, b; aba^{-1}b^{-1}\rangle$ onto $G = \langle a, b; R(a, b)\rangle$ given by $a \to a$, $b \to b$. Since the free Abelian group of rank two is Hopfian, the mapping is an isomorphism. Thus by Theorem 4.11, $R(a, b)$ is a conjugate of $aba^{-1}b^{-1}$ or its inverse.

As a further application of Theorem 4.10 and its method of proof, we characterize those groups with one defining relator which have elements of finite order.

Theorem 4.12. *Let* $G = \langle a_1, \ldots, a_n; R(a_1, \ldots a_n)\rangle$. *Then* G *has an element* ($\neq 1$) *of finite order if and only if* R *is the* k*th power,* $k > 1$, *of some non-empty word* V *in the free group on* $a_1, \ldots, a_n$.

Proof. Suppose $G = \langle a_1, \ldots, a_n; V^k\rangle$, where V is non-empty and $k > 1$. Then by the Corollary 4.11, V defines an element of order k in G; hence, G has an element ($\neq 1$) of finite order.

To show the converse, suppose

$$G = \langle a_1, \ldots, a_n; R(a_1, \ldots, a_n)\rangle$$

has an element ($\neq 1$) of finite order. We use induction on the length of R to show that R is a true power in the free group on $a_1, \ldots, a_n$. If R has length two or involves only one generator, then the assertion easily follows.

Suppose next that the assertion holds for any group with one defining relator whose length is less than the length of R. Clearly then, we may assume R is cyclically reduced.

Case 1. Suppose R has exponent sum zero on some generator involved in it, say, a_n; R also involves at least one other generator, say a_1. We again represent a_n by t and $a_1, a_2, \ldots$ by $b, c, \ldots$. Now every relator in G is a product of conjugates of R and R^{-1} and so has exponent sum zero on t. In particular, if U defines an element ($\neq 1$) of finite order d in G, then $\sigma_t(U^d) = d\sigma_t(U) = 0$. Thus U defines an element in the normal subgroup N of G generated by $b, c, \ldots$. If τ is the Reidemeister-Schreier rewriting process using t^m as representatives, and $b_m, c_m, \ldots$ represent $t^mbt^{-m}, t^mct^{-m}, \ldots$ and $P_m = \tau(t^mRt^{-m})$, m any integer, then

$$N = \langle \ldots, b_{-1}, b_0, b_1, \ldots, c_{-1}, c_0, c_1, \ldots; \ldots, P_{-1}, P_0, P_1, \ldots \rangle.$$

Moreover, if μ and M are the minimum and maximum subscripts that occur on b in P_0, we define

$$N_i = \langle \ldots, c_{-1}, c_0, c_1, \ldots, b_{\mu+i}, \ldots, b_{M+i}; P_i \rangle. \tag{18}$$

Then we construct

$$N_{0,1} = \langle \ldots, c_{-1}, c_0, c_1, \ldots, b_\mu, \ldots, b_{M+1}; P_0, P_1 \rangle,$$

which is the free product of N_0 and N_1 with the free subgroups in each, freely generated by

$$\ldots, c_{-1}, c_0, c_1 \ldots, b_{\mu+1}, \ldots, b_M,$$

amalgamated under the identity mapping. Continuing in this way, we construct the chain of groups

(19)

$$N_0 \subset N_{0,1} \subset N_{-1,1} \subset N_{-1,2} \subset \ldots \subset N_{-i+1,i} \subset N_{-i,i} \subset N_{-i,i+1} \subset \ldots.$$

Finally, we obtain N as the union of the constructed groups. Since N has an element ($\neq 1$) of finite order, one of the groups in (19) has an element ($\neq 1$) of finite order. But the groups in (19) are obtained from the groups in (18) by a repeated free product with amalgamation construction. By Corollary 4.4.5, one of the groups in (18) must have an element ($\neq 1$) of finite order. Since N_0 is isomorphic to N_i (under the mapping $b_m \rightarrow b_{m+i}$, for $\mu \leq m \leq M$ and $c_m \rightarrow c_{m+i}, \ldots$ for all integers m), N_0 must have an element ($\neq 1$) of finite order. By inductive hypothesis, since the length of P_0 is less than the length of R,

$$P_0(\ldots, b_m, \ldots, c_m, \ldots) \approx W^k(\ldots, b_m, \ldots, c_m, \ldots)$$

for some $k > 1$. But then, by a property of the rewriting process τ,

$$\begin{aligned} R(b, c, \ldots, t) &\approx P_0(\ldots, t^mbt^{-m}, \ldots, t^mct^{-m}, \ldots) \\ &\approx W^k(\ldots, t^mbt^{-m}, \ldots, t^mct^{-m}, \ldots) \\ &= V^k(b, c, \ldots, t), \end{aligned}$$

where

$$V(b, c, \ldots, t) = W(\ldots, t^mbt^{-m}, \ldots, t^mct^{-m}, \ldots).$$

Hence, our assertion holds.

CASE 2. There remains the case where each exponent sum on a generator involved in $R(b, c, \ldots, t)$ is different from zero. Again we introduce the group

$$(20) \qquad \langle x, c, \ldots, t; R(x^\alpha, c, \ldots, t)\rangle$$

where $\alpha = \sigma_t(R)$. The group in (20) is simply the amalgamated product of G and the infinite cyclic group on x, with the subgroups generated by x^α and b amalgamated under $x^\alpha \rightarrow b$. Since G has an element ($\neq 1$) of finite order, (20) must also have such an element. Applying a Tietze transformation, (20) can be presented as

$$(21) \qquad \langle x, c, \ldots, y; R(x^\alpha, c, \ldots, yx^{-\beta})\rangle$$

where $\beta = \sigma_b(R)$, and in (20), t was replaced by $yx^{-\beta}$. Since the defining relator in (21) has zero exponent sum on x, if we consider the normal subgroup of (21) generated by $c, \ldots, y$, we can show as in the previous case that $R(x^\alpha, c, \ldots, yx^{-\beta}) = V^k(x, c, \ldots, y)$, for some $k > 1$, in the free group on $x, c, \ldots, y$. Letting $y = tx^\beta$, $R(x^\alpha, c, \ldots, t) = S^k(x, c, \ldots, t)$ in the free group on $x, c, \ldots, t$. It remains then to show that $S(x, c, \ldots, t)$ is actually some $W(x^\alpha, c, \ldots, t)$. Since $R(b, c, \ldots, t)$ is cyclically reduced, $R(x^\alpha, c, \ldots, t)$ is cyclically reduced. Hence, since we may assume $S(x, c, \ldots, t)$ is freely reduced, it must by cyclically reduced and an initial and terminal segment of $R(x^\alpha, c, \ldots, t)$. Clearly, each x^η which occurs in S surrounded by generators other than x must have η divisible by α. Moreover, the beginning of S and the end of S are the beginning and end of $R(x^\alpha, c, \ldots, t)$, and so any power x^η occurring at the beginning or end of S has η divisible by α. Thus,

$$R(x^\alpha, c, \ldots, t) = Q^k(x^\alpha, c, \ldots, t)$$

and finally,

$$R(b, c \ldots, t) = Q^k(b, c, \ldots, t)$$

in the free group on $b, c, \ldots, t$, ◀

It should be noted that in the free group on $a_1, \ldots, a_n$, we can decide if a given $R(a_1, \ldots, a_n)$ is a true power. For $R = TST^{-1}$ where S is cyclically reduced, and R is a true power if and only if S is. In the case of the cyclically reduced word S, inspection of the initial segments of S whose lengths divide the length of S will reveal the possible roots of S.

Moreover, Theorem 4.12 may be used even when the word R cannot be specifically examined. Thus, if $R = V^k$, then k divides all exponent sums of R on each generator. In particular then, the word

$$b \cdot [U(b, c, \ldots, t), W(b, c, \ldots, t)],$$

where $[U, W] = U^{-1}W^{-1}UW$, cannot be a true power; for k must divide

$\sigma_b(R) = 1$. Hence,

$$\langle b, c, \ldots t; b[U, W]\rangle$$

has no element ($\neq 1$) of finite order.

We shall now characterize the elements of finite order in a group with one defining relator.

THEOREM 4.13. *Let* $G = \langle a_1, \ldots, a_n; V^k(a_1, \ldots, a_n)\rangle$ *where* $k > 1$, *and* $V(a_1, \ldots, a_n)$ *is not a true power in the free group on* $a_1, \ldots, a_n$. *Then every element of finite order in* G *is defined by a conjugate of a power of* V.

PROOF. Again we use induction on the length of V. If V has length one or involves only one generator, then the result follows from Corollary 4.1.4.

Assume then that the assertion holds for all groups with one defining relator U^k, where U is not a true power and has length less than the length of V.

Clearly, we may assume that V is cyclically reduced; otherwise, we replace V by a cyclically reduced conjugate of smaller length, and so obtain the assertion by inductive hypothesis.

Again we represent $a_1, \ldots, a_n$ by $b, c, \ldots, t$.

CASE 1. Suppose first that V has exponent sum zero on some generator, say t, and involves some other generator, say b. Since the defining relator V^k has zero exponent sum on t, all relators do, and hence, all elements of finite order are in the normal subgroup N of G generated by $b, c, \ldots$. Presenting N as in the previous proof, we have

$$N = \langle \ldots, b_{-1}, b_0, b_1, \ldots, c_{-1}, c_0, c_1, \ldots; \ldots, P_{-1}^k, P_0^{\,k}, P_1^{\,k}, \ldots\rangle,$$

where $P_i(\ldots, b_m, \ldots, c_m, \ldots) = \tau(t^i V t^{-i})$. Moreover, if μ and M are the smallest and largest subscripts that occur on b in P_0, then we define

$$N_i = \langle \ldots, c_{-1}, c_0, c_1, \ldots, b_{\mu+i}, \ldots, b_{M+i}; P_i^{\,k}\rangle; \tag{22}$$

using N_i we construct (as in the proof of the preceding theorem) the chain of groups

$$N_0 \subset N_{0,1} \subset N_{-1,1} \subset \ldots \subset N_{-i+1,i} \subset N_{-i,i} \subset N_{-i,i+1} \subset \ldots \tag{23}$$

and find that N is the union of this chain. Every element of finite order in N is therefore in one of the groups of the chain (23). But each group in the chain (23) is constructed from the groups in (22) by a repeated free product with amalgamation construction. Hence, by Corollary 4.4.5 each element of finite order in the groups of the chain (23), and so in G, is a conjugate of an element of finite order in the groups (22). Now, by a property of τ,

(24) $$t^i V(b, c, \ldots, t)t^{-i} \approx P_i(\ldots, t^m bt^{-m}, \ldots, t^m ct^{-m}).$$

Hence, if V is not a true power in the free group on $b, c, \ldots, t$, then P_i cannot be a true power in the free group on $\ldots, c_{-1}, c_0, c_1, \ldots, b_{\mu+i}, \ldots, b_{M+i}$. Thus, by inductive hypothesis, the elements of finite order in (22) are conjugates of powers of P_i. Since in G, the element defined by P_i, i.e.,

$$P_i(\ldots, t^m bt^{-m}, \ldots, t^m ct^{-m}, \ldots),$$

is by (24) a conjugate of V, it follows that every element of finite order in G is a conjugate of a power of V.

CASE 2. In the remaining case, V has non-zero exponent sum on all generators involved in V. If $\alpha = \sigma_t(V)$ and $\beta = \sigma_b(V)$, then we construct

(25) $$\langle x, c, \ldots, t; V^k(x^\alpha, c, \ldots, t)\rangle$$

and its Tietze transform, obtained by letting $t = yx^{-\beta}$,

(26) $$\langle x, c, \ldots, y; V^k(x^\alpha, c, \ldots, yx^{-\beta})\rangle.$$

Since $V(b, c, \ldots, t)$ is cyclically reduced, we can show that if $V(x^\alpha, c, \ldots, yx^{-\beta})$ is a true power, then $V(x^\alpha, c, \ldots, t)$ is a true power, and also $V(b, c, \ldots, t)$ is a true power. Hence, $V(x^\alpha, c, \ldots, yx^{-\beta})$ is not a true power. Using the method of proof in the preceding case, where the normal subgroup of (26) generated by $c, \ldots, y$ is taken as N, we can show that the elements of (26), of finite order, are conjugates in (26) of powers of $V(x^\alpha, c, \ldots, yx^{-\beta})$. Applying the Tietze transformation $y = tx^\beta$ to (26), we obtain that the elements in (25) of finite order are conjugates in (25) of powers of $V(x^\alpha, c, \ldots, t)$. Thus an element $W(\neq 1)$ of finite order in G, which is the subgroup of (25) generated by $x^\alpha, c, \ldots, t$ must be a conjugate of a power of $V(x^\alpha, c, \ldots, t)$. Hence, in (25),

$$W(x^\alpha, c, \ldots, t) = TV^q(x^\alpha, c, \ldots, t)T^{-1} \text{ where } T \text{ is some word in } x, c, \ldots, t.$$

We must show that T is in fact a word in $x^\alpha, c, \ldots, t$, i.e., that T defines an element of G.

Now, (25) is the free product of X, the infinite cyclic group on x, and the group G, with the infinite cyclic subgroups H, K, generated by x^α, b in X and G, respectively, amalgamated under φ: $x^\alpha \to b$. Since V^q has finite order and H is infinite cyclic, V^q is not in a conjugate of H. Hence, our problem reduces to showing that in $*(X, G, H, K, \varphi)$, which is the group in (25), if g and sgs^{-1} are in G, and g is not in a conjugate of H, then s is in G. The proof is immediate; for if s is not in G, s has the form $s_1 \ldots s_p$ where s_i alternate out of X and G, are not in H or K, and s_p is in X, or s_p is in G and s_{p-1} is in X. But, the representative length of

$$sgs^{-1} = s_1 \ldots s_{p-1}s_p g s_p^{-1} s_{p-1}^{-1} \ldots s_1^{-1}$$
$$= s_1 \ldots s_{p-1}(s_p g s_p^{-1}) s_{p-1}^{-1} \ldots s_1^{-1}$$

cannot be one if s_p is in X, or if s_p is in G and s_{p-1} is in X, and so sgs^{-1} cannot be in G.

Thus T defines an element of G, and so W which is an element of finite order in G must be a conjugate of a power of V in G. ◀

COROLLARY 4.13.1. *If* $G = \langle a_1, a_2, \ldots, a_n; V^k(a_1, \ldots, a_n)\rangle$ *is isomorphic to*

$$H = \langle a_1, a_2, \ldots, a_n; U^k(a_1, \ldots, a_n)\rangle$$

then

$$K = \langle a_1, a_2, \ldots, a_n; V(a_1, \ldots, a_n)\rangle$$

is isomorphic to

$$L = \langle a_1, a_2, \ldots, a_n; U(a_1, \ldots, a_n)\rangle.$$

PROOF. (Note that Theorem 4.11 need not apply since we are not given that $a_\nu \to a_\nu$ under the isomorphism.) Let $V = W^r$ where W is not a true power. Then

$$G = \langle a_1, \ldots, a_n; W^{kr}\rangle,$$

and so, by Corollary 4.11, W has order kr in G. Since every other element of finite order in G is a conjugate of a power of W, kr is the maximum finite order of an element in G. But $H \simeq G$; therefore, if $U = Y^s$, where Y is not a true power, then $s = r$. The elements in G of order dividing k are just the conjugates of powers of $W^r = V$. Similarly in H, the elements of order dividing k are just the conjugates of powers of $Y^r = U$. Hence, under the isomorphism between G and H, the normal subgroup N of G generated by V must correspond to the normal subgroup M of H generated by U. But then $G/N \simeq H/M$; since $G/N = K$ and $H/M = L$, we have our result. ◀

COROLLARY 4.13.2. *Let R and S be in the free group F on $a_1, \ldots, a_n$. If S^k is in the normal subgroup of F generated by R^k, then S is in the normal subgroup of F generated by R. In other words, if S^k is derivable from R^k, then S is derivable from R.*

PROOF. Suppose $R = W^r$ where W is not a true power. In the group G

$$G = \langle a_1, \ldots a_n; R^k\rangle = \langle a_1, \ldots, a_n; W^{rk}\rangle$$

S^k is the identity. Hence, S has finite order dividing k in G. Thus, in G, S must be a conjugate of $(W^r)^t = R^t$. Since R^k is the defining relator in

G, S as an element of the free group F must be freely equal to the product of a conjugate of R^t and conjugates of R^k and R^{-k}. Thus S is a product of conjugates of R and R^{-1}; hence, we have our result. ◀

The converse of Corollary 4.13.2 is false. For example, $a \cdot bab^{-1}$ is derivable from a but $(abab^{-1})^2$ is not derivable from a^2; indeed, $abab^{-1}abab^{-1}$ has syllable length eight in $\langle a, b; a^2 \rangle$. (The converse of Corollary 4.13.1 would be true if a conjecture on the isomorphism problem for groups with one defining relator were true; see Section 6.1. However, this is still unknown.)

We turn finally to the solution of the word problem for groups with a single defining relator. An essential part of the method will be to go to a normal subgroup as in the proof of Theorem 4.10. Since this normal subgroup is the union of a chain of groups each of which is constructed as an amalgamated product, we shall require the following lemma on amalgamated products.

LEMMA 4.9. *Let* $G = *(A_1, A_2, H_1, H_2, \varphi)$. *Suppose that we have a process for deciding if an element of* A_i *is in* H_i, *and if it is, to write it as an element of* H_i; *and suppose* φ *and* φ^{-1} *are each given by a specific process. Then we can decide if an element of* G *is in* A_i, *and if it is, we can write it as an element of* A_i.

PROOF. Let g be given as the product $g_1 g_2 \cdots g_n$, where the g_j are elements alternating out of the A_i, $i = 1, 2$. We show by induction on n that we can decide if g is in A_i, and if so write it as an element in A_i.

If $n = 1$, then $g = g_1$ is in either A_1 or A_2. Moreover, it is in both if and only if g_1 is in A_1 and in H_1, or g_1 is in A_2 and in H_2. Since this can be decided by assumption, and φ is specifically given, we can decide if g_1 is in A_i, and if so write it as an element of A_i.

Assume the decision can be made for all elements defined by words of apparent syllable length less than n, and suppose g has apparent syllable length n, $n > 1$, $g = g_1 g_2 \cdots g_n$. If none of the g_j is in H_1 or H_2, then g will have representative length $n > 1$ in G and so cannot be in A_i. If some g_j is in A_1 and H_1, or A_2 and H_2, say A_1 and H_1, then $\varphi(g_j)$ is in A_2 and H_2. In G,

$$\begin{aligned} g &= g_1 g_2 \cdots g_{j-1}\varphi(g_j)g_{j+1} \cdots g_n \\ &= g_1 g_2 \cdots g_{j-2}(g_{j-1}\varphi(g_j)g_{j+1})g_{j+2} \cdots g_n; \end{aligned}$$

since g_{j-1} and g_{j+1} are in A_2, g has apparent syllable length less than n, and so we can decide if g is in A_i, and if it is we can express g as an element of A_i. ◀

To solve the word problem for a group with a single defining relator, we shall actually solve a more general problem, *the extended word problem: Given any proper subset (possibly empty) of the generators, decide if an element can be expressed in terms of this subset, and if so, express the element in terms of this subset in at least one way.* For groups with one defining relator, Theorem 4.10 enables the solution of the extended word problem to be reduced to the problem of deciding whether an element, in a group which has one defining relator involving all the generators, can be expressed without one particular generator, and if so, to express it as such in at least one way. For suppose first that

$$G = \langle a_1, \ldots, a_n; R(a_1 \ldots, a_n)\rangle$$

and R involves all the generators. Given the proper subset of generators $a_{i_1}, \ldots, a_{i_r}$, suppose that, say a_n, is left out. To decide if an element g can be expressed in terms of $a_{i_1}, \ldots, a_{i_r}$, first decide if g can be expressed in terms of $a_1, \ldots, a_{n-1}$. If g can be, express g as such, and then freely reduce the resulting word. Since $a_1, \ldots, a_{n-1}$ freely generate a free group by the Freiheitssatz, g can be expressed in terms of $a_{i_1}, \ldots, a_{i_r}$ if and only if the freely reduced word for g in $a_1, \ldots, a_{n-1}$ involves only $a_{i_1}, \ldots, a_{i_r}$. Thus the case of the extended word problem for any proper subset of generators is reduced to the extended word problem for maximal proper subsets, if all generators are involved in $R(a_1, \ldots, a_n)$; in particular, the ordinary word problem has been reduced to the extended word problem for maximal proper subsets of generators in such a group.

Suppose next $R(a_1, \ldots, a_n)$ involves only $a_{k+1}, \ldots, a_n$. If $a_{k+1}, \ldots, a_n$ are all included among $a_{i_1}, \ldots, a_{i_r}$, then the missing generators must come from among $a_1, \ldots, a_k$. Suppose $a_1, \ldots, a_s$ are the generators missing from the subset $a_{i_1}, \ldots, a_{i_r}$. Then G is the free product of the free group F on $a_1, \ldots, a_s$ and the group

$$G' = \langle a_{s+1}, \ldots, a_n; R(a_{k+1}, \ldots, a_n)\rangle.$$

To decide if an element of $F * G'$ is in G' it suffices by our lemma, taking $A_1 = F$, $A_2 = G'$, $H_1 = H_2 = 1$, $\varphi =$ identity, to decide if a given element of F is 1 and if a given element of G' is 1, i.e., to solve the word problems for F and G'. Since F is a free group on free generators, the word problem is solvable in F. Moreover, G' is the free product of the free group F' on $a_{s+1}, \ldots, a_k$ and the group

$$G'' = \langle a_{k+1}, \ldots, a_n; R(a_{k+1}, \ldots, a_n)\rangle.$$

Hence, the word problem for $G' = F' * G''$ can be reduced to the word problem for F' which is solvable, and the word problem for G'' which, by the above, can be reduced to the extended word problem for maximal proper subsets of generators in G''.

Suppose, finally, that R involves $a_{k+1}, \ldots, a_n$ and that one of these generators, say, a_n, is missing from $a_{i_1}, \ldots, a_{i_r}$. Now G is the free product of the free group F on $a_1, \ldots, a_k$ and $G' = \langle a_{k+1}, \ldots, a_n; R(a_{k+1}, \ldots, a_n)\rangle$; moreover, $a_1, \ldots, a_{n-1}$ freely generate a free group in G. To decide if an element g of $G = F * G'$ can be expressed in terms of $a_1, \ldots, a_{n-1}$, we first write the element g in canonical form with respect to $F * G'$. This depends upon solving the word problem for F a free group, and the word problem for G', which can be reduced to the extended word problem for maximal proper subsets in G'. Moreover, we can decide if the G' syllables of g can be expressed without a_n, by use of the extended word problem for maximal subsets in G'. Thus we can decide if an element of G is in the free group F' freely generated by $a_1, \ldots, a_{n-1}$, and if so, write it as such. Since $a_{i_1}, \ldots, a_{i_r}$ are among $a_1, \ldots, a_{n-1}$ which freely generate F', we can decide easily if g can be expressed as a word in $a_{i_1}, \ldots, a_{i_r}$ by freely reducing the word in $a_1, \ldots, a_{n-1}$ for g.

Thus the extended word problem for groups with one defining relator can be reduced to the extended word problem for subsets which leave out one generator; moreover, we may assume that all the generators are involved in the one defining relator.

THEOREM 4.14. *Given a group*

$$G = \langle a_1, \ldots, a_n; R(a_1, \ldots, a_n)\rangle$$

then the word problem for G is solvable; moreover, the extended word problem for G is solvable, i.e., *given any proper subset $a_{i_1}, \ldots, a_{i_r}$ of generators we can decide if an element of G is in the subgroup generated by this subset, and, if so, we can express the element in terms of this subset.*

(*Note*: G may be infinitely generated.)

PROOF. As the remarks preceding the statement of Theorem 4.14 show, we need only show that the extended word problem can be solved for maximal proper subsets of generators. We do this by induction on the length of the relator R.

If R has length one or involves only one generator, then G is a free product of a free group and a cyclic group. In this case, the solution to the extended word problem follows easily from the solution of the word problem in free products.

Assume then that the assertion holds for all groups with one defining relator of length less than that of R, and suppose, moreover, that R involves at least two generators. We again represent $a_1, a_2, \ldots, a_n$ by $b, c, \ldots, t$. Now in the arguments preceding the statement of Theorem 4.14 we showed that we may assume that R involves all the generators; for otherwise, we reduce the extended word problem for G to the extended word problem for the subgroup of G generated by the generators involved in R.

CASE 1. R has exponent sum zero on one of the generators, say, t. We shall go to N, the normal subgroup of G generated by $b, c, \ldots,$ denoting as before the rewriting process involved by τ.

To see that the extended word problem for maximal subsets can be reduced to a question in N, suppose first that the maximal proper subset of generators given excludes t. Then we must decide if a word W which apparently involves t can be expressed as a word involving only $b, c, \ldots$; since $\sigma_t(R) = 0$, if W can be so expressed in G then $\sigma_t(W) = 0$. Hence, W defines an element of N. The question thus becomes one of expressing an element of N in terms of $b_0, c_0, \ldots,$ if possible.

On the other hand, if t is not excluded from the maximal proper subset but, say, b is, then W may not define an element of N. However, W can be expressed in G without using b if and only if $Wt^{-\alpha}$, $\alpha = \sigma_t(W)$, can be expressed in G without using b. The word $Wt^{-\alpha}$ defines an element of N, and the question thus becomes one of expressing an element of N without using any b_k, if possible.

By cyclically permuting R, we can assume that R begins with a b-symbol. Thus $\tau(R)$ involves b_0, and if

$$N_i = \langle \ldots, c_{-1}, c_0, c_1, \ldots, b_{\mu+i}, \ldots, b_{M+i}; P_i \rangle$$

then b_0 occurs in P_0, and in both situations, whether t or b is missing from the maximal proper subset of generators of G, we wish to decide if a given element of N can be expressed in terms of a subset of the generators in

$$N_0 = \langle \ldots, c_{-1}, c_0, c_1, \ldots, b_\mu, \ldots, b_M; P_0 \rangle,$$

and if so, we wish to express it in terms of these generators. But once an element is expressed as an element of N_0 (since the length of P_0 is less than the length of R), we can decide if the element is expressible in terms of a proper subset of the generators of N_0, and if so, we can so express it.

Thus our problem is reduced to determining if an element of N is in N_0, and if so, to express it as such. We shall, in fact, show how to determine if an element of N is in N_i, and, if so, how to express it as such.

Now N is the union of the chain of groups

$$Q_1 = N_0, \qquad Q_2 = N_{0,1}, \qquad Q_3 = N_{-1,1}, \qquad \ldots,$$

as defined by (18) and (19). Hence, each element of N is in some term Q_j. We show by induction on j that in each Q_j if the generators of N_i are among the generators of Q_j then we can decide if an element of Q_j is in N_i, and if so, can so express it. If $j = 1$, then $Q_1 = N_0$ and the only N_i with generators among those of Q_j is N_0. Clearly then the result holds for $j = 1$.

Assume the result for Q_s. Now Q_{s+1} is the free product of Q_s and some N_p, with the subgroup K of N_p generated by all the generators of N_p except some b_k generator amalgamated under the identity mapping. Now any N_i with generators included among the generators of Q_{s+1} is either in Q_s or is N_p.

The hypothesis of Lemma 4.9 is satisfied with $A_1 = Q_s$, $A_2 = N_p$, $H_1 = H_2 = K$, $\varphi =$ identity. For, since P_i has length less than the length of R, the extended word problem can be solved for N_p; and so we can decide if an element of N_p is in K. Now K is contained in some N_u whose generators are among those of Q_s; by inductive hypothesis, we can determine if an element of Q_s is in this N_u; if the element of Q_s is in N_u then we can decide if it is in K. Hence, we can decide if an element of Q_s is in K. Moreover, $\varphi =$ identity is specified. Thus we can decide by the lemma if an element of Q_{s+1} is in Q_s or in N_p. Whether $N_i \subset Q_s$ or $N_i = N_p$ we can therefore tell if an element of Q_{s+1} is in N_i, and if so, express it as such. This completes our induction on j to show that for any N_i whose generators are among those of Q_j, we can decide if an element of Q_j defines an element of N_i, and if so, can so express the element.

Therefore, we can decide if any element of N is an element of N_i, and if so, can so express it; in particular, we can decide if an element of N is an element of N_0, and if so, can so express it.

Finally, then, we can decide if an element of N is in N_0, and, if so, whether it can be expressed in terms of $b_0, c_0, \ldots$, or whether it can be expressed without using any of the b_k generators. Thus the extended word problem for maximal proper subsets of generators of G can be solved in Case 1.

CASE 2. All the generators involved in R have non-zero exponent sums. R must involve at least two generators say b, t, and R involves all the generators.

We may assume that we wish to decide in G if an element can be expressed without t, and to do so if possible.

As in Theorem 4.10, we go to the group

$$E = \langle x, c, \ldots, t; R(x^\alpha, c, \ldots, t)\rangle \tag{27}$$

where $\alpha = \sigma_t(R)$, and then by a Tietze transformation to

$$E = \langle x, c, \ldots, y; R(x^\alpha, c, \ldots, yx^{-\beta})\rangle \tag{28}$$

where $\beta = \sigma_b(R)$. Then, since x has zero exponent sum in (28), by applying Case 1 with N, the normal subgroup of E generated by $c, \ldots, y$, we can solve the extended word problem for (28).

Let $W(b, c, \ldots, t)$ be any word in G. The mapping $b \to x^\alpha$, $c \to c$, $\ldots, t \to t$ gives an isomorphism of G into E in (27). Moreover, the mapping

$x \to x$, $c \to c, \ldots, t \to yx^{-\beta}$ gives an isomorphism of (27) into (28). Thus

$$b \to x^{\alpha},\ c \to c,\ \ldots,\ t \to yx^{-\beta}$$

gives an isomorphism of G into (28). Therefore, $W(b, c, \ldots, t)$ can define the same element as $V(b, c, \ldots)$ in G if and only if $W(x^{\alpha}, c, \ldots, yx^{-\beta})$ can define the same element as $V(x^{\alpha}, c, \ldots)$ in E. Thus, to decide if W can be expressed without t in G, we must decide if $W(x^{\alpha}, c, \ldots, yx^{-\beta})$ can be expressed in terms of $x^{\alpha}, c, \ldots,$ in E, and, if so, to so express it. Since the extended word problem is solvable in (28), we can decide if $W(x^{\alpha}, c, \ldots, yx^{-\beta})$ can be expressed in terms of $x, c, \ldots,$ without using y, and if so, can so express it. Finally, since $x, c, \ldots,$ freely generate a free group in (28) we can decide if a word in $x, c, \ldots,$ can be expressed in (28) as a word in $x^{\alpha}, c, \ldots,$ and if so, can so express it. Thus we can decide in (28) if $W(x^{\alpha}, c, \ldots, yx^{-\beta})$ can be expressed in terms of $x^{\alpha}, c, \ldots,$ in E, and if so, can so express it. Therefore, we can decide if $W(b, c, \ldots, t)$ can be expressed in G without t, and if so, can so express it.

This completes the proof of Theorem 4.14. ◀

As an application of Theorem 4.14 and Theorem 4.10 to groups with more than one defining relator we have the following:

Corollary 4.14.1. *Let $R(a_1, \ldots, a_n)$, $S(b_1, \ldots, b_m)$ be cyclically reduced words involving a_n and b_m, respectively; let $U(a_1, \ldots, a_{n-1})$, $V(b_1, \ldots, b_{m-1})$ be freely reduced non-empty words. Then the word problem for*

$$G = \langle a_1, \ldots, a_n, b_1, \ldots, b_m;\ R, S, U = V\rangle$$

is solvable.

Proof. Let $A = \langle a_1, \ldots, a_n; R\rangle$, $B = \langle b_1, \ldots, b_m; S\rangle$, $H =$ the cyclic subgroup generated by U, and $K =$ the cyclic subgroup generated by V. Then, by Theorem 4.10, $a_1, \ldots, a_{n-1}$ freely generate a free group in A, $b_1, \ldots, b_{m-1}$ freely generate a free group in B. Hence, $U(a_1, \ldots, a_{n-1})$ and $V(b_1, \ldots, b_{m-1})$ have infinite order, and $H \simeq K$ under the mapping φ: $U \to V$. Thus $G = *(A, B, H, K, \varphi)$. To decide if $W(a_1, \ldots, a_n, b_1, \ldots, b_m)$ defines 1 in G, decide first if it defines an element of A, and if so write it as this element. To show this can be done, we show the hypotheses of Lemma 4.9 are satisfied with $A_1 = A$, $A_2 = B$, $H_1 = H$, $H_2 = K$, $\varphi = \varphi$. The extended word problem in A and B can be solved by Theorem 4.14. Hence, we can decide if an element of A or B is generated by $a_1, \ldots, a_{n-1}$ or $b_1, \ldots, b_{m-1}$, respectively. We can then decide by examination if an element expressible in $a_1, \ldots, a_{n-1}$ or $b_1, \ldots, b_{m-1}$ is a power of U or V, respectively; this can be done since $a_1, \ldots, a_{n-1}$ and $b_1, \ldots, b_{m-1}$ freely generate free groups. Moreover, φ is specified by replacing U by V, and φ^{-1} by replacing V by U.

Once $W(a_1, \ldots, a_n, b_1, \ldots, b_m)$ is written as an element in A we may use Theorem 4.14 to solve the word problem in A, and so decide if $W = 1$ in G. ◀

COROLLARY 4.14.2. *Let $R_i(x_i, y_1, \ldots, y_n)$ be a cyclically reduced word involving x_i, $i = 1, \ldots, r$. Then in*

$$G = \langle x_1, \ldots, x_r, y_1, \ldots, y_n;\ R_1, R_2, \ldots, R_r \rangle$$

the word problem is solvable.

PROOF. By Theorem 4.14, the word problem is solvable for

$$G_i = \langle x_i, y_1, \ldots, y_n;\ R_i \rangle;$$

we can also decide if a given element of G_i can be expressed in terms of $y_1, \ldots, y_n$ alone. If

$$H_j = \langle x_1, \ldots, x_j, y_1, \ldots, y_n;\ R_1, \ldots, R_j, \rangle$$

then H_j is the free product of H_{j-1} and G_j with the free subgroup in each, freely generated by $y_1, \ldots, y_n$, amalgamated under the identity mapping. We can then prove by induction on j that in H_j we can decide if an element can be expressed in terms of $y_1, \ldots, y_n$, and if so, can so express it. Since $y_1, \ldots, y_n$, freely generate a free group in $H_r = G$, we can solve the word problem in G. ◀

Further applications of the results and methods of this section are given in the exercises.

References and Remarks. The systematic theory of free products was created by Kurosh, 1934, 1937 (see also Kurosh, 1955). His subgroup theorem (proved in slightly simplified form in Section 4.3, Corollary 4.9.1) has been proved since by many authors and by a variety of methods. We mention here only the paper by MacLane, 1958, which also contains references to earlier proofs. The proof given in Section 4.3 is based on a paper by Karrass and Solitar, 1958a. The method used there has the advantage of giving results for free products with amalgamations as well.

In addition to the papers and monograph mentioned in the introductory remarks of Sections 4.1 and 4.2, we list here briefly some of the many papers relevant to the topics treated in Chapter 4. (Others are mentioned in Section 6.1.)

Artin, 1947a, gave a new definition of a free product. Subgroup theorems and other results concerning free products have been found by R. Baer and F. Levi, 1936; M. Hall, 1953; Takahasi, 1944; and by Karrass and Solitar, 1958b. Commutators in free products have been studied by Griffiths, 1956.

The contents of Section 4.4 are based on papers by Magnus, 1931, 1932, and by Karrass, Magnus, and Solitar, 1960. Theorem 4.1 has been generalized by Greendlinger, 1961.

There are important results in the theory of groups with a single defining relator which are due to R. Lyndon, 1950, and have not been included in Section 4.4. We merely mention here that Lyndon determines a set of free generators for the free Abelian group R/R', where F/R is the presentation of a group with a single defining relator as a quotient group of a free group F, and where R' is the commutator subgroup of R. For much more general results of a related type see Lyndon, 1962.

Problems for Section 4.4

1. Show that the words $x^3y^3x^{-3}y^{-3}$, $(xy)^2x^{-2}y^{-2}$, and $xyx^{-1}yxy^{-1}x^{-1}y^{-1}$ are not relators in

$$G = \langle x, y; x^2y^2 = y^2x^2\rangle.$$

[*Hint:* Let N be the normal subgroup of G generated by y. If $y_i = x^iyx^{-i}$, then show that N is the free product of the subgroup H_0 generated by all the y_{2n} and the subgroup H_1 generated by all the y_{2n+1}. Show that every element of H_0 can be written uniquely as a product $y_0{}^{2k}y_{i_1}y_{i_2}\cdots y_{i_m}$, where k is an integer and i_j is an even integer, $i_j \neq i_{j+1}$. Similarly, every element of H_1 can be written uniquely as a product $y_1{}^{2k}y_{i_1}y_{i_2}\cdots y_{i_m}$, where k is an integer and i_j is an odd integer, $i_j \neq i_{j+1}$.]

2. Show how to solve the word problem for

$$G = \langle x, y; x^2y^2 = y^2x^2\rangle.$$

[*Hint:* If N is as in the hint to Problem 1, then every relator of G must be in N. Write the supposed relator in terms of the y_i and use the hint to Problem 1.]

3. Show how to solve the word problem for

$$G = \langle x, y; x^3y^3 = y^3x^3\rangle.$$

[*Hint:* If N is the normal subgroup of G generated by y, then N is the free product of the subgroup H_0 generated by all y_{3n}, the subgroup H_1 generated by all y_{3n+1}, and the subgroup H_2 generated by all y_{3n+2}. Moreover, for $r = 0, 1, 2$, every element in H_r can be uniquely represented as a product $y_r{}^{3k}y_{i_1}^{\alpha_1}y_{i_2}^{\alpha_2}\cdots y_{i_m}^{\alpha_m}$, where k is an integer, i_j is an integer $3n + r$, $i_j \neq i_{j+1}$, and $\alpha_j = 1$ or 2.]

4. Show how to solve the word problem for

$$G = \langle x, y; x^ry^r = y^rx^r\rangle, \qquad r \geq 1.$$

[*Hint:* See the hint to Problems 2 and 3.]

5. Show how to solve the word problem for

$$G = \langle x, y; x^r y^s = y^s x^r \rangle, \qquad r, s \geq 1.$$

[*Hint:* See the hint to Problems 2 and 3.]

6. Show how to solve the word problem for

$$G = \langle x, y; x^r y^s = y^t x^r \rangle, \qquad r, s \geq 1, t \text{ any integer.}$$

[*Hint:* If $t = 0$, then G is the free product of the infinite cyclic group on x and the cyclic group of order s on y. If $t \neq 0$, use the hint to Problem 2. Moreover show that N is the free product of subgroups H_i, $i = 0, 1, \ldots, r - 1$, where

$$H_i = \langle \ldots y_{-r+i}, y_i, y_{r+i}, \ldots; \ldots, y_i^{\,s} = y_{-r+i}^{\,t}, y_{r+i}^{\,s} = y_i^{\,t}, \ldots \rangle.$$

Since any subgroup of H_i, generated by finitely many of the generators of H_i can be obtained by successively forming amalgamated free products, where infinite cyclic groups are amalgamated, the word problem is particularly simple to solve in H_i.]

7. Show that $[y, x] = y^{-1}x^{-1}yx$ and y do not define the same element in

$$G = \langle x, y; xy^2 = y^3x \rangle.$$

[*Hint:* Let N be the normal subgroup of G generated by y. Rewrite both $[y, x]$ and y in N. Use that the subgroup generated by y_0 and y_{-1} is $\langle y_{-1}, y_0; y_0^{\,2} = y_{-1}^{\,3} \rangle$.]

8. Let

$$H_n = \langle a_1, \ldots, a_n; a_1^{\alpha_1} = a_2^{\beta_1}, a_2^{\alpha_2} = a_3^{\beta_2}, \ldots, a_{n-1}^{\alpha_{n-1}} = a_n^{\beta_{n-1}} \rangle,$$

where $\alpha_i \neq 0$ and $\beta_j \neq 0$, $1 \leq i, j \leq n - 1$. Show that H_n is the free product of H_{n-1} and the infinite cyclic group generated by a_n with the infinite cyclic subgroups generated by $a_{n-1}^{\alpha_{n-1}}$ and $a_n^{\beta_{n-1}}$ amalgamated under the identity mapping. [*Hint:* Use induction on n to show the assertion.]

9. Let

$$H = \langle a_1, a_2, \ldots, a_n; a_1^{\alpha_1} = a_2^{\beta_1}, a_2^{\alpha_2} = a_3^{\beta_2}, \ldots, a_{n-1}^{\alpha_{n-1}} = a_n^{\beta_{n-1}} \rangle$$

where $|\alpha_i|, |\beta_j| > 1$. Then

$$W = a_{i_1}^{\epsilon_1} a_{i_2}^{\epsilon_2} \ldots a_{i_r}^{\epsilon_r},$$

where $|\epsilon_j| = 1$ and $i_j \neq i_{j+1}$, does not define the identity in H. Moreover, if $r \geq 2$ then W does not define a power of a_k, $k = 1, \ldots, n$, in H. [*Hint:* To prove this result, we use double induction on r and n. If $n = 2$, and $W = a_{i_1}^{\epsilon_1} \ldots a_{i_r}^{\epsilon_r}$, where $|\epsilon_j| = 1$ and $i_j \neq i_{j+1}$, then W has representative length r in the amalgamated product $\langle a_1, a_2; a_1^{\alpha_1} = a_2^{\beta_1} \rangle$, and so W does not define the identity; and if $r > 1$, W does not define a power of a_1 or a_2. Assume the assertion holds for all words W for all groups H with fewer generators than n, and for all words W of length less than r for any group H with n generators. Consider

$$H = \langle a_1, \ldots, a_n; a_1^{\alpha_1} = a_2^{\beta_1}, \ldots, a_{n-1}^{\alpha_{n-1}} = a_n^{\beta_{n-1}} \rangle$$

and $W = a_{i_1}^{\epsilon_1}a_{i_2}^{\epsilon_2} \dots a_{i_r}^{\epsilon_r}$. Now, by Problem 8, H is the free product of

$$K = \langle a_1, \dots, a_{n-1}; a_1^{\alpha_1} = a_2^{\beta_1}, \dots, a_{n-2}^{\alpha_{n-2}} = a_{n-1}^{\beta_{n-2}}\rangle$$

and the infinite cyclic group on a_n with the infinite cyclic subgroups generated by $a_{n-1}^{\alpha_{n-1}}$ and $a_n^{\beta_{n-1}}$ identified.

If $i_j \neq n$ for all j then W defines a word of K and so the inductive hypothesis yields the required assertion. Otherwise, W involves a_n and $W = U_1 \dots U_s$, where the U_i are subwords of W alternating from K and from the infinite cyclic group on a_n. By inductive hypothesis each U_i in K of letter length greater than one does not define an element of the cyclic subgroup of K generated by $a_{n-1}^{\alpha_{n-1}}$. Moreover, if U_i is in K and has letter length one, it cannot define a power of $a_{n-1}^{\alpha_{n-1}}$. Hence, $W = U_1 \dots U_s$ cannot define the identity. Moreover, if the letter length of W is greater than one, then W does not define a power of any a_k, $k = 1, \dots, n$.]

10. Show that if G is as in Problem 7, and $u_0 = y$, $u_1 = [y, x]$ and $u_s = [u_{s-1}, x]$ then u_s does not define the identity in G. Moreover, u_s and u_t do not define the same element of G if $s \neq t$. [*Hint:* Let N be as in the hint to Problem 7, and rewrite u_s in N. Show by induction on s that $u_s = y_{i_1}^{\epsilon_1} \dots y_{i_{2^s}}^{\epsilon_{2^s}}$, where

$$i_1 = -s + 1, \qquad i_{2^s} = -s, \qquad i_j \leq 0, i_j \neq i_{j+1}, \qquad \text{and} \qquad |\epsilon_j| = 1.$$

Hence, if $s \neq t$,

$$u_s u_t^{-1} = y_{k_1}^{\eta_1} \dots y_{k_m}^{\eta_m},$$

where $k_j \neq k_{j+1}$ and $|\eta_j| = 1$. Since the subgroup of N generated by y_0, $y_{-1}, \dots, y_{-n}$ has defining relations

$$y_0^2 = y_{-1}^3, y_{-1}^2 = y_{-2}^3, \dots, y_{-n+1}^2 = y_{-n}^3,$$

by Problem 9 we know that u_s and $u_s u_t^{-1}$ do not define the identity in N.]

11. Let

$$G = \langle x, y; x^r y^p = y^q x^r\rangle, \qquad r \geq 1, |p|, |q| > 1.$$

Show that $u_0 = y$, $u_1 = [y, x], \dots, u_s = [u_{s-1}, x]$, all define distinct elements of G, different from the identity. [*Hint:* Use the hint to Problem 10, and that N, in this case, is a free product of the subgroups H_k, $0 \leq k < r$, generated by all y_i, where i is congruent to k mod r.]

12. Let

$$G = \langle x, y; x^r y^p = y^q x^r\rangle, \qquad r \geq 1, |p|, |q| > 1.$$

If $W = x^{\alpha_1}y^{\epsilon_1}x^{\alpha_2}y^{\epsilon_2} \dots x^{\alpha_t}y^{\epsilon_t}x^{\alpha_{t+1}}$, where α_i is an integer, non-zero except possibly for α_1 or α_{t+1}, and $|\epsilon_i| = 1$, then W does not define the identity in G. [*Hint:* Let N be the normal subgroup generated by y. If W is a relator, it has zero x-exponent sum. Rewrite W in N and use Problem 9.]

13. Let $G = \langle x, z; (x^{-1}zxz)^2\rangle$. Show that none of the following words are relators in G:

$$z^3xzx^{-1}, xz^{-1}x^{-1}z^2xzxz^{-1}x^{-1}z^{-1}x^{-1}, z^7xzx^{-1}zxzx^{-1}zxzx^{-1}.$$

Show how to solve the word problem in G. [*Hint:* Let N be the normal subgroup of G generated by z; and let $z_i = x^i z x^{-i}$. Show that the subgroup N_k generated by $z_0, z_1, \ldots, z_k$ is

$$N_k = \langle z_0, z_1, \ldots, z_k; (z_0z_1)^2, (z_1z_2)^2, \ldots, (z_{k-1}z_k)^2 \rangle.$$

Introduce the generators $t_0 = z_0$, $t_1 = z_0z_1$, $t_2 = z_1z_2, \ldots, t_k = z_{k-1}z_k$. Then by a Tietze transformation

$$N_k = \langle t_0, t_1, t_2, \ldots, t_k; t_1^2, t_2^2, \ldots, t_k^2 \rangle.$$

Thus N_k is a free product of cyclic groups; the word problem for this group is particularly easy to solve. Now any relator W in G must have zero x-exponent sum. Moreover, there is some x^p such that every initial segment of x^pWx^{-p} will have non-negative x-exponent sum, and so x^pWx^{-p} will define an element of N_k. To express a word in $z_0, \ldots, z_k$ as a word in $t_0, \ldots, t_k$, we replace z_0 by t_0, z_1 by $t_0^{-1}t_1$, z_2 by $t_1^{-1}t_0t_2$, and in general z_{i+1} by $t_i^{-1}Z_{i-1}t_{i+1}$, where Z_{i-1} is the word in $t_0, \ldots, t_k$ which replaces z_{i-1}.]

14. Show how to solve the word problem for

$$G = \langle x, y; (x^{-2}y^2)^2 \rangle.$$

[*Hint:* Introduce a new generator z such that $y = xz$. When expressed in terms of x and z the defining relator in G will have zero x-exponent sum. Then use Problem 13.]

15. Show how to solve the word problem for

$$G = \langle x, y; (x^{-2}y^2)^r \rangle, \qquad r \geq 1.$$

[*Hint:* See the hint in Problem 14.]

16. Show how to solve the word problem for

$$G = \langle x, z; (x^{-2}zxzxz)^r \rangle, \qquad r \geq 1.$$

[*Hint:* If N is the normal subgroup of G generated by z, then the subgroup N_k generated by $z_0, \ldots, z_k$ is $\langle z_0, z_1, z_2, \ldots, z_k; (z_0z_1z_2)^r, \ldots, (z_{k-2}z_{k-1}z_k)^r \rangle$. Introduce the generators $t_0 = z_0$, $t_1 = z_1$, $t_2 = z_0z_1z_2, \ldots, t_k = z_{k-2}z_{k-1}z_k$. Then by a Tietze transformation

$$N_k = \langle t_0, t_1, t_2, \ldots, t_k; t_2^r, t_3^r, \ldots, t_k^r \rangle.$$

Thus N_k is a free product of cyclic groups whose word problem is particularly easy to solve. Moreover, to express a word in $z_0, \ldots, z_k$ as a word in $t_0, \ldots, t_k$ we replace z_0 by t_0, z_1 by t_1, z_2 by $t_1^{-1}t_0^{-1}t_2$ and, in general, z_{i+1} by $t_i^{-1}Z_{i-2}t_{i+1}$, where Z_{i-2} is the word in $t_0, \ldots, t_k$ which replaces z_{i-2}.]

17. Show how to solve the word problem for

$$G = \langle x, y; (x^{-3}y^3)^r \rangle, \qquad r \geq 1.$$

[*Hint:* See the hint to Problems 14 and 16.]

18. Show how to solve the word problem for

$$G = \langle x, y; (x^{-s}y^s)^r \rangle, \qquad r \geq 1, \quad s \geq 1.$$

[*Hint:* See the hint to Problems 14 and 16.]

19. Let F be the free group on a and b. Show that if $U(a, b)$ and $V(a, b)$ generate F, then $UVU^{-1}V^{-1}$ is freely equal to a conjugate of $aba^{-1}b^{-1}$ or $bab^{-1}a^{-1}$. [*Hint:* If U and V generate F, then $\langle a, b; UVU^{-1}V^{-1}\rangle$ is the free Abelian group on two generators.]

20. Let F be the free group on $a_1 \ldots, a_n$. Suppose that

$$G = \langle a_1, \ldots, a_n; R(a_\nu)\rangle$$

is a free group H on $n - 1$ generators. Then R is a primitive element of F. [*Hint:* Let $b_1, \ldots, b_{n-1}$ be free generators of H. If a_i equals $A_i(b_\mu)$, then $A_1(b_\mu), \ldots, A_n(b_\mu)$ generate H. Hence, by a Nielsen transformation we may reduce $A_1, \ldots, A_n$ to $b_1, \ldots, b_{n-1}, 1$. Applying this same transformation to $a_1, \ldots, a_n$ we obtain free generators $W_1(a_\nu), \ldots, W_n(a_\nu)$ for F. Since $W_n(a_\nu)$ defines 1 in G, the normal subgroup of R in F contains the primitive element W_n. Thus, if R is written in terms of $W_1, \ldots, W_n$, then R must involve W_n. Moreover, $R = V(W_1, \ldots, W_{n-1}) \cdot N$ where N is in the normal subgroup of F generated by W_n. But then since N is derivable from W_n, which is derivable from R, $V(W_1, \ldots, W_{n-1})$ is derivable from R and must be empty. Thus R and W_n are derivable from one another, and R, a conjugate of W_n or W_n^{-1}, is primitive.]

21. Show that if $aba^{-1}b^{-1}$ can be derived from $R(a, b)$, then $R(a, b)$ is primitive in the free group F on a and b, or $R(a, b)$ is a conjugate of $aba^{-1}b^{-1}$ or $bab^{-1}a^{-1}$ in F. [*Hint:* If $aba^{-1}b^{-1}$ can be derived from $R(a, b)$ then $G = \langle a, b; R(a, b)\rangle$ is an Abelian group. Moreover, by applying a Nielsen transformation to a, b we can find generators $U(a, b)$ and $V(a, b)$ for F such that $R(a, b) = U^kC(U, V)$ where C is a product of commutators in U and V. Since G is an Abelian group, in $G = \langle U, V; U^kC(U, V)\rangle$, U has finite order if $k \neq 0$. But by Theorem 4.10 U has infinite order in G if $C(U, V)$ involves V. Hence (conjugating by some power of V if necessary) either $k = 0$, or $k \neq 0$ and $C(U, V)$ does not involve V. If $k = 0$, then $R(a, b) = C(U, V)$ and then $R(a, b)$ and $aba^{-1}b^{-1}$ are derivable from one another; thus $R(a, b)$ is a conjugate of $aba^{-1}b^{-1}$ or $bab^{-1}a^{-1}$. If $k \neq 0$ and $C(U, V)$ does not involve V then $R(a, b) = U^k$. But $G = \langle U, V; U^k\rangle$ is not Abelian unless $k = 1$ or -1. Thus $R(a, b) = U$ or U^{-1} and so $R(a, b)$ is primitive.]

22. Show that the groups $\langle a, b, c; a^2b^3[U(a, b, c), V(a, b, c)]\rangle$ and $\langle a, b, c; a^6b^{10}c^{15}[U(a, b, c), V(a, b, c)]\rangle$ have no elements $(\neq 1)$ of finite order, where $[U, V] = U^{-1}V^{-1}UV$. [*Hint:* If $R = L^k$, then k divides $\sigma_a(R)$, $\sigma_b(R)$, $\sigma_c(R)$.]

23. Let $G = \langle a, b; ab^2a^{-1} = b^3\rangle$ and let N be the normal subgroup of G generated by b. Show that

$$N = \langle \ldots, b_{-1}, b_0, b_1, \ldots; \ldots, b_0^2 = b_{-1}^3, b_1^2 = b_0^3, \ldots\rangle$$

where b_i defines a^iba^{-i} in G. Show that

$$K = \langle a, \ldots, b_i, \ldots; \ldots, ab_ia^{-1} = b_{i+1}, \ldots, b_1^2 = b_0^3\rangle,$$

where $i = \ldots, -1, 0, 1, \ldots,$ is isomorphic to G. [*Hint:* Map the last presentation for K into the first for G by sending $a \to a$, $b_i \to a^iba^{-i}$. Show that every element in K can be written as a word $a^j W(b_i)$. If $a^j W(b_i)$ goes into the identity in G then $j = 0$ and $W(b_i)$ is 1 in N and so in K.]

24. Show that $H = \langle a, b; ab^2a^{-1} = b^3, ba^2b^{-1} = a^3\rangle$ is the identity group. [*Hint:* To show H is the identity, it suffices to show that if we divide

$$G = \langle a, b; ab^2a^{-1} = b^3\rangle$$

by the normal subgroup of G generated by $ba^2b^{-1}a^{-3}$ we get the identity. But we can just as well show that if

$$K = \langle a, \ldots, b_i, \ldots; \ldots, ab_ia^{-1} = b_{i+1}, \ldots, b_1{}^2 = b_0{}^3\rangle$$

is divided by the normal subgroup generated by $b_0b_2{}^{-1}a^{-1}$, the identity group results. Thus we show

$$L = \langle a, \ldots, b_i, \ldots; \ldots, ab_ia^{-1} = b_{i+1}, \ldots, b_1{}^2 = b_0{}^3, b_0b_2{}^{-1} = a\rangle$$

is the identity group. Consider the following equations which hold in L. Since $ab_2a^{-1} = b_3$ and $a = b_0b_2{}^{-1}$, we have $b_0b_2{}^{-1}(b_2)(b_2b_0{}^{-1}) = b_0b_2b_0{}^{-1} = b_3$. Hence, $b_0b_2{}^nb_0{}^{-1} = b_3{}^n$. We now try to choose n so that $b_2{}^n$ and $b_3{}^n$ are powers of b_0. Since

$$b_1{}^2 = b_0{}^3,\ b_2{}^2 = b_1{}^3,\ b_3{}^2 = b_2{}^3,$$

we choose $n = 8$. Thus

$$b_3{}^8 = (b_3{}^2)^4 = (b_2{}^3)^4 = (b_2{}^4)^3 = (b_2{}^2)^6 = (b_1{}^3)^6 = (b_1{}^2)^9 = (b_0{}^3)^9 = b_0{}^{27}.$$

Moreover,

$$b_2{}^8 = (b_2{}^2)^4 = (b_1{}^3)^4 = (b_1{}^2)^6 = (b_0{}^3)^6 = b_0{}^{18}.$$

Thus from $b_0b_2{}^8b_0{}^{-1} = b_3{}^8$ it follows that $b_0b_0{}^{18}b_0{}^{-1} = b_0{}^{18} = b_0{}^{27}$ and so $b_0{}^9 = 1$. But

$$1 = b_0{}^9 = (b_0{}^3)^3 = (b_1{}^2)^3 = (b_1{}^3)^2 = (b_2{}^2)^2 = b_2{}^4.$$

Since $b_2 = a^2b_0a^{-2}$, we have $b_0{}^4 = 1$. But since 9 and 4 are coprime and $b_0{}^9 = b_0{}^4 = 1$, it follows $b_0 = 1$. Hence, $b_i = a^ib_0a^{-i} = 1$, and also $a = b_0b_2{}^{-1} = 1$. Thus $L = 1$.]

25. Show that $U = ab^2a^{-1}b^{-3}$ and $V = ba^2b^{-1}a^{-3}$ do not generate the free group F on a and b. Show that the normal subgroup generated by U and V is all of F. [*Hint:* See Problems 19 and 24.]

26. Show that

$$L = \langle a, b; b = [a, b^3], a = [b^r, a^{s+1}]\rangle,$$

where $[u, v] = u^{-1}v^{-1}uv$ and r and s are non-negative integers, defines the identity group. [*Hint:* Note that $b = [a, b^3]$ is equivalent as a relation to $ab^2a^{-1} = b^3$. Consider the group $G = \langle a, b; ab^2a^{-1} = b^3\rangle$. Present the group as

$$K = \langle a, \ldots, b_i, \ldots; \ldots, ab_ia^{-1} = b_{i+1}, \ldots, b_1{}^2 = b_0{}^3\rangle.$$

In K, the element $b^ra^sb^{-r}a^{-s-1}$ of G is written $b_0{}^rb_s{}^{-r}a^{-1}$. Thus we must show

$$L = \langle a, \ldots, b_i, \ldots; \ldots, ab_ia^{-1} = b_{i+1}, \ldots, b_1{}^2 = b_0{}^3, a = b_0{}^rb_s{}^{-r}\rangle$$

is the identity group. Again we compute in L.

$$ab_s a^{-1} = b_{s+1}$$

and so

$$(b_0{}^r b_s{}^{-r}) b_s (b_s{}^r b_0{}^{-r}) = b_0{}^r b_s b_0{}^{-r} = b_{s+1}.$$

Thus $b_0{}^r b_s{}^n b_0{}^{-r} = b_{s+1}^n$. To express $b_s{}^n$ and b_{s+1}^n in terms of b_0, we use induction on q to show $b_q{}^{2^q} = b_0{}^{3^q}$. For $q = 0$ it holds. Moreover, if it holds for q, then

$$b_{q+1}^{2^{q+1}} = (b_{q+1}^2)^{2^q} = (b_q{}^3)^{2^q} = (b_q{}^{2^q})^3 = (b_0{}^{3^q})^3 = b_0^{3^{q+1}}.$$

Thus, letting $n = 2^{s+1}$, we have

$$b_{s+1}^{2^{s+1}} = b_0^{3^{s+1}} = b_0{}^r b_s^{2^{s+1}} b_0{}^{-r} = b_0{}^r (b_s{}^{2^s})^2 b_0{}^{-r} = b_0{}^r (b_0{}^{3^s})^2 b_0{}^{-r} = b_0^{2\cdot 3^s}.$$

Thus $b_0{}^{3^s} = 1$; hence $b_s{}^{2^s} = 1$. Since $b_s = a^s b_0 a^{-s}$, also $b_0{}^{2^s} = 1$. From 2^s and 3^s coprime and $b_0{}^{2^s} = b_0{}^{3^s} = 1$, it follows $b_0 = 1$. Thus $L = 1$.]

27. Show that

$$L = \langle a, b; b = [a, b^{p+1}], a = [b^r, a^{s+1}]\rangle$$

where p, r, and s are non-negative integers, defines the identity group. [*Hint:* See the hint to Problem 26.]

28. Show that

$$L = \langle a, b; b = [a, b^{p+1}], a^t = [b^r, a^{s+t}]\rangle$$

where p, t, r and s are non-negative integers, defines the identity group. [*Hint:* See the hint to Problem 26.]

29. Give a reason why each of the following pairs of groups cannot be isomorphic:

(a) $\langle x, y; (x^2y^2x^{-2}y^{-2})^3\rangle$

and

(a′) $\langle x, y; (x^3y^3x^{-3}y^{-3})^2\rangle$;

(b) $\langle x, y; (xyx^{-1}y^{-1})^6\rangle$

and

(b′) $\langle x, y; (xy^2x^{-1}y^{-2})^6\rangle$;

(c) $\langle x, y; (xyxy^2)^5\rangle$

and

(c′) $\langle x, y; (x^2y^3)^5\rangle$.

[*Hint:* Use Theorem 4.13 and Corollary 4.13.1.]

30. (G. Baumslag) Let F be a free group on $a_1, \ldots, a_m$. Show that if $W(a_\nu)$ is not a true power in F, then $W(a_\nu)$ is not a true power in $F/F^{(n)}$ for some $n > 1$, where $F^{(n)}$ is the nth derived group defined inductively by

$$F^{(1)} = (F, F) \qquad \text{and} \qquad F^{(n)} = (F^{(n-1)}, F^{(n-1)}).$$

(See Definition 5.3, Section 5.1.) [*Hint:* We use induction on the length of W. If W has length 1 then $W = a_\nu{}^\epsilon$, $\epsilon = 1$ or -1. Now if $W = V^k \cdot U$ where $k > 1$ and U is in $F^{(n)}$, then the a_ν-exponent sum of W is divisible by k.

Thus W is not a true power mod any term of the derived series. Suppose then the result is true for any word of length less than r, and suppose $W(a_\nu)$ has length $r > 1$. If $W(a_\nu)$ has exponent sum zero on some a_ν, say a_1, let N be the normal subgroup of F generated by $a_2, \ldots, a_m$. Then $W(a_\nu)$ is in N and moreover, using the free generators $a_1{}^i a_\nu a_1{}^{-i}$, $\nu \neq 1$, for N, $W(a_\nu)$ has smaller length than r in N. Hence, $W(a_\nu)$ is not a true power in N mod some $N^{(n)}$. Suppose $W(a_\nu)$ is a true power mod $F^{(n+1)}$. Then $WF^{(n+1)} = V^k F^{(n+1)}$. Since W has a_1-exponent sum zero, so does V. Hence, V is in N; since $F^{(1)} \subset N$ and so $F^{(n+1)} \subset N^{(n)}$, we have $WN^{(n)} = V^k N^{(n)}$, and so W is a true power in N mod $N^{(n)}$. If $W(a_\nu)$ has exponent sum different from zero on all generators involved in it (there are at least two since $r > 1$ and W is not a true power in F, say a_1 and a_2), let H be the free group on $b_1, b_2, \ldots, b_m$ and imbed F into H so that W has zero exponent sum on b_1, as in the proof of Theorem 4.10. Then use the above argument on W as an element of H together with $F^{(n)} \subset H^{(n)}$.]

31. Let $R(b_\lambda, c_\mu, d_\nu)$ be a cyclically reduced word in b_λ, c_μ, d_ν, λ, μ, ν ranging over all integers. Suppose that the b-subscripts in R range from 0 to r, and that those of c range from p to $p + s$. If

$$R_i(b_\lambda, c_\mu, d_\nu) = R(b_{\lambda+i}, c_{\mu+i}, d_{\nu+i})$$

where i is an integer, show that in the group

$$G = \langle \ldots, b_\lambda, c_\mu, d_\nu, \ldots; \ldots, R_i(b_\lambda, c_\mu, d_\nu), \ldots \rangle$$

the elements $b_0, \ldots, b_r, c_q, \ldots, c_{q+s}$, and all d_ν freely generate a free group, if $q \neq p$. [*Hint:* Define the group

$$G_i = \langle b_i, \ldots, b_{r+i}, c_q, \ldots, c_{q+s}, c_{p+i}, \ldots, c_{p+i+s}, \ldots, d_\nu, \ldots; R_i(b_\lambda, c_\mu, d_\nu) \rangle$$

Since b_i is in R_i but not in R_{i-1}, and b_{i+1+s} is in R_{i+1} but not in R_i, use Theorem 4.10 to show that G can be obtained as the union of groups each of which is an amalgamated product of the groups G_i. In particular, G contains G_0 as a subgroup. Since $R_0 = R$ contains c_p and c_{p+s}, and one of these is not in $c_q, \ldots, c_{q+s}$, it follows by Theorem 4.10 that $b_0, \ldots, b_r, c_q, \ldots, c_{q+s}$, and all d_ν freely generate a free group of G_0, and hence, of G.]

Chapter 5

Commutator Calculus

5.1. Introduction

A simple method for obtaining information about a group G which is presented in terms of generators and defining relators consists in Abelianizing G. The quotient group G/G' of G with respect to its commutator subgroup G' is an Abelian group independent of the particular presentation of G, and if G is finitely generated, G/G' is "computable" in the sense that we can determine explicitly a complete set of numbers which determines G/G' up to an isomorphism. Of course, the scope of the method is a very limited one. It does not allow us to distinguish between two groups on the same number of generators if the defining relators of both groups are products of commutators. It would seem natural to study next the Abelianized commutator subgroup G'/G''. This, however, leads to difficult problems even if G is finitely generated; for, although G'/G'' is Abelian, it will, in general, be infinitely generated and may have a very complicated structure. To bypass this difficulty we investigate the groups G_n of the lower central series of G (see section 5.3), which, like G', G'', etc., are fully invariant in G but have the advantage that G_n/G_{n+1} is a finitely generated Abelian group whenever G is finitely generated. The theory of the groups of the lower central series has been fully developed by P. Hall, 1933, and it is the far-reaching scope of his work which makes the commutator calculus a systematic theory.

P. Hall uses purely group theoretical methods, and the difficulties he has to overcome are in part due to the fact that commutation (i.e., the composition of group elements defined by forming their commutator) is an operation which is not connected in a simple manner to the composition of group elements (multiplication) which determines the group. Theorem 5.1 (Section 5.2), and Theorem 5.3 (Section 5.3) show that commutation and multiplication in a group are related "almost" like multiplication and addition in a Lie algebra. This remark, however, is not found in the paper by P. Hall, 1933. It is suggested by an observation arising from the theory of Lie groups, or of linear operators, where one is accustomed to inverting an operator L "close" to the identity 1 by putting $L = 1 + u$ and expanding L^{-1} in a power series in u, obtaining $L^{-1} = 1 - u + u^2 - u^3 + \ldots + (-1)^n u^n + \ldots$. (The Neumann series for the solution of a Fredholm integral equation of the second kind arises in this way.) To use this method in general, we need a sufficiently abstract meaning for the word "close." Of course, if $u^n = 0$ for some n, then the power series terminates, and we need make no quantitative restriction on u to guarantee "convergence" of the power series for L^{-1}; examples of such linear operators u are given by finite matrices in which the elements on and below the main diagonal vanish. (Problem 5.4.1 indicates that infinite triangular matrices could be used for our development.) However, questions of convergence can be completely sidestepped by using the abstract algebraic construction of a ring A of formal power series in non-commuting variables. Such a ring A was constructed by H. F. Baker, 1904, and by Hausdorff, 1905, for the purpose of investigating Lie groups, whose elements were represented as certain elements of the form $1 + u$ in A. Our account of the commutator calculus is based on the use of such a ring A. It can be read by starting with Section 5.4 and returning to parts of Sections 5.2 and 5.3 as they are needed and referred to in the text.

5.2. Commutator Identities

Let a, b be any elements in a group G. The element

(1) $$a^{-1}b^{-1}ab$$

is called the commutator of a and b (in this order) and is denoted by (a, b)

We shall take (1) as the definition of a new binary composition in G, which we shall call *commutation*.

Commutation is ordinarily not an associative composition, i.e., in general, $(a, (b, c)) \neq ((a, b), c)$. In order to describe the various ways in which commutation can be applied to a sequence of n elements, we use the concept of a bracket arrangement of weight n.

Definition 5.1. A *bracket arrangement* β^n *of weight* n, $n = 1, 2, 3, \ldots$ is defined recursively as a certain sequence of asterisks (which act as place holders) and parentheses (which indicate the order in which commutation is performed) in the following manner:

There is only one bracket arrangement of weight one,

$$\beta^1 = (*).$$

A bracket arrangement β^n of weight $n > 1$ is obtained by choosing bracket arrangements β^k and β^l of weights k and l respectively such that $k + l = n$ and setting

(2) $$\beta^n = (\beta^k\beta^l),$$

that is, juxtaposing the sequences β^k and β^l and enclosing the resulting sequence in a pair of parentheses.

In accordance with this definition, the (sole) bracket arrangement of weight two is $((*)(*))$, and the (two) bracket arrangements of weight three are

(3) $$(((*)(*))(*)), \qquad ((*)((*)(*))).$$

It is both customary and expedient not to use parentheses if they enclose a single asterisk. For example, we shall simply write $*$ for the bracket arrangement of weight one, $(**)$ for the bracket arrangement of weight two, and

(4) $$((**)*), \qquad (*(**))$$

for the two bracket arrangements of weight three.

We can now introduce the following definitions:

Definition 5.2. Let G be a group, let $a_1, a_2, \ldots, a_n$ be a finite sequence of elements of G and let β^n be a bracket arrangement of weight n. We define the elements

(5) $$\beta^n(a_1, a_2, \ldots, a_n)$$

of G, recursively as follows:

$$\beta^1(a_1) = a_1,$$

and if $n > 1$ and $\beta^n = (\beta^k\beta^l)$, then

$$\beta^n(a_1, \ldots, a_n) = (\beta^k(a_1, \ldots, a_k), \beta^l(a_{k+1}, \ldots, a_n)).$$

We call (5) a *commutator of weight n in the components* $a_1, \ldots, a_n$.

Definition 5.3. Let $A_1, \ldots, A_n$ be normal subgroups of a group G. Then the set of all elements

$$\beta^n(a_1, \ldots, a_n)$$

where $a_\rho \in A_\rho$, $\rho = 1, 2, \ldots, n$, generates a subgroup of G. This subgroup, which is normal in G, is denoted by

(6) $$\beta^n(A_1, \ldots, A_n)$$

and is called a *commutator of weight n in the components* $A_1, \ldots, A_n$.

There exist numerous identities between commutators. Some of these will be of great importance later on. We collect these in Theorem 5.1. To simplify notation we use

Definition 5.4. Let a, b be any elements of a group G. Then

$$a^b = b^{-1}ab, \qquad a^{-b} = b^{-1}a^{-1}b. \tag{7}$$

THEOREM 5.1. (*Witt-Hall identities*). *For any three elements a, b, c of a group G:*

$$(a, b) \cdot (b, a) = 1 \tag{8}$$
$$(a, b \cdot c) = (a, c) \cdot (a, b) \cdot ((a, b), c) \tag{9}$$
$$(a \cdot b, c) = (a, c) \cdot ((a, c), b) \cdot (b, c) \tag{10}$$
$$((a, b), c^a) \cdot ((c, a), b^c) \cdot ((b, c), a^b) = 1 \tag{11}$$
$$\begin{aligned}((a, b), c) \cdot ((b, c), a) \cdot ((c, a), b) \\ = (b, a) \cdot (c, a) \cdot (c, b)^a \cdot (a, b) \cdot (a, c)^b \cdot (b, c)^a \cdot (a, c) \cdot (c, a)^b\end{aligned} \tag{12}$$

PROOF. The proof is straightforward and consists in expressing each commutator explicitly as a product of group elements and then simplifying. ◀

Formulas (8) to (11) are due to P. Hall (see P. Hall, 1933, and M. Lazard, 1954, p. 107). Formula (12) is due to Witt, 1937. The significance of Theorem 5.1 will become clear in Section 5.3 and 5.7.

For commutators involving normal divisors we have the rather obvious

LEMMA 5.1. *If A and B are normal subgroups of group G, then (A, B) is again a normal subgroup of G which is contained in the intersection of A and B.*

Not so obvious is the following result due to P. Hall, 1933.

THEOREM 5.2. *Let A, B, C be any three normal divisors of a group G. Then each one of the three normal divisors*

$$((A, B), C), \quad ((B, C), A), \quad ((C, A), B), \tag{13}$$

is contained in the product of the other two.

PROOF. It suffices to show that $((A, B), C)$ is contained in the product

$$L = ((B, C), A) \cdot ((C, A), B).$$

Let a, b, c be elements of A, B, C, respectively. Then $b^c \in B$, $a^b \in A$ and therefore we have from (11),

$$((a, b), c^a) \in L.$$

Since c^a runs through all elements of C if c does, this proves Theorem 5.2. ◀

Without proof, we mention the following theorem due to Auslander and Lyndon, 1955:

Let A, B be normal divisors of a non-Abelian free group F. Then

$$(A, A) \subset (B, B)$$

implies

$$A \subset B.$$

The conditions for the validity of this theorem can be somewhat weakened according to a comment made by Cartan (*Mathematical Reviews*, **17**, p. 709). For different proofs and for generalizations see H. Neumann, 1962, and B. H. Neumann, 1962.

Problems for Section 5.2

1. Let A and B be normal divisors of a group G. Show that $(A, B) \neq A^{-1}B^{-1}AB$ unless $A = B$. Show that $(A, B) = (B, A)$.

2. Show that if $a_1, a_2, \ldots, a_n$ are free generators of a free group F and β^n is a bracket arrangement of weight n then

$$\beta^n(a_1, a_2, \ldots, a_n)$$

is a freely reduced word in F which ends in a_n.

3. Let $a_1, a_2, \ldots, a_n$ freely generate a free group F and let β^n be a bracket arrangement of weight n. If L denotes the letter length of a word in a_1, $a_2, \ldots, a_n$, show that

$$L(\beta^n(a_1, a_2, \ldots, a_n)) \leq 3 \cdot 2^{n-1} - 2, \qquad n \geq 1.$$

Moreover,

$$L(\beta^n(a_1, a_2, \ldots, a_n)) = 3 \cdot 2^{n-1} - 2$$

if and only if

$$\beta^n = (\ldots(((**)*)*)\ldots*) \text{ or } \beta^n = (*\ldots(*(*(**)))\ldots)$$

4. Let $a_1, a_2, \ldots, a_n$ freely generate a free group F and let β^n, γ^n be bracket arrangements of weight n. Show that

$$\beta^n(a_1, a_2, \ldots, a_n) = \gamma^n(a_1, a_2, \ldots, a_n)$$

if and only if $\beta^n = \gamma^n$.

5. Show that there are

$$\frac{1}{n-1}\binom{2n-2}{n}$$

bracket arrangements of weight n, $n > 1$. [*Hint:* Let p_n be the number of bracket arrangements of weight n and consider the formal power series

$$P(x) = \sum_{n=1}^{\infty} p_n x^n.$$

Show that

$$P(x)^2 - P(x) + x = 0 \quad \text{and} \quad P(0) = 0,$$

Hence, obtain p_n by expanding

$$\tfrac{1}{2}[1 - \sqrt{1 - 4x}]$$

in a power series.]

6. Let $a_1, a_2, \ldots, a_n$ freely generate a free group F. Show that the set of

$$\beta^n(a_{i_1}, a_{i_2}, \ldots, a_{i_n}),$$

where β^n is any bracket arrangement of weight $n > 1$ and $i_1, i_2, \ldots, i_n$ is any permutation of $1, 2, \ldots, n$, has

$$2^{n-1} \cdot (2n - 3) \cdot (2n - 5) \ldots \cdot 3 \cdot 1$$

elements in it.

7. Let G be the subgroup of one-one onto transformations of the integers given by $n \to \epsilon n + k$, k an integer, $\epsilon = \pm 1$. Show that

$$(G, G) = \{n \to n + 2k, k \text{ any integer}\}$$

and that $((G, G), G) = \{n \to n + 4k, k \text{ any integer}\}$. Conclude that if $A = (G, G)$, $B = G$ and $C = G$ then $((A, B), C) \neq (A, (B, C))$.

8. Let $A \,\theta\, B$ be a commutative but non-associative operation and let β^n be a bracket arrangement of weight n. Define $\beta^n(A_1, A_2, \ldots, A_n)$ inductively by

$$\beta^1(A_1) = A_1 \quad \text{and}$$
$$\beta^n(A_1, A_2, \ldots, A_n) = \beta^k(A_1, \ldots, A_k) \,\theta\, \beta^l(A_{k+1}, \ldots, A_n)$$

where $\beta^n = (\beta^k\beta^l)$.

Show that in general the set of

$$\beta^n(A_{i_1}, A_{i_2}, \ldots, A_{i_n}),$$

where β^n is any bracket arrangement of weight $n > 1$ and $i_1, i_2, \ldots, i_n$ is any permutation of $1, 2, \ldots, n$, has

$$(2n - 3) \cdot (2n - 5) \cdot \ldots \cdot 3 \cdot 1$$

elements in it.

5.3. The Lower Central Series

Let G be a group and let β^n be a bracket arrangement of weight n. The subgroup (see Definition 5.3)

$$\beta^n(G, G, \ldots, G) \tag{1}$$

is fully invariant in G. Special names have been given to two particular sequences of such fully invariant subgroups.

The nth *derived group* $G^{(n)}$ of G is defined recursively by

$$G^{(0)} = G, \qquad G^{(n+1)} = (G^{(n)}, G^{(n)}), n = 0, 1, 2, \ldots .$$

A group G for which $G^{(N)} = 1$ for some positive integer N is called *solvable*. The quotient groups $G/G^{(1)}$, $G/G^{(2)}$ were used in Chapter II to obtain tests for isomorphism.

The subgroups G_n of the *lower central series* of G are defined recursively by

$$G_1 = G, \qquad G_{n+1} = (G_n, G), n = 1, 2, \ldots .$$

A group G for which $G_N = 1$ for some positive integer N is called *nilpotent*.

We may show that G_n is the verbal subgroup of G generated by the word

$$((\ldots((X_1, X_2), X_3), \ldots), X_n)$$

i.e., G_n is generated by the set of all commutators

$$((\ldots((g_1, g_2), g_3), \ldots), g_n) \tag{2}$$

where $g_i \in G$, $i = 1, 2, \ldots, n$ (see Problem 3). A commutator of the type (2) with all of its open parentheses to the left of all the elements occurring is called a *simple n-fold commutator* and is denoted by

$$(g_1, g_2, \ldots, g_n). \tag{3}$$

Thus (5.2.11) can be written

$$(a, b, c^a) \cdot (c, a, b^c) \cdot (b, c, a^b) = 1$$

Modulo the terms of the lower central series, the commutator identities established in the preceding section take on a particularly simple form.

THEOREM 5.3. *Let a, b, c be elements of a group G. Let k, m, n be positive integers such that $a \in G_k$, $b \in G_m$, $c \in G_n$. Then*

$$a \cdot b \equiv b \cdot a \bmod G_{k+m}. \tag{4}$$

$$(a, b \cdot c) \equiv (a, b) \cdot (a, c) \bmod G_{k+m+n}. \tag{5}$$

$$(a \cdot b, c) \equiv (a, c) \cdot (b, c) \bmod G_{k+m+n}. \tag{6}$$

$$(a, b, c) \cdot (b, c, a) \cdot (c, a, b) \equiv 1 \bmod G_{k+m+n+1}. \tag{7}$$

PROOF. Clearly $a \cdot b = b \cdot a \cdot (a, b)$. Hence (4) is equivalent to $(a, b) \in G_{k+m}$, i.e.,

$$(G_k, G_m) \subset G_{k+m} \tag{8}$$

[Note that (G_k, G_m) need not be equal to G_{k+m} (see Problem 1).] We prove (8) by induction on m. For $m = 1$ the result holds by definition of the lower central series. Assuming (8) for m we note that by Theorem 5.2,

$$(G_k, G_{m+1}) = (G_k, (G_m, G_1)) \subset ((G_k, G_m), G_1) \cdot ((G_k, G_1), G_m).$$

By the inductive hypothesis

$$((G_k, G_m), G_1) \subset (G_{k+m}, G_1) = G_{k+m+1}$$

and

$$((G_k, G_1), G_m) = (G_{k+1}, G_m) \subset G_{k+1+m}.$$

Hence, $(G_k, G_{m+1}) \subset G_{k+m+1}$ and we have (8) and (4).

To prove (5) we use (5.2.9), For,

$$(a, b \cdot c) = (a, c)(a, b) \cdot ((a, b), c) = (a, b)(a, c) \cdot ((a, c), (a, b))((a, b), c).$$

Since $G_r \subset G_s$ if $r \geq s$, by (8)

$$(a, bc) \equiv (a, b)(a, c) \bmod G_{k+m+n}.$$

Similarly, one can show (6) using (5.2.10).

Finally, to show (7) we use (5.2.11) and (5). For,

$$((a, b), c^a) = ((a, b), c \cdot (c, a)) \equiv ((a, b), c) \cdot ((a, b), (c, a)) \bmod G_{2k+m+2n}.$$

Since $((a, b), (c, a)) \in G_{2k+m+n}$,

$$((a, b), c^a) \equiv ((a, b), c) \bmod G_{2k+m+n}.$$

Similarly, $((c, a), b^c) \equiv ((c, a), b) \bmod G_{k+m+2n}$ and

$$((b, c), a^b) \equiv ((b, c), a) \bmod G_{k+2m+n}.$$

Since, clearly, congruence mod G_r implies congruence mod G_s for $r \geq s$, each of these congruences holds mod $G_{k+m+n+1}$. Hence,

$$((a, b), c) \cdot ((c, a), b) \cdot ((b, c), a) \equiv 1 \bmod G_{k+m+n+1}.$$

This completes the proof of Theorem 5.3. ◀

Corollary 5.3. *If* $g_1, \ldots, g_p \in G_k$, $g \in G_m$ *and* $\epsilon_i = \pm 1$, *then*

(5.3.9) $$\left(\prod_{i=1}^{p} g_i^{\epsilon_i}, g\right) \equiv \prod_{i=1}^{p} (g_i, g)^{\epsilon_i} \bmod G_{2k+m}$$

and

(5.3.10) $$\left(g, \prod_{i=1}^{p} g_i^{\epsilon_i}\right) \equiv \prod_{i=1}^{p} (g, g_i)^{\epsilon_i} \bmod G_{2k+m}.$$

Proof. Use of induction on p and (6) yields

$$\left(\prod_{i=1}^{p} g_i^{\epsilon_i}, g\right) \equiv \prod_{i=1}^{p} (g_i^{\epsilon_i}, g) \bmod G_{2k+m}.$$

Since

$$(g_i^{-1}, g) \cdot (g_i, g) \equiv (g_i^{-1} g_i, g) = 1 \bmod G_{2k+m},$$

it follows that $(g_i^{\epsilon_i}, g) \equiv (g_i, g)^{\epsilon_i} \bmod G_{2k+m}$. Hence, (9) holds and similarly for (10). ◀

The significance of formulas (4) to (7) lies in their similarity to the formal laws of a Lie ring. Indeed, if (4) to (7) were equalities rather than congruences modulo certain terms of the lower central series, then the group G would be a Lie ring, whose ring addition is given by the group multiplication and whose ring multiplication is given by group commutation. (For a formal definition of a Lie ring see Section 5.4.)

For, under ring addition we have a group which by (4) is Abelian. Moreover, (5) and (6) show that ring multiplication is distributive over ring addition. The anti-commutative law for ring multiplication is given by (5.2.8). The Jacobi identity for ring multiplication is supplied by (7).

We shall use Theorem 5.3 later (see Section 5.7) to give precise information about the structure of the factor groups G_n/G_{n+1} in the case that G is a finitely generated free group. However, at this stage we can prove G_n/G_{n+1} is a finitely generated Abelian group, if G is finitely generated.

THEOREM 5.4. *Let the group G be generated by $a_1, a_2, \ldots, a_r$. Then G_n/G_{n+1} is abelian and is generated by the cosets of the simple n-fold commutators*

(11) $$(a_{\rho_1}, \ldots, a_{\rho_n})$$
where $$\rho_i \in \{1, 2, \ldots, r\}.$$

PROOF. Clearly, G_n/G_{n+1} is Abelian, for $G_{n+1} \supset G_{2n} \supset (G_n, G_n)$. We use induction on n to obtain the generators. When $n = 1$, (11) yields the given generators of G and hence, the cosets of (11) generate G_1/G_2. Assume then that the cosets of (11) generate G_n/G_{n+1}. Since $G_{n+1} = (G_n, G)$ is generated by (h, g) where $h \in G_n$, $g \in G$, clearly G_{n+1}/G_{n+2} is generated by the cosets of these elements. Moreover, by the inductive hypothesis,
$$h = \prod_{i=1}^{k} h_i^{\epsilon_i} \cdot h', \qquad \epsilon_i = \pm 1.$$
where $h' \in G_{n+1}$, $h_i \in G_n$ and h_i has the form (11). By (9),
$$(h, g) = \left(\left[\prod_{i=1}^{k} h_i^{\epsilon_i}\right] \cdot h', g\right) \equiv \left[\prod_{i=1}^{k} (h_i, g)^{\epsilon_i}\right] \cdot (h', g) \bmod G_{2n+1}.$$
Since
$$(h', g) \in G_{n+2}, \qquad (h, g) \equiv \prod_{i=1}^{k} (h_i, g)^{\epsilon_i} \bmod G_{n+2}.$$
Now,
$$g = \prod_{j=1}^{s} a_{\rho_j}^{\eta_j}, \qquad \eta_j = \pm 1, \qquad \rho_j \in \{1, 2, \ldots, r\}.$$
Hence,
$$(h_i, g) = \left(h_i, \prod_{j=1}^{s} a_{\rho_j}^{\eta_j}\right) \equiv \prod_{j=1}^{s} (h_i, a_{\rho_j})^{\eta_j} \bmod G_{n+2}$$
by (10). Thus,
$$(h, g) \equiv \prod_{i=1}^{k} \left[\prod_{j=1}^{s} (h_i, a_{\rho_j})^{\eta_j}\right]^{\epsilon_i} \bmod G_{n+2}.$$
Since (h_i, a_{ρ_j}) has the form (11) with n replaced by $n + 1$, we have completed the proof of Theorem 5.4. ◀

The generators (11) for G_n/G_{n+1} are always dependent if $n > 1$. For the particular case where G is a free group of finite rank r we shall show that G_n/G_{n+1} is a free Abelian group whose rank can be computed explicitly (see Witt's formula, Section 5.6 and Theorem 5.12).

We now apply Theorem 5.4 to *residually nilpotent* groups, i.e., to groups whose lower central series intersects in the identity. [For the general use of the term "residually", see Section 6.5.]

THEOREM 5.5. *A finitely generated, residually nilpotent group G is Hopfian, i.e., G cannot be isomorphic to any of its quotient groups G/K unless $K = 1$.*

PROOF. To prove Theorem 5.5, assume that

$$G \simeq G/K = G^*.$$

Since $G \simeq G^*$ and the terms of the lower central series are verbal subgroups,

$$G_n/G_{n+1} \simeq G_n{}^*/G^*_{n+1},$$

and $G_n{}^* = KG_n/K$, $G^*_{n+1} = KG_{n+1}/K$ (see Problem 2.3.11). If $K \neq 1$, there is a smallest integer n such that

$$K \subset G_n \quad \text{but } K \not\subset G_{n+1}.$$

Then

$$\begin{aligned} G_n/G_{n+1} \simeq G_n{}^*/G^*_{n+1} &= (KG_n/K)/(KG_{n+1}/K) \simeq KG_n/KG_{n+1} \\ &= G_n/KG_{n+1} \simeq (G_n/G_{n+1})/(KG_{n+1}/G_{n+1}). \end{aligned}$$

Since G_n/G_{n+1} is a finitely generated Abelian group (Theorem 5.4), it cannot be isomorphic to a proper quotient group of itself (see Problem 3.3.20 or 3.3.21). Therefore, $K = 1$ and G is Hopfian. ◀

Theorem 5.5 can be used to give another proof that finitely generated free groups are Hopfian (see Section 5.5).

Problems for Section 5.3

1. Show that if G is the group of transformations of the integers given by $x \to \epsilon x + k$ where $\epsilon = \pm 1$ and k is an integer, then the commutator

$$(x \to \epsilon_1 x + k_1, x \to \epsilon_2 x + k_2) = [x \to x + \epsilon_1(\epsilon_2 - 1)k_1 + \epsilon_2(1 - \epsilon_1)k_2].$$

Show that $\quad G^{(1)} = \{x \to x + 2k, k = 0, \pm 1, \pm 2, \ldots\}$;

show that $G^{(2)} = 1$ and so G is solvable. On the other hand show that for $n > 1$

$$G_n = \{x \to x + 2^{n-1}k, k = 0, \pm 1, \pm 2, \ldots\},$$

so that G is not nilpotent. Show that G is residually nilpotent. Note in particular that

$$(G_2, G_2) \neq G_4.$$

2. Let G be the group of non-singular matrices with entries which are rational functions of the four variables s, t, u, v. Show that the subgroup H generated by

$$a = \begin{pmatrix} s & u \\ 0 & 1 \end{pmatrix}, \qquad b = \begin{pmatrix} t & v \\ 0 & 1 \end{pmatrix}$$

has $H^{(2)} = 1$, but that H is not nilpotent. Show that H is residually nilpotent.

3. Show that G_n is the verbal subgroup of G generated by

$$((\ldots((X_1, X_2), X_3), \ldots), X_n).$$

[*Hint:* Show that if a is a simple n-fold commutator then a^b is also. Show that $(a \cdot b, c) = (a^b, c^b) \cdot (b, c)$ and that $(a^{-1}, b) = (a, b^a)^{-1}$. Show by induction on k that if $g_1, \ldots, g_k$ are simple n-fold commutators then $\left(\prod_{i=1}^{k} g_1^{\epsilon_i}, g\right)$ is the product of simple $(n + 1)$-fold commutators or their inverses.]

4. Show that if G is nilpotent (solvable) then any subgroup or factor group of G is nilpotent (solvable).

5. Show that a non-cyclic free group is neither nilpotent nor solvable. [*Hint:* Use Problem 5.2.2.]

6. Show that if K is any field and G is a group of triangular matrices

$$\begin{pmatrix} k_1 & k_2 \\ 0 & k_3 \end{pmatrix}, \qquad k_1, k_2, k_3 \in K,$$

then a subgroup of G cannot be a non-cyclic free group. [*Hint:* Show that $G^{(2)} = 1$ so that G is solvable.]

7. Show that in a nilpotent group, elements of coprime order must commute. In particular, a finite nilpotent group is the direct product of its Sylow subgroups. [*Hint:* Let a, b be elements of coprime order, p, q respectively. If H is the subgroup they generate then H_2 is the normal subgroup generated by (a, b). Now $1 = (a^p, b) \equiv (a, b)^p \bmod H_3$ and $1 = (a, b^q) \equiv (a, b)^q \bmod H_3$. Hence, $(a, b) \in H_3$ and $H_2 = H_3$. Since H is nilpotent, $H_N = H_2 = 1$.]

8. Let $G = \langle a, b; a^2b^{-3}\rangle$. Show that $G_N = G_2$ for $N \geq 2$ and hence, G is not nilpotent. Show that $G^{(1)} \neq G^{(2)}$. [*Hint:* Use the hint to Problem 7 to show that $G_N = G_2$. To show that $G^{(1)} \neq G^{(2)}$ use the symmetric group of degree three given by $\langle a, b; a^2, b^3, (ab)^2\rangle$ as a factor group.]

9. Show that the group of matrices generated by

$$a = \begin{pmatrix} -1 & 0 \\ 0 & 1 \end{pmatrix} \quad \text{and} \quad b = \begin{pmatrix} 1 & 1 \\ 0 & 1 \end{pmatrix}$$

over the field of integers mod 3 is not nilpotent.

10. Show that if G is a group of n-by-n triangular matrices over a field F, then G is solvable. [*Hint:* Show that $G^{(1)}$ consists of triangular matrices M whose main diagonal entries are all one. Such a matrix M can be written as $I + A$ where I is the n-by-n identity matrix and A is a triangular matrix whose main diagonal entries are all zero. Show that $A^n = 0$ and that $M^{-1} = I - A + A^2 - \ldots + (-1)^{n-1}A^{n-1}$. Next show that if $M_1 = I + A_1$ and $M_2 = I + A_2$ then $(M_1, M_2) = I + B$ where B has more upper diagonals consisting of zeros than either A_1 or A_2. Thus a matrix M in $G^{(n)}$ is of the form $I + A$ where A has zeros in at least n upper diagonals, i.e., $A = 0$ and $G^{(n)} = 1$.]

11. Show that if G is a group of n-by-n triangular matrices over a field F, each matrix having equal main diagonal entries, then G is nilpotent. [*Hint:* Every such matrix can be written as $\lambda(I + A)$ where A is a triangular matrix with main diagonal entries all zero. Observe that $(I + A_1, I + A_2) = I + B$ where B is a sum of products, each containing at least one factor A_1 and at least one factor A_2. Hence, if A_1 has k upper diagonals all zeros and A_2 has its main diagonal all zeros then B will have at least $k + 1$ upper diagonals all zeros. Show that if $M \in G_n$, $n > 1$, then $M = I + A$ where A has at least n upper diagonals all zeros, i.e., $A = 0$. Hence, $G_n = 1$.]

12. Show that a group of infinite triangular matrices is residually solvable, but not necessarily residually nilpotent, unless each matrix has equal main diagonal entries. [*Hint:* Use Problem 9 and the hints to Problems 10 and 11.]

5.4. Some Freely Generated, Graded Algebras

For the investigation of the lower central series of a group (especially a free group) we need some algebraic aids which are not directly group-theoretical concepts. We shall introduce the concepts needed in the simplest possible manner. For a systematic and thorough discussion see Chevalley, 1956. In addition to a few definitions given in this section, we shall also introduce notations that will be used throughout the remainder of this chapter.

We start with a ring R, called the ring of coefficients. In this chapter, R will always be an integral domain with an identity element 1 ($\neq 0$) under multiplication; in fact, R will be one of the following three rings:

Z, the ring of integers
Q, the field of rational numbers
G_q, the Galois field of order q, where q is a power of a prime number.

The various algebras we need will be R-modules (modules over R) with respect to addition. With respect to multiplication, we shall assume the validity of both distributive laws in all cases. Also, for $r_1, r_2 \in R$ and for any two elements u, v of our algebras, we shall require that

$$(r_1u)(r_2v) = (r_1r_2)(uv),$$

and

$$1u = u,$$

where juxtaposition denotes product. An R-module with a multiplication satisfying both distributive laws, and the above properties will be called an *R-algebra*. (Note that we do not require commutativity or even associativity of multiplication in the algebra, although multiplication in R has these properties.)

We can now define

I. The algebra A_0 (R, r)

This may be described as the associative R-algebra for which the formal power products or monomials

$$x_{n_1}^{e_1}x_{n_2}^{e_2}\cdots x_{n_k}^{e_k} \tag{1}$$

(with $n_j \neq n_{j+1}$ and $n_1, n_2, \ldots, n_k \in \{1, 2, \ldots, r\}$, and k, $e_1, e_2, \ldots, e_k$ positive integers) of r associative non-commutative indeterminates or "variables" $x_1, \ldots, x_r$, form basis elements with R as ring of coefficients.

The basis is completed by adjoining an identity (unit element) 1 to the algebra which may be identified with the identity element of R and which, by definition, will also be considered as the zero-th power of all indeterminates x_ρ, ($\rho = 1, \ldots, r$), and as the empty monomial.

Multiplication of the basis elements is defined in the obvious manner, by juxtaposition of the basis elements and amalgamation of the last power of an indeterminate appearing in the first factor with the first power appearing in the second factor, provided that it is a power of the same indeterminate. Multiplication of sums of R-multiplies of monomials is defined using the distributive laws and $(r_1u)(r_2v) = (r_1r_2)uv$, where $r_1, r_2 \in R$ and u, v are monomials.

Verification of the associative law presents no difficulty. It is also clear that any associative algebra over R with an identity and with the property that it is generated by r of its elements (in addition to the identity) must necessarily be a homomorphic image of $A_0(R, r)$; for, the distributive and associative laws without commutativity guarantee that every element can be expressed as a linear combination (with coefficients in R) of power products of the type (1). For this reason we shall call $A_0(R, r)$ the *freely generated associative algebra of rank* r, and we shall call the x_ρ a *set of free generators*.

Concepts like *degree of a basis element, degree in an indeterminate (generator)*, (and hence, *homogeneous terms*, and *homogeneous component of a given degree* of an element of $A_0(R, r)$) are defined in the natural manner, by adding the exponents on all, or on one generator appearing in the monomial.

Problem 1 at the end of this section will show how the construction of $A_0(R, r)$ can be carried out by using merely ordinary, commuting indeterminates together with the concept of an infinite matrix. (We can write down a faithful representation of $A_0(R, r)$ in terms of triangular infinite matrices whose elements are from the ordinary polynomial ring over R in several commuting and associating variables.)

The second type of algebra we need is a Lie algebra. We define

II. The algebra $\Lambda_0(R, r)$

First of all, Λ_0 is an R-algebra. Multiplication is (in general) nonassociative. Instead, for any three elements ω, φ, ψ of Λ_0, we have (with $\circ$ denoting multiplication)

$$\omega \circ \omega = 0 \tag{2}$$

$$(\omega \circ \varphi) \circ \psi + (\psi \circ \omega) \circ \varphi + (\varphi \circ \psi) \circ \omega = 0. \tag{3}$$

From (2) and the distributive laws it follows that

$$\omega \circ \varphi + \varphi \circ \omega = 0 \tag{4}$$

since

$$\begin{aligned} 0 &= (\omega + \varphi) \circ (\omega + \varphi) = \omega \circ \omega + \omega \circ \varphi + \varphi \circ \omega + \varphi \circ \varphi \\ &= \omega \circ \varphi + \varphi \circ \omega. \end{aligned}$$

Equation (3) is called the *Jacobi identity*. Clearly, Λ_0 cannot have an identity element. So far, our definitions amount to saying that Λ_0 is a *Lie algebra over R*. In addition, we now postulate that Λ_0 can be generated by r of its elements, ξ_ρ, $(\rho = 1, \ldots, r)$ and that any other Lie algebra over R generated by r elements ξ_ρ' is a homomorphic image of $\Lambda_0(R, r)$ under the mapping $\xi_\rho \to \xi'_\rho$. [Alternatively, we might say that any relation between the ξ_ρ is derivable from the laws of an R-algebra together with the laws (2) and (3).]

We shall call $\Lambda_0(R, r)$ the *free Lie algebra of rank r over R*, and we shall call the ξ_ρ *free generators* of $\Lambda_0(R, r)$. [For the existence of Λ_0 see Problem 7(d).] Although we cannot as yet describe Λ_0 in the same manner as A_0 by writing down a basis for it as for A_0, we can express every element in Λ_0 as a linear combination of monomials. These we can write as expressions

$$\beta^n(\xi_{\rho_1}, \xi_{\rho_2}, \ldots, \xi_{\rho_n}), \qquad (\rho_1, \ldots, \rho_n \in \{1, 2, \ldots, r\}) \tag{5}$$

in the ξ_ρ by choosing a bracket arrangement β^n (see Section 5.2) and replacing commutation by multiplication in Λ_0. The weight n of the bracket arrangement will then be called the *degree of the monomial* (5) in the ξ_ρ, and we can also count the number l of times a particular ξ_ρ occurs in such a monomial and say that it is of *degree l in ξ_ρ*. This concept of degree allows us to introduce *homogeneous terms* and *homogeneous component of a given degree* in the natural way. Clearly homogeneous terms of a given degree form an R-submodule. In fact, as the homogeneous nature of the defining identities (2) and (3) for a Lie algebra allows us to prove,

LEMMA 5.1. *As an R-module, $\Lambda_0(R, r)$ is the direct sum of its homogeneous submodules, i.e., every element of Λ_0 can be written uniquely as a sum of homogeneous elements of Λ_0. In particular:*

(A) *If $P(\xi_1, \ldots, \xi_r)$ is an element of Λ_0 and $P = 0$, then each of the homogeneous components of P is zero.*

(B) *If H is a sum of monomials of type* (5) *which are all of degree at least n, and if H' is a sum of monomials of degree less than n, then $H = H'$ implies $H = 0$.*

PROOF. From the properties of any R-algebra generated by $\xi_1, \ldots, \xi_r$, it follows that every element of Λ_0 can be expressed as a sum of R-multiples of monomials of type (5). Moreover, using associativity and commutativity of addition, these may be grouped into homogeneous elements.

To show that Λ_0 is the direct sum of its homogeneous submodules, or equivalently, to show (A), we must make use of the "freeness" of Λ_0, in some direct manner. We therefore leave the completion of the proof for the exercises where a constructive procedure for obtaining $\Lambda_0(R, r)$ is given in outline (see Problems 6 and 7). For a more detailed development see Chevalley, 1956. ◀

Lemma 5.1 shows that the problem of finding a basis for Λ_0 can be reduced to the construction of a basis for each set of elements of a fixed degree. We shall carry out such a construction recursively in Section 5.6.

We need two extensions of our algebras $A_0(R, r)$ and $\Lambda_0(R, r)$, to allow infinitely many variables, and to allow infinite sums.

III. The algebras $A_0(R, \infty)$ and $A(R, r)$

The algebra $A_0(R, \infty)$ is simply the free associative algebra generated by countably many indeterminates y_ρ ($\rho = 1, 2, 3, \ldots$). We shall restrict the elements of $A_0(R, \infty)$ to finite sums, so that every element of the algebra is contained in a finitely generated subalgebra $A_0(R, r)$ for some finite r.

The algebra $A(R, r)$ arises from $A_0(R, r)$ by admitting infinite sums. We shall write the general element v of $A(R, r)$ in the form of a "*formal power series in non-commuting variables*"

$$v = \sum_{n=0}^{\infty} u_n, \tag{6}$$

where u_n is a homogeneous element of degree n belonging to $A_0(R, r)$. We define addition and multiplication in the natural manner and we observe that multiplication can be carried out no matter what the u_n are, since in a product of two infinite sums of this type only a finite number of terms in each sum can contribute to the component of a given degree in the product. [We could define $A(R, r)$ as the closure of $A_0(R, r)$ in the topology defined by a suitable distance function (valuation) in $A_0(R, r)$. However, we do not need this and refer the matter to the problems at the end of this section (see Problem 4).]

It will be convenient later on to use the following:

Definition 5.5. The ideal X generated by the x_ρ $(\rho = 1, \ldots, r)$ in $A(R, r)$ will be called the *basic ideal*. The statement that an element u of $A(R, r)$ is contained in the nth power of X,

$$u \equiv 0 \bmod X^n$$

is equivalent to saying that the homogeneous components of u are all of a degree exceeding $n - 1$. If $u \equiv 0 \bmod X^n$ for all $n = 1, 2, 3, \ldots$, then $u = 0$.

IV. The algebras $\Lambda_0(R, \infty)$ and $\Lambda(R, r)$

We define $\Lambda_0(R, \infty)$ as we defined $A_0(R, \infty)$, simply by admitting a denumerable infinity of generators ξ_ρ but allowing only finite sums. And, as $A(R, r)$ from $A_0(R, r)$, so we derive $\Lambda(R, r)$ from $\Lambda_0(R, r)$, by admitting infinite sums

$$\psi = \sum_{n=1}^{\infty} \omega_n \tag{7}$$

where ω_n is a finite sum of monomials of degree n in the generators $\xi_\rho(\rho = 1, \ldots, r)$ (and, of course, each monomial has a coefficient that is an element of R). Since the product of terms of degree n and m consists of terms of degree $n + m$ (unless it vanishes), we can apply formal multiplication to sums of this type; for, under formal multiplication of two sums of the type (7), only a finite number of terms of each sum contributes to the terms of a given degree in the product.

Problems for Section 5.4

1. Let $A_0(Z, 2)$ be freely generated by x, y. Show that we obtain an isomorphic representation of $A_0(Z, 2)$ by mapping the identity on the unit matrix I of infinitely many rows and columns and x, y respectively onto matrices X, Y, defined by

$$X = \begin{pmatrix} 0 & x_1 & 0 & 0 & 0 & 0 & . & . & . \\ 0 & 0 & x_2 & 0 & 0 & 0 & . & . & . \\ 0 & 0 & 0 & x_3 & 0 & 0 & . & . & . \\ 0 & 0 & 0 & 0 & x_4 & 0 & . & . & . \\ 0 & 0 & 0 & 0 & 0 & x_5 & . & . & . \\ 0 & 0 & 0 & 0 & 0 & 0 & . & . & . \\ . & . & . & . & . & . & . & . & . \end{pmatrix}$$

$$Y = \begin{pmatrix} 0 & y_1 & 0 & 0 & 0 & 0 & . & . & . \\ 0 & 0 & y_2 & 0 & 0 & 0 & . & . & . \\ 0 & 0 & 0 & y_3 & 0 & 0 & . & . & . \\ 0 & 0 & 0 & 0 & y_4 & 0 & . & . & . \\ 0 & 0 & 0 & 0 & 0 & y_5 & . & . & . \\ 0 & 0 & 0 & 0 & 0 & 0 & . & . & . \\ . & . & . & . & . & . & . & . & . \end{pmatrix}$$

where the x_ν, y_ν, $(\nu = 1, 2, 3, \ldots)$ are ordinary commuting indeterminates (for instance, real variables). Show in particular that the basis element

$$b = x^{\alpha_1}y^{\beta_1}x^{\alpha_2}y^{\beta_2} \ldots x^{\alpha_n}y^{\beta_n}$$

of $A_0(Z, 2)$ is mapped onto a matrix B defined as follows:

In its first row, all but one of the elements of B are zero. The non-zero element b^* occurs in the $(d + 1)$st column where

$$d = \alpha_1 + \beta_1 + \alpha_2 + \beta_2 + \ldots + \alpha_n + \beta_n.$$

The element b^* is the product

$$b^* = x_1 x_2 \ldots x_{\alpha_1} y_{\alpha_1+1} \ldots y_{\alpha_1+\beta_1} x_{\alpha_1+\beta_1+1} \ldots x_{\alpha_1+\beta_1+\alpha_2} \ldots$$

which may be described as follows: If the kth factor $(k = 1, \ldots, d)$ of b is an x, then b^* has the factor x_k exactly once; if the kth factor of b is a y, then b^* contains the factor y_k exactly once.

In the lth row of B all elements but the one in the $(d + l)$th column vanish. The element in the $(d + l)$th column arises from b^* by adding $l - 1$ to each one of the subscripts of the x_k, y_k occurring in b^*. Show that $A(Z, 2)$ can also be represented in this manner.

2. Let Q be the field of rational numbers and let $A_0(Q, 2)$ be the freely generated associative algebra over Q on two free generators x, y. Consider

in A_0 the two sided ideal J generated by x^2 and y^2. Then the quotient ring $A^* = A_0/J$ is an associative algebra again; it arises from A_0 by simply adding the relations $x^2 = y^2 = 0$. Show that A^* is isomorphic to an algebra M of matrices, the elements of which are polynomials in two ordinary commuting variables s, t, with rational coefficients. M consists of two-by-two matrices; it is generated (over Q) by the matrices

$$I = \begin{pmatrix} 1 & 0 \\ 0 & 1 \end{pmatrix}, \quad S = \begin{pmatrix} 0 & s \\ 0 & 0 \end{pmatrix}, \quad T = \begin{pmatrix} 0 & 0 \\ t & 0 \end{pmatrix}$$

and the isomorphism between A^* and M is established by the mapping

$$1 \to I, \quad x \to S, \quad y \to T.$$

3. Let R be an integral domain with an identity element. Show that $A_0(R, r)$ and $A(R, r)$ do not have any divisors of zero.

4. For any element u of $A_0(R, r)$ define the absolute value of u by $|u| = 0$ if $u = 0$ and by $|u| = 2^{-n}$ if n is the lowest degree of (non-vanishing) multiples of basis elements occurring in u. Show that for $u, v \in A_0(R, r)$:

$$|uv| = |u|\,|v|$$

$$|u + v| \leqq \max(|u|, |v|).$$

[In other words, show that $|u|$ is a non-Archimedean valuation of $A_0(R, r)$.] Show that the distance function $|u - v|$ makes a metric space out of $A_0(R, r)$. Extend this metric to $A(R, r)$, and show that $A(R, r)$ is a complete metric space in which $A_0(R, r)$ is dense.

5. Define a distance function Δ in $\Lambda_0(R, r)$ as follows: For any two elements $\varphi, \psi \in \Lambda_0$, their distance $\Delta(\varphi, \psi)$ is defined as 2^{-n} if $\varphi - \psi \neq 0$ and if the non-vanishing homogeneous component of lowest degree in $\varphi - \psi$ is of degree n. If $\varphi - \psi = 0$, $\Delta(\varphi, \psi) = 0$. Prove that for any three elements $\varphi, \psi, \omega \in \Lambda_0$:

$$\Delta(\varphi, \psi) + \Delta(\psi, \omega) \geqq \Delta(\varphi, \omega).$$

Show that Λ_0 is a metric space under this distance function. Extend this metric to $\Lambda(R, r)$, and show that $\Lambda(R, r)$ is a complete metric space in which $\Lambda_0(R, r)$ is dense.

6. Let R be an integral domain with infinitely many elements. Suppose that an element φ in $\Lambda_0(R, r)$ (which is freely generated by $\xi_1, \ldots, \xi_r$) is zero.

(a) Show that if $\varphi = \varphi_1 + \ldots + \varphi_n$, where the terms of φ_i all have degree i on ξ_1, then each φ_i is zero.

(b) Show that if $\varphi = \varphi_1 + \ldots + \varphi_n$, where the terms of φ_i all have total degree i on some fixed subset of $\xi_1, \ldots, \xi_r$, then each φ_i is zero. In particular, if φ_i is the homogeneous component of degree i in φ, then $\varphi_i = 0$.

(c) State and prove analogues of (a) and (b) for the free associative algebra $A_0(S, r)$, where S is any integral domain (possibly finite).

[*Hint:* For (a) and (b), show that if $\xi_{\rho_1}, \ldots, \xi_{\rho_k}$ is a subset of $\xi_1, \ldots, \xi_r$, then the free Lie algebra can be mapped into itself by sending $\xi_j \to \xi_j$ if $j \neq \rho_i$, and $\xi_{\rho_i} \to \alpha\xi_{\rho_i}$, where α is an element in R. Now φ must go into zero, and so

$$\alpha\varphi_1 + \alpha^2\varphi_2 + \ldots + \alpha^n\varphi_n = 0,$$

for all α in R. Since R is infinite, we can find n elements $\alpha_1, \ldots, \alpha_n$ in R such that the determinant whose rows are $\alpha_i, \alpha_i^2, \ldots, \alpha_i^n$, where $i = 1, \ldots, n$ does not vanish. Hence, $\varphi_1 = \ldots = \varphi_n = 0$. For (c), use the linear independence of distinct monomials in A_0.]

7. The *free R-algebra $B_0(R, r)$ freely generated by $x_1, x_2, \ldots, x_r$* is obtained by taking as a basis $\beta^n(x_{\rho_1}, \ldots, x_{\rho_n})$ where $\rho_i \in \{1, \ldots, r\}$ and β^n is a bracket arrangement of weight $n \geqslant 0$, and defining the product of the basis elements

$$\beta^k(x_{\rho_1}, \ldots, x_{\rho_k}), \qquad \beta^l(x_{\sigma_1}, \ldots, x_{\sigma_l})$$

to be the basis element

$$\beta^n(x_{\rho_1}, \ldots, x_{\rho_k}, x_{\sigma_1}, \ldots, x_{\sigma_l}),$$

where $n = k + l$ and $\beta^n = (\beta^k\beta^l)$. (Multiplication is extended in an obvious manner to all elements of $B_0(R, r)$.)

(a) Show that any R-algebra generated by $a_1, \ldots, a_r$ is a homomorphic image of $B_0(R, r)$.

(b) Show that the ideal of $B_0(R, r)$ (i.e., the smallest R-subalgebra of B_0 which is closed under right and left multiplication by all elements of B_0) generated by an element θ in B_0, is given by all finite sums of R-multiples of elements

$$\beta^n(x_{\rho_1}, x_{\rho_2}, \ldots, x_{\rho_j}, \theta, x_{\rho_{j+2}}, \ldots, x_{\rho_n}),$$

where $\rho_i \in \{1, \ldots, r\}$. Generalize this to the ideal of B_0 generated by any set of elements $\{\theta\}$ in B_0.

(c) Suppose that the *degree* of a basis element $\beta^n(x_{\rho_1}, \ldots, x_{\rho_n})$ is n and *homogeneous elements, homogeneous components of an element* of B_0, etc., are defined in the obvious manner. Show that if each element in $\{\theta\}$ is homogeneous (two elements may have different degrees) and y is in the ideal C of B_0 generated by the set $\{\theta\}$, then each of the homogeneous components of y are in C.

(d) Show that the free Lie algebra $\Lambda_0(R, r)$ freely generated by $\xi_1, \xi_2, \ldots, \xi_r$ is isomorphic to the quotient algebra of B_0 by the ideal C generated by all elements of B_0 of the form $\alpha\beta - \beta\alpha$ and $(\alpha\beta)\gamma + (\gamma\alpha)\beta + (\beta\gamma)\alpha$, where α, β, γ are basis elements of B_0.

(e) Show that an element of $\Lambda_0(R, r)$ is zero implies all of its homogeneous components are zero.

[*Hint:* For (b), if $y_1, \ldots, y_n$ are any elements of B_0, then define $\beta^n(y_1, \ldots, y_n)$ inductively on n, using the product in B_0. Show that the elements given are closed under right or left multiplication by basis elements. For (c), use that the homogeneous component of a sum of R-multiples of elements is the sum of the R-multiples of the homogeneous components of the elements; also use that

$$\beta^n(x_{\rho_1}, \ldots, x_{\rho_j}, \theta, x_{\rho_{j+2}}, \ldots, x_{\rho_n})$$

is homogeneous if θ is. For (d), show that B_0/C can be mapped into any Lie algebra with r generators. For (e), use (c) and (d).]

8. Let $B_0(R, r)$ be as in Problem 7.
   (a) Show that if each element in a set $\{\theta\}$ is a sum of terms of the same total degree on a fixed subset $x_{\rho_1}, \ldots, x_{\rho_k}$, of $x_1, \ldots, x_r$ (two elements may have different total degree), then if an element y of B_0 is in C the ideal of B_0 generated by $\{\theta\}$, and $y = y_1 + \ldots + y_n$, where y_i has total degree i on $x_{\rho_1}, \ldots, x_{\rho_k}$, then each y_i is in C.
   (b) Show that if $\Lambda_0(R, r)$ is as in Problem 7(d), then if $\varphi = \varphi_1 + \ldots + \varphi_n$, where φ_i has total degree i on $\xi_{\rho_1}, \ldots, \xi_{\rho_k}$, and $\varphi = 0$ then each φ_i is zero. [*Hint:* See the hint to Problem 7.]

9. Let R be the real number field and let V be the set of three-dimensional real vectors $ai + bj + ck$, where i, j, k are the unit vectors in the positive x, y, z directions, respectively. Show that if multiplication is taken to be the vector (or cross) product, then V is a Lie algebra over R. Find the kernel of the homomorphism of $\Lambda_0(R, 3)$ onto V given by

$$\xi_1 \to i, \quad \xi_2 \to j, \quad \xi_3 \to k.$$

[*Hint:* Under the given homomorphism, clearly

$$\xi_1\xi_2 - \xi_3, \quad \xi_2\xi_3 - \xi_1, \quad \text{and} \quad \xi_3\xi_1 - \xi_2$$

go into zero. If C is the ideal of Λ_0 generated by these three elements, then show by induction on its degree that every product of ξ_1, ξ_2, ξ_3 is a linear combination of ξ_1, ξ_2, and ξ_3 in Λ_0/C. Conclude $V \simeq \Lambda_0/C$.]

10. Let B be any associative R-algebra. Define a new operation $[x, y]$ for any two elements of B by

$$[x, y] = xy - yx.$$

Show that the R-module B is a Lie algebra under the "bracket multiplication." [*Hint:* Verify (2) and (3), and show that B is an R-algebra.]

11. Let B be an associative algebra, and suppose that C and D are subalgebras of B such that $C \cap D = 0$. Show that if all elements of C commute under multiplication with all elements of D, then $C + D$ is a Lie algebra under the bracket multiplication of Problem 10; moreover, the Lie algebra $C + D$ is isomorphic to the direct sum of the Lie algebras C and D.

12. Show that an associative R-algebra B is isomorphic to $A_0(R, r)$ if and only if there exists r elements $y_1, \ldots, y_r$ in B such that any mapping of $y_1, \ldots, y_r$ into an associative R-algebra C can be extended to a homomorphism of B into C. (This extendability property is often used to define the notion of the freeness of a general algebraic system, and of a set of generators for this system.)

13. State and prove a result analogous to that in Problem 12, but for $\Lambda_0(R, r)$ rather than $A_0(R, r)$.

14. Let $A_0(R, r)$ be the associative algebra freely generated by $x_1, \ldots, x_r$, $r \geq 2$, where some element of R has infinite multiplicative order (for example, if R has characteristic zero then 2 has infinite multiplicative order). Suppose an operation $*$ is defined in A_0 by means of a "polynomial expression" i.e., for all P, Q in A_0

$$P * Q = \sum_{i=1}^{k} \alpha_i P^{l_1} Q^{l'_1} \ldots P^{l_{m_i}} Q^{l'_{m_i}},$$

where α_i is in R, and l_j, l_j' are positive integers except that l_1 or l'_{m_i} may be zero.

(a) Show that if

$$P * \alpha Q = \alpha P * Q = \alpha(P * Q)$$

for all α in R, and P, Q in A_0, then

$$P * Q = \alpha_1 P + \alpha_2 Q + \alpha_3 PQ + \alpha_4 QP.$$

(b) Show that if $*$ is both left and right distributive over $+$, and satisfies (a), then

$$P * Q = \gamma PQ + \delta QP.$$

(c) Show that if in addition to (a) and (b), $*$ is anti-commutative, i.e., $P * Q = -Q * P$ then

$$P * Q = \beta[P, Q].$$

(d) Conclude that an anti-commutative R-algebra whose multiplication is given by a "polynomial expression" in a free associative algebra is automatically a Lie algebra, i.e., satisfies the Jacobi identity.

[*Hint:* For (a), if α has infinite multiplicative order in k, consider $\alpha^t x_1 * x_2$ and $x_1 * \alpha^t x_2$. For (b), show that $(P_1 + P_2) * Q = P_1 * Q + P_2 * Q$ implies α_2 in (a) is zero. For (c) use $x_1 * x_2 = -x_2 * x_1$.]

15. Let M be the algebra over the integers having the linear basis

$$x, y, xx, xy, yx, yy$$

with a multiplication given by juxtaposition except that if a monomial of degree three or more occurs then that product is zero.

(a) Show that M is an associative algebra.

(b) Show that M has a multiplication which does not satisfy the nilpotent condition (2) but does satisfy the Jacobi identity (3), though it is associative and not identically zero.

[*Hint:* Show that if the Jacobi identity holds for all monomials, it holds for all elements.]

16. Let $G_3[x, y, z]$ be the associative and commutative algebra of polynomials in x, y, and z over the field of integers modulo three. Define for each P and Q in $G_3[x, y, z]$

$$P * Q = P^3Q - PQ^3.$$

(a) Show that $\alpha P * Q = P * \alpha Q = \alpha(P * Q)$ for all α in G_3 and P, Q in $G_3[x, y, z]$.
(b) Show that $*$ is left and right distributive over $+$.
(c) Show that $*$ satisfies the nilpotent condition (2) but not the Jacobi identity (3).
(d) Show that in any algebra with a linear basis of two elements the nilpotent condition (2) implies (3).

17. Let R be any integral domain with characteristic different from two, and let L be an R-algebra.

(a) Show that the nilpotent condition (2) is equivalent to the anti-commutative law

$$\varphi \cdot \psi = -\psi \cdot \varphi.$$

(b) Show that assuming the nilpotent condition (2), the Jacobi identity (3) is equivalent to the "derivative law," the φ-multiple of a product is the φ-multiple of the first factor times the second factor plus the first factor times the φ-multiple of the second factor, i.e.,

$$\varphi \cdot (\psi \cdot \xi) = (\varphi \cdot \psi) \cdot \xi + \psi \cdot (\varphi \cdot \xi),$$
$$(\psi \cdot \xi) \cdot \varphi = (\psi \cdot \varphi) \cdot \xi + \psi \cdot (\xi \cdot \varphi).$$

18. Let the Lie algebra C be generated by $\eta_1, \eta_2, \ldots, \eta_r$ (but not necessarily freely). Show that every element of C is a linear combination of the "simple" monomials

$$(\ldots((\eta_{i_1} \cdot \eta_{i_2}) \cdot \eta_{i_3}) \ldots)\eta_{i_k}.$$

[*Hint:* Use induction on the weight of a bracket arrangement; and also if $\beta^n = (\beta^p\beta^q)$, where β^p, β^q are "simple" use induction on q and Problem 17(b).]

5.5. A Mapping of a Free Group into $A(Z, r)$

The associative Z-algebra $A_0(Z, r)$ in the non-commuting variables $x_1, \ldots, x_r$ consist of "polynomials" in $x_1, \ldots, x_r$ with integer coefficients. As such, besides the operations of addition and multiplication in A_0, one may consider "substitution" of one element into another; e.g., if $P(x_\rho) = x_1x_2^2 + x_2x_3x_1$ and $Q_1(x_\rho)$, $Q_2(x_\rho)$, $Q_3(x_\rho)$ are any elements of A_0, then $Q_1Q_2^2 + Q_2Q_3Q_1$ is a well-defined element of A_0. If we try to extend this idea of "substitution" to $A(Z, r)$, the associative Z-algebra of formal power series in non-commuting variables $x_1, \ldots, x_r$ with integer coefficients, some difficulties present themselves; e.g., if $P(x_1) =$

$1 + x_1 + x_1^2 + \ldots + x_1^n + \ldots$, then $P(2)$ should mean

$$1 + 2 + 2^2 + \ldots + 2^n + \ldots,$$

which makes no sense as an element in $A(Z, r)$. If, however, we try to compute $P(x_2 + x_2^2)$, using the "power series"

$$1 + (x_2 + x_2^2) + (x_2 + x_2^2)^2 + (x_2 + x_2^2)^3 + \ldots + (x_2 + x_2^2)^n + \ldots,$$

we obtain

$$1 + x_2 + 2x_2^2 + 3x_2^3 + 5x_2^4 + \ldots + \left(\sum_{n/2 \le k \le n} \binom{k}{n-k}\right) x_2^n + \ldots;$$

for $(x_2 + x_2^2)^k = x_2^k(1 + x_2)^k$ has the term $\binom{k}{n-k} x^n$ in its expansion, if $k \le n \le 2k$. Since only finitely many of the terms in $P(x_1)$ can contribute a term of nth degree to $P(x_2 + x_2^2)$, a definite element of $A(Z, r)$ is arrived at.

In general, if $P(x_1, \ldots, x_r)$ is an element of $A(Z, r)$ and $Q_1(x_\rho), \ldots, Q_r(x_\rho)$ are elements of $A(Z, r)$ with zero constant term, then the *value* $P(Q_1, \ldots, Q_r)$ *obtained by substituting* $Q_1, \ldots, Q_r$ *into* $P(x_1, \ldots, x_r)$ is the unique element of $A(Z, r)$ whose homogeneous component of nth degree is the homogeneous component of nth degree of $H(Q_1, \ldots, Q_r)$, where $H(x_1, \ldots, x_r)$ is the sum of the homogeneous components of degree $\le n$ of $P(x_1, \ldots, x_r)$.

This definition is well-defined and agrees with the intuitive notion of $P(Q_1, \ldots, Q_r)$; for, if

$$P(x_\rho) = P_0(x_\rho) + P_1(x_\rho) + P_2(x_\rho) + \ldots + P_n(x_\rho) + \ldots,$$

where $P_n(x_\rho)$ is homogeneous of degree n, then $P_i(Q_\rho)$, for $i > n$, can contribute no term of degree n to $P(Q_\rho)$, since Q_ρ has no constant term.

Substitution of $Q_1, \ldots, Q_r$ for $x_1, \ldots, x_r$ in each element of $A(Z, r)$, provided it is defined (i.e., Q_ρ has a zero constant term), behaves much like substitution into polynomials. Specifically, we have

LEMMA 5.2. *Let* $Q_1(x_\rho), \ldots, Q_r(x_\rho)$ *be elements of* $A(Z, r)$ *with zero constant term. Then the mapping*

$$x_\rho \to Q_\rho, \qquad \rho = 1, \ldots, r$$

determines a homomorphism of the Z*-algebra* A *into itself.*

PROOF. The proof is straightforward and is based upon obtaining the nth degree homogeneous component of a sum, Z-multiple, and product of elements of $A(Z, r)$, in terms of the homogeneous components of the individual elements. The details are left as an exercise for the reader. ◀

This lemma allows us to show that $A(Z, r)$ contains a large multiplicative group; within this group we shall find a useful representation for the free group of rank r.

LEMMA 5.3. *The set M of all elements g in $A(Z, r)$ with constant term 1 is a group under multiplication. Moreover, if $g = 1 + h$, then*

$$g^{-1} = 1 - h + h^2 - h^3 + \ldots + (-1)^n h^n + \ldots. \tag{1}$$

PROOF. Since the constant term of a product is the product of the individual constant terms, M is closed under multiplication; since multiplication is associative in $A(Z, r)$, and 1 is in M, associativity and the existence of an identity holds. To show M is a group under multiplication therefore, it suffices to produce a multiplicative inverse for $g = 1 + h$ in M. It is natural to try to show that $g^{-1} = 1 - h + h^2 - h^3 + \ldots + (-1)^n h^n + \ldots$, exists and is the required inverse. Indeed, since h has a zero constant term and

$$\begin{aligned}(1 + x_1)(1 - x_1 + x_1{}^2 - x_1{}^3 + \ldots + (-1)^n x_1{}^n + \ldots)\\ = (1 - x_1 + x_1{}^2 - x_1{}^3 + \ldots + (-1)^n x_1{}^n + \ldots)\\ + (x_1 - x_1{}^2 + x_1{}^3 - x_1{}^4 + \ldots + (-1)^{n+1} x_1{}^n + \ldots) = 1,\end{aligned}$$

it follows from Lemma 5.2 that

$$g \cdot g^{-1} = (1 + h)(1 - h + h^2 - h^3 + \ldots + (-1)^n h^n + \ldots) = 1.$$

Since g^{-1} is in M, we have our result. ◀

THEOREM 5.6. *If $A(Z, r)$ is freely generated by $x_1, \ldots, x_r$ then the elements*

$$a_\rho = 1 + x_\rho, \qquad \rho = 1, \ldots, r$$

of $A(Z, r)$ are generators of a free group $F(r)$ of rank r. Moreover,

$$a_\rho{}^{-1} = 1 - x_\rho + x_\rho{}^2 - x_\rho{}^3 + \ldots, + (-1)^n x_\rho{}^n + \ldots.$$

PROOF. In view of Lemma 5.3, all we must show is that a freely reduced word in $a_1, \ldots, a_r$ is not 1 unless it is the empty word. Consider the word

$$W = a_{\rho_1}^{e_1} a_{\rho_2}^{e_2} \ldots a_{\rho_k}^{e_k},$$

where e_j and ρ_j are integers, $1 \leq \rho_j \leq r$, for $j = 1, \ldots, k$, and $\rho_j \neq \rho_{j+1}$. Now it is easily shown by induction on n that

$$a_\rho{}^n = 1 + n x_\rho + x_\rho{}^2 h(x_\rho),$$

where $h(x_\rho)$ is an infinite power series in x_ρ alone, for n any integer. Hence, W is

$$(1 + e_1 x_{\rho_1} + x_{\rho_1}^2 h_1(x_{\rho_1})) \ldots (1 + e_k x_{\rho_k} + x_{\rho_k}^2 h_k(x_{\rho_k}))$$

which contains the unique monomial of degree k and syllable length k

$$e_1 \dots e_k x_{\rho_1} \dots x_{\rho_k}.$$

Since

$$e_1 \dots e_k \neq 0,$$

we have $W \neq 1$. ◀

The representation of the free group $F(r)$ given by Theorem 5.6 is particularly useful for exhibiting an important sequence of fully invariant subgroups of $F(r)$.

THEOREM 5.7. *Let $D_n(F)$ be the set of elements C_n of $F(r)$ such that*

$$C_n = 1 + h_n(x_1, \dots, x_r).$$

where h_n is an element of $A(Z, r)$ with no term of degree less than n (i.e., h_n *is in the ideal X^n, the nth power of the basic ideal X of elements with zero constant term*). *Then $D_n(F)$ is an invariant subgroup of $F(r)$ and the qoutient group*

$$D_n(F)/D_{n+1}(F)$$

is a finitely generated (if r is finite) free Abelian group. Moreover, clearly the intersection of all $D_n(F)$ is the identity.

PROOF. To show that $D_n(F)$ is a fully invariant subgroup of $F(r)$, since $D_n(F)$ is clearly a subgroup, it suffices to show that if

$$W(a_1, \dots, a_r) = 1 + h_n(x_1, \dots, x_r), \tag{2}$$

where h_n is in X^n, then for any $b_1, \dots, b_r$ in $F(r)$,

$$W(b_1, \dots, b_r)$$

is in $D_n(F)$. Since b_ρ is in $F(r)$, $b_\rho = 1 + y_\rho$, where y_ρ is in X. By Lemma 5.2, the mapping

$$x_\rho \to y_\rho, \qquad \rho = 1, \dots, r$$

determines a homomorphism of $A(Z, r)$ into itself. Under this homomorphism,

$$a_\rho \to b_\rho,$$

and hence, by (1),

$$a_\rho^{-1} \to b_\rho^{-1},$$

and therefore,

$$W(a_1, \dots, a_r) \to W(b_1, \dots, b_r).$$

Thus, from (2),

$$W(b_1, \dots, b_r) = 1 + h_n(y_1, \dots, y_r),$$

and so $D_n(F)$ is fully invariant.

To show that $D_n(F)/D_{n+1}(F)$ is a finitely generated free Abelian group, write each element C_n of $D_n(F)$ as

$$C_n = 1 + P_n(x_1, \ldots, x_r) + h_{n+1}(x_1, \ldots, x_r),$$

where $P_n(x_\rho)$ is homogeneous of degree n and $h_{n+1}(x_\rho)$ is in X^{n+1}. Then the mapping

$$(3) \qquad C_n \to P_n(x_1, \ldots, x_r),$$

is a homomorphism of $D_n(F)$ onto an additive group H_n of homogeneous polynomials of degree n in the non-commuting variables $x_1, \ldots, x_r$, with integral coefficients; for $C_nC_n' = 1 + P_n + P_n' + h_{n+1}$, where h_{n+1} is in X^{n+1}. Since H_n is a subgroup of the additive group of all homogeneous polynomials of degree n in $x_1, \ldots, x_r$ with integral coefficients, which is a finitely generated (if r is finite) free Abelian group, then H_n is a finitely generated free Abelian group. Moreover, the kernel of the mapping (3) is $D_{n+1}(F)$, so that

$$D_n(F)/D_{n+1}(F) \simeq H_n,$$

which completes the proof. ◀

Theorem 5.7 can be used directly to show that a finitely generated free group is Hopfian (see Problem 5). This may also be done by showing that Theorem 5.5 applies since

Corollary 5.7. *If F_n is the nth term in the lower central series for $F(r)$, then $D_n(F)$ contains F_n.*

Proof. (It should be noted that $D_n(F)$ is actually equal to F_n; but this is not easy to prove and we shall first establish it in Section 5.7.)

We use induction on n to show that $D_n(F) \supset F_n$; for $D_1(F) = F(r) = F_1$. Moreover, it is easy to show that $D_n(F)/D_{n+1}(F)$ is in the center of $F(r)/D_{n+1}(F)$. Indeed, let

$$C = 1 + P(x_\rho) + h(x_\rho)$$

and

$$C_n = 1 + Q_n(x_\rho) + h'(x_\rho),$$

where C is in $F(r)$, C_n is in $D_n(F)$, $P(x_\rho)$ has terms of degree between 1 and n, $Q_n(x_\rho)$ is homogeneous of degree n, and both $h(x_\rho)$ and $h'(x_\rho)$ are in X^{n+1}. Then

$$CC_n = 1 + P(x_\rho) + Q_n(x_\rho) + h''(x_\rho),$$

and

$$C_nC = 1 + Q_n(x_\rho) + P(x_\rho) + h'''(x_\rho),$$

where $h''(x_\rho)$ and $h'''(x_\rho)$ are in X^{n+1}. Thus CC_n and C_nC are equal modulo X^{n+1}. But then from (1) it follows that $(CC_n)^{-1}$ and $(C_nC)^{-1}$ are equal modulo X^{n+1}. Thus $(CC_n)(C_nC)^{-1}$ is equal to 1 modulo X^{n+1} and so is in $D_{n+1}(F)$. Hence, $CC_nD_{n+1}(F) = C_nCD_{n+1}(F)$, and $C_nD_{n+1}(F)$ is in the center of $F(r)/D_{n+1}(F)$.

It therefore follows that the commutator

$$(D_n(F), F(r)) \subset D_{n+1}(F).$$

Since by inductive hypothesis, $F_n \subset D_n(F)$, we have

$$F_{n+1} = (F_n, F(r)) \subset (D_n(F), F(r)) \subset D_{n+1}(F),$$

and we have our result. ◀

The calculations in the last proofs suggest the following:

Definition 5.6. Let W be a word in the free generators $a_\rho = 1 + x_\rho$ of $F(r)$. Then the *deviation* $\delta(W)$ *of* W is defined by

$$\delta(W) = 0,$$

if $W = 1$, and otherwise

$$\delta(W) = u_n, \tag{4}$$

where u_n is of degree n and is the non-vanishing homogeneous component of lowest positive degree in W.

The following formulae for δ are easily derived.

LEMMA 5.4. *Let U and V be words ($\neq 1$) in the $a_\rho = 1 + x_\rho$, and let*

$$\delta(U) = u_n, \quad \delta(V) = v_m.$$

Then, for all integers l,

$$\delta(U^l) = lu_n. \tag{5}$$

If $n < m$, then

$$\delta(UV) = \delta(VU) = u_n. \tag{6a}$$

If $n = m$ and $u_n + v_n \neq 0$, then

$$\delta(UV) = \delta(VU) = u_n + v_n. \tag{6b}$$

If $n = m$ and $u_n + v_n = 0$, then $UV = 1$, or $\delta(UV) = \delta(VU)$ is in X^{n+1}.

If $u_nv_m - v_mu_n \neq 0$, then

$$\delta(U^{-1}V^{-1}UV) = u_nv_m - v_mu_n. \tag{7}$$

If $u_nv_m - v_mu_n = 0$, then $UV = VU$, or $\delta(U^{-1}V^{-1}UV)$ is in X^{n+m+1}.

Finally,

$$\delta(U^{-1}VU) = v_m. \tag{8}$$

PROOF. For (5), we can establish the "binomial expansion"

$$(1 + u)^l = \sum_{s=0}^{\infty} \binom{l}{s} u^s,$$

where

$$\binom{l}{s} = l(l-1)\ldots(l-s+1)/s!,$$

by using induction on l. Since u is in X, it is clear that the lowest degree non-constant terms of $(1 + u)^l$ are contained precisely in lu; therefore, (5) follows.

Formulae (6a) and (6b) follow from multiplication. Formulae (7) and (8) will follow from the equations

$$(1 + u)^{-1}(1 + v)(1 + u) = 1 + v + \sum_{s=0}^{\infty} (-1)^s u^s(vu - uv) \tag{8a}$$

and

$$(1 + u)^{-1}(1 + v)^{-1}(1 + u)(1 + v) = 1 + \sum_{s,t=0}^{\infty} (-1)^{s+t} u^s v^t (uv - vu). \tag{7a}$$

The first equation (8a) follows from (1) by multiplication; the second equation (7a) follows from (8a) and (1) by multiplication.

From equation (7a), it is clear that the lowest degree non-constant terms of the left hand side of (7a) occur precisely in $uv - vu$. Similarly, from equation (8a), the lowest degree non-constant terms of the left-hand side of (8a) occur precisely in v. Thus, we have formulae (7) and (8), and hence, have completed our proof. ◀

Corollary 5.7 can be proved easily by means of Lemma 5.4; for, if we assume that $F_n \subset D_n(F)$, and that C_n is in F_n, C in $F(r)$, then C_n and C commute, i.e., $C_n^{-1}C^{-1}C_nC = 1$, or $\delta(C_n^{-1}C^{-1}C_nC)$ has degree at least $n + 1$, since $\delta(C_n)$ has degree at least n. Thus, in any event $C_n^{-1}C^{-1}C_nC$ is in $D_{n+1}(F)$, and so is F_{n+1}.

References and Remarks. Related methods for representing a group in terms of elements of rings of power series have been developed by Chen, 1954a, b.

The result that a free group of finite rank is Hopfian was derived here from Theorem 5.7. It was proved first by J. Nielsen, 1921, in an entirely different manner; essentially he proved Theorem 3.3 from which the Hopfian property of a finitely generated free group follows. For the proof given here see Magnus, 1935. A related proof was given by Fuchs-Rabinovitch, 1940a, b; the connection is discussed by Lyndon, 1953. Still another proof was given by M. Hall, 1950a.

Problems for Section 5.5

1. Let C be the ideal of $A(Z, r)$ which is generated by $x_1^2, \ldots, x_r^2$, where $x_1, \ldots, x_r$ freely generate $A(Z, r)$.

(a) Show that $A(Z, r)/C$ is isomorphic to the algebra of "infinite sums" of integral multiples of simple monomials consisting of 1 and

$$x_{\rho_1}x_{\rho_2} \cdots x_{\rho_k},$$

where ρ_j is in $\{1, \ldots, r\}$ and $\rho_j \neq \rho_{j+1}$, under a multiplication in which the product of two monomials

$$x_{\rho_1}x_{\rho_2} \cdots x_{\rho_k}, \quad x_{\sigma_1}x_{\sigma_2} \cdots x_{\sigma_n},$$

is obtained by juxtaposition if $\rho_k \neq \sigma_1$, and is zero otherwise.

(b) Show that in $A(Z, r)/C$, $1 + x_\rho$ and $1 - x_\rho$ are multiplicative inverses.

(c) Show that the free group $F(r)$ is faithfully represented in $A(Z, r)/C$.

2. Let G_p be the galois field of integers modulo p, where p is a prime integer; let C be the ideal of $A(G_p, r)$ generated by $x_1^p, \ldots, x_r^p$, where $x_1, \ldots, x_r$ freely generate $A(G_p, r)$.

(a) Show that $A(G_p, r)/C$ is isomorphic to the G_p-algebra of "infinite sums" of G_p-multiples of monomials consisting of 1 and

$$x_{\rho_1}^{e_1}x_{\rho_2}^{e_2} \cdots x_{\rho_k}^{e_k},$$

where ρ_j is in $\{1, \ldots, r\}$, $\rho_j \neq \rho_{j+1}$, and e_j is an integer,

$$1 \leq e_j \leq p - 1,$$

under a multiplication in which the product of two monomials

$$x_{\rho_1}^{e_1} \cdots x_{\rho_k}^{e_k}, \quad x_{\sigma_1}^{f_1} \cdots x_{\sigma_n}^{f_n},$$

is obtained by juxtaposition if $\rho_k \neq \sigma_1$ or $e_k + f_1 < p$, and is zero otherwise.

(b) Show that the free product of r cyclic groups of order p is faithfully represented in $A(G_p, r)/C$.

[*Hint:* Use the reduced form of a word in

$$E = \langle a_1, \ldots, a_r; a_1^p, \ldots, a_r^p \rangle$$

to show that under the mapping

$$a_\rho \to 1 + x_\rho,$$

an element $\neq 1$ in E is not mapped into 1 in $A(G_p, r)/C$; to do this also use the proof of Theorem 5.6.]

3. Let $F(r)$ be the free group of $A(Z, r)$ generated by $a_\rho = 1 + x_\rho$. Show that if $W \neq 1$ is an element of the nth derived group of $F(r)$, then $\delta(W)$ has degree at least 2^n. [*Hint:* Use induction on n and Lemma 5.4.]

4. Let Q be the rational number field. Show that the set of elements M in $A(Q, r)$ which have the constant term 1 form a group under multiplication, in which each element has a unique nth root for any positive integer n. [*Hint:* Consider the "binomial expansion" for the function

$$(1 + x)^{\frac{1}{n}} = 1 + \frac{1}{n} x + \dots + \binom{\frac{1}{n}}{k} x^k + \dots,$$

of a real variable x, which results from expansion x in a Taylor series about 0. Since multiplication of power series is done formally, it follows that if x is replaced by x_1 then the nth power of the right hand side in the above equation will be $1 + x_1$. Hence, by Lemma 5.2, an nth root of $1 + h$ is

$$1 + \frac{1}{n} h + \dots + \binom{\frac{1}{n}}{k} h^k + \dots.$$

Moreover, if $P_i(x_1, \dots, x_r)$, $i = 1, 2, \dots$ ranges over all non-empty monomials, the equality

$$\left(1 + \sum_{i=1}^{\infty} a_i P_i(x_\rho)\right)^n = 1 + \sum_{i=1}^{\infty} b_i P_i(x_\rho),$$

where a_i and b_i are in Q, leads to an infinite number of linear equations in the a_i which must have a unique solution (if any) because when the b_i are all zero, the only solution is $a_i = 0$, for all i.]

5. (a) Let G be a group such that a verbal subgroup V of G is Hopfian and G/V is Hopfian. Show that G is Hopfian.
   (b) Show that $D_n(F)$ is a verbal subgroup of $F(r)$, and that $F(r)/D_n(F)$ is Hopfian.
   (c) Conclude from Problems 2.4.20(e) and 2.4.21 that since $\bigcap_{n=1}^{\infty} D_n(F) = 1$, $F(r)$ is Hopfian.

[*Hint:* For (a), note that if α: $G \to G$ is onto, then, since V is verbal, α: $V \to V$ is onto, and α: $G/V \to G/V$ is onto. Since G/V is Hopfian, the kernel of α must be in V; since V is Hopfian, the kernel of α is therefore 1. For (b), use induction on n, Theorem 5.7, and (a). We have $D_n(F)$ is verbal since it is a fully invariant subgroup of a free group and hence, Theorem 2.2 applies.]

6. (a) Complete the details of the proof of Lemma 5.2.
   (b) State and prove an analogue of Lemma 5.2 for the Lie algebra $\Lambda(Z, r)$. Is there an analogue for Lemma 5.3?

7. Let $F(r)$ be the free group of $A(Z, r)$ generated by $a_\rho = 1 + x_\rho$, and let

$$W(a_\rho) = a_{\rho_1}^{l_1} a_{\rho_2}^{l_2} \dots a_{\rho_k}^{l_k},$$

where l_i is a non-zero integer and ρ_i is in $\{1, \ldots, r\}$, $\rho_i \neq \rho_{i+1}$, be a freely reduced word in $F(r)$.

(a) Show that the monomials occurring in $W(a_\rho)$ have a maximum syllable length of k, and that the monomial

$$x_{\rho_1} \cdots x_{\rho_{j-1}} x^n_{\rho_j} x_{\rho_{j+1}} \cdots x_{\rho_k}$$

occurs in $W(a_\rho)$ with the coefficient (possibly zero)

$$l_1 \cdots l_{j-1} \cdot \binom{l_j}{n} \cdot l_{j+1} \cdots l_k.$$

(b) Show that $W(a_\rho)$ has infinitely many monomials of syllable length k occurring if and only if some l_j is negative; moreover, if some l_j is negative, then infinitely many monomials of syllable length k have negative coefficients, and infinitely many have positive coefficients.

(c) Show that each of the monomials

$$x_{\rho_1} \cdots x_{\rho_{j-1}} x_{\rho_{j+1}} \cdots x_{\rho_k}$$

(ρ_{j-1} may equal ρ_{j+1}) occurs in $W(a_\rho)$ with coefficient

$$l_1 \cdots l_{j-1} l_{j+1} \cdots l_k.$$

(d) Show how to recover the word $W(a_\rho)$ from its power series representation in $A(Z, r)$.

8. State a reason why each of the following power series in $A(Z, r)$ is not in $F(r)$:

(a) a finite sum with some negative coefficient;
(b) an infinite sum with finitely many negative coefficients;
(c) an infinite sum with finitely many positive coefficients;
(d) a finite or infinite sum with at most one monomial of any degree, and having some monomial of syllable length at least two;
(e) a finite sum whose coefficients do not add up to a power 2^k, where k is a non-negative integer;
(f) an infinite sum in which the coefficients take on finitely many values, and 1 and -1 occur only finitely often.

5.6. Lie Elements and Basis Theorems

The properties of the function $\delta(u)$ in Lemma 5.4 at the end of Section 5.5 show a relationship between commutation in a group (see Section 5.2) and the so-called "bracket multiplication" in an associative algebra, that is, the composition $[u, v]$ defined by

(1) $$[u, v] = uv - vu$$

for elements u, v of an associative algebra. The expression $[u, v]$ is called the *bracket product* (also called the "commutator" by some) of u and v. For any three elements u, v, w of an associative ring we have

$$[u, u] = 0, [[u, v]w] + [[w, u], v] + [[v, w], u] = 0,$$

(compare Problem 5.4.10). Therefore, bracket multiplication in an associative algebra satisfies the relations for multiplication in a Lie ring (the distributive laws are, of course, immediate). This implies the following:

LEMMA 5.5. *There exists a mapping μ of the Lie algebra $\Lambda_0(R, r)$ on free generators ξ_ρ into the associative algebra $A_0(R, r)$ on free generators x_ρ with the following properties:*

1. *Every element $\varphi \in \Lambda_0$ has exactly one image $\mu(\varphi)$ in A_0.*
2. $\mu(\xi_\rho) = x_\rho$, $(\rho = 1, 2, , , , . r)$
3. *If φ, ψ are in Λ_0 and if $\alpha \in R$,* then

$$\mu(\alpha\varphi) = \alpha\mu(\varphi)$$
$$\mu(\varphi + \psi) = \mu(\varphi) + \mu(\psi)$$
$$\mu(\varphi \circ \psi) = [\mu(\varphi), u(\psi)].$$

PROOF. Since every relation among the generators of Λ_0 are a consequence of the laws of a Lie algebra, and A_0 is a Lie algebra under the bracket multiplication, we have our result. ◀

We shall use:

Definition 5.7. An element of $A_0(R, r)$ that is the image of some element of $\Lambda_0(R, r)$ under the mapping μ of Lemma 5.5 is called a *Lie element* of $A_0(R, r)$ or a *Lie element in the x_ρ*.

It should be mentioned that not every element of A_0 is a Lie element; it is easy to show, for instance, that no power $x_\rho{}^m$ for $m > 1$ can be a Lie element (see the problems at the end of this section for details). However, all elements of A_0 can be expressed in terms of Lie elements in a unique manner. The resulting "basis theorem" (Theorem 5.8) will permit us to prove that the mapping μ is a one-one mapping and this, in turn, will permit us to find basis elements for Λ_0. In carrying out this program, we start with

LEMMA 5.6. *Let $A_0(R, \infty)$ be an associative algebra freely generated by the elements*

$$x, y_1, y_2, y_3, \ldots.$$

Let S_x be the associative subalgebra (containing 1) *of A_0 generated by the particular Lie elements defined inductively by*

(2) $$y_\lambda^{(0)} = y_\lambda, \qquad y_\lambda^{(k+1)} = [y_\lambda^{(k)}, x], \qquad k = 0, 1, 2, \ldots.$$

Then the $y_\lambda^{(k)}$ *are free generators of* S_x. *Moreover, any element* u *of* A_0 *can be expressed in the form*

$$(3) \qquad u = \alpha_0 x^m + x^{m-1}s_1 + x^{m-2}s_2 + \ldots + s_m$$

where $\alpha_0 \in R$ *and* $s_1, \ldots, s_m \in S_x$, *and this expression for* u *is unique, i.e.,* $u = 0$ *implies* $\alpha_0 = 0$, $s_1 = \ldots = s_m = 0$. *Finally, if* u *is homogeneous of degree* m *and a Lie element, then either* $m = 1$ *or* $\alpha_0 = s_1 = \ldots = s_{m-1} = 0$. (Note: We shall call the $y_\lambda^{(k)}$ the *elements arising by elimination of* x *from* A_0.)

PROOF. We first show that the monomials

$$(4) \qquad z(e_\nu, \lambda_\nu) = y_{\lambda_1}^{(e_1)} y_{\lambda_2}^{(e_2)} \ldots y_{\lambda_n}^{(e_n)},$$

(where $\nu = 1, \ldots, n$ and λ_ν, e_ν are positive and non-negative integers respectively, with repetitions permitted) are linearly independent over R when viewed as elements of S_x.

For this purpose, it is easily proved by induction on k that

$$(5) \quad y_\lambda^{(k)} = y_\lambda x^k - \binom{k}{1} x y_\lambda x^{k-1} + \binom{k}{2} x^2 y_\lambda x^{k-2} - \ldots + (-1)^k \binom{k}{k} x^k y_\lambda$$
$$= \sum_{j=0}^{k} (-1)^j \binom{k}{j} x^j y_\lambda x^{k-j}.$$

Hence, when (4) is expressed in A_0, the result will be a sum of integral multiples of monomials in x and the y_λ such that the degree on x is

$$e = e_1 + e_2 + \ldots + e_n,$$

and if x is replaced in the monomial by 1 then

$$y_{\lambda_1} y_{\lambda_2} \cdots y_{\lambda_n}$$

results. Since x and the y_λ freely generate the associative algebra A_0, it is clear that if a non-trivial linear relation exists among different elements in (4), then there is one using non-zero coefficients of R between elements of (4) in which the same n and the same e occurs [see Problem 5.4.6(c)]. Now we may lexicographically order monomials in x and the y_λ which have the same degree on x, and the same total degree, by comparing corresponding factors starting from the left and using

$$x < y_1 < y_2 < \ldots;$$

for example, if $e = 2$ and $n = 2$, then

$$x y_1 x y_3 < x y_1 y_1 x < x y_1 y_2 x.$$

As may be easily verified, this ordering is preserved under multiplication. Since the largest monomial in (5) is $y_\lambda x^k$, the largest monomial in (4) will be

$$y_{\lambda_1}x^{e_1}y_{\lambda_2}x^{e_2}\ldots y_{\lambda_n}x^{e_n}, \tag{6}$$

and as can be seen from (5), it will occur with coefficient 1. Since distinct elements in (4) have distinct largest terms (6), in any linear expression with non-zero coefficients from R involving finitely many elements of (4), there will be a non-zero coefficient on the largest of the monomials from (6). Hence, the elements of (4) are linearly independent over R.

We now show that any homogeneous element u in A_0, and hence, any element in A_0, can be expressed in the form (3). Clearly we must show that in any monomial in x and the y_λ, the x factors may be collected to the left by introducing suitable $y_\lambda^{(k)}$. During the collection process we shall have to deal with monomials in x and the $y_\lambda^{(k)}$. We therefore introduce the "deficiency" of a monomial in x and the $y_\lambda^{(k)}$; the *deficiency* for such a monomial is the sum over all $y_\lambda^{(k)}$ occurring of the number of x factors to the right of $y_\lambda^{(k)}$; for example, the deficiencies of

$$xy_1^{(2)}xxy_2^{(1)}, \qquad y_2^{(3)}xy_1^{(1)}xxx$$

are two and seven, respectively.

We shall prove by induction on its deficiency that any monomial in x and the $y_\lambda^{(k)}$ can be expressed in the form (3); for in A_0

$$y_\lambda^{(k)}x = xy_\lambda^{(k)} + y_\lambda^{(k+1)},$$

and so we may replace any monomial in x and the $y_\lambda^{(k)}$ by the sum of two such monomials, each with smaller deficiency than the original. Repeating this process on all the new monomials introduced we obtain (after no more than the original deficiency number of steps) only monomials which have deficiency zero. Moreover, during the replacement process the degree of a monomial in x is never increased.

Hence, if u is a homogeneous element of degree m in A_0, and the above process is applied to the monomials in u, we obtain monomials of deficiency zero and degree in x of no more than m. Moreover, if some monomial in u has degree m in x, then it must be an R-multiple of x^m, and hence remains unchanged. Thus u can be expressed in the form (3), and so can every element of A_0.

To show that u can be represented uniquely by (3), it suffices to show that if

$$\alpha_0 x^m + x^{m-1}s_1 + x^{m-2}s_2 + \ldots + s_m = 0 \tag{7}$$

in A_0, then

$$\alpha_0 = s_1 = s_2 = \ldots = s_m = 0.$$

We shall make use of the freeness of x and the y_λ to show that equation (7) is valid when x is replaced by $x + h$, where h is an indeterminate commuting with every element of A_0 but otherwise independent of A_0. More precisely, let $A_0[h]$ be the domain of polynomials in h with coefficients in $A_0(R, \infty)$; clearly then h commutes with all elements in A_0. Since $A_0[h]$ is an associative algebra, the free associative algebra A_0 may be mapped homomorphically into it by sending

$$(8) \qquad x \to x + h, \quad y_\lambda \to y_\lambda.$$

It is easily shown by induction on k that $y_\lambda^{(k)}$ is left fixed by (8). Hence, under (8) the equation (7) becomes the equation

$$(9) \qquad \alpha_0(x + h)^m + (x + h)^{m-1}s_1 + (x + h)^{m-2}s_2 + \ldots + s_m = 0,$$

in $A_0[h]$. Collecting terms according to their degree in h and starting with the highest degree first, we obtain successively

$$\alpha_0 = 0, \quad s_1 = 0, \quad s_2 = 0, \quad \ldots, \quad s_m = 0,$$

and thus we have our result.

Finally, we consider the case that u is a homogeneous Lie element not involving an R-multiple of x, and show that u must be left fixed under (8). For this purpose let

$$(10) \qquad \beta^n(z_1, z_2, \ldots, z_n),$$

where z_ν is one of the generators x or y_λ and β^n is a bracket arrangement of weight n interpreted as acting by means of bracket multiplication in $A_0[h]$, be different from x; we show by induction on n that (10) is left fixed under (8). In the case $n = 1$, (10) reduces to x or y_λ, and so if it is not x, it must be left fixed by (8). If $n > 1$ and $\beta^n = (\beta^k\beta^l)$, then

$$(11) \qquad \beta^n(z_1, z_2, \ldots, z_n) = [P, Q],$$

where

$$P = \beta^k(z_1, \ldots, z_k), \qquad Q = \beta^l(z_{k+1}, \ldots, z_n).$$

If both P and Q are x, then (11) reduces to 0; and if neither P nor Q is x, then the inductive hypothesis applies and (11) is left fixed by (8). If P alone is x, then the inductive hypothesis applies to Q, and so, under (8), (11) becomes

$$[x + h, Q] = xQ + hQ - Qx - Qh = [x, Q],$$

since Q is in A_0 and h commutes with all elements in A_0; similarly if Q alone is x, (11) is left fixed by (8).

Suppose now that u is a Lie element which when expressed as in (3) has $m > 1$. Then since u has degree m it cannot involve any R-multiple of x, and so is left fixed by (8). Applying (8) to (3), and collecting powers of h starting with the highest we obtain successively

$$\alpha_0 = 0, \quad s_1 = 0, \quad s_2 = 0, \quad \ldots, \quad s_{m-1} = 0,$$

as desired. ◀

Although the Lie elements (not involving any R-multiple of x) of $A_0(R, \infty)$ must be in S_x, this is only a necessary condition; for example, $y_1{}^2$ is in S_x but is not a Lie element, since we can interchange the roles of x and y_1 in Lemma 5.6. We shall now establish a necessary and sufficient condition for an element of S_x to be a Lie element.

Lemma 5.7. *A necessary and sufficient condition for an element of* S_x *to be a Lie element in* x *and the* y_λ *is for it to be a Lie element in the free generators* $y_\lambda^{(k)}$ *of the associative algebra* S_x.

The proof of Lemma 5.7 is an immediate consequence of the following:

LEMMA 5.8. *Let* $\Lambda_0(R, \infty)$ *be the Lie algebra freely generated by* ξ *and* $\eta_1, \eta_2, \ldots$. *Let* Σ_ξ *be the subalgebra of* Λ_0 *generated by the elements* $\eta_\lambda^{(0)}$, $\eta_\lambda^{(1)}, \eta_\lambda^{(2)}, \ldots$ *which are defined inductively by*

$$\eta_\lambda^{(0)} = \eta_\lambda, \qquad \eta_\lambda^{(k+1)} = \eta_\lambda^{(k)} \circ \xi.$$

Then Σ_ξ *is an ideal in* Λ_0 *and every element* ω *in* Λ_0 *has the property*

$$\omega \equiv \alpha_0 \xi \text{ modulo } \Sigma_\xi, \qquad \alpha_0 \in R. \tag{12}$$

PROOF. We first show that Lemma 5.7 follows from Lemma 5.8. For, the Lie elements of $A_0(R, \infty)$ are the images of elements of Λ_0 under the mapping μ of Λ_0 into A_0 given by

$$\mu(\xi) = x, \qquad \mu(\eta_\lambda) = y_\lambda$$

(see Lemma 5.5 and Definition 5.7). Since

$$\mu(\eta_\lambda^{(k)}) = y_\lambda^{(k)}$$

the images of the elements of Σ_ξ under μ are all in S_x, and are Lie elements in the $y_\lambda^{(k)}$ (because of condition 3 in Lemma 5.5). Moreover, from (12) and Lemma 5.6, μ maps Σ_ξ onto all Lie elements of $A_0(R, \infty)$ which involve no term $\alpha_0 x$, where α_0 is in R. Hence, we have proved Lemma 5.7.

To prove Lemma 5.8, it suffices to show that every homogeneous element in Λ_0 of degree > 1 in ξ and the η_λ is in Σ_ξ. For then Σ_ξ is clearly an ideal; since linear combinations of the η_λ are in it, (12) follows immediately. Moreover, we may restrict ourselves to monomials.

Since any monomial of degree 2 is clearly in Σ_ξ $(\xi \circ \eta = -\eta \circ \xi)$, we may use induction on the degree of a monomial to establish our result. Moreover, since Σ_ξ consists of polynomials in $\eta_\lambda^{(k)}$, our problem is reduced to showing that if γ is in Σ_ξ then $\gamma \circ \xi$ is also in Σ_ξ. Since the elements of Σ_ξ are linear combinations of monomials in $\eta_\lambda^{(k)}$, it suffices to show that such monomials when multiplied by ξ yield elements of Σ_ξ. For this purpose we may use induction on the degree of such a monomial in $\eta_\lambda^{(k)}$.

From the definition of the $\eta_\lambda^{(k)}$, we have our result for degree 1. Any monomial of degree > 1 is a product of two monomials, each of smaller degree. Hence, if $\gamma = \phi \circ \psi$, we may assume by inductive hypothesis that $\phi \circ \xi$ and $\psi \circ \xi$ are in Σ_ξ. From the Jacobi identity we have

$$(\phi \circ \psi) \circ \xi = -(\psi \circ \xi) \circ \phi + (\phi \circ \xi) \circ \psi.$$

This completes the proof of Lemma 5.8 and hence of Lemma 5.7 also. ◀

We can now prove the following fundamental result.

THEOREM 5.8 (*Basis Theorem for Lie elements*). *In $A_0(R, r)$, $r \geq 2$, there exists a sequence $z_1, z_2, z_3, \ldots$ of homogeneous Lie elements with non-decreasing degrees in the free generators of A_0, such that*

(i) *The elements z_ν form a linear basis (over R) for the Lie elements of A_0.*

(ii) *The products*

$$(13) \qquad z_{\nu_1}^{e_1} z_{\nu_2}^{e_2} \ldots z_{\nu_k}^{e_k}, \qquad 1 \leq \nu_1 < \nu_2 < \ldots < \nu_k, \qquad k \geq 1$$

with positive integral exponents $e_1, \ldots, e_k$, together with the identity element 1 of R form a linear basis (over R) for all elements of A_0.

(iii) *All lie elements z_ν of degree ≥ 2 in the free generators of A_0 can be written in the form*

$$(14) \qquad z_\nu = [z_\lambda, z_\mu], \qquad 1 \leq \mu < \lambda.$$

(iv) *Moreover, every element z_ν of degree ≥ 2 can be written in the form (14) in such a way that for each τ with $\mu \leq \tau < \lambda$, the element*

$$[[z_\lambda, z_\mu], z_\tau] = [z_\nu, z_\tau]$$

occurs in the sequence $z_1, z_2, z_3, \ldots$.

(*Note:* We call the elements $z_1, z_2, z_3, \ldots$ *basic Lie elements.*)

PROOF. We shall construct the sequence of basic Lie elements inductively. For this purpose we shall also define inductively the basic Lie elements of a given level.

Let $A_0(R, r)$ be the associative algebra freely generated by $x_1, x_2, \ldots, x_r$ where $r \geq 2$. Then the *basic Lie elements of the first level* are

$$x_1, x_2, \ldots, x_r,$$

and the *first basic Lie element* is

$$z_1 = x_1.$$

The *basic Lie elements of the second level* are the elements arising by the elimination of z_1 from A_0 (in the sense of Lemma 5.6). In particular, the basic Lie elements of the second level are the elements of the form

$$x_2, \ldots, x_r, [x_2, x_1], \ldots, [x_r, x_1], [[x_2, x_1], x_1], \ldots, [[x_r, x_1], x_1], \ldots.$$

We now select a basic Lie element of the second level which has minimum degree in A_0, say, x_2, and define the *second basic Lie element* by

$$z_2 = x_2.$$

Now the basic Lie elements of the second level are free generators of the associative subalgebra S_1 which they generate (Lemma 5.6). Hence, we may define the *basic Lie elements of the third level* to be the elements arising by the elimination of z_2 from S_1 (in the sense of Lemma 5.6). In particular, the basic Lie elements of the third level are those in the second level, excluding x_2, together with the elements of the form

$$[x_3, x_2], \ldots, [x_r, x_2], [[x_2, x_1], x_2], [[x_3, x_1], x_2], \ldots, [[x_r, x_1], x_2], \ldots.$$

We now select a basic Lie element of the third level which has minimum degree in A_0, say, x_3, and define the *third basic Lie element* by

$$z_3 = x_3.$$

The basic Lie elements of the third level are free generators of the associative subalgebra S_2 which they generate (Lemma 5.6). Hence, we may define the *basic Lie elements of the fourth level* to be the elements arising by the elimination of z_3 from S_2 (in the sense of Lemma 5.6), and then select a basic Lie element of the fourth level which has minimum degree in A_0 and set it equal to the *fourth basic Lie element.*

Suppose that we have defined L_n, *the set of basic Lie elements of the* nth *level*, and have also defined z_n, *the* nth *basic Lie element*. Moreover, suppose that z_n is in L_n, has minimum degree (as an element of A_0) in L_n, and also that the elements in L_n freely generate their associative algebra S_{n-1} in A_0. Then we define L_{n+1}, *the set of basic Lie elements of the* $(n + 1)$st *level* to be the set of elements that arise by elimination of z_n from S_{n-1} (in the sense of Lemma 5.6). Moreover, we select an element in L_{n+1} which has minimum degree in A_0, and set it equal to z_{n+1}, *the* $(n + 1)$st *basic Lie element*.

Any element in some L_n is called a *basic Lie element*.

Clearly, the basic Lie elements are homogeneous Lie elements of A_0 with degrees ≥ 1. In particular, if we examine the degrees of elements in L_{n+1} we note that the elements of minimum degree in L_{n+1} are also in L_n. For, an element v of L_{n+1} which is not in L_n has the form

$$v = [[. . . [u, z_n], . . .], z_n,$$

where u is in L_n and also in L_{n+1}; clearly degree $u <$ degree v. Thus degree $z_n \leq$ degree z_{n+1} since z_{n+1} is in L_n. Moreover, if v is in L_{n+1} but not in L_n, then degree $z_n <$ degree v. Hence, once z_n is eliminated from L_n it cannot occur in any higher level.

Now, L_n has only a finite number M_n of elements of minimum degree m_n (in fact, since A_0 is finitely generated there are only a finite number of "monomial" Lie elements, i.e., images of monomials in Λ_0, with a given degree, and the basic Lie elements are all monomial Lie elements). If z_n and z_{n+1} have the same degree then L_{n+1} has one less element of this degree than L_n. Thus

$$L_{n+M_n}$$

has no element of degree m_n, and so as n goes to infinity, so must m_n. Since any basic Lie element u has finite degree, it cannot be in all L_n; if p is the maximum level containing u, then

$$u = z_p.$$

Thus the sequence $z_1, z_2, z_3, . . .$ exhausts all basic Lie elements.

We may now show that the purported bases in (i) and (ii) are linearly independent over R. Since the z_ν are included among the elements in (13), it suffices to show that the identity element of R and the elements in (13) are linearly independent. Consider any linear relation between 1 and the elements in (13). Since each z_n is homogeneous of degree ≥ 1, so is every element in (13), and so the coefficient on 1 in the linear relation is 0. To show that the coefficients on the elements of (13) are also 0, we use induction on the difference between the maximum and minimum subscript occurring on any z_ν in any element of (13) in the linear relation. If this difference is zero, then all the elements of (13) which are involved are powers of the same z_ν; since z_ν is in a set of free generators for the associative subalgebra $S_{\nu-1}$, its powers are linearly independent. Moreover, if the difference in maximum and minimum subscript is positive, and μ is the minimum subscript occurring, we may arrange the elements in (13) according to the power of z_μ they contain, factor out the common power of

z_μ, and then use the uniqueness of the form (7) in Lemma 5.6 for $x = z_\mu$ and $S_x = S_\mu$ (clearly S_μ contains S_ν for all $\nu \geq \mu$, and hence z_ν is in S_μ for all $\nu > \mu$), to obtain by inductive hypothesis that all the coefficients are 0.

We now turn our attention to showing that the elements in (i) and (13) span the required spaces.

We shall show by induction on n that any monomial u in A_0 can be written as a linear combination of elements

(15) $$z_1^{e_1} z_2^{e_2} \dots z_n^{e_n} Q,$$

where the exponents e_ν are non-negative integers and Q is an element of S_n. For $n = 1$, the result is immediate from Lemma 5.6 with $x = z_1$. Since Q is in S_n it follows from Lemma 5.6 with $x = z_{n+1}$ that Q is a linear combination of elements

$$z_{n+1}^{e_{n+1}} Q',$$

where Q' is in S_{n+1}. Thus, assuming our result for n, we have it for $n + 1$. A given element of A_0 can only involve monomials of some bounded degree. If the minimum degree of an element in L_n exceeds this bound then all elements of S_n which are not in R have degrees exceeding this bound. Thus when the given element of A_0 is written as a linear combination of the elements in (15), Q must be an element of R. Hence, the elements (13) and the identity element of R span all of A_0.

Suppose now that u is a homogeneous Lie element of A_0 with degree $d \geq 1$. If the minimum degree in L_n exceeds d, then u cannot be in S_n. Hence, u is in some S_n with a maximum subscript n (of course $S_0 = A_0$). But if u is not a linear combination of the z_ν (the free generators of each S_μ are among the z_ν), then it follows from Lemma 5.6 that u is a Lie element in each S_ν, and hence must be in S_{n+1}. This contradiction shows that each homogeneous Lie element, and hence each Lie element, of A_0 is a linear combination of some of the z_ν.

We have thus established (i) and (ii).

To establish (iii) we note that if z_ν is a basic Lie element of degree ≥ 2 then z_ν first occurs in some level $\mu + 1$, where $\mu \geq 1$. The basic Lie elements of the μth level consist of z_μ and z_β, where β ranges over a subset of integers $> \mu$; for all z_β with $\beta < \mu$ are eliminated before the μth level, and cannot reoccur. The elements of $L_{\mu+1}$ which are not in L_μ, have the form

$$[[\dots[z_\beta, z_\mu], \dots], z_\mu].$$

Since $[\dots[z_\beta, z_\mu], \dots]$ is an element of $L_{\mu+1}$ it is some z_λ with $\lambda > \mu$. Thus,

$$z_\nu = [z_\lambda, z_\mu],$$

where $\lambda > \mu$, and (iii) is established. Moreover, since z_ν is in $L_{\mu+1}$ the element

$$[[z_\lambda, z_\mu], z_\mu] = [z_\nu, z_\mu]$$

is also in $L_{\mu+1}$ and so is in the sequence $z_1, z_2, z_3, \ldots$. Also since z_ν is not eliminated till the $(\nu + 1)$st level, it is in all τth levels for $\mu < \tau < \nu$. Hence, when z_τ is eliminated, the element

$$[[z_\lambda, z_\mu], z_\tau] = [z_\nu, z_\tau]$$

occurs in $L_{\tau+1}$, and hence occurs in the sequence $z_1, z_2, z_3, \ldots$. ◀

As an application of Theorem 5.8 and its proof we shall now describe a linear basis for the free Lie algebra $\Lambda_0(R, r)$ which is freely generated by $\xi_1, \xi_2, \ldots, \xi_r$.

THEOREM 5.9. *Let $\Lambda_0(R, r)$ be the free Lie algebra freely generated by $\xi_1, \xi_2, \ldots, \xi_r$, and let μ be the natural mapping of Λ_0 into A_0 (as in Lemma 5.5). Then μ is a one-one mapping, and the unique preimages $\zeta_1, \zeta_2, \ldots$ of the basic Lie elements $z_1, z_2, \ldots$ of A_0 are a linear basis for Λ_0 over R.*

PROOF. The construction of the basic Lie elements in Theorem 5.8 can be mirrored in the Lie algebra Λ_0, with the ordinary product in Λ_0 replacing the bracket product in A_0, and the elements $\xi_1, \ldots, \xi_r$ replacing the elements $x_1, \ldots, x_r$ in the first level. Clearly, the image of the sequence $\zeta_1, \zeta_2, \zeta_3, \ldots$ under μ is the sequence $z_1, z_2, z_3, \ldots$ of basic Lie elements in A_0. Since μ is a linear mapping of the R-module Λ_0 into the R-module A_0 (this is contained in condition 3 of Lemma 5.5), and $z_1, z_2, z_3, \ldots$ are linearly independent over R, $\zeta_1, \zeta_2, \zeta_3, \ldots$ are also.

To show that $\zeta_1, \zeta_2, \zeta_3, \ldots$ span all of Λ_0, we note that if

$$\zeta_n = \zeta_{\nu_1}, \zeta_{\nu_2}, \zeta_{\nu_3}, \ldots$$

are the basic Lie elements of the nth level in Λ_0, then under the mapping of $\Lambda_0(R, \infty)$ into $\Lambda_0(R, r)$ given by

$$\xi \to \zeta_n, \eta_1 \to \zeta_{\nu_2}, \ldots, \eta_\lambda \to \zeta_{\nu_{\lambda+1}}, \ldots,$$

the elements in Lemma 5.8

$$\eta_\lambda^{(0)}, \eta_\lambda^{(1)}, \eta_\lambda^{(2)}, \ldots,$$

where λ ranges over all positive integers, are precisely mapped onto the basic Lie elements of $(n + 1)$st level in Λ_0. Hence, it follows from (12) of Lemma 5.8 that any element in the Lie subalgebra Σ_{n-1} generated (not necessarily freely) by the basic Lie elements of nth level in Λ_0, is the sum of an R-multiple of ζ_n and an element in Σ_n, the Lie subalgebra generated by the basic Lie elements of $(n + 1)$st level in Λ_0. Since the degree of an element of Σ_n in the free generators $\xi_1, \xi_2, \ldots, \xi_r$ of Λ_0 is at least as large as the degree of its image under μ in S_n, the minimum degree of an element in Σ_n tends to infinity as n does. Select an n so that the minimum degree of any element of Σ_n exceeds the maximum degree of any term in a given element ψ of Λ_0. Then

$$\psi = \beta_1\zeta_1 + \beta_2\zeta_2 + \ldots + \beta_n\zeta_n + \delta,$$

where δ is in Σ_n, by a repeated application of the above remarks. Since the ζ_ν are homogeneous and have non-decreasing degrees, if k is the largest subscript such that degree $\zeta_k \leq$ the maximum degree of a term in ψ, then from Lemma 5.1 in Section 5.4 it follows that

$$\psi = \beta_1\zeta_1 + \beta_2\zeta_2 + \ldots + \beta_k\zeta_k,$$

and since ψ was any element of Λ_0, the basic Lie elements of Λ_0 span all of it.

Since the linear mapping μ sends the basis $\zeta_1, \zeta_2, \ldots$ of Λ_0 into the linear independent subset $z_1, z_2, \ldots$ of A_0, μ must be one-one. ◀

COROLLARY 5.9. *If R_1 is a subintegral domain of R_2, then the free Lie algebra $\Lambda_0(R_1, r)$ is naturally isomorphic to a subring of the free Lie algebra $\Lambda_0(R_2, r)$ (as a ring, not as an algebra). Moreover, the R_1-multiple of an element goes into the same R_1-multiple of the image element.*

PROOF. Clearly the free associative algebra $A_0(R_1, r)$ is a subring of the free associative algebra $A_0(R_2, r)$, because of the linear basis of monomials that both possess. Moreover, the Lie elements of $A_0(R_1, r)$ under bracket product are a subring of the Lie elements of $A_0(R_2, r)$ under bracket product. Our result follows easily then from the representation of $\Lambda_0(R_1, r)$ and $\Lambda_0(R_2, r)$ as the Lie elements of $A_0(R_1, r)$ and $A_0(R_2, r)$ respectively, under bracket product. ◀

Our representation of a free Lie algebra in the free associative algebra with the same R and r allows us to prove the following theorem which will be used later on:

THEOREM 5.10. *Let ϕ and ψ be homogeneous elements of $\Lambda_0(R, r)$, a free Lie algebra. Then $\phi \circ \psi = 0$ if and only if ϕ and ψ are linearly dependent (over R).*

PROOF. If ϕ and ψ are linearly dependent over R then

$$\alpha\phi + \beta\psi = 0,$$

where we may assume by symmetry that $\alpha \neq 0$; hence,

$$\alpha(\phi \circ \psi) = (\alpha\phi + \beta\psi) \circ \psi = 0 \circ \psi = 0.$$

Since R is an integral domain and $\Lambda_0(R, r)$ has a linear basis over R, and $\alpha \neq 0$, it follows that $\phi \circ \psi = 0$.

Next, suppose that $\phi \circ \psi = 0$. If F is the field of quotients of R, then according to Corollary 5.9 we may view $\Lambda_0(R, r)$ as a subring of $\Lambda_0(F, r)$; moreover, if a set of elements of $\Lambda_0(R, r)$ are linearly dependent over F, by clearing of fractions, we see they are linearly dependent over R. Thus, we may assume that R itself is a field. We next consider the elements u and v of $A_0(R, r)$ which are the images of ϕ and ψ, respectively, under the mapping μ of Theorem 5.9. Since u and v are Lie elements of A_0, each can be expressed as a linear combination of basic Lie elements of A_0. Let

$$u = \alpha_1 z_{\nu_1} + \ldots + \alpha_s z_{\nu_s}, \qquad v = \beta_1 z_{\mu_1} + \ldots + \beta_t z_{\mu_t},$$

where α_i, β_j are in R, $\alpha_1, \beta_1 \neq 0$, and the ν_i and the μ_j are strictly increasing sequences of positive integers, and z_{ν_i}, z_{μ_j} are basic Lie elements of A_0, as given in Theorem 5.8. Since u and v are homogeneous, all the z_{ν_i} have the same degree and all the z_{μ_j} have the same degree, and by symmetry, we may assume that $\nu_1 \leq \mu_1$. Thus all basic Lie elements occurring in u must be in the ν_1-level and be part of a set of free generators for the free associative algebra $S_{\nu_1 - 1}$ generated by all basic Lie elements of the ν_1-level. All the basic Lie elements involved in v are in $S_{\nu_1 - 1}$. Since R is a field, we may select u as part of a set of free generators for the free associative algebra $S_{\nu_1 - 1}$ (see Problem 3). Our result will be established once we show that in a free associative algebra any element commuting with a free generator x must be a linear combination of powers of x (we show this result in Problem 5). For since $\phi \circ \psi = 0$, it follows that $[u, v] = 0$, i.e., u and v commute. Thus v is a sum of R-multiples of powers of u, and since u and v are homogeneous in A_0, u must be an R-multiple of a single power of u. But v is a Lie element of A_0, and hence, since it is in $S_{\nu_1 - 1}$, v must be a Lie element of $S_{\nu_1 - 1}$ (Lemma 5.7), and so an R-multiple of u (Lemma 5.6). ◀

Although the construction of a basis for $\Lambda_0(R, r)$ or, equivalently, for the Lie elements in $A_0(R, r)$, is a complicated matter, the number of basis elements for the homogeneous terms of a given degree in Λ_0 has been computed by E. Witt, 1937, in a simple manner.

Let $\mu(d)$ be the *Moebius function*, defined for all positive integers by $\mu(1) = 1$, $\mu(p) = -1$ if p is a prime number, $\mu(p^k) = 0$ for $k > 1$, and

$$\mu(b \cdot c) = \mu(b) \cdot \mu(c)$$

if b, c are coprime integers. Then we have

THEOREM 5.11. *(Witt's Formulae) In $\Lambda_0(R, r)$ there exist*

$$N_n = \frac{1}{n} \sum_{d|n} \mu(d) r^{n/d}$$

basis elements of degree n in the free generators ξ_ρ. The number of basis elements exactly of degree n_ρ in ξ_ρ, $\rho = 1, \ldots, r$, is given by

$$\frac{1}{n} \sum_{d|n_\rho} \mu(d) \frac{(n/d)!}{(n_1/d)!\,(n_2/d)! \ldots (n_r/d)!}$$

$$(n = n_1 + n_2 + \ldots + n_r)$$

In the first formula, d runs through all divisors of n, and in the second one through all common divisors of $n_1, n_2, \ldots, n_r$.

PROOF. To prove Theorem 5.11, we observe first that the number of possibilities of selecting m objects (repetitions allowed) out of a collection of N different ones equals the coefficient of t^m in the power series expansion of

$$(1 - t)^{-N}$$

(where t is an ordinary variable). Assume now that we are given N_l basic Lie elements of degree l (where, of course, $N_0 = 1$, $N_1 = r$). Then the number of possibilities of forming products

$$z_1^{e_1} z_2^{e_2} \ldots z_k^{e_k}$$

which are exactly of degree n (and in which the first N_1 of the z_ν are of first degree, the next N_2 of second degree, and so on) equals the coefficient of t^n in the infinite product

$$(1 - t)^{-N_1}(1 - t^2)^{-N_2}(1 - t^3)^{-N_3} \ldots$$

Now we happen to know the number of these products if N_k is just the number of linearly independent Lie elements of degree k in r indeterminates: According to Theorem 5.8 it is equal to the number of linearly independent elements of degree n in $A_0(R, r)$. Since this number is obviously r^n, we have the relation

$$\prod_{l=1}^{\infty} (1 - t^l)^{-N_l} = \sum_{n=0}^{\infty} r^n t^n = (1 - rt)^{-1}.$$

Taking logarithms on both sides and differentiating with respect to t, we find (after multiplication by t):

$$\sum_{l=1}^{\infty} \frac{lN_l t^l}{1 - t^l} = \frac{rt}{1 - rt} = \sum_{n=1}^{\infty} (rt)^n.$$

By expanding the denominators on the left-hand side in power series and collecting terms of the same degree n on both sides, we find

$$\sum_{l|n} lN_l = r^n$$

where l runs through all divisors of n. By an application of the Moebius inversion formula (see, for example, Hardy and Wright, 1954), we obtain the formula in Theorem 5.11 for N_n. The second formula in Theorem 5.11 can be derived in a similar manner. ◀

References and Remarks. The proof of a theorem equivalent to conditions (i) and (ii) of Theorem 5.8 was first given by P. Hall, 1933, in a form which involves only commutators in a free group. A much simpler proof for his statements was found by Meier-Wunderli, 1952, who proved also the group-theoretical equivalent of Witt's formulae (Theorem 5.11). A proof and a formulation operating entirely in $\Lambda_0(R, r)$ was given by M. Hall, 1950b. For his presentation, see also M. Hall, 1959. The proofs given in Section 5.6 essentially follow Magnus, 1937, with improvements due to M. Hall. Since our main interest is group theory, we have developed our results by using the relationship between A_0 and Λ_0, which we shall require in later sections.

Theorems 5.8 and 5.9 are closely related to a much more general result due to Birkhoff, 1937 and Witt, 1937. We have the

Birkhoff-Witt Theorem. Let Λ be a Lie algebra with coefficients in a field F. Then there exists exactly one associative algebra A over F and a mapping μ of Λ into A with the properties of the mapping in Lemma 5.5, such that A is generated by the images of elements of Λ and such that μ is a one-one mapping.

Theorem 5.9 is a very special case of the Birkhoff-Witt Theorem if R is a field. It should be noted that we merely assumed that R was an integral domain with an identity. Lyndon, 1955, has given an example of a Lie algebra over an integral domain R consisting of the ring of polynomials in three (commuting) indeterminates for which the Birkhoff-Witt Theorem is not true. Of course, Lyndon's algebra is not free.

For the subalgebras of free Lie algebras there exists a result which has no analog for freely generated non-commutative, associative algebras. We have the

Širšov-Witt Theorem. Let F be a field and Λ be a Lie algebra over F which has a set of free generators. Then every subalgebra of Λ has the same property.

Proofs were given independently by Širšov, 1953, and Witt, 1956. For generalizations see Širšov, 1954.

For a systematic theory of Lie algebras see N. Jacobson, 1962.

Problems for Section 5.6

1. Let $A_0(R, r)$ be the associative algebra freely generated by $x_1, \ldots, x_r$. Suppose B is an associative and commutative R-algebra, and A_0 is mapped homomorphically into B by sending

$$x_\lambda \to b_\lambda.$$

(a) Show that the Lie elements of A_0 are mapped into linear combinations of the b_λ over R.

(b) In particular, if $B = R[h]$, the polynomial domain in one variable, and $b_\lambda = h$ for all λ, then all Lie elements in A_0 go into R-multiples of h.

(c) Conclude that the sum of the coefficients on all monomials of a given degree $\neq 1$ in a Lie element of A_0 must be zero, and hence, if $k > 1$, then $x_\lambda{}^k$ is not a Lie element of A_0.

2. Let $A_0(Z, 2)$ be freely generated by x and y, where Z is the ring of integers. Express each of the following elements in the form (3) of Lemma 5.6 and determine which are Lie elements of A_0:

(a) $x^3 + 3xyx^2 + x^2y^2$
(b) $yx^3 + 3yx - 3xy - yxy$
(c) $xy^2 - 2yxy + y^2x$
(d) $xyxy - xy^2x - yx^2y + yxyx$

3. Let $A_0(F, \infty)$ be the associative algebra freely generated by $x_1, x_2, x_3, \ldots$ over a field F. Show that the mapping

$$x_1 \to \alpha_1 x_1 + \ldots + \alpha_n x_n, \quad x_\lambda \to x_\lambda, \quad \lambda > 1,$$

where α_i is in F and $\alpha_1 \neq 0$, determines an automorphism of A_0 onto itself. Hence, conclude that

$$\alpha_1 x_1 + \ldots + \alpha_n x_n, x_2, x_3, \ldots$$

freely generate A_0. [*Hint:* Since A_0 is free, the given mapping induces an endomorphism of A_0. Moreover, the mapping

$$x_1 \to \frac{1}{\alpha_1}(x_1 - \alpha_2 x_2 - \ldots - \alpha_n x_n),$$

$$x_\lambda \to x_\lambda, \quad \lambda > 1,$$

induces the inverse endomorphism.]

4. Let $A_0(R, r)$ be the associative algebra freely generated by $x_1, \ldots, x_r$.

(a) Show that all "linear" mappings

$$x_\lambda \to \alpha_{\lambda 1}x_1 + \alpha_{\lambda 2}x_2 + \ldots + \alpha_{\lambda r}x_r + \beta_\lambda,$$

where $\alpha_{\lambda\mu}, \beta_\lambda$ are in R and the determinant $|a_{\lambda\mu}|$ is a unit in R, i.e., has a multiplicative inverse in R, determines an automorphism of A_0.

(b) Show that if $r = 1$, then all automorphisms of A_0 are determined by "linear" mappings.

(c) Show that if $r = 2$, then the mappings

$$x_1 \to x_1{}^2 + x_2, \quad x_2 \to x_1$$

and

$$x_1 \to x_2, \qquad x_2 \to x_1 - x_2{}^2$$

determine inverse automorphisms of A_0, and are "non-linear" mappings.

[*Hint:* For (a), see the hint to Problem 3. For (b), use that the degree of $P(Q(x))$ is the product of the degrees of P and of Q; hence, if $x = P(Q(x))$, then $Q(x)$ is "linear".]

5. Let $A_0(R, \infty)$ be the associative algebra freely generated by x and $y_1, y_2, y_3, \ldots$. Show that if u is in A_0 and $ux = xu$, then u is a polynomial in x with coefficients in R. [*Hint:* If $u = u_1 + u_2 + \ldots + u_n$, where each u_i consists of the terms of u with the same total degree and same degree on x, then $u_i x = x u_i$. Now monomials with the same total degree and same degree on x were ordered lexicographically in the proof of Lemma 5.6. Since the ordering is preserved under multiplication, if v is the largest monomial occurring in u_i and $v = x^m w$, where w is a monomial not beginning with x, then the largest monomial occurring in $u_i x$ is $vx = x^m wx$, which is larger than $xv = x^{m+1}w$, the largest monomial occurring in xu_i, unless of course $w = 1$. Thus $w = 1$, and x^m is the largest monomial occurring in u_i, and hence, is the only monomial occurring in u_i.]

6. Let $A_0(R, \infty)$ be the associative algebra freely generated by $x, y_1, y_2, y_3, \ldots$. Show that if u is a homogeneous element of degree m in A_0, then in the representation of u given by (3) of Lemma 5.6, the element s_j is the sum of R-multiples of monomials (4) with $e_1 + e_2 + \ldots + e_n + n = j$. [*Hint:* Show that if a monomial in x and the $y_\lambda^{(k)}$ has m as the sum of its degree on x added to the superscripts on the y_λ each increased by one, then under the substitution

$$y_\lambda^{(k)} x = x y_\lambda^{(k)} + y_\lambda^{(k+1)}$$

each new monomial introduced also has sum m. To begin with, u is a sum of R-multiples of monomials in x and the $y_\lambda = y_\lambda^{(0)}$ with total sum m.]

7. Let $A_0(Z, 2)$ be the associative algebra freely generated by x and y, and suppose that the basic Lie elements of Theorem 5.8 have been so selected that

$$z_1 = x, \quad z_2 = y.$$

(a) Find all basic Lie elements of degree ≤ 5 in level two, but not in level one.

(b) Find all basic Lie elements of degree ≤ 5 in level three but not in level two, and show that z_3 must be $[y, x]$.

(c) Find all basic Lie elements of degree ≤ 5 in level four but not in level three.

(d) Show that there are no basic Lie elements of degree ≤ 5 in level five but not in level four.

(e) Show that $[[[y, x], y], x]$ is not a basic Lie element.

(f) Show that we may select

$$\begin{aligned} z_4 &= [y, x, x];\ z_5 = [y, x, y]; \\ z_6 &= [y, x, x, x];\ z_7 = [y, x, x, y]; \\ z_8 &= [y, x, y, y];\ z_9 = [y, x, x, x, x]; \\ z_{10} &= [y, x, x, x, y];\ z_{11} = [y, x, x, y, y]; \\ z_{12} &= [y, x, y, y, y];\ z_{13} = [[y, x, x], [y, x]]; \\ z_{14} &= [[y, x, y], [y, x]]; \end{aligned}$$

where we have used the abbreviation

$$[a_1, a_2, \ldots, a_n]$$

for

$$[[\ldots[a_1, a_2], \ldots], a_n].$$

(g) Verify Witt's formulae (Theorem 5.12) for total degree 2, 3, 4, and 5, and for the various degrees on x and y with such total degrees.

(h) Assuming $z_1 = x$ and $z_2 = y$ find all ways of selecting z_3 through z_{13}.

8. Let $z_1, z_2, z_3, \ldots$ be a sequence of basic Lie elements of $A_0(R, r)$ as in Theorem 5.8. Show that if

$$[z_\lambda, z_\mu] = [z_\tau, z_\mu],$$

where $\lambda > \mu$ and $\tau > \mu$, then $z_\lambda = z_\tau$. [*Hint:* If $S_{\mu-1}$ is the associative subalgebra of A_0 freely generated by the basic Lie elements of the μth level, then z_μ is a free generator of $S_{\mu-1}$ and z_τ and z_λ are in S_μ and $S_{\mu-1}$. Since $[z_\lambda - z_\tau, z_\mu] = 0$, $z_\lambda - z_\tau$ commutes with z_μ, and hence, by Problem 5, $z_\lambda - z_\tau$ is a polynomial in z_μ. Since $z_\lambda - z_\tau$ is in S_μ, which is obtained by eliminating z_μ from $S_{\mu-1}$, by the uniqueness of the representation (3) in Lemma 5.6, $z_\lambda - z_\tau$ is in R, although it is a Lie element. Therefore, $z_\lambda - z_\tau = 0$.]

9. Let $z_1, z_2, z_3, \ldots$ be as in Problem 8. If z_ν has degree ≥ 2 in A_0 and first occurs as a basic Lie element in the $(\mu + 1)$st level and $z_\nu = [z_\lambda, z_\mu]$, $\lambda > \mu$ we call z_λ and z_μ *the factors in the standard form of* z_ν.

(a) Show that each z_ν has a unique pair of factors in its standard form.

(b) Show that if z_ν has the pair z_λ, z_μ as the factors in its standard form, then $[z_\nu, z_\tau]$ is the standard form of a basic Lie element if and only if $\mu \leq \tau < \nu$.

[*Hint:* For (a), use Problem 8 and the proof of (iii) in Theorem 5.8. For (b), use the proof of (iv) in Theorem 5.8.]

10. The (*formal*) *standard basic Lie elements on the symbols* $x_1, x_2, \ldots, x_r$ and their *standard ordering* are defined inductively as follows: The *standard basic Lie elements of degree one* in their *standard ordering* are

$$x_1 < x_2 < \ldots < x_r.$$

The *standard basic Lie elements of degree 2 are*

$$[x_\lambda, x_\mu]$$

where $x_\mu < x_\lambda$, and their *standard ordering* is given by having any standard basic Lie element of degree 1 less than any of degree 2, and also for those of degree 2,

$$[x_\lambda, x_\mu] < [x_\sigma, x_\tau]$$

if $x_\mu < x_\tau$, or if $x_\mu = x_\tau$ and $x_\lambda < x_\sigma$.

Assuming we have constructed all standard basic Lie elements of degree less than n, where $n \geq 3$, and assigned a standard ordering to them, we shall now construct those of degree n and order them. A *standard basic Lie element z of degree n* is obtained by selecting a standard basic Lie element $t = [u, v]$ of degree k, $2 \leq k < n$, where u and v are themselves standard basic Lie elements, and also selecting a standard basic Lie element w of degree $n - k$ such that $v \leq w < t$, and setting

$$z = [t, w].$$

Moreover, the *standard ordering* is extended to degree n by having any standard basic Lie element of degree less than n precede one of degree n, and for two of degree n we set

$$[t, w] < [p, q]$$

if $w < q$, or if $w = q$ and $t < p$.

(a) Show that the standard basic Lie elements are linearly ordered under their standard ordering.

(b) Show that we may construct the sequence $z_1, z_2, z_3, \ldots$ of Theorem 5.8 so that the elements of the sequence correspond precisely to the (formal) standard basic Lie elements (if the formal bracket is interpreted as bracket multiplication in $A_0(R, r)$, the associative algebra freely generated by $x_1, x_2, \ldots, x_r$).

(c) Verify that the elements in Problem 7 are standard basic Lie elements.

[*Hint:* Show by induction on the level that the basic Lie elements on the nth level correspond to standard basic Lie elements, provided that to obtain the nth level, the basic Lie element corresponding to the least standard basic Lie element of the $(n - 1)$st level is eliminated. Also use Problem 9.]

11. Let $A_0(R, 2)$ be the associative algebra freely generated by x and y. If $y^{(k)}$ is defined inductively by

$$y^{(0)} = y, \quad y^{(k+1)} = [y^{(k)}, x],$$

show that

$$[[y^{(1)}, y], x] = [[y^{(1)}, x], y] + [y^{(1)}, [y, x]]$$
$$= [y^{(2)}, y] + [y^{(1)}, y^{(1)}] = [y^{(2)}, y];$$

hence, if z_ν is as in Problem 7 then

$$[z_5, z_1] = [z_4, z_2] = z_7.$$

In particular, a basic Lie element can be represented in several ways as the bracket product of basic Lie elements, although its standard form (as in Problem 9) is unique. [*Hint:* Use the Jacobi identity and the anti-commutativity of bracket multiplication; see Problem 5.4.17(b).]

12. (Friedrichs) Let R be an integral domain of characteristic zero and let B be the associative R-algebra freely generated by $x_1, \ldots, x_r$. Let C be the associative R-algebra freely generated by $h_1, \ldots, h_r$, and let D be the associative B-algebra freely generated by $h_1, \ldots, h_r$ (i.e., x_λ commutes with all h_μ). Show that $P(x_1, \ldots, x_r)$ is a Lie element of B if and only if in D

$$P(x_1 + h_1, \ldots, x_r + h_r) = P(x_1, \ldots, x_r) + P(h_1, \ldots, h_r).$$

(Compare Theorem 5.18 of Section 5.9 and Magnus, 1954.) [*Hint:* If $z_1, z_2, z_3, \ldots$ are a sequence of basic Lie elements of B as given in Theorem 5.8, show by induction on n that if

$$z_n = P_n(x_1, \ldots, x_r)$$

then

$$P_n(x_1 + h_1, \ldots, x_r + h_r) = P_n(x_1, \ldots, x_r) + P_n(h_1, \ldots, h_r).$$

For $n = 1, \ldots, r$ the result is clear; if

$$z_n = [z_\lambda, z_\mu],$$

then we may use the inductive hypothesis on z_λ and z_μ. Since each Lie element of B is a linear combination of basic Lie elements, the additive property follows for any Lie element (without using R has characteristic zero). To show that the additive property implies that $P(x_1, \ldots, x_r)$ is a Lie element, it is convenient to think of $z_1, z_2, z_3, \ldots$ as the basic Lie elements of B and $z_1', z_2', z_3', \ldots$ as the corresponding basic Lie elements of C, and to order the monomials of (13) in Theorem 5.8 lexicographically assuming $z_\lambda < z_\mu$ if $\lambda < \mu$, and using the sum of the exponents $e_1 + \cdots + e_k$ as a "length" and putting smaller length monomials first; similarly for the z_ν'. If

$$P(x_1, \ldots, x_r) = \Sigma\, \alpha z_{\nu_1}^{l_1} \ldots z_{\nu_k}^{l_k}$$

where α is in R and $z_{\nu_1}^{e_1} \ldots z_{\nu_k}^{e_k}$ is the largest monomial in P, then by the first part, and the additivity of P,

$$(16) \qquad P(x_1 + h_1, \ldots, x_r + h_r) = \Sigma\alpha(z_{\nu_1} + z'_{\nu_1})^{l_1} \ldots (z_{\nu_k} + z'_{\nu_k})^{l_k}$$

$$(17) \qquad = \Sigma\alpha z_{\nu_1}^{l_1} \ldots z_{\nu_k}^{l_k} + \Sigma\alpha z'^{l_1}_{\nu_1} \ldots z'^{l_k}_{\nu_k}.$$

But, $e_k z_{\nu_k} \cdot z'^{e_1}_{\nu_1} \ldots z'^{e_{k-1}}_{\nu_{k-1}} z'^{e_k - 1}_{\nu_k}$ occurs in the right hand side of (16) but not in (17), unless $k = 1$ and $e_k = 1$, where we consider (16) and (17) as elements of D. Thus P is a Lie element.]

5.7. The Lower Central Series of Free Groups

We are now able to utilize the identities of Section 5.2 and Theorem 5.3 of Section 5.3 for a thorough analysis of the lower central series of free groups.

To summarize our set of notations and definitions, we recall here the following symbols and concepts:

The associative algebra $A_0(Z, r)$ is freely generated by $x_1, \ldots, x_r$ (and contains a multiplicative identity), and the Lie algebra $\Lambda_0(Z, r)$ is freely generated by $\xi_1, \ldots, \xi_r$, where Z is the ring of integers. Moreover, the mapping μ of Λ_0 onto the Lie elements of A_0, defined by Lemma 5.5 and $\mu(\xi_\rho) = x_\rho$, is a one-one mapping according to Theorem 5.9 in Section 5.6. Therefore, μ has an inverse, denoted by μ^{-1}. The free group $F(r)$ on r free generators $a_1, \ldots, a_r$ can be represented in terms of elements of A_0 by setting $a_\rho = 1 + x_\rho$ (Theorem 5.6). The nth term of the lower central series of $F(r)$ is denoted by $F_n(r)$ or simply by F_n, with $F_1 = F(r)$. The mapping $\delta(W)$ of the words W in the a_ρ onto homogeneous elements of A_0 is described in Definition 5.6 and Lemma 5.4 of Section 5.5. Now we have:

THEOREM 5.12. *The quotient groups F_n/F_{n+1}, $n = 1, 2, 3, \ldots$, of the lower central series of a free group F freely generated by $a_1, \ldots, a_r$ are isomorphic, as Abelian groups, to the submodules Λ_n of homogeneous elements of degree n in the Lie algebra $\Lambda_0(Z, r)$ freely generated by $\xi_1, \ldots, \xi_r$. Explicitly, the isomorphism is given as follows:*

Let $\beta^n(\xi_{\rho_1}, \ldots, \xi_{\rho_n})$ be a Lie monomial of degree n. Then the mapping

(1) $$\beta^n(\xi_{\rho_1}, \ldots, \xi_{\rho_n}) \to \beta^n(a_{\rho_1}, \ldots, a_{\rho_n})F_{n+1}$$

is an isomorphism of Λ_n under addition onto F_n/F_{n+1} under multiplication where the bracket arrangement β^n applies to the ξ_ρ by Lie multiplication and to the a_ρ by commutation.

PROOF. We first use the quotient groups F_n/F_{n+1} to construct a Lie algebra over Z. Since $C_n = F_n/F_{n+1}$ is an Abelian group under multiplication we may form C the restricted direct sum of $C_1, C_2, \ldots, C_n, \ldots$, i.e., form the set of formal sums

$$c_1 + c_2 + \ldots + c_n + \ldots$$

where c_n is in C_n and all but finitely many c_n are 1 (the "zero" in C_n), and define an operation *pseudo-addition in C* by

(2) $$g \oplus h = g_1h_1 + g_2h_2 + \ldots + g_nh_n + \ldots,$$

where

(3a) $$g = g_1 + g_2 + \ldots + g_n + \ldots,$$

(3b) $$h = h_1 + h_2 + \ldots + h_n + \ldots.$$

Under pseudo-addition, C is an Abelian group; indeed,

$$1 + 1 + \ldots + 1 + \ldots$$

is the pseudo-zero, and

$$\ominus g = g_1^{-1} + g_2^{-1} + \ldots, + g_n^{-1} + \ldots$$

is the pseudo-negative of g.

We may introduce a *pseudo-multiplication* $\odot$ *in* C, using commutation, so that $(C, \oplus, \odot)$ is a Lie ring. Indeed, we note that it follows from (5.3.5), (5.3.6), and (5.3.8) of Theorem 5.3 that if a is in F_n, b is in F_{n+1}, c is in F_m, and d is in F_{m+1}, then

$$(ab, cd)F_{n+m+1} = (a, c)F_{n+m+1}.$$

Hence, if

$$g_n = aF_{n+1}, \qquad g_m = cF_{m+1},$$

then

(4) $$g_n \odot g_m = (a, c)F_{n+m+1}$$

is a unique element of

$$C_{n+m} = F_{n+m}/F_{n+m+1}.$$

Moreover, we extend *pseudo-multiplication* to all of C as follows:

(5) $$g \odot h = \sum_{i,j=1}^{\infty} g_i \odot h_j,$$

where g and h are as in (3), and Σ denotes pseudo-addition (since all but finitely many g_i and h_j are 1, Σ is actually a finite pseudo-sum). Clearly, the component of (5) which is in C_n is precisely the pseudo-sum of all

$$g_i \odot h_j,$$

where $i + j = n$.

To show that $(C, \oplus, \odot)$ is a Lie ring, we first show that $\odot$ distributes over $\oplus$. Let

(6) $$f = f_1 + f_2 + \ldots + f_n + \ldots$$

and let g and h as in (3) be in C, and consider

(7) $$f \odot (g \oplus h), (f \odot g) \oplus (f \odot h).$$

The left-hand side of (7) is by definition

$$\sum_{i,j=1}^{\infty} f_i \odot (g_j h_j).$$

If $f_i = aF_{i+1}$, $g_j = bF_{j+1}$, $h_j = cF_{j+1}$, then

$$\begin{aligned} f_i \odot (g_j h_j) &= aF_{i+1} \odot bcF_{j+1} = (a, bc)F_{i+j+1} \\ &= (a, b)(a, c)F_{i+j+1} = (a, b)F_{i+j+1} \cdot (a, c)F_{i+j+1} \\ &= (aF_i \odot bF_j)(aF_i \odot cF_j) \\ &= (f_i \odot g_j) \cdot (f_i \odot h_j) \end{aligned}$$

(where we have made use of (5.3.5), and so

$$\sum_{i,j=1}^{\infty} f_i \odot (g_j h_j) = \sum_{i,j=1}^{\infty} (f_i \odot g_j)\cdot(f_i \odot h_j) = (f \odot g) \oplus (f \odot h).$$

Thus both elements in (7) are equal, and $\odot$ is left-distributive over $\oplus$. Similarly, $\odot$ is right-distributive over $\oplus$. Thus, C is a ring. To show the nilpotency and Jacobi identity for pseudo-multiplication, first let g be as in (3). Then

$$g \odot g = \sum_{i,j=1}^{\infty} g_i \odot g_j.$$

Let $g_i = aF_{i+1}$ and $g_j = bF_{j+1}$; then if $i < j$,

$$\begin{aligned} g_i \odot g_j &= (a, b)F_{i+j+1} = (b, a)^{-1}F_{i+j+1} \\ &= \ominus(g_j \odot g_i), \end{aligned}$$

and if $i = j$,

$$g_i \odot g_i = (a, a)F_{i+j+1} = F_{i+j+1}$$

(the identity element of C_{i+j}). Thus,

$$g \odot g = 1 + 1 + \ldots + 1 + \ldots$$

and we have the nilpotency requirement for $\odot$.

Next, let f, g, and h be as in (6) and (3). Then if $f_i = aF_{i+1}$, $g_j = bF_{j+1}$, and $h_k = cF_{k+1}$,

$$\begin{aligned} &[(f \odot g) \odot h] \oplus [(h \odot f) \odot g] \oplus [(g \odot h) \odot f] \\ &\quad = \left[\sum_{i,j,k=1}^{\infty} (a, b, c)F_{i+j+k+1}\right] \oplus \left[\sum_{i,j,k=1}^{\infty} (c, a, b)F_{i+j+k+1}\right] \\ &\qquad \oplus \left[\sum_{i,j,k=1}^{\infty} (b, c, a)F_{i+j+k+1}\right] \\ &\quad = \sum_{i,j,k=1}^{\infty} (a, b, c)(c, a, b)(b, c, a)F_{i+j+k+1} \\ &\quad = \sum_{i,j,k=1}^{\infty} F_{i+j+k+1} = 1 + 1 + \ldots + 1 + \ldots \end{aligned}$$

(where we have made use of (5.3.7) of Theorem 5.3). Therefore, $\odot$ satisfies the Jacobi identity, and $(C, \oplus, \odot)$ is a Lie ring.

Since any ring is automatically an algebra over the ring of integers, $(C, \oplus, \odot)$ is a Lie algebra over Z, and we may map $\Lambda_0(Z, r)$ homomorphically into C by sending

(8) $$\xi_\rho \rightarrow a_\rho F_2.$$

Clearly, since $\odot$ was defined by (4) and (5), under the mapping (8), a monomial

$$\beta^n(\xi_{\rho_1}, \ldots, \xi_{\rho_n})$$

goes into

$$\beta^n(a_{\rho_1}, \ldots, a_{\rho_n})F_{n+1}.$$

Thus, to show that (1) is an isomorphism of Λ_n onto $F_n/F_{n+1} = C_n$, it suffices to show that (8) is an isomorphism of Λ_0 onto C (clearly Λ_n goes into C_n).

According to Theorem 5.4 of Section 5.3, the simple n-fold commutators

$$((\ldots((a_{\rho_1}, a_{\rho_2}), a_{\rho_3}), \ldots), a_{\rho_n})$$

generate C_n. Hence, (8) is clearly onto C.

To show that (8) is one-one, we shall construct its inverse. Every element a in F_n, when expressed as an element of the power series ring $A(Z, r)$ using the mapping

$$a_\rho \to 1 + x_\rho,$$

has the form

$$1 + u_n + u_{n+1} + \ldots u_k + \ldots$$

where u_k has degree k or is zero and $k \geq n$, according to Corollary 5.7 of Section 5.5. Moreover, if we define

$$\delta_n(aF_{n+1}) = \delta_n(a) = u_n,$$

then by Corollary 5.7, δ_n is well defined on C_n, and maps C_n into A_n the set of homogeneous polynomials of degree n in $A_0(Z, r)$. Moreover, because of (5.5.5) and (5.5.6) in Lemma 5.4, δ_n is a linear mapping of C_n into A_n. Thus the mapping

$$\delta(g) = \delta_1(g_1) + \delta_2(g_2) + \ldots + \delta_n(g_n) + \ldots$$

is a linear mapping of C into $A_0(Z, r)$ (all but finitely many g_n are 1). Now the mapping μ^{-1} of $A_0(Z, r)$ onto $\Lambda_0(Z, r)$, where μ is given as in Lemma 5.5 of Section 5.6 (μ is one-one by Theorem 5.9 of Section 5.6), is a linear mapping.

Hence, the mapping

$$\mu^{-1}\delta(g) = \mu^{-1}\delta_1(g_1) + \mu^{-1}\delta_2(g_2) + \ldots + \mu^{-1}\delta_n(g_n) + \ldots \tag{9}$$

is a linear mapping of C into $\Lambda_0(Z, r)$.

To show that $\mu^{-1}\delta$ is the inverse of the mapping in (8), we show that

$$\beta^n(a_{\rho_1}, \ldots, a_{\rho_n})$$

which is the image of

$$\beta^n(\xi_{\rho_1}, \ldots, \xi_{\rho_n})$$

under (8), goes back into it under $\mu^{-1}\delta$. It suffices to show by induction on n that

$$\delta\beta^n(a_{\rho_1}, \ldots, a_{\rho_n}) = \beta^n(x_{\rho_1}, \ldots, x_{\rho_n}),$$

where β^n applies to the x_ρ by bracket multiplication in A_0. For $n = 1$,

$$\delta(a_\rho) = \delta_1(a_\rho) = x_\rho.$$

Moreover, if $\beta^n = (\beta^k\beta^l)$ and so $n = k + l$, then

$$\begin{aligned}\delta(\beta^n(a_{\rho_1}, \ldots, a_{\rho_n})) &= \delta_n((\beta^k(a_{\rho_1}, \ldots, a_{\rho_k}), \beta^l(a_{\rho_{k+1}}, \ldots, a_{\rho_n}))) \\ &= [\delta_k(\beta^k(a_{\rho_1}, \ldots, a_{\rho_k})), \delta_l(\beta^l(a_{\rho_{k+1}}, \ldots, a_{\rho_n}))],\end{aligned}$$

by (5.5.7) of Lemma 5.4 in Section 5.5. Thus the inductive hypothesis implies that

$$\delta(\beta^n(a_{\rho_1}, \ldots, a_{\rho_n})) = \beta^n(x_{\rho_1}, \ldots, x_{\rho_n}),$$

and hence

$$\mu^{-1}\delta(\beta^n(a_{\rho_1}, \ldots, a_{\rho_n})) = \beta^n(\xi_{\rho_1}, \ldots, \xi_{\rho_n}).$$

Therefore, the mapping (8) has the left-inverse $\mu^{-1}\delta$, and so (8) must be one-one of Λ_0 onto C. Then, since (8) has a unique inverse, it must be $\mu^{-1}\delta$.

We have thus established our result. ◀

For later use we shall now list some consequences easily derived from Theorem 5.12.

COROLLARY 5.12. *If F is the free group on $a_1, \ldots, a_r$, F_n is the nth term of the lower central series of F, $A_0(Z, r)$ is the free associative algebra on $x_1, \ldots, x_r$, and $\Lambda_0(Z, r)$ is the free Lie algebra on $\xi_1, \ldots, \xi_r$, then the following statements hold:*

(i) *If u_n is any homogeneous Lie element of degree n in $A_0(Z, r)$, then there exists an element W in F_n, uniquely determined mod F_{n+1}, such that*

$$\delta(W) = u_n.$$

(ii) *If W is in F_n but not in F_{n+1}, then W is an mth power* mod F_{n+1}, i.e.,

$$WF_{n+1} = V^mF_{n+1},$$

where V is in F_n, if and only if the coefficients used to write

$$\delta(W) = u_n$$

as a linear combination of basic Lie elements, or of distinct monomials in $A_0(Z, r)$, are all divisible by m.

(iii) *If U is in F_m and V is in F_k, then unless $m = k$, and also U, V determine powers of the same element in F_m/F_{m+1}*, i.e.,

$$UF_{m+1} = W^p F_{m+1}, \; VF_{m+1} = W^q F_{m+1},$$

where p, q are integers, the commutator (U, V) is in F_{m+k} but not in F_{m+k+1}.

(iv) *The group F_n/F_{n+1} is a free Abelian group of rank N_n, the number of basic Lie elements of degree n.*

(v) *The module Λ_n of elements of degree n in $\Lambda_0(Z, r)$ is generated by the simple Lie products of n factors:*

$$(10) \qquad (\ldots((\xi_{\rho_1} \circ \xi_{\rho_2}) \circ \xi_{\rho_3}) \ldots) \circ \xi_{\rho_n},$$

where $\rho_1, \ldots, \rho_n$ range independently over $1, \ldots, r$.

Proof. The mapping (1) is an isomorphism of Λ_n onto F_n/F_{n+1} whose inverse is the mapping $\mu^{-1}\delta$. Since μ is an isomorphism (under addition) of Λ_n onto the homogeneous Lie elements of degree n in $A_0(Z, r)$, the mapping

$$W \rightarrow \delta(W)$$

is an isomorphism of F_n/F_{n+1}, under group multiplication, onto the homogeneous Lie elements of degree n in $A_0(Z, r)$, under addition.

Thus (i) follows. Moreover, since the basic Lie elements of degree n form a set of free Abelian generators for the homogeneous Lie elements of degree n in $A_0(Z, r)$, and since distinct monomials (5.4.1) form a set of free Abelian generators for all of $A_0(Z, r)$, we must have (ii).

The statement (iii) follows immediately from the correspondence in the proof of Theorem 5.12 between commutation among the groups F_n/F_{n+1} and multiplication among the modules Λ_n, with the help of Theorem 5.10 in Section 5.6.

Statement (iv) follows from Theorem 5.12, and Theorem 5.9 in Section 5.6. (Theorem 5.11 in Section 5.6 gives the explicit form of N_n.)

Statement (v) follows from the fact that the simple n-fold commutators

$$(a_{\rho_1}, \ldots, a_{\rho_n})$$

of Theorem 5.4 in Section 5.3 generate F_n/F_{n+1}. Hence, their pre-images under the isomorphism (1) of Λ_n onto F_n/F_{n+1} must generate Λ_n; but the simple Lie products with n factors are precisely the pre-images under (1) of the simple n-fold commutators. (For another proof of (v) see Problem 5.4.18.) ◀

We shall now use Theorem 5.12 to prove two important results which were first proved by P. Hall, 1933, by purely group theoretic methods.

THEOREM 5.13A (*P. Hall's Basis Theorem*). *In a free group* $F(r)$, *there exists an infinite sequence of commutators* C_ν, $(\nu = 1, 2, 3, \ldots)$ *of non-decreasing weights in the generators such that, for* $n = 1, 2, 3, \ldots$ *and for any element* W *of* $F(r)$,

$$(11) \qquad W = C_1^{e_1} C_2^{e_2} \ldots C_{k(n)}^{e_{k(n)}} V_{n+1},$$

where $V_{n+1} \in F_{n+1}$ (the $(n + 1)$st group of the lower central series of $F(r)$), *and where the integers* $e_1, \ldots, e_{k(n)}$ *and* V_{n+1} *are uniquely determined by* W. *The number* $k(n)$ *is the number of* C_ν *of a weight* $\leq n$.

We shall call (11) an nth *commutator-power product* for W.

THEOREM 5.13B. *Let* $q = p^m$ *be a power of the prime number* p. *Then the exponents* $e_1, \ldots, e_{k(p-1)}$ *in* (11) *are divisible by* q *if and only if there exists an element* U *of* F *such that*

$$(12) \qquad W \equiv U^q \bmod F_p.$$

(If q divides $e_1, \ldots, e_{k(p-1)}$, we shall say that W is *q-divisible mod* F_p.)

PROOF OF 5.13A. By Theorem 5.12, each element of F_n/F_{n+1} corresponds to a unique element of Λ_n under the mapping (1). If $\zeta_1, \zeta_2, \zeta_3, \ldots$ is the sequence of basic Lie elements of $\Lambda_0(Z, r)$ given by Theorem 5.9 in Section 5.6, then their images under (1) form a sequence $C_1, C_2, C_3, \ldots$ of commutators of non-decreasing weight. Moreover, this sequence of commutators has the properties required by our theorem, as we show now by induction on n in (11).

For $n = 1$, the basic Lie elements of $\Lambda_0(Z, r)$ of degree 1 are $\xi_1, \ldots, \xi_r$ and $C_1 = a_1, \ldots, C_r = a_r$; every element W of $F(r)$ is a product mod F_2

$$a_1^{e_1} \ldots a_r^{e_r},$$

and the exponents are determined uniquely. Assuming by inductive hypothesis that

$$W = C_1^{e_1} \ldots C_{k(n)}^{e_{k(n)}} V_{n+1}$$

where V_{n+1} is in F_{n+1}, then $V_{n+1}F_{n+2}$ is an element of F_{n+1}/F_{n+2}. Hence, under (1), $V_{n+1}F_{n+2}$ has a pre-image in Λ_{n+1}, which can be written uniquely in the form

$$e_t\zeta_t + \ldots + e_s\zeta_s$$

where $t, \ldots, s$ are $k(n) + 1, \ldots, k(n + 1)$. Thus

$$V_{n+1} = C_t^{e_t} \ldots C_s^{e_s} V_{n+2},$$

where V_{n+2} is in F_{n+2}. Thus,

$$W = C_1^{e_1} \ldots C_{k(n)}^{e_{k(n)}} C_t^{e_t} \ldots C_s^{e_s} V_{n+2},$$

where $t, \ldots, s$ is $k(n) + 1, \ldots, k(n+1)$ and V_{n+2} is in F_{n+2}. To show uniqueness of the exponents, suppose

$$W = C_1^{e'_1} \ldots C_{k(n)}^{e'_{k(n)}} C_t^{e'_t} \ldots C_s^{e'_s} V'_{n+2},$$

where V'_{n+2} is in F_{n+2}. Since $C_t, \ldots, C_s, V_{n+2}, V'_{n+2}$ are each in F_{n+1}, the inductive hypothesis implies

$$e_1 = e_1', \ldots, e_{k(n)} = e'_{k(n)},$$

and that

$$C_t^{e_t} \ldots C_s^{e_s} V_{n+2} = C_t^{e'_t} \ldots C_s^{e'_s} V'_{n+2}$$

Considering this last equation mod F_{n+2}, and taking pre-images under (1), we obtain

$$e_t\zeta_t + \ldots + e_s\zeta_s = e_t'\zeta_t + \ldots + e_s'\zeta_s.$$

Since the ζ_ν are a linear basis for Λ_0,

$$e_t = e_t', \ldots, e_s = e_s'.$$

Now, by cancelling common factors from the two representations for W, we obtain $V_{n+2} = V'_{n+2}$. Thus we have proved Theorem 5.13A.

Proof of 5.13B. We first show that if an element V of F_n is a product of qth powers, mod F_{n+1}, where $n < p$, then if the pre-image of V in Λ_n is

$$f_t\zeta_t + \ldots + f_s\zeta_s,$$

where $t = k(n-1) + 1, \ldots, s = k(n)$, then all f_j are divisible by q. For, if

$$V \equiv U_1^{qd_1} \ldots U_m^{qd_m} \bmod F_{n+1}$$

and we represent the elements in $A(Z, r)$, then

$$1 + v = (1 + u_1)^{qd_1} \ldots (1 + u_m)^{qd_m}(1 + v')$$

where v' has only monomials of degree greater than n; since $\binom{qd_i}{l}$ is divisible by q if $l < p$, if follows that all monomials in $A(Z, r)$ of degree $\leq n$ occur in v with coefficient divisible by q. Replacing each such monomial with its representation as a polynomial in basic Lie elements $z_1, z_2, \ldots$ as given in (ii) of Theorem 5.8 of Section 5.6, we see that each monomial in the basic Lie elements which has degree in $x_1, \ldots, x_r$ no more than n, has a coefficient divisible by q. In particular, the basic Lie elements of degree n have coefficients divisible by q. Applying μ^{-1} to these, gives us the result we want.

Suppose now that

$$W \equiv U^q \bmod F_p,$$

where U is any element of F. Then

$$W \equiv U_1{}^q \bmod F_2$$

where U_1 is in F; thus, by the above result and the construction (in the proof of Theorem 5.13A) to express W in the form (11), we have that

$$W \equiv C_1{}^{e_1} \ldots C_{k(1)}^{e_{k(1)}} \bmod F_2,$$

where q divides $e_1, \ldots, e_{k(1)}$. Since

$$U_1{}^{-q}W \equiv (U_1{}^{-1})^q U^q \bmod F_3$$

and $U_1{}^{-q}W$ is in F_2, it follows by the result above and the construction in the proof of Theorem 5.13A that

$$W \equiv C_1{}^{e_1} \ldots C_{k(1)}^{e_{k(1)}} C_{k(1)+1}^{e_{k(1)+1}} \ldots C_{k(2)}^{e_{k(2)}} \bmod F_3$$

where q divides all exponents e_j.
Continuing in this way we obtain that W is q-divisible mod F_p.

Suppose, finally, that W is q-divisible mod F_p. We use induction on n to show that

$$W \equiv U_n{}^q \bmod F_{n+1}$$

for all $n < p$.

For $n = 1$, since F/F_2 is Abelian, we may let

$$U_1 = C_1{}^{f_1} \ldots C_{k(1)}^{f_{k(1)}},$$

where $f_j = e_j/q$. Assuming by inductive hypothesis that

$$W \equiv U_n{}^q \bmod F_{n+1},$$

let $n + 1 < p$. Then

$$V_{n+1} \equiv U_n{}^{-q}W \equiv U_n{}^{-q}C_1{}^{e_1} \ldots C_{k(p-1)}^{e_{k(p-1)}} \bmod F_{n+2}.$$

Since $q, e_1, \ldots, e_{k(p-1)}$ are all divisible by q, and V_{n+1} is in F_{n+1}, we may apply the above result to obtain that when V_{n+1} is represented in Λ_{n+1}, all coefficients on basic Lie elements are divisible by q. Hence, by (ii) of Corollary 5.12,

$$V_{n+1} \equiv V^q \bmod F_{n+2},$$

where V is in F_{n+1}. Thus,

$$W \equiv U_n{}^q V^q \bmod F_{n+2},$$

and since $U_n V \equiv V U_n \bmod F_{n+2}$,

$$W \equiv (U_n V)^q \bmod F_{n+2}.$$

Thus we have completed the induction, and the proof of Theorem 5.13B. ◀

The problems at the end of this section give some indication of the use of Theorems 5.13A and 5.13B. For other applications, see P. Hall, 1933, and Section 5.11.

Reference and Remarks. Theorems 5.13A and B were first proved by P. Hall, 1933, using purely group theoretical methods. His procedure (which he called a "collecting process") was used again by M. Hall, 1950b, to construct a basis for $\Lambda_0(R, r)$, operating exclusively within the Lie algebra. The simplifications due to Meier-Wunderli, 1950, have already been mentioned. Theorem 5.13B was used by P. Hall, 1933, in the theory of groups of prime power order. We shall apply it later (after having derived some refinements from Jacobson, Artin, and Zassenhaus in Section 5.11) to the Burnside problem (Section 5.12). Various applications of the results of the present Section 5.7 will be given in the next section.

Theorems 5.12 and 5.13A indicate that the associative ring $A_0(Z, r)$ is larger than necessary to describe the lower central series of a free group, since we actually need only Lie elements and commutators for this purpose. Therefore, a Lie algebra would seem to be the appropriate tool. We shall see in Section 5.10 that there exists an associative composition in $\Lambda(Q, r)$, where Q is the field of rational numbers, which permits us to represent a free group completely in terms of elements of this Lie algebra.

Problems for Section 5.7

1. Let G be a group such that the cosets of the elements $g_1, g_2, g_3, \ldots$ generate G/G_2, the cosets of the elements $h_1, h_2, h_3, \ldots$ generate $G_2/G_3, \ldots,$ the cosets of the elements $k_1, k_2, k_3, \ldots$ generate G_n/G_{n+1}. Show that mod G_{n+1} every element of G can be expressed in the form

$$W_1(g_\nu)W_2(h_\nu)\ldots W_n(k_\nu), \qquad \nu = 1, 2, 3, \ldots.$$

[*Hint:* Use induction on n, and the method of proof of Theorem 5.13A.]

2. Let G be a group on r generators and suppose that $G_1, G_2, G_3, \ldots$ is the lower central series of G, and that

$$G_n/G_{n+1} \simeq F_n/F_{n+1}$$

for all n; then $G \simeq F(r)$. Show that if

$$G/G_n \simeq F/F_n$$

for infinitely many n then $G \simeq F(r)$. [*Hint:* Suppose that under a homomorphism α of $F(r)$ onto G the freely reduced word $W(a_1, \ldots, a_r) \neq 1$ goes into the identity in G. Since $\bigcap_{n=1}^{\infty} F_n = 1$, W is in some F_n, but not in F_{n+1}.

The verbal subgroups F_n, F_{n+1} are mapped by α onto G_n, G_{n+1} and hence, F_n/F_{n+1} is mapped onto G_n/G_{n+1} by α. But $\alpha(W) = 1$ although $W \neq 1$, and W is not in F_{n+1}. This contradicts that F_n/F_{n+1}, a finitely generated Abelian group, is Hopfian.

If $G/G_n \simeq F/F_n$, then

$$G_k/G_n \simeq F_k/F_n$$

for all $k \leq n$, and hence, the first part applies.]

3. Generalize Problem 2 to show that if H is residually nilpotent, i.e., $\bigcap_{n=1}^{\infty} H_n = 1$, and is a finitely generated group which has G as a homomorphic image, and

$$G_n/G_{n+1} \simeq H_n/H_{n+1}$$

for all n, or

$$G/G_n \simeq H/H_n$$

for infinitely many n, then $G \simeq H$.

4. Let $F(r)$ be the free group of rank r, $r \geq 2$. Show that F_{m+1}/F_{2m+1} is an Abelian subgroup of F/F_{2m+1}. Show that if F_m/F_{2m+1} is Abelian, then $r = 2$ and $m = 2$. [*Hint:* If F_m/F_{2m+1} is Abelian and C_μ and C_λ are in F_m, then (C_μ, C_λ) is in F_{2m+1}. Hence, under the isomorphism of the direct sum of the F_n/F_{n+1}, and the modules Λ_n, in which commutation among the F_n/F_{n+1} is reflected by Lie multiplication among the Λ_n, it follows that for any two elements ξ, η in Λ_m, $\xi \circ \eta$ is in Λ_{2m} and Λ_{2m+1} and so is 0. Hence, ξ and η are linearly dependent, and Λ_m has a basis consisting of one element. But Λ_m has a basis consisting of the basic Lie elements of degree m. If $m > 2$, then

$$(\ldots((\xi_2 \circ \xi_1) \circ \xi_1) \circ \ldots) \circ \xi_1, \qquad (\ldots((\xi_3 \circ \xi_1) \circ \xi_1) \circ \ldots) \circ \xi_1,$$

where $m - 1$ factors ξ_1 are used, are two different basic Lie elements of degree r. Moreover, if $r = 2$ but $m > 2$, then

$$(\ldots((\xi_2 \circ \xi_1) \circ \xi_1) \circ \ldots) \circ \xi_1, \qquad (\ldots((\xi_2 \circ \xi_1) \circ \xi_2) \circ \ldots) \circ \xi_2,$$

where $m - 1$ factors ξ_1 are used in the first, and $m - 1$ factors ξ_2 are used in the second, are two different basic Lie elements of degree m.]

5. Let $F(r)$ be the free group of rank r, $r \geq 2$. Show that the center of F/F_{n+1} is precisely F_n/F_{n+1}. [*Hint:* If an element u not in F_n has its coset uF_{n+1} in the center of F/F_{n+1} then (a_1, u) and (a_2, u) are in F_{n+1}. Hence, if u is in F_k but not in F_{k+1}, then

$$\xi_1 \circ \eta = 0, \qquad \xi_2 \circ \eta = 0,$$

where η is the element of Λ_k corresponding to uF_{k+1}. But then

$$\alpha_1 \xi_1 + \beta_1 \eta = 0, \qquad \alpha_2 \xi_2 + \beta_2 \eta = 0,$$

where $\alpha_1, \alpha_2, \beta_1, \beta_2$ are non-zero integers. Then

$$\beta_2\alpha_1\xi_1 - \beta_1\alpha_2\xi_2 = 0$$

contradicting the linear independence of ξ_1 and ξ_2.]

6. In (5.7.11) of Theorem 5.13A, let $n = 2$ and denote $k(2)$ by k. Show that if $e_1, \ldots, e_k, f_1, \ldots, f_k$ is any sequence of $2k$ integers, then

$$C_1^{e_1} \ldots C_k^{e_k} C_1^{f_1} \ldots C_k^{f_k} \equiv C_1^{g_1} \ldots C_k^{g_k} \bmod F_3 \tag{13}$$

where $g_1, \ldots, g_k$ are integers. Show that $g_1, \ldots, g_k$ are each polynomials in $e_1, \ldots, e_k, f_1, \ldots, f_k$ with integer coefficients. (This is true for $n > 2$, but is much harder to prove.) [*Hint:* If $C_1 = a_1, \ldots, C_r = a_r$, then we may choose $C_{r+1}, \ldots, C_k$ to be of the form (a_i, a_l), $i > l$. By reducing (13) mod F_2 one easily obtains $g_j = e_j + f_j$, $1 \leq j \leq r$. Thus,

$$C_1^{-e_1-f_1} \ldots C_r^{-e_r-f_r} C_1^{e_1} \ldots C_k^{e_k} C_1^{f_1} \ldots C_k^{f_2} \tag{14}$$

is in F_2. To expand it in a power series in $A(Z, r)$ we let

$$C_i = 1 + x_i, \qquad 1 \leq i \leq r$$
$$C_j \equiv 1 + x_{i_j}x_{l_j} - x_{l_j}x_{i_j} \bmod D_3, \qquad r + 1 \leq j \leq k$$

where D_3 consists of the power series with monomials of degree three or more only. The second-degree monomial $x_q x_p$, where $p > q$ has contributions from

$$C_q^{-e_q-f_q} C_p^{-e_p-f_p},\ C_q^{-e_q-f_q} C_p^{e_p},\ C_q^{-e_q-f_q} C_p^{f_p},$$
$$C_q^{e_q} C_p^{e_p},\ C_q^{e_q} C_p^{f_p},\ C_q^{f_q} C_p^{f_p},\ C_s^{e_s},\ C_s^{f_s},$$

where $C_s = (a_p, a_q)$. Similarly, the monomial $x_p x_q$ has contributions from

$$C_p^{-e_p-f_p} C_q^{e_q},\ C_p^{-e_p-f_p} C_q^{f_q},\ C_p^{e_p} C_q^{f_q},\ C_s^{e_s},\ C_s^{f}.$$

Hence, $x_p x_q - x_q x_p$ has the coefficient

$$e_s + f_s - e_q e_p - e_q f_p - f_q f_p$$

in (14). Thus,

$$g_s = e_s + f_s - e_q e_p - e_q f_p - f_q f_p.]$$

7. Suppose that G is a group generated by finitely many elements, each of finite order. Show that if some term G_{n+1} of the lower central series of G is 1, i.e., G is nilpotent, then G is a finite group. [*Hint:* Show that there are only finitely many commutators of weight k for $k \leq n$. Show by induction on k that each commutator of weight k has finite order mod G_{k+1}. Conclude from the expression (11) for any element W in the free group, that every element of G can be expressed in the form (11) mod $G_{n+1} = 1$.]

8. Let p be a prime number, and let H be a group for which H_p, the pth term of the lower central series for H, is 1. Show that the elements of H which are pth powers form a subgroup of H, i.e., the product of two pth powers is again a pth power. [*Hint:* Use the method of proof of Theorem 5.13A and show by induction on n that

$$U^p \cdot V^p \equiv W^p \bmod F_n,$$

for $n \leq p$.]

9. Let G be the *reduced free group on r generators with exponent p and nilpotent of class $p - 1$*, i.e.,

$$G = \langle a_1, \ldots, a_r; X^p, (X_1, X_2, \ldots, X_p)\rangle,$$

where p is a prime number and $(X_1, X_2, \ldots, X_p)$ is a simple p-fold commutator. Show that G has order p^s, where

$$s = \sum_{n=1}^{p-1} N_n,$$

and N_n is given by Witt's formula in Theorem 5.11 of Section 5.6. [*Hint:* Let Z_p be the Galois field of integers mod p; we shall show first that G is isomorphic to a subgroup of $A(Z_p, r)/D_p(Z_p)$, where $D_p(Z_p)$ is the ideal of polynomials having no monomial of degree less than p. Indeed, we first map $F(r)$, the free group on $a_1, \ldots, a_r$, into $A(Z, r)/D_p(Z)$ by sending

$$a_\rho \to 1 + x_\rho. \tag{15}$$

Since F_p, the pth term of the lower central series goes into $1 + u$, where u is in $D_p(Z)$, under the mapping (15) (Corollary 5.7 of Section 5), the mapping (15) is a homomorphism of F/F_p into $A(Z, r)/D_p(Z)$. If W is in F_t but not in F_{t+1}, $t < p$, then under (15), W will go into a power series with some monomial $\neq 1$ of degree less than p. Hence, the mapping (15) is one-one from F/F_p into $A(Z, r)/D_p(Z)$. Next we map $A(Z, r)/D_p(Z)$ homomorphically into $A(Z_p, r)/D_p(Z_p)$ by using the homomorphism of Z onto Z_p. The kernel of the mapping of

$$F/F_p \to A(Z_p, r)/D_p(Z_p)$$

contains precisely those cosets WF_p such that in $A(Z, r)$, each monomial of degree less than p in the power series representing W has coefficient divisible by p. Hence, as in the proof of Corollary 5.13B, we can show that $WF_p = V^pF_p$. Thus G is isomorphic to the multiplicative subgroup of $A(Z_p, r)/D_p(Z_p)$ generated by $1 + x_1, 1 + x_2, \ldots, 1 + x_r$. Moreover, if W is in G_t, $t < p$, then

$$WG_{t+1} \to \delta_t(W)$$

is an isomorphism of G_t/G_{t+1} onto the homogeneous Lie elements of degree t in $A_0(Z_p, r)$. Hence, G_t/G_{t+1} has order p^{N_t} and G has order p^s.]

5.8. Some Applications

In this section we shall apply the commutator calculus to groups which are not necessarily free.

For any group G, we denote the nth group in the lower central series of G by G_n; in particular $G_1 = G$ and $G_2 = (G, G)$. If $G_{n+1} = 1$, we call G *nilpotent of class n*. If the intersection of all the terms of the lower central series of G is the identity, we call G *residually nilpotent*.

Lemma 5.9. *Suppose that a group G is generated by a set of elements $a_1, a_2, a_3, \ldots$, together with a set of elements $b_1, b_2, b_3, \ldots$ which lie in G_2. Then there exists a set of elements $c_1, c_2, c_3, \ldots$ in G_n such that*

$$a_1, a_2, a_3, \ldots, \quad c_1, c_2, c_3, \ldots$$

generates G. In particular, if G is nilpotent (of any class), then G is generated by $a_1, a_2, a_3, \ldots$ alone.

Proof. We shall show by induction on n that $a_1, a_2, a_3, \ldots$ together with some elements of G_n generate G.

For $n = 1$ the result is trivial, and for $n = 2$ it is the hypothesis. Assume that $a_1, a_2, a_3, \ldots, c_1, c_2, c_2, \ldots$ generate G, where $c_1, c_2, c_3, \ldots$ are in G_n, $n \geq 2$. Now, by Problem 5.7.1, mod G_{n+1} every element of G has the form

$$g_1 g_2 \ldots g_n G_{n+1} \tag{1}$$

where g_i is a word in a set of elements whose cosets generate G_i/G_{i+1}. Since $a_1, a_2, a_3, \ldots, b_1, b_2, b_3, \ldots$ generate G and $b_1, b_2, b_3, \ldots$ are in G_2, the cosets of $a_1, a_2, a_3, \ldots$ generate G/G_2; thus we may choose

$$g_1 = W_1(a_1, a_2, a_3, \ldots).$$

Since $a_1, a_2, a_3, \ldots, c_1, c_2, c_3, \ldots$ generate G, by Theorem 5.4 of Section 5.3, we may generate the group G_i/G_{i+1} by the cosets of the simple monomials of weight i in $a_1, a_2, a_3, \ldots, c_1, c_2, c_3, \ldots$. Moreover, if $i \geq 2$, any simple monomial of weight i which involves c_ν must be in G_{n+1}, since c_ν itself is in G_n. Thus mod G_{n+1}, each element g_i may be chosen as a word $W_i(a_\nu)$. Thus any element of G is in some coset

$$\begin{aligned} &W_1(a_\nu)W_2(a_\nu) \ldots W_n(a_\nu)G_{n+1}; \qquad \text{in particular,} \\ &c_j = W_1^{(j)}(a_\nu)W_2^{(j)}(a_\nu) \ldots W_n^{(j)}(a_\nu)\, d_j, \end{aligned} \tag{2}$$

where d_j is in G_{n+1}. Thus,

$$a_1, a_2, a_3, \ldots, \quad d_1, d_2, d_3, \ldots$$

generate G, where $d_1, d_2, d_3, \ldots$ are in G_{n+1}. This completes the induction.

If G is nilpotent of class n, then $G_{n+1} = 1$ and so

$$d_1 = d_2 = d_3 = \ldots = 1,$$

and $a_1, a_2, a_3, \ldots$ generate G. ◀

To guarantee that generators of G Abelianized are necessarily generators of G, it does not suffice to assume that G is residually nilpotent instead of nilpotent. Indeed, the elements a^2b^3 and a^3b^4 generate the

Abelianized free group on a and b, but do not generate the free group itself (Problem 4.4.19); the free group on a and b is residually nilpotent.

Theorem 4.10 (which states that in a group with one defining relator any subset of generators which omits a generator involved in the defining relator freely generates its subgroup) cannot be generalized, without restrictions, to a group with more than one defining relator. For example, in the group G,

$$G = \langle a, b, c; b = a^2c, c = a^3b \rangle$$

the subgroup generated by a is not free; indeed,

$$a^2 = bc^{-1} = (cb^{-1})^{-1} = a^{-3},$$

so that $a^5 = 1$. Theorem 5.14 provides a restricted generalization of Theorem 4.10.

THEOREM 5.14. *Let G have the presentation on $r + m$ generators and m defining relators given by*

$$(3) \quad G = \langle a_1, \ldots, a_r, b_1, \ldots, b_m; b_1 = C_1(a_\rho, b_\mu), \ldots, b_m = C_m(a_\rho, b_\mu)\rangle,$$

where $C_i(a_\rho, b_\mu)$ has zero exponent sum on both a_ρ and b_μ. Then $a_1, \ldots, a_r$ freely generate a subgroup A of G.

PROOF: Let F be the free group on $a_1, \ldots, a_r$, and let Φ be the free group on $a_1, \ldots, a_r, c_1, \ldots, c_m$. We shall show by induction on n that G has a presentation

$$(4) \quad \langle a_1, \ldots, a_r, c_1, \ldots, c_m; c_1 = K_1(a_\rho, c_\mu), \ldots, c_m = K_m(a_\rho, c_\mu)\rangle,$$

where $K_i(a_\rho, c_\mu)$ is in the commutator subgroup of Φ and $K_i(a_\rho, 1)$ is in F_n, the nth term of the lower central series of F.

For $n = 1$ or 2, the result follows immediately from (3), with $c_\mu = b_\mu$. Suppose then that G has the presentation (4) with the $K_i(a_\rho, c_\mu)$ in Φ_2 and the $K_i(a_\rho, 1)$ in F_n. Since $K_i(a_\rho, c_\mu)$ is in Φ_2,

$$(5) \qquad K_i(a_\rho, c_\mu) = L_1(a_\rho, c_\mu) \cdot \ldots \cdot L_s(a_\rho, c_\mu) \cdot V(a_\rho, c_\mu),$$

where $V(a_\rho, c_\mu)$ is in Φ_{n+1} and $L_j(a_\rho, c_\mu)$ is a simple commutator in a_ρ and c_μ of weight at least two [this follows easily from Problem 5.7.1, Theorem 5.4 of Section 5.3, and the fact that $K_i(a_\rho, c_\mu)$ is in Φ_2]. Collect to the front those $L_j(a_\rho, c_\mu)$ which involve only the a_ρ, by introducing commutators of the form

$$(6) \qquad (U(a_\rho, c_\mu), L_{j_s}(a_\rho) \cdot L_{j_{s+1}}(a_\rho) \cdot \ldots \cdot L_{j_t}(a_\rho))$$

where $U(a_\rho, c_\mu)$ is a commutator (not necessarily simple) with some c_μ as a component, and

$$L_{j_1}(a_\rho), \ldots, L_{j_s}(a_\rho), \ldots, L_{j_t}(a_\rho)$$

are those $L_j(a_\rho, c_\mu)$ which involve only the a_ρ. Then

$$K_i(a_\rho, c_\mu) = W_i(a_\rho) \cdot \prod_k M_k(a_\rho, c_\mu) \cdot V(a_\rho, c_\mu) \tag{7}$$

where $M_k(a_\rho, c_\mu)$ is a commutator which has more than one component, and has some c_μ as one of its components.

Since $K_i(a_\rho, 1)$ is in F_n, and moreover, $M_k(a_\rho, 1) = 1$, it follows from (7) that $W_i(a_\rho)$ is in F_n. Introduce the new generator d_i for G given by

$$d_i = W_i^{-1}(a_\rho)c_i,$$

or equivalently,

$$c_i = W_i(a_\rho)d_i.$$

Under the Tietze transformation which replaces c_i by $W_i(a_\rho)d_i$, the defining relation

$$c_i = K_i(a_\rho, c_\mu)$$

becomes the defining relation

$$W_i(a_\rho)d_i = K_i(a_\rho, W_\mu(a_\rho)d_\mu),$$

or equivalently

$$d_i = W_i^{-1}(a_\rho)K_i(a_\rho, W_\mu(a_\rho)d_\mu).$$

Using (7), this last defining relation becomes

$$d_i = \prod_k M_k(a_\rho, W_\mu(a_\rho)d_\mu) \cdot V(a_\rho, W_\mu(a_\rho)d_\mu). \tag{8}$$

Since M_k and V are in the commutator subgroup of the free group on $a_1, \ldots, a_r, d_1, \ldots, d_m$, clearly the right-hand side of (8) is also. Moreover, if we replace each d_μ on the right-hand side of (8) with 1, we obtain

$$\prod_k M_k(a_\rho, W_\mu(a_\rho)) \cdot V(a_\rho, W_\mu(a_\rho)). \tag{9}$$

Since $W_\mu(a_\rho)$ is in F_n, it follows easily that (9) is in F_{n+1}. Hence, if we present G on the generators $a_1, \ldots, a_r, d_1, \ldots, d_m$ with the defining relations (8), we satisfy the inductive requirement for $n + 1$, and have completed our induction.

Suppose now that G is presented as in (4), with $K_i(a_\rho, c_\mu)$ in Φ_2 and $K_i(a_\rho, 1)$ in F_n. Let N be the normal subgroup of G generated by $c_1, \ldots, c_m$ and all simple commutators

$$(a_{\rho_1}, a_{\rho_2}, \ldots, a_{\rho_n}).$$

Then G/N is generated by $a_1, \ldots, a_r, c_1, \ldots, c_m$ and has the defining relators

$$c_1, \ldots, c_m, \tag{10a}$$

$$c_1 = K_1(a_\rho, c_\mu), \ldots, c_m = K_m(a_\rho, c_\mu), \tag{10b}$$

$$(a_{\rho_1}, a_{\rho_2}, \ldots, a_{\rho_n}), \ldots . \tag{10c}$$

Using Tietze transformations to eliminate the generators $c_1, \ldots, c_m$ and the defining relators (10a) we obtain that G/N is generated by $a_1, \ldots, a_r$ and has the defining relators

$$K_1(a_\rho, 1), \ldots, K_m(a_\rho, 1) \tag{11a}$$

$$(a_{\rho_1}, a_{\rho_2}, \ldots, a_{\rho_n}), \ldots \tag{11b}$$

Since $K_i(a_\rho, 1)$ is in F_n, the defining relators (11a) can be eliminated because of (11b). Hence, G/N has the generators $a_1, \ldots, a_r$ and the defining relators (11b), and is therefore isomorphic to F/F_n. But since (10c) are in N, we can map A/A_n homomorphically onto G/N, and since A has r generators, we can map F/F_n homomorphically onto A/A_n. Thus

$$F/F_n \xrightarrow[\text{onto}]{} A/A_n \xrightarrow[\text{onto}]{} G/N \xrightarrow[\text{onto}]{} F/F_n.$$

Since F/F_n is Hopfian (it is nilpotent and hence, residually nilpotent so that Theorem 5.5 of Section 5.3 applies), each of these homomorphisms must be one-one, and

$$F/F_n \simeq A/A_n.$$

Thus by Problem 5.7.2, A is the free group on $a_1, \ldots, a_r$ and is freely generated by them. ◀

Corollary 5.14.1. *If G is as in Theorem 5.14, then*

$$G/G_n \simeq F/F_n,$$

where F is the free group on $a_1, \ldots, a_r$.

Proof. If N is the normal subgroup of G generated by (10a) and (10c), then G_n is contained in N; indeed, G_n is generated by simple commutators

$$(g_1, \ldots, g_n),$$

and mod the normal subgroup of G generated by (10a), each such element is equal to 1 or to a commutator in (10c). Thus

$$G/G_n \xrightarrow[\text{onto}]{} G/N \xrightarrow[\text{onto}]{} F/F_n.$$

Since G is generated by $a_1, \ldots, a_r, b_1, \ldots, b_m$, where $b_1, \ldots, b_m$ are in G_2, it follows that G is generated by $a_1, \ldots, a_r$ and elements of G_n; thus, G/G_n has at most r generators and F/F_n can be mapped homomorphically onto G/G_n. From the Hopfian property of F/F_n, it follows that

$$G/G_n \simeq F/F_n,$$

which completes our proof. ◀

COROLLARY 5.14.2. *If G has some presentation with r generators, and has another presentation with $r + m$ generators and m defining relators, then G is the free group on r generators.*

PROOF. By Theorem 3.5 in Section 3.3, we may assume that G has the presentation

$$\langle a_1, \ldots, a_r, b_1, \ldots, b_m; b_1^{d_1}C_1, \ldots, b_m^{d_m}C_m \rangle,$$

where C_j has zero exponent sum on the a_ρ and b_μ. Since G has a presentation with r generators it follows that G/G_2 must have no more than r generators. Hence, $d_1 = \ldots = d_m = 1$, and G has the presentation

$$\langle a_1, \ldots, a_r, b_1, \ldots, b_m; b_1 = C_1^{-1}, \ldots, b_m = C_m^{-1} \rangle.$$

Thus Corollary 5.14.1 applies and

$$G/G_n \simeq F/F_n,$$

where F is the free group on $a_1, \ldots, a_r$. But G has a presentation with r generators and so, by Problem 5.7.2, $G \simeq F$, and we have our result. ◀

We have not yet connected the quotient groups G_n/G_{n+1} of the lower central series of a non-free group G to any Lie algebras. To do this, we first represent G as the quotient group F/N of a free group F. If F has r generators, then we define the nth *relation module* M_n *of* $G = F/N$ to be the set of homogeneous elements of $\Lambda_0(Z, r)$ of degree n which correspond to cosets of F_n/F_{n+1} (under the mapping in Theorem 5.12 of Section 5.7) which are defined by elements in $F_n \cap N$. Specifically, M_n consists of the elements

$$\mu^{-1}\delta_n(W),$$

where W ranges over all elements of $F_n \cap N$, μ is the natural mapping of $\Lambda_0(Z, r)$ onto the Lie elements of $A_0(Z, r)$, and $\delta_n(W)$ is the deviation of W if it has degree n and is zero otherwise. Now F_n/F_{n+1} is isomorphic to the submodule Λ_n of $\Lambda_0(Z, r)$ consisting of all homogeneous elements of degree n. Moreover, under the homomorphism of F onto G given by

$$F \to F/N,$$

the subgroups F_n and F_{n+1} are mapped onto F_nN/N and $F_{n+1}N/N$. Thus the mapping of F_n/F_{n+1} onto G_n/G_{n+1} induced by the homomorphism of F onto G has as its kernel the cosets of F_{n+1} determined by elements of $F_{n+1}N$, which are also in F_n. Since $F_{n+1} \subset F_n$, the kernel of the natural homorphism of F_n/F_{n+1} onto G_n/G_{n+1} is precisely the F_{n+1} cosets of $F_n \cap N$. But under the isomorphism $\mu^{-1}\delta_n$ of F_n/F_{n+1} onto Λ_n, the elements of $F_n \cap N$ correspond to M_n. Thus M_n is a submodule of Λ_n and

$$\Lambda_n/M_n \simeq G_n/G_{n+1}.$$

The relation submodules M_n of a group $G = F/N$ allow us to relate automorphisms of G to non-singular integral linear substitutions if N is in the commutator subgroup of F. Specifically, we have

THEOREM 5.15. *Let $F = F(r)$ be a free group of finite rank r, let N be a normal divisor of F such that $N \subset F_2$, and let $G = F/N$. Then any automorphism α of G induces in M_1 an automorphism $S(\alpha)$ which can be written as a linear substitution, with integral coefficients and determinant ± 1, of the basis elements ξ_ρ of M_1. This substitution $S(\alpha)$ is further restricted by the condition that it maps each M_n, $n = 2, 3, \ldots$, onto itself.*

PROOF. Let a_ρ, $\rho = 1, \ldots, r$, be the generators of F. Let $b_\rho = \alpha(a_\rho)$ be their maps under α. Since $N \subset F_2$, we must have

$$b_\sigma \equiv \prod_{\rho=1}^{r} a_\rho^{s_{\sigma,\rho}} \bmod F_2, \qquad (\sigma = 1, \ldots, r)$$

where the $s_{\sigma,\rho}$ are integers. The determinant of the r-by-r matrix $S(\alpha)$ with elements $s_{\sigma,\rho}$ must be ± 1 since α [and, therefore, $S(\alpha)$] must have an inverse. We have, from $\delta(a_\rho) = x_\rho$,

$$y_\rho = \delta(b_\sigma) = \sum_{\rho=1}^{r} s_{\sigma,\rho} x_\rho.$$

Since each of the y_ρ is homogeneous of degree 1, and since the inverse of a linear substitution with integral coefficients and determinant ± 1 is another such, it follows that each of these substitution maps $A_0(Z, r)$ in a one-one manner onto itself, preserving the degree of homogeneous elements. Therefore, if $W_n(a_\rho)$ is a word in the a_ρ and $\delta(W_n)$ is of degree n, then $\delta_n(W_n(b_\rho))$ arises from $\delta(W_n(a_\rho))$ by substituting the y_ρ for the x_ρ, since doing this does not increase the degree of $\delta(W_n)$. Now if $W_n(a_\rho) \in N$, we also have $W_n(b_\rho) \in N$ since α is an automorphism of $G = F/N$. Therefore μM_n is mapped into, and since α is invertible, onto μM_n by the mapping

$$x_\rho \to \sum_{\rho=1}^{r} s_{\sigma,\rho} x_\rho.$$

If we now go, by means of μ^{-1}, from the x_ρ to the free generators ξ_ρ of $\Lambda_0(Z, r)$, we obtain Theorem 5.15. ◄

As a further illustration of the use of the Lie algebra to restrict automorphisms of a group, we prove

COROLLARY 5.15. *Let Φ be the fundamental group of a closed two dimensional manifold of genus g. [Φ may be defined by $2g$ generators a_ρ, $\rho = 1, \ldots, 2g$, and the defining relator*

$$(a_1, a_{g+1})(a_2, a_{g+2}) \ldots (a_g, a_{2g}).]$$

Let α be an automorphism of Φ and assume that

$$\alpha(a_\sigma) \equiv \prod_{\rho=1}^{2g} a_\rho^{s_{\sigma,\rho}} \bmod \Phi_2, \qquad (\sigma = 1, \ldots, 2g).$$

Then the matrix $S = (s_{\sigma,\rho})$ must be symplectic, that is we must have

$$S'JS = \pm J,$$

where S' is the transpose of S and where J is the $2g$-by-$2g$ matrix

$$\begin{pmatrix} 0 & I \\ -I & 0 \end{pmatrix}$$

in which 0 stands for the g-by-g null matrix and I stands for the g-by-g unit matrix.

Proof. The relation module M_2 of Φ consists of the multiples of

$$\sum_{\sigma=1}^{g} \xi_\sigma \circ \xi_{\sigma+g}$$

If it is to be mapped onto itself by S, we must have,

$$\sum_{\sigma=1}^{g}\left[\left(\sum_{\rho=1}^{2g} s_{\sigma,\rho}\xi_\rho\right) \circ \left(\sum_{\rho=1}^{2g} s_{\sigma+g,\rho}\xi_\rho\right)\right] = \pm \sum_{\sigma=1}^{g} \xi_\sigma \circ \xi_{\sigma+g}$$

By collecting and equating the coefficients of $\xi_\rho \circ \xi_\tau$, $\rho < \tau$, on both sides (observing that $\xi_\tau \circ \xi_\rho = -\xi_\rho \circ \xi_\tau$), we arrive at the desired result. ◀

The fact that all symplectic substitutions S can be induced by automorphisms of Φ has to be proved by a direct construction of such automorphisms. See Theorem N13 of Section 3.7 and the references given at the end of its proof.

Theorem 5.15 shows that the automorphisms induced in G/G_2 by an automorphism of G will, in general, be restricted by any relation for the generators of G, no matter in which group of the lower central series the relators for G lie. Also, Theorem 5.15 suggests the problem of finding the submodules M_n of Λ_n which are left invariant under all unimodular substitutions S of the generators ξ_ρ of $\Lambda_0(Z, r)$. If such a module M_n is of rank one, its basis element J_n is called a *Lie invariant* of degree n *of the unimodular group*. Such Lie invariants can exist only if n is a proper multiple of r or $n = 2$. They do not exist if $r = 2$, $n = 4$ or if $r = 3$ and $n = 6$. For $r = 2$, $n = 2, 6, 8, \ldots$ they have been studied by Wever, 1949, who also found (Wever, 1947) a Lie invariant for $r = 3$, $n = 9$. Burrow, 1958, showed that they exist in all cases not excluded above. It can be shown that if J_n is a Lie invariant in the r indeterminates ξ_ρ and of degree n, then the corresponding product $W_n(a_\rho)$ of commutators of weight n in the r group elements a_ρ has the following property: Whenever we substitute for the a_ρ elements g_ρ of a group G which can be generated by fewer then r elements, then $W_n(g_\rho) \in G_{n+1}$.

References and Remarks. It can be shown that the union of the relation modules M_n of a group $G = F/N$ is an ideal M in $\Lambda_0(Z, r)$, and the quotient Lie ring

$$\Lambda_0/M$$

is independent of the presentation of G in the form F/N (Magnus, 1939). The connection between nilpotent groups and Lie rings has been developed systematically by Lazard, 1954; the results of Section 5.8 are very special cases of theorems that can be found in Lazard's work, which also, in part, summarizes the results of earlier papers on automorphisms, as for instance the work of Kaloujnine, 1949, 1950, 1953.

Problems for Section 5.8

1. Show that the union M of the relation modules M_n of $G = F/N$ is an ideal in $\Lambda_0(Z, r)$. In other words, if $\varphi \in M$ and $\omega \in \Lambda_0$, then $\omega \circ \varphi \in M$. [*Hint:* Use the facts that, for any $\varphi \in M$, the homogeneous components of φ are also in M and that, for any $W \in F$ and $W^* \in N$, the commutator (W, W^*) is in N.]

2. Let $G = F/N$, where the free generators of F are a_ρ, $\delta(a_\rho) = x_\rho$, and $\mu^{-1}(x_\rho) = \xi_\rho \in \Lambda_0(Z, r)$. Show that if $a_\rho{}^m \in G_2$, and if ω is any homogeneous element of degree n in Λ_0 which is a linear combination of monomials involving the factor ξ_ρ, then $m\omega \in M_n$, the nth relation module of F/N. Use this result to prove that if all but one of the generators of F are of finite order modulo N, then G_n/G_{n+1} is finite for all $n > 1$.

3. Let F be a free group, and let A be the group of those automorphisms of F/F_n which induce the identical automorphism in F/F_2. Show that if a_ρ, $\rho = 1, \ldots, r$, is a set of generators for F/F_n, and if the c_ρ form any set of elements which are products of commutators then there exists an automorphism in A which maps a_ρ onto $a_\rho c_\rho$, $\rho = 1, \ldots, r$. Prove (in the case $r = 2$) that not all of these automorphisms can be induced by automorphisms of the free group F itself if n is sufficiently large. [*Hint:* Use Theorem 3.9 and Lemma 5.9.]

5.9. Identities

Our construction of a basis for Lie elements in an associative algebra $A_0(R, r)$ and in a free Lie algebra $\Lambda_0(R, r)$ permit us to decide whether a given element of A_0 is a Lie element or whether two elements in Λ_0 are equal. However, our constructions are of a recursive nature and, in practice, rather clumsy ones. It is helpful for many purposes to have more explicit formulas, and we shall derive first some useful identities in Lie algebras.

Let Λ be a Lie algebra over a ring of coefficients which has a unit element. Let η_ν, $\nu = 1, \ldots, n$, be any elements of Λ. By definition, we have the "anti-commutative" law

$$\eta_1 \circ \eta_2 + \eta_2 \circ \eta_1 = 0 \tag{1}$$

and the Jacobi identity:

$$(\eta_1 \circ \eta_2) \circ \eta_3 + (\eta_3 \circ \eta_1) \circ \eta_2 + (\eta_2 \circ \eta_3) \circ \eta_1 = 0. \tag{2}$$

The left-hand sides of these identities are of a type which may be described as follows: We take a product of several factors, permute them in various ways and add the resulting products. In order to obtain a convenient description for this procedure in a more general situation, we introduce the following notations:

Let Σ_n be the group of permutations of n symbols $1, 2, \ldots, n$. We introduce the *group ring* $J\Sigma_n$ as the associative algebra with basis elements σ_k, where $k = 1, \ldots, n!$ and where σ_k runs through the different elements of Σ_n. As coefficients of $J\Sigma_n$ we choose the integers. Multiplication of two elements

$$\sum_{k=1}^{n!} i_k\sigma_k, \quad \sum_{l=1}^{n!} j_l\sigma_l \qquad (i_k, j_l = 0, \pm 1, \pm 2, \ldots)$$

of $J\Sigma_n$ is, of course, defined by

$$\left(\sum_{k=1}^{n!} i_k\sigma_k\right)\left(\sum_{l=1}^{n!} j_l\sigma_l\right) = \sum_{k,l} i_k j_l \sigma_{k(l)}$$

where $\sigma_{k(l)}$ is the element $\sigma_k\sigma_l$ of Σ_n.

Now let $\Lambda^{(n)}$ be the subset of elements of Λ which can be written as a sum of integral multiples of products involving n factors (in any arrangement) $\eta_1, \ldots, \eta_n$. A multiple $m\zeta$ of an element $\zeta \in \Lambda$ is, of course, the same as $\zeta + \zeta + \ldots + \zeta$ (m times), if m is positive.

We can use the elements of $J\Sigma_n$ as operators acting on the elements of $\Lambda^{(n)}$. First, let ζ be any product of $\eta_1, \ldots \eta_n$, where ζ is defined by using any bracket arrangement of weight n applied to any order of the η_ν. For any element $\sigma \in \Sigma_n$, we define $\sigma\zeta$ as the product arising from ζ by an application of σ to the subscripts of the η_ν. Next, let τ_1, τ_2 be any elements of $J\Sigma_n$ and ζ_1, ζ_2 be any elements of $\Lambda^{(n)}$. We complete the definition of the action of $J\Sigma_n$ on $\Lambda^{(n)}$ by requiring that, for any integers $i_1, i_2, \ldots, i_k$, and $\tau_1, \tau_2, \ldots, \tau_k$ in Σ_n, and $\zeta_1, \zeta_2, \ldots, \zeta_k$ in $\Lambda^{(n)}$,

$$(i_1\tau_1 + i_2\tau_2 + \ldots + i_k\tau_k)\zeta_1 = i_1\tau_1\zeta_1 + i_2\tau_2\zeta_1 + \ldots + i_k\tau_k\zeta_1 \tag{3}$$

$$\tau_1(i_1\zeta_1 + i_2\zeta_2 + \ldots + i_k\zeta_k) = i_1\tau_1\zeta_1 + i_2\tau_1\zeta_2 + \ldots + i_k\tau_1\zeta_k. \tag{4}$$

Since we are writing the operators on the left hand side, $(\tau_1 \cdot \tau_2)\zeta$ means $\tau_1 \cdot (\tau_2 \zeta)$. (Moreover, the product of two permutations must be computed from the right, e.g., (12)(123) = (23) and not (13) as it would be if we computed from the left.)

We shall need some special elements of $J\Sigma_n$ which we may define as follows. Let k, l be integers such that $1 \leq k < l \leq n$, and let

$$c_{k,l} = \begin{pmatrix} k, & k+1, & \ldots, & l-1, & l \\ l, & k & , \ldots, & l-2, & l-1 \end{pmatrix} \tag{5}$$

by the cyclic permutation of the symbols $k, k+1, \ldots, l$ described by (5) which leaves the remaining symbols of the set $1, 2, \ldots, n$ fixed. In terms of the $c_{k,l}$ we define the elements $\Omega_{k,m}$ of $J\Sigma_n$ for $k = 1, 2, \ldots, m-1$ and $k < m \leq n$ by

$$\Omega_{k,m} = (1 - c_{k,k+1})(1 - c_{k,k+2}) \ldots (1 - c_{k,m-1})(1 - c_{k,m}), \tag{6}$$

where 1 denotes the identical permutation. We shall also write Ω_n for $\Omega_{1,n}$.

Finally, we shall use the notation

$$(\eta_1, \eta_2, \eta_3, \ldots, \eta_n) \tag{7}$$

for the "*simple products*" [Corollary 5.12, (5.7.10)]

$$(\ldots((\eta_1 \circ \eta_2) \circ \eta_3) \ldots) \circ \eta_n \tag{8}$$

which arise from the bracket arrangement with all open parentheses to the left of all elements occurring. [Compare this with (5.3.2), (5.3.3).]

Now we shall prove

THEOREM 5.16. (*Wever's Formulae*) *For* $1 < k < n$,

$$\Omega_{k,n}(\eta_1, \ldots, \eta_n) = (\eta_1, \ldots, \eta_{k-1}) \circ (\eta_k, \ldots, \eta_n). \tag{9}$$

Moreover, for $k = 1$, *we have* $(\Omega_n = \Omega_{1,n})$

$$\Omega_n(\eta_1, \ldots, \eta_n) = n(\eta_1, \ldots, \eta_n). \tag{10}$$

PROOF. We may note that for $n = 2$ and $n = 3$ the expression (10) yields respectively

$$\eta_1 \circ \eta_2 - \eta_2 \circ \eta_1 = 2\eta_1 \circ \eta_2,$$

and

$$(\eta_1 \circ \eta_2) \circ \eta_3 - (\eta_3 \circ \eta_1) \circ \eta_2 - (\eta_2 \circ \eta_1) \circ \eta_3 + (\eta_3 \circ \eta_2) \circ \eta_1 = 3(\eta_1 \circ \eta_2) \circ \eta_3$$

These relations are easily seen to be equivalent to the relations (1) and (2). Hence, the Wever formulae are indeed a generalization of (1) and (2).

To prove (9), we shall use induction on $m - k$, where $1 \leq k < m \leq n$, to show that if the η_i and the η_j are such that i and j are outside the interval from k to m inclusive, i.e., η_i and n_j are left fixed by $\Omega_{k,m}$, and if there are some η_i (though η_j may be empty), then

$$\Omega_{k,m}(\ldots, \eta_i, \eta_k, \ldots, \eta_m, \eta_j, \ldots) = ((\ldots, \eta_i) \circ (\eta_k, \ldots, \eta_m), \eta_j, \ldots).$$

The case $m - k = 1$ holds, for $\Omega_{k,m} = \Omega_{m-1,m} = 1 - C_{m-1,m}$ and

$$\begin{aligned}
&(1 - C_{m-1,m})(\ldots, \eta_i, \eta_{m-1}, \eta_m, \eta_j, \ldots) \\
&= (\ldots, \eta_i, \eta_{m-1}, \eta_m, \eta_j, \ldots) - (\ldots, \eta_i, \eta_m, \eta_{m-1}, \eta_j, \ldots) \\
&= (((\ldots, \eta_i) \circ \eta_{m-1}) \circ \eta_m, \eta_j, \ldots) - (((\ldots, \eta_i) \circ \eta_m) \circ \eta_{m-1}, \eta_j, \ldots) \\
&= ((\ldots, \eta_i) \circ (\eta_{m-1} \circ \eta_m), \eta_j, \ldots)
\end{aligned}$$

by the definition of the simple Lie product of n factors, and by (1) and (2), and the distributive law for Lie multiplication.

Suppose that we have proved

$$\Omega_{l,p}(\ldots, \eta_i, \eta_l, \ldots, \eta_p, \eta_j, \ldots) = ((\ldots, \eta_i) \circ (\eta_l, \ldots, \eta_p), \eta_j, \ldots)$$

if $p - l < m - k$; consider $\Omega_{k,m}$. Since

$$\Omega_{k,m} = \Omega_{k,m-1}(1 - C_{k,m}),$$

we have

$$\begin{aligned}
&\Omega_{k,m}(\ldots, \eta_i, \eta_k, \ldots, \eta_m, \eta_j, \ldots) \\
&\quad = \Omega_{k,m-1}(\ldots, \eta_i, \eta_k, \ldots, \eta_m, \eta_j, \ldots) \\
&\quad\quad - \Omega_{k,m-1}(\ldots, \eta_i, \eta_m, \eta_k, \ldots, \eta_{m-1}, \eta_j, \ldots) \\
&\quad = (((\ldots, \eta_i) \circ (\eta_k, \ldots, \eta_{m-1})) \circ \eta_m, \eta_j, \ldots) \\
&\quad\quad - (((\ldots, \eta_i) \circ \eta_m) \circ (\eta_k, \ldots, \eta_{m-1}), \eta_j, \ldots) \\
&\quad = ((\ldots, \eta_i) \circ ((\eta_k, \ldots, \eta_{m-1}) \circ \eta_m), \eta_j, \ldots) \\
&\quad = ((\ldots, \eta_i) \circ (\eta_k, \ldots, \eta_{m-1}, \eta_m), \eta_j, \ldots),
\end{aligned}$$

by inductive hypothesis, by definition of the simple Lie product of n factors, by (1) and (2), and by the distributive law for Lie multiplication. This completes our induction.

Applying the above to the left-hand side of (9) (when $k > 1$, there are some η_i preceding η_k though the η_j are empty), we obtain the desired result.

To prove (10), we show by induction on m, where $1 < m \leq n$ that if j is outside the interval from 1 to m, then ($\Omega_m = \Omega_{1,m}$)

$$\Omega_m(\eta_1, \ldots, \eta_m, \eta_j, \ldots) = m(\eta_1, \ldots, \eta_m, \eta_j, \ldots)$$

(where the η_j are possibly empty).

The case $m = 2$ follows easily since $\Omega_2 = (1 - C_{1,2})$ and

$$\begin{aligned}(1 - C_{1,2})(\eta_1, \eta_2, \eta_j, \ldots) &= (\eta_1, \eta_2, \eta_j, \ldots) - (\eta_2, \eta_1, \eta_j, \ldots)\\ &= (\eta_1 \circ \eta_2, \eta_j, \ldots) - (\eta_2 \circ \eta_1, \eta_j, \ldots)\\ &= 2(\eta_1, \eta_2, \eta_j, \ldots),\end{aligned}$$

by the anti-commutative law and the distributive law for Lie multiplication.

Suppose that we have proved

$$\Omega_p(\eta_1, \ldots, \eta_p, \eta_j, \ldots) = p(\eta_1, \ldots, \eta_p, \eta_j, \ldots)$$

for all $p < m$; consider Ω_m, $m > 2$.

Since $\Omega_m = \Omega_{m-1}(1 - C_{1,m})$,

$$\begin{aligned}&\Omega_m(\eta_1, \ldots, \eta_m, \eta_j, \ldots)\\ &= \Omega_{m-1}(\eta_1, \ldots, \eta_{m-1}, \eta_m, \eta_j, \ldots) - \Omega_{m-1}(\eta_m, \eta_1, \ldots, \eta_{m-1}, \eta_j, \ldots)\\ &= (m-1)(\eta_1, \ldots, \eta_{m-1}, \eta_m, \eta_j, \ldots) - (\eta_m \circ (\eta_1, \ldots, \eta_{m-1}), \eta_j, \ldots),\end{aligned}$$

where in the first term we used the inductive hypothesis, and in the second term we used the first part (with $\eta_i = \eta_m$). Applying the anti-commutative law to this last term we obtain

$$\Omega_m(\eta_1, \ldots, \eta_m, \eta_j, \ldots) = m(\eta_1, \ldots, \eta_m, \eta_j, \ldots),$$

and so complete our induction.

This proves Theorem 5.16. ◀

As an immediate consequence of Theorem 5.16 we have

COROLLARY 5.16. *In a free Lie ring on free generators ξ_ρ, every element of degree n is a linear combination of simple Lie products in which each factor is a generator. The product of two simple products of degrees k and l respectively can be written as a linear combination of at most 2^l or 2^k simple products.*

PROOF. The second statement of Corollary 5.16 is part of the statement of Theorem 5.16 since $\Omega_{k,n}$ is a linear combination of at most 2^{n-k} permutations. The first statement of Corollary 5.16 follows from the second by induction. (Of course, we have proven the first statement of Corollary 5.16 before; see Statement (v) of Corollary 5.12, Section 5.7.)

We shall apply Theorem 5.16 to derive two tests which will be useful in determining whether a given element in $A_0(Z, r)$ is a Lie element. To do this, we first introduce

Definition 5.8. Let $A_0(R, r)$ be a freely generated associative algebra with free generators x_ρ, $\rho = 1, \ldots, r$. The *Curly Bracket Operator* $\{\ \}$ in

A_0 is defined as follows: Let y_ν, $\nu = 1, \ldots, n$ be any n elements out of the set of the x_ρ (repetitions allowed), and let

(11) $$z = \alpha y_1 y_2 y_3 \ldots y_n, \qquad \alpha \in R.$$

Then we define $\{z\} = \alpha y_1$ if $n = 1$ and, for $n > 1$,

(12) $$\{z\} = \alpha[[\ldots[[y_1, y_2], y_3]\ldots], y_n]$$

We also define $\{\alpha\} = 0$, and require that $\{\ \}$ be additive, i.e., if $z_1, z_2, \ldots, z_k$ are as in (11), then

$$\{z_1 + z_2 + \ldots + z_k\} = \{z_1\} + \{z_2\} + \ldots + \{z_k\}.$$

Clearly, $\{\ \}$ maps $A_0(R, r)$ onto its Lie elements. We have

THEOREM 5.17. (*Dynkin-Specht-Wever Theorem*) *Let R be an integral domain with an identity and of characteristic zero. Let u_n be a homogeneous element of degree n in $A_0(R, r)$. Then u_n is a Lie element if and only if*

(13) $$\{u_n\} = nu_n.$$

PROOF. If (13) holds, nu_n is a Lie element. By embedding the integral domain R into its field of quotients $\bar{R}$, we see that u_n itself is a Lie element in $A_0(\bar{R}, r)$; but since the basic Lie elements in $A_0(\bar{R}, r)$ form part of a basis for $A_0(R, r)$ (Theorem 5.8 of Section 5.6), it follows that u_n must be a Lie element in $A_0(R, r)$. It remains to show that (13) is true for all homogeneous Lie elements of $A_0(R, r)$. Letting the operator Ω_n of Theorem 5.16 act on the subscripts of the y_ν, it is easily shown by induction on n that

(14) $$\Omega_n(y_1 y_2 \ldots y_n) = \{y_1 y_2 \ldots y_n\}$$

Since the bracket product [ ] in $A_0(R, r)$ satisfies the same laws as multiplication in a Lie algbera, it follows from (10) and (12) that

(15) $$\Omega_n\{y_1 y_2 \ldots y_n\} = n\{y_1 y_2 \ldots y_n\}$$

By its definition, the curly bracket operator commutes with all permutations of the y_ν. Therefore,

$$\Omega_n\{y_1 y_2 \ldots y_n\} = \{\Omega_n(y_1 y_2 \ldots y_n)\},$$

and using (14) and (15) this equation may be written

$$n\{y_1 y_2 \ldots y_n\} = \{\{y_1 y_2 \ldots y_n\}\}.$$

or equivalently as

(16) $$\{[y_1, y_2, \ldots, y_n]\} = n[y_1, y_2, \ldots, y_n].$$

Since Corollary 5.16 shows that any homogeneous Lie element may be written as a linear combination of terms of type (12), the equation in (16) proves (13) and Theorem 5.17 completely. ◄

We shall derive a second characterization of Lie elements from Theorem 5.17 which will be used in Section 5.10. Again we need a new concept. Let $A_0(R, r)$ be a freely generated associative algebra on free generators x_ρ, $\rho = 1, \ldots, r$, and $A_0'(R, r)$ be an isomorphic replica of $A_0(R, r)$ on free generators x_ρ'. We define the *tensor product*

$$A_0 \otimes A_0'$$

as an associative algebra over R on generators x_ρ, x_ρ' in which

$$x_\rho x_\sigma' = x_\sigma' x_\rho, \qquad \sigma, \rho = 1, \ldots, r$$

and which has as basis elements exactly the products of the basis elements of A_0 and A_0' namely the unit element (which may be considered as the common unit element of A_0 and A_0') and the products

$$y_1 y_2 \cdots y_n y_1' y_2' \cdots y_m' \tag{17}$$

where the $y_\nu (\nu = 1, \ldots, n)$ belong to the set of the x_ρ and the y_μ' ($\mu = 1, \ldots, m$) belong to the set of the x_ρ'. If $n = m = 0$, (17) means the unit element. (For more systematic information about tensor products see, for example, Chevalley, 1956, or Jacobson, 1962.) Now we can state

THEOREM 5.18. (*Friedrichs' Theorem*) *Let R be a field of characteristic zero. An element $P(x_\rho)$ in $A_0(R, r)$ is a Lie element if and only if as elements of $A_0 \otimes A_0'$ we have*

$$P(x_\rho + x_\rho') = P(x_\rho) + P(x_\rho'). \tag{18}$$

PROOF. Clearly a Lie element has a zero constant term. Moreover, since R has characteristic zero, (18) implies that $P(x_\rho)$ has a zero constant term.

If $P(x_\rho)$ is a Lie element, then $P(x_\rho)$ is a linear combination of the simple Lie products

$$[x_{\rho_1}, x_{\rho_2}, \ldots, x_{\rho_k}].$$

Hence, it suffices to show that (18) holds for such simple Lie products. This is easily done by induction on k, since x_{ρ_i} commutes with x_{ρ_j}'.

Conversely, suppose $P(x_\rho)$ satisfies (18). To show that $P(x_\rho)$ is a Lie element, it is convenient to introduce an algebra of operators very similar to $A_0 \otimes A_0'$ [indeed, the algebra to be introduced is isomorphic to $A_0 \otimes A_0'$ (see Problem 6)]. Let $A_0(R, r + 1)$ be the associative algebra freely generated by t and $x_1, x_2, \ldots, x_r$, and let T be the submodule of $A_0(R, r+1)$ of polynomials of degree one in t. If X_ρ is the operator which multiplies an element of T by x_ρ on the right, and X_ρ' is the operator that multiplies an element of T on the left by x_ρ, then clearly

$$P(X_\rho)u = uP^*(x_\rho), \qquad P(X_\rho')u = P(x_\rho)u$$

where u is an element of T, and P^* is the polynomial in x_ρ obtained by replacing each monomial in P with its mirror image [e.g., $(x_1x_2x_1^3)^* = x_1^3x_2x_1$]; we have implicitly assumed that the sum of two operators and product of two operators have been defined in their customary manners, i.e.,

$$(P + Q)u = Pu + Qu, \qquad (P \cdot Q)u = P(Q(u)), \qquad (\alpha P)u = \alpha(P(u)),$$

where P, Q are operators and α is in R, and u is in T.

Returning now to $P(x_\rho)$ which satisfies (18), since if $P(x_\rho)$ is homogeneous of degree d, then $P(x_\rho + x_\rho')$, $P(x_\rho')$ are also, we may assume that $P(x_\rho)$ is homogeneous of degree d and satisfies (18). The operators X_ρ and X_σ' commute, and hence from (18) it follows that

$$(19) \quad P(X_\rho - X_\rho') = P(X_\rho + (-X_\rho')) = P(X_\rho) + (-1)^d P(X_\rho').$$

Applying the operator in (19) to t, it is easily shown that we have

$$P(X_\rho - X_\rho')t = \{tP^*(x_\rho)\},$$

where $tP^*(x_\rho)$ is regarded as a polynomial in t and $x_1, x_2, \ldots, x_r$ (see Problem 5). Hence, from (19) it follows that

$$(20) \qquad \{tP^*(x_\rho)\} = tP^*(x_\rho) + (-1)^d P(x_\rho)t.$$

Applying the curly bracket to both sides of (20), and using Theorem 5.17, we obtain

$$\{\{tP^*(x_\rho)\}\} = (d + 1)\{tP^*(x_\rho)\} = \{tP^*(x_\rho)\} + (-1)^d\{P(x_\rho)t\}$$

and therefore,

$$(21) \qquad \{tP^*(x_\rho)\} = (-1)^d d^{-1}\{P(x_\rho)t\}.$$

Expressing both sides of (21) as polynomials in t and $x_1, x_2, \ldots, x_r$, and equating the monomials which have t on the extreme left, we obtain

$$(22) \qquad P^*(x_\rho) = (-1)^{d+1}d^{-1}\{P(x_\rho)\}$$

(see Problem 5). Applying the "star" to both sides of (22), we obtain

$$(23) \qquad P(x_\rho) = (-1)^{d+1}d^{-1}\{P(x_\rho)\}^*.$$

Since the right-hand side of (23) is a Lie element (see Problem 5), we have that $P(x_\rho)$ is a Lie element. ◀

References and Remarks. Theorem 5.16 was first proved by Wever, 1947. We have used his proof in the text. Theorem 5.17 was found independently by Dynkin, 1947, Specht, 1948, and Wever, 1947. Theorem 5.18 was formulated by Freidrichs, 1952. The proof given here is due to Lyndon, 1955a. Others proof were given almost simultaneously by P. M. Cohn, 1954a, Finkelstein, 1955, Magnus, 1954, Zassenhaus, 1955.

Problems for Section 5.9

1. Show by induction on $m - k$ that if $1 \leq k < m \leq n$, and the η_j are such that j is outside the interval from k to m (possibly the j are empty) inclusive, then

$$\Omega_{k,m}(\eta_k, \ldots, \eta_m, \eta_j, \ldots) = (m - k + 1)(\eta_k, \ldots, \eta_m, \eta_j, \ldots).$$

2. Let $J\Sigma_n$ by the group ring of permutations of n objects with integer coefficients, and let $A_0(R, n)$ be the associative algebra freely generated by $x_1, \ldots, x_n$. Let an element of $J\Sigma_n$ act naturally on a homogeneous element of degree n in A_0.

(a) Show that if Γ, Γ' are in $J\Sigma_n$ and

$$\Gamma(x_1x_2 \ldots x_n) = \Gamma'(x_1x_2 \ldots x_n)$$

then $\Gamma = \Gamma'$ in $J\Sigma_n$.

(b) Show by induction on n that

$$\Omega_n(x_1x_2 \ldots x_n) = \{x_1x_2 \ldots x_n\}.$$

(c) Show that

$$\Omega_n^2(x_1x_2 \ldots x_n) = n\Omega_n(x_1x_2 \ldots x_n),$$

and hence, that

$$\Omega_n^2 = n\Omega_n$$

in $J\Sigma_n$.

3. Let $A_0(R, n)$ be as in Problem 2, and suppose that R has characteristic zero. Let $P(x_v)$ be a homogeneous element of degree n in A_0, such that each monomial in $P(x_v)$ has syllable length n.

(a) Show that if $\Gamma P(x_v) = P(x_v)$ for each Γ in Σ_n then

$$P(x_v) = \alpha(\Gamma_1(x_1x_2 \ldots x_n) + \Gamma_2(x_1x_2 \ldots x_n) + \ldots + \Gamma_{n!}(x_1x_2 \ldots x_n)),$$

where $\Gamma_1, \Gamma_2, \ldots, \Gamma_{n!}$ are the elements of Σ_n and α is in R.

(b) Show that if $P(x_v)$ is a Lie element of degree $n > 1$, and $\Gamma P(x_v) = P(x_v)$ for all Γ in Σ_n, then $P(x_v) = 0$.

(c) Show that if $\Lambda_0(R, n)$ is the free Lie algebra freely generated by $\xi_1, \xi_2, \ldots, \xi_n$, then if the simple Lie product

$$(\ldots((\xi_1 \circ \xi_2) \circ \xi_3) \ldots) \circ \xi_n = (\xi_1, \xi_2, \ldots, \xi_n)$$

is represented by a polynomial in A_0 then each monomial occurring has degree n and syllable length n.

(d) If $\Lambda_0(R, n)$ is as in (c), and $n > 1$, show that the sum of all possible permutations of the simple Lie product $(\xi_1, \xi_2, \ldots, \xi_n)$ is zero, i.e.,

$$\Gamma_1(\xi_1, \xi_2, \ldots, \xi_n) + \Gamma_2(\xi_1, \xi_2, \ldots, \xi_n) + \ldots + \Gamma_{n!}(\xi_1, \xi_2, \ldots, \xi_n) = 0,$$

where $\Gamma_1, \Gamma_2, \ldots, \Gamma_{n!}$ are the elements of Σ_n.

[*Hint:* For (b), map A_0 into $R[x]$, the polynomial ring in one variable x, by mapping each x_ν into x. Then by (a), $P(x_\nu)$ goes into $n!\, \alpha x^n$; since each Lie element of degree >1 goes into zero, $\alpha = 0$ and so $P(x_\nu) = 0$. For (c), use induction on n. For (d), use (b) and (c).]

4. Show that if $\Lambda_0(R, n)$ is as in Problem 3(c), then

$$\Delta_1(\xi_1, \xi_2, \ldots, \xi_n) + \Delta_2(\xi_1, \xi_2, \ldots, \xi_n) + \ldots + \Delta_n(\xi_1, \xi_2, \ldots, \xi_n),$$

where $\Delta_1, \Delta_2, \ldots, \Delta_n$ are the powers of $C_{1,n}$, is zero for $n = 2$ and $n = 3$, but is not zero for $n = 4$. [*Hint:* For $n = 4$, represent $\Lambda_0(R, 4)$ in $A_0(R, 4)$ and show that the given sum has a coefficient of -1 on the monomial $x_3x_1x_2x_4$.]

5. Let X_ρ, X_ρ', P^* be as in the proof of Theorem 5.18.

(a) Show by induction on the degree of a monomial Q in the x_ρ that

$$Q(X_\rho - X_\rho')t = \{tQ^*(x_\rho)\},$$

and hence, for any polynomial Q the result also holds.

(b) Show by induction on the degree of a monomial Q in the x_ρ, that the sum of the monomials in

$$\{tQ(x_\rho)\}, \quad \{Q(x_\rho)t\},$$

which have t on the extreme left is

$$Q(x_\rho), \quad -\{Q(x_\rho)\},$$

respectively, and hence, the result is true for any polynomial Q.

(c) Show that
$$(P^*)^* = P, \qquad (P + Q)^* = P^* + Q^*,$$
$$(\alpha P)^* = \alpha P^*, \qquad (P \cdot Q)^* = Q^* \cdot P^*,$$
$$[P, Q]^* = [Q^*, P^*], \qquad \{P\}^* = \langle P^* \rangle,$$

where $\langle x_1 \rangle = x_1$ and $\langle x_1x_2 \ldots x_n \rangle = [x_1, \langle x_2 \ldots x_n \rangle]$, i.e., $\langle\ \rangle$ "brackets from the right."

[*Hint:* For (c), show the results for monomials.]

6. Let A_0 be the associative algebra freely generated by $x_1, x_2, \ldots, x_r$ and similarly for A_0'. Let X_ρ, X_ρ' be as in the proof of Theorem 5.18, and let B be the algebra of operators generated by all the X_ρ and X_ρ' over R. Show that the algebra $A_0 \otimes A_0'$ is isomorphic to the algebra B under the mapping

$$x_\rho \to X_\rho, \quad x_\rho' \to X_\rho'.$$

[*Hint:* Since B is an associative algebra, and X_ρ commutes with X_σ', the given mapping is a homomorphism of $A_0 \otimes A_0'$ onto B. Moreover, suppose the element

$$\alpha_1P_1(x_\rho)Q_1(x_\rho') + \ldots + \alpha_kP_k(x_\rho)Q_k(x_\rho')$$

goes into zero in B, where $P_i(x_\rho)$, $Q_i(x_\rho')$ are monomials and

$$P_i(x_\rho) \neq P_j(x_\rho) \qquad \text{or} \qquad Q_i(x_\rho') \neq Q_j(x_\rho') \qquad \text{if } i \neq j.$$

Then

$$\alpha_1Q_1(x_\rho)tP_1^*(x_\rho) + \ldots + \alpha_kQ_k(x_\rho)tP_k^*(x_\rho)$$

must be zero. Since

$$Q_i(x_\rho)tP_i^*(x_\rho), \quad Q_j(x_\rho)tP_j^*(x_\rho),$$

are different monomials in t and $x_1, x_2, \ldots, x_r$, each α_k is zero. Thus the mapping is one-one.]

7. Let Q be the field of rational numbers, and let $A(Q, 2)$ be the algebra of formal power series freely generated by x and y over Q. Define e^x in the natural manner by

$$e^x = \sum_{n=0}^{\infty} x^n/n!$$

Prove that

$$u = e^{-x}ye^x$$

is a Lie element. Show that u can be expressed as a Lie element explicitly by

$$u = y + \sum_{n=1}^{\infty} \{yx^n\}/n!.$$

[*Hint:* Show that if x and x' commute, then $e^{-x}x'e^x = x'$ and $e^{x+x'} = e^x \cdot e^{x'}$; then use Theorem 5.18 to show that u is a Lie element. Show that the homogeneous component of degree $n + 1$, $n \geq 1$, in u is

$$u_{n+1} = \sum_{i+j=n} (-x)^i yx^j/i!\, j!,$$

which is a Lie element since u is. Hence, by (13) of Theorem 5.17,

$$\begin{aligned} u_{n+1} &= \frac{1}{n+1} \sum_{i+j=n} \{(-x)^i yx^j\}/i!\, j! \\ &= \frac{1}{n+1}\left(\frac{\{(-x)yx^{n-1}\}}{(n-1)!} + \frac{\{yx^n\}}{n!}\right) \\ &= \frac{1}{n+1}\left(\frac{\{yx^n\}}{(n-1)!} + \frac{\{yx^n\}}{n!}\right) \\ &= \{yx^n\}/n!.] \end{aligned}$$

8. Let F be a free group of rank r, and let A be the subgroup of F/F_{n+1}

$$A = F_n/F_{n+1},$$

and let B be the subgroup of A

$$B = (F_k, F_{n-k})/F_{n+1}.$$

Show that there is a homomorphism of A onto B such that each element of B is mapped into its $(n - k)$th power. [*Hint:* Represent the Abelian group A by the submodule Λ_n of $\Lambda_0(Z, r)$ of homogeneous elements of Λ_0 with degree n. If $J\Sigma_n$ is the group ring of permutations of n objects with integral coefficients, then each element of $J\Sigma_n$ induces a homomorphism of the homogeneous elements of $A_0(Z, r)$ of degree n by applying the permutations in Σ_n to the subscripts of a monomial $z_1 z_2 \ldots z_n$, where each z_ν is some x_{ρ_ν}. Such a homorphism of the homogeneous elements of degree n in A_0 maps Lie elements into Lie elements, and hence, induces a homomorphism in a natural manner on Λ_n, and hence, on A. In particular, $\Omega_{k+1,n}$ induces a homomorphism of A. From (9) in Theorem 5.16, it follows that $\Omega_{k+1,n}$ maps A onto B. Moreover, since $\Omega_{k+1,n}$ acts on $(\eta_{k+1}, \ldots, \eta_n)$ in the same way that $\Omega_{1,n-k}$ acts on $(\eta_1, \ldots, \eta_{n-k})$, $\Omega_{k+1,n}$ acts on the elements of Λ_n corresponding to B by multiplying them by $n - k$.]

5.10. The Baker-Hausdorff Formula

We shall prove, and comment on, a formula which has stimulated investigations related to the commutator calculus although its direct applicability has been surprisingly small, at least so far. We shall prove

THEOREM 5.19. *(Baker-Hausdorff Formula) Let Q be the field of rational numbers, and let $A(Q, 2)$ be freely generated by x, y. Let*

$$e^x = \sum_{n=0}^{\infty} x^n/n! \tag{1}$$

Then

$$e^x e^y = e^{\Phi(x,y)} \tag{2}$$

where $\Phi(x, y)$ is an (infinite) sum of homogeneous Lie elements. Φ can be computed explicitly by putting

$$v = e^x e^y - 1 = \sum_{\substack{n,m=0\\ n+m>0}}^{\infty} x^n y^m/(n!\, m!), \tag{3}$$

collecting the homogeneous terms u_n of

$$\sum_{k=1}^{\infty} (-1)^{k+1} \frac{v^k}{k} = \sum_{n=1}^{\infty} u_n, \tag{4}$$

and computing

$$\Phi(x, y) = \sum_{n=1}^{\infty} \{u_n\}/n. \tag{5}$$

(The first terms of $\Phi(x, y)$ are (up to and including terms of degree five):

$$x + y + \frac{1}{2}[x, y] + \frac{1}{12}[[x, y]x] + \frac{1}{12}[[x, y]y]$$
$$- \frac{1}{24}[[[x, y]y]x] - \frac{1}{720}[[[[x, y]y]y]y] - \frac{1}{720}[[[[y, x]x]x]x]$$
$$+ \frac{1}{180}[[\Delta_1, x]y] - \frac{1}{180}[[\Delta_1, y]y] - \frac{1}{120}[\Delta_1, \Delta] - \frac{1}{360}[\Delta_2, \Delta]$$

where

$$\Delta = [x, y], \Delta_1 = [\Delta, x], \Delta_2 = [\Delta, y].)$$

PROOF. It is clear that $\Phi = \log(1 + v)$ exists and that it is an element of $A(Q, 2)$, since both the exponential series and the power series for $\log(1 + v)$ have rational coefficients. If Φ is a Lie element, then Theorem 5.17 also shows that it must have the form (5). Therefore, it remains to prove only that Φ is a Lie element. For this purpose, we shall use Theorem 5.18, where we put $x_1 = x$, $x_2 = y$, $x_1' = x'$, $x_2' = y'$. We find

$$e^{x+x'}e^{y+y'} = e^x e^y e^{x'} e^{y'} = e^{\Phi(x,y)}e^{\Phi(x',y')},$$

since

$$e^{x+x'} = e^x e^{x'}, \qquad e^{y+y'} = e^y e^{y'}, \qquad e^{x'}e^y = e^y e^{x'}$$

because the ordinary multiplicative property holds for the exponential function if the indeterminates involved commute. For this reason, we also have

$$e^{x+x'}e^{y+y'} = e^{\Phi(x+x',y+y')} = e^{\Phi(x,y)}e^{\Phi(x',y')}$$
$$= e^{\Phi(x,y)+\Phi(x',y')}.$$

Putting

$$\Phi(x + x', y + y') = w \qquad \Phi(x, y) + \Phi(x', y') = z,$$

we must now show that

$$e^w = e^z$$

implies $w = z$. Then Theorem 5.18 shows that $\Phi(x, y)$ is indeed a Lie element, and our proof of Theorem 5.19 is complete. Since both w and z lack a constant term, the relation

$$w + w^2/2! + w^3/3! + \ldots = z + z^2/2! + z^3/3! + \ldots$$

shows at once that the terms of lowest degree in w and z agree. Denoting these by w_1, and putting $w = w_1 + w^*$, $z = w_1 + z^*$, we find

$$w^* + (w_1 + w^*)^2/2! + \ldots = z^* + (w_1 + z^*)^2/2! + \ldots$$

Since the powers of w_1 on both sides cancel, we find again that the terms of lowest degree in w^* and z^* agree, and, by continuing in this manner, we can prove that $w = z$. ◀

There are two aspects of Theorem 5.19 which may be mentioned. First, we may say that (2) defines an associative composition in a free Lie algebra. In fact, since $\Phi(x, y)$ is a Lie element, we can define its pre-image $\mu^{-1}\Phi = \Psi(\xi, \eta)$ in a Lie algebra Λ over R_0 containing ξ, η, if $\mu^{-1}(x) = \xi$, $\mu^{-1}(y) = \eta$ and μ is defined as in Lemma 5.5, Section 5.6. Then the composition $\xi * \eta$ defined by

$$(6) \quad \xi * \eta = \Psi(\xi, \eta)$$
$$= \xi + \eta + \tfrac{1}{2}\xi \circ \eta + \tfrac{1}{12}(\xi \circ \eta) \circ \xi + \tfrac{1}{12}(\xi \circ \eta) \circ \eta + \ldots$$

will be associative provided that the right hand side in (6) exists. This will always exist if Λ is a free Lie algebra $\Lambda(R_0, r)$ of formal power series. It will also exist if Λ arises from $\Lambda(R_0, r)$ by taking the quotient ring with respect to an ideal containing all terms of a degree $> n$, ($n = 1, 2, 3, \ldots$).

Second, we may use Theorem 5.19 to produce a representation of a free group that involves either Lie elements only or elements of a Lie algebra. We summarize the situation by stating

COROLLARY 5.19. *Let $F(r)$ be a free group on free generators a_ρ, $\rho = 1, \ldots, r$. Let Q be the field of rationals and let $A(Q, r)$ and $B(Q, r)$ be isomorphic, freely generated associative algebras, the sets of free generators being respectively, x_ρ and y_ρ. Let $\Lambda(Q, r)$ be the free Lie algebra on free generators η_ρ, and let U, V be any elements in $F(r)$. Then we obtain an isomorphic mapping E of A onto B by*

$$Ex_\rho = \sum_{n=1}^{\infty} y_\rho{}^n/n! = e^{y_\rho} - 1$$

and a one-one mapping of $F(r)$ into A or B by the mappings α, β generated by the equation

$$\alpha a_\rho = 1 + x_\rho, \qquad \beta a_\rho = e^{y_\rho} \qquad (\beta = E\alpha).$$

If

$$\alpha U = u, \qquad \beta U = e^{u^*}, u^* \in B$$

then u^* is a Lie element of $B(Q, r)$. The non-vanishing homogeneous terms of lowest degree in u^* arise from $\delta(U) \in A$ if we replace each x_ρ by y_ρ.

Let (as in Lemma 5.5) μ be the mapping of Λ into B defined by $\mu(\eta_\rho) = y_\rho$. Then the mapping γ defined by

$$\gamma U = \omega = \mu^{-1}u^*$$

is a one-one mapping of F into Λ. If $\gamma U = \omega$, $\gamma V = \varphi$, then

$$\gamma(UV) = \Psi(\omega, \varphi) = \omega * \varphi$$

where Ψ (or the composition $*$) is defined as in (6).

PROOF. The proof of Corollary 5.19 does not require more than the repetition of a few arguments used before and the application of Theorems 5.8, 5.9 and 5.19. We shall not go into the rather obvious details. However, we shall prove, as an application:

THEOREM 5.20. *Let $F(r)$ be a free group on r free generators a_ρ. There exists a group $\tilde{F}$ containing F with the following properties:*

(i) *Every element W of $\tilde{F}$ has a unique* n*th root for $n = 1, 2, 3, \ldots$ in $\tilde{F}$, that is, there exists exactly one element $V^{(n)}$ such that*

$$(V^{(n)})^n = W.$$

(ii) $\tilde{F}$ *can be generated by the generators* a_ρ *of* F *and the operation of taking the* nth *roots for all* n.

(iii) *Any two elements* W_1, W_2 *of* $\tilde{F}$ *either generate a free subgroup of rank two, or there exist integers* p, q *such that*

$$W_1{}^p = W_2{}^q$$

(iv) $\tilde{F}$ *is residually nilpotent.*

PROOF. We shall represent F in terms of elements of $\Lambda(Q, r)$ as in Corollary 5.19. Then, if $W \in F$, $\omega \in \Lambda$ and $W \to \omega$ under our mapping,

$$V^{(n)} \to \omega/n$$

gives us the desired (and unique) nth root of W and, similarly, of every element of $\tilde{F}$ obtained from the elements of F by composition and the process of taking nth roots. This proves statements (i) and (ii) of Theorem 5.10. Statement (iii) follows from the Širšov-Witt Theorem (end of Section 5.6) and Theorem 5.10 (Section 5.6); for, if $W_1 \to \omega_1$ and $W_2 \to \omega_2$ under our mapping of $F \to \Lambda$, then ω_1, ω_2 generate a free subalgebra in Λ and therefore ω_1, ω_2 generate a free group. If this group is not of rank 2, we must have $\omega_1 \circ \omega_2 = 0$. From Theorem 5.10 it follows then that the homogeneous components (in the generators of Λ) $\bar{\omega}_1$, $\bar{\omega}_2$ of lowest degree in ω_1, ω_2 must satisfy a relation

$$p\bar{\omega}_1 - q\bar{\omega}_2 = 0.$$

Therefore, homogeneous components of lowest degree in

$$\omega_3 = p\omega_1 - q\omega_2$$

must have a higher degree than the component of lowest degree in ω_1. Since $\omega_1 \circ \omega_3 = 0$, this implies $\omega_3 = 0$ according to Theorem 5.10. Therefore, Statement (iii) is true. Statement (iv) follows from the remark that if $W \in \tilde{F}_n$, then the components of lowest degree (with respect to the generators of Λ) in the map $\omega \in \Lambda$ of W must be at least of degree n. ◀

If in Theorem 5.20 we replace the group F with the free nilpotent group $F^{(c)}$ of class c, we can construct a group $\tilde{F}^{(c)}$ which satisfies (i) and (ii) of Theorem 5.20 and, instead of (iv), is nilpotent of class c. This group is obtained by replacing $\Lambda(Q, r)$ by its quotient ring with respect to the ideal generated by the elements of a degree $> c$.

Using the methods developed by Baumslag, 1961, it can be shown that the group $\tilde{F}^{(c)}$ thus obtained is free in the following sense: It is generated by r of its elements a_ρ and the process of taking nth roots for $n = 2, 3, 4, \ldots$; any nilpotent group of class c having properties (i) and (ii) of Theorem 5.20 is a quotient group of $\tilde{F}^{(c)}$.

A formula that is somewhat the converse of the Baker-Hausdorff formula can be derived from it in the same way as Theorem 5.13A from Theorem 5.12. We state it as

THEOREM 5.21. (*Zassenhaus' Formula*) *Let x, y be free generators of $A(Q, 2)$. Then*

$$e^{x+y} = e^x e^y \prod_{n=2}^{\infty} e^{c_n(x,y)}$$

where $c_n(x, y)$ is a homogeneous Lie element in x, y of degree n. The first few terms of the formula are given by

$$c_2 = -\tfrac{1}{2}[x, y], \qquad c_3 = -\tfrac{1}{3}[[x, y]y] - \tfrac{1}{6}[[x, y]x]$$

References and Remarks. The Baker-Hausdorff Formula was discovered by Campbell, 1898. However, Campbell assumed that the elements x, y of $A(R, 2)$ were matrices with real numbers as elements, and this leads to convergence difficulties, since the infinite series $\Phi(x, y)$ will, in general, not converge. Campbell had started his investigation for the purpose of constructing a Lie group directly from the Lie algebra of its "infinitesimal transformations." Baker, 1904, and Hausdorff, 1906, proved Theorem 5.18 independently, operating in a sufficiently abstract algebra where convergence questions are trivial. This, however, destroys the direct applicability to Lie groups. Later, the Baker-Hausdorff formula was used in the theory of groups defined by generators and relations by Adelsberger, 1930, Magnus, 1950, Lyndon, 1954, and, above all, Lazard 1954. For the properties of the numerical coefficients in $\Phi(x, y)$ see Goldberg, 1956. For generalizations, see Dieudonne, 1956.

For a continuous analog of the Baker-Hausdorff formula and the Zassenhaus formula, see respectively, papers by Magnus, 1954, and Fer, 1958, where these continuous analogs appear in the theory of linear differential equations.

Our method of constructing the group $\tilde{F}$ of Theorem 5.20 is a special case of a construction due to Mal'cev, 1949.

Problems for Section 5.10

1. Show that if x and y are free generators of $A(Q, 2)$, then

$$e^{-x}e^y e^x = e^u,$$

where

$$u = \sum_{n=0}^{\infty} \{yx^n\}/n!.$$

[*Hint:* Show that $\exp(e^{-x}ye^x) = e^{-x}e^ye^x$, and use Problem 5.9.7.]

2. Show that every two-by-two matrix A with real entries and determinant one can be written in the form $A = \exp \Omega$, where

$$\Omega = \begin{pmatrix} \omega & \varphi \\ \psi & -\omega \end{pmatrix}$$

is a real two-by-two matrix with trace zero. Show that

$$A = (\cosh \Delta)I + \left(\frac{\sinh \Delta}{\Delta}\right)\Omega,$$

where I is the two-by-two unit matrix and

$$\Delta = (\omega^2 + \varphi\psi)^{1/2}.$$

3. Show that if B is a two-by-two matrix with real entries and determinant one and A is as in the preceding problem, and if $B = \exp \Omega'$ and Ω'' is defined by

$$A \cdot B = \exp \Omega'',$$

then the elements of Ω'' cannot be written as one-valued analytic functions of the elements of Ω and Ω'. (This problem indicates some of the limitations of the Baker-Hausdorff formula as mentioned under "References and Remarks".)

5.11. Power Relations and Commutator Relations

The properties of the lower central series of a free group F make it possible to classify the relations satisfied by the generators of a group G which is presented as a quotient group $G = F/N$ of F.

Let a_ρ, $\rho = 1, \ldots, r$, be the generators of F and let W be a word in the a_ρ which belongs to the normal divisor N of F but is not freely equal to 1. Then $W = 1$ is a relation for G. Assume that $W \in F_n$ but that W is not in F_{n+1}. If there exists an element $V \in F_n$ which is not contained in N such that, for an integer $m > 1$,

$$W \equiv V^m \bmod F_{n+1}$$

then $W = 1$ is called a *power relation* mod F_{n+1} or simply a *power relation for the generators* a_ν of G. All other relations are called *commutator relations*. In a group G without power relations, the quotient groups G_n/G_{n+1} of the lower central series are *torsion free*, that is, they contain no elements of finite order > 1.

Not much is known about the question whether the generators of a group G can satisfy power relations if G itself is defined by commutator relations only. For free groups, free nilpotent groups and a few other generalized free groups the answer to the question is in the negative and can be given immediately on the basis of our knowledge of these groups. The weaker question whether a group G defined by commutator relations can have elements of finite order has been answered for free products whenever it can be solved for the factors (see the Kurosh Subgroup Theorem in Chapter 4) and has also been answered for groups with a single defining relation as well as for certain quotient groups of the groups Φ as defined in Corollary 5.15 of Section 5.8. (See Karrass, Magnus, Solitar, 1960.) It seems that there is no case known where commutator relations imply power relations if all commutator relations are in F_n and linearly independent mod F_{n+1}.

The converse question, whether power relations will imply commutator relations has an affirmative answer even in very simple cases. For instance, it has been shown by Koehler, 1947, that in a group G on generators a, b with the defining relator a^2 the word W defined by

$$W = (((a, b), a), (a, b))$$

is a relator. $W = 1$ is the first commutator relation occurring in G. The groups G_n/G_{n+1} for $n \leqq 4$ all arise from the corresponding groups F_n/F_{n+1} for a free group F of rank two by taking their quotient groups with respect to all squares. However, W is not a square mod F_6, according to Corollary 5.12 (ii), Section 5.7, since $\delta(W)$ is a basic Lie element.

No general formulae are known which would permit us to state which commutator relations follow, say, from a single power relation. However, if we have sufficiently many power relations, then there exist commutator relations derivable from them which can be exhibited explicitly. In particular, this will be true for the groups B_p, where p is a prime number and where $W^p = 1$ for every $W \in B_p$. We shall study these groups in some detail in the next two sections. Right now, we shall prove two basic results which will be applied later. First, we shall sharpen Theorem 5.13B of Section 5.7. For this purpose, we need a new concept which we may introduce by

Definition 5.9. (*Hausdorff Differentiation*) Let $A_0(R, r)$ be an associative algebra on free generators x_ρ, $\rho = 1, 2, \ldots, r$. Let k, l be any two fixed values of the set $1, 2, \ldots, r$. We define the *Hausdorff derivative*

$$x_k \frac{\partial}{\partial x_l}$$

first for a monomial $u = z_1 z_2 \ldots z_n$ (where z_ν for $\nu = 1, \ldots, n$, denotes any one of the x_ρ) as follows: Let $\nu = l_1, l_2, \ldots, l_s$ be the values of ν for which $z_\nu = x_l$. In u, we replace z_{l_σ} by x_k, calling the result u_σ. Then we define

(1) $$\left(x_k \frac{\partial}{\partial x_l}\right) u = \sum_{\sigma=1}^{s} u_\sigma.$$

We extend the definition of the left hand side in (1) to all elements $v \in A_0$ by requiring that, for any two elements $u_1, u_2 \in A_0$ and $\alpha_1, \alpha_2 \in R$,

$$x_k \frac{\partial}{\partial x_l}(\alpha_1 u_1 + \alpha_2 u_2) = \alpha_1\left(x_k \frac{\partial}{\partial x_l}\right)u_1 + \alpha_2\left(x_k \frac{\partial}{\partial x_l}\right)u_2.$$

Definition 5.9 looks rather artificial. However, it can be shown easily (by starting with the case where u is a monomial) that our definition of a derivative is exactly the one needed for having a Taylor theorem. We may state this result as

LEMMA 5.10. *Let $u(x_1, \ldots, x_l, \ldots, x_r)$ be an element of $A_0(R, r)$ involving terms of a degree not exceeding N. Then, for every $t \in R$,*

$$(2)\quad u(x_1, \ldots, x_l + tx_k, \ldots x_r) = \sum_{\nu=0}^{N} \frac{t^\nu}{\nu!}\left(x_k \frac{\partial}{\partial x_l}\right)^\nu u(x_1, \ldots, x_l, \ldots, x_r)$$

where

$$\left(x_k \frac{\partial}{\partial x_l}\right)^0 = 1.$$

Now we can improve on Theorem 5.13B as follows:

THEOREM 5.22. *Let x, y be free generators of an associative algebra $A(Z, 2)$ and let F be the free group generated by $a = 1 + x$, $b = 1 + y$. Let C_λ, $\lambda = 1, 2, \ldots, L$, be the commutators defined in Theorem 5.13A of a weight w such that $2 \leq w \leq p - 1$ where p is a prime number. Then*

$$(3)\qquad (ab)^p = a^p b^p C_1^{pd_1} C_2^{pd_2} \ldots C_L^{pd_L} C$$

with integral exponents $d_1, \ldots d_L$, $C \in F_p$ and

$$(4)\qquad \delta(C) \equiv \sum_{\nu=1}^{p-1} \frac{1}{\nu!}\left(x \frac{\partial}{\partial y}\right)^{\nu-1} \{xy^{p-1}\} \bmod p.$$

(The improvement over Theorem 5.13B consists, of course, in formula (4) which gives us information about the "remainder term" C in (3). It should be noted that the fractions $1/\nu!$ on the right hand side of (3) are, of course, integers mod p since $\nu < p$.)

PROOF. By substituting on both sides of (3) the expressions for the factors in terms of elements of $A(Z, 2)$ and comparing the terms of degree p in x, y on both sides, we find as in the proof of Theorem 5.13B:

$$(5)\qquad (x + y)^p \equiv x^p + y^p + \delta(C) \bmod p.$$

Since $\delta(C)$ is a Lie element, we must be able to rewrite

$$(6)\qquad D = (y + x)^p - y^p - x^p$$

as a sum of Lie elements by changing D mod p. We do this first for the terms in D which are linear in x by proving

$$(7) \qquad xy^{p-1} + xyx^{p-2} + y^2xy^{p-3} + \ldots + y^{p-1}x \equiv \{xy^{p-1}\} \bmod p.$$

From Definition 5.8 (Section 5.9) it follows easily that

$$(8) \qquad \{xy^{p-1}\} = xy^{p-1} - \binom{p-1}{1}yxy^{p-2} + \binom{p-1}{2}y^2xy^{p-3} \pm \cdots.$$

Therefore, we must prove the following congruence for binomial coefficients:

$$(9) \qquad \binom{p-1}{k} \equiv (-1)^k \bmod p, \qquad (k = 1, 2, \ldots, p-1).$$

The well-known relation for binomial coefficients

$$\binom{p}{k} = \binom{p-1}{k} + \binom{p-1}{k-1}$$

and the fact that, for $k = 1, 2, \ldots, p-1$

$$\binom{p}{k} \equiv 0 \bmod p,$$

proves (9) immediately, and therefore we have proved (7). Now we apply (2) of Lemma 5.10 to D in (6) and find

$$(10) \qquad (y+x)^p - y^p - x^p = \sum_{\nu=1}^{p-1} \frac{1}{\nu!}\left(x\frac{\partial}{\partial y}\right)^{\nu} y^p$$

This means that we can obtain the terms in D of degree $\nu + 1$ in x from those of degree ν by applying the operator

$$(11) \qquad \frac{\nu!}{(\nu+1)!}\, x \frac{\partial}{\partial y}$$

Since, according to (7), the terms in D of first degree in x can be put into the form

$$\{xy^{p-1}\}$$

if we calculate mod p, we obtain now (4). It should be noted that Hausdorff differentiation does not introduce denominators. Therefore, the operators (11) are well-defined mod p, for $\nu + 1 \leqq p - 1$. ◀

We can use Theorem 5.22 for proving a result that has found an important application to the Burnside Problem (Section 5.13). In order to formulate it concisely, we shall introduce a new concept. In the algebra

$A_0(Z, r)$ on free generators x_ρ, $\rho = 1, \ldots, r$, we shall call a Lie element an *Engel element* if it can be derived from the p-fold bracket product

(12) $$[[\ldots[[x_1, x_2], x_2]\ldots], x_2] = \{x_1x_2^{p-1}\}.$$

by the following procedures:

(i) Application (any number of times) of the Hausdorff differentiation operators

(13) $$x_k\frac{\partial}{\partial x_l}, \qquad k, l = 1, \ldots, r.$$

(ii) Substitution of any homogeneous Lie elements for the x_ρ in all elements derived from (12) by repeated application of the operators (13).

(iii) Bracket multiplication of the elements thus derived by an arbitrary homogeneous Lie element in $A_0(Z, r)$.

The module (with coefficients in Z) generated by these Engel elements will be called the *Engel module of class* $p - 1$ *in* $A_0(Z, r)$.

Of course, the Engel module of class $p - 1$ in A_0 has a well defined pre-image in $\Lambda_0(Z, r)$ which will be called the *Engel module of class* $p - 1$ *in* Λ_0. It is easily seen to be an ideal in $\Lambda_0(Z, r)$.

Now we can state:

THEOREM 5.23. *Let p be a prime number, and let $B(p, r) = F/N$ be the quotient group of the free group $F(r) = F$ of rank r with respect to the normal divisor N generated by all* pth *powers of elements of F. Let F be represented in terms of elements of $A_0(Z, r)$, and let, for any $n \geqq p$, W be an element of F_n. If $\delta(W)$ is in the Engel module of class $p - 1$ in A_0, then W belongs to the* $(n + 1)$st *group B_{n+1} of the lower central series of $B(p, r)$.*

Using the notation introduced in Section 5.8 before Theorem 5.15, we could also say:

The nth *relation module M_n of F/N contains the elements of degree n of the Engel module of class $p - 1$ in $\Lambda_0(Z, r)$.*

(Since we shall operate in A_0 rather than in Λ_0, we shall denote the map μM_n of M_n in A_0 by $M_n{}^*$. Then $M_n{}^*$ consists of all homogeneous Lie elements of degree n such that $W \in B_{n+1}$ if $\delta(W) \in M_n{}^*$.)

PROOF. The main difficulty consists in proving that $u_1 \in M_p{}^*$, where

(14) $$u_1 = \{x_1x_2^{p-1}\}.$$

We know from Theorem 5.22 that

(15) $$\sum_{\nu=1}^{p-1} u_\nu \in M_p{}^*$$

where, for $\nu > 1$,

$$u_\nu = \frac{1}{\nu!}\left(x_1 \frac{\partial}{\partial x_2}\right)^{\nu-1} u_1$$

If we replace, in Theorem 5.22, $(ab)^p$ by $(a^k b)^p$, where $k = 1, 2, \ldots, p - 1$, we see that

$$k \sum_{\nu=1}^{p-1} k^{\nu-1} u_\nu \in M_p^*, \qquad (k = 1, 2, \ldots, p - 1). \tag{16}$$

Now for every homogeneous Lie element v of any degree n, $pv \in M_n^*$. Therefore, since M_p^* is a module, all u_ν will be in M_p^* if the Vandermonde determinant D of $p - 1$ rows and columns defined by

$$D = \begin{vmatrix} 1 & 1 & \cdots & 1 \\ 1 & 2 & \cdots & 2^{p-2} \\ 1 & 3 & \cdots & 3^{p-2} \\ \cdot & \cdot & \cdot & \cdot \\ \cdot & \cdot & \cdot & \cdot \\ \cdot & \cdot & \cdot & \cdot \\ 1 & p-1 & \cdots & (p-1)^{p-2} \end{vmatrix}$$

is not divisible by p. This is easily seen to be true and proves $u_\nu \in M_p^*$.

So far, we have proved that the Engel element (12) belongs to M_p^* and also all the elements arising from it by applying (repeatedly) the operator (13) for $k = 1$, $l = 2$. In order to generalize this result, let F have the generators $a_\rho = 1 + x_\rho$, replace b in (3) by $a_2 a_3^k$ and put $a = a_1$. Then we find as in the proof of Theorem 5.22 that

$$\delta(C) \equiv (x_1 + x_2 + kx_3)^p - x_1^p - (x_2 + kx_3)^p \bmod p.$$

Therefore, on the right-hand side of (4) we may replace x with x_1 and y with $x_2 + kx_3$ and obtain an element of M_p^*. But then we will also obtain an element of M_p^* if in any one of the u_ν we replace x_2 with $x_2 + kx_3$. Applying the Taylor theorem of Lemma 5.10, we find

$$u_\nu(x_1, x_2 + kx_3) = \sum_{s=0}^{p-\nu} k^s \left(x_3 \frac{\partial}{\partial x_2}\right)^s u_\nu(x_1, x_2)$$

In the last equation, both sides are elements of M_p^* for $k = 0, 1, \ldots, p - 1$ By using the same argument which showed that (16) implies that the u_ν belong to M_p^* we can now show that all of the elements

$$\nu! \left(x_3 \frac{\partial}{\partial x_2}\right)^s u_\nu(x_1, x_2) = \left(x_3 \frac{\partial}{\partial x_2}\right)^s \left(x_1 \frac{\partial}{\partial x_2}\right)^{\nu-1} u_1$$

belong to M_p^*. A repetition of this argument, and the remark that our definition of N allows us to permute the x_ρ arbitrarily, proves that all Engel elements listed under (i) (before (13)) belong to M_p^*. To prove also that the Engel elements listed under (ii) belong to some M_n^*, we use the fact that N is a verbal subgroup. Let $W(a_\rho) \in F_p$ be a word belonging to N. Let b_ρ be any elements in F such that $\delta(b_\rho) = y_\rho$ are preassigned Lie elements of $A_0(Z, r)$. Then $W(b_\rho) \in N$. Let the Lie element v of degree p be defined by

$$\delta(W(a_\rho)) = v(x_\rho).$$

Then either $v(y_\rho) = 0$ or

$$\delta(W(b_\rho)) = v(y_\rho).$$

If $v(y_\rho)$ is of degree n, it belongs to M_n^* since $W(b_\rho) \in N$.

It remains to show that the Engel elements defined under (iii) belong to some M_n^*. This follows from the fact that the commutator of an element in N and any element in F belongs to N. Therefore, we have proved Theorem 5.23 completely. ◀

References and Remarks. Theorem 5.23 has been stated explicitly in the literature by Magnus, 1950, and by Sanov, 1952. It is implicit in the results of P. Hall, 1933, and Zassenhaus, 1939, combined with Magnus, 1940.

Problems for Section 5.11

1. Prove Lemma 5.10.

5.12. Burnside's Problem, Exponents 3 and 4

Burnside, 1902, posed the following problem which has had a very stimulating effect on the development of the methods discussed in this chapter. We state:

Burnside's Problem. Let $B(e, r)$ be the quotient group F/N of the free group F on r free generators, where N is the normal subgroup of F generated by all eth *powers in F,* i.e., $N = F(X^e)$. *For which values of e and r is $B(e, r)$ a finite group?*

We shall call e the *exponent* and r the *rank* of $B(e, r)$.

Clearly, if e is divisible by a prime number p, then since $B(e, r)$ has $B(p, r)$ as a factor group, $B(e, r)$ is infinite if $B(p, r)$ is infinite, and $B(p, r)$ will be finite if $B(e, r)$ is finite. Therefore, the case $e = p$ deserves special attention. In this case, (and, similarly, in the case where e is a power of p), we know from the theory of finite groups that if q is a power of a prime number p, and if $B(q, r)$ is finite then $B(q, r)$ is nilpotent, i.e., its lower central series terminates with 1 after a finite number of steps (see, for example, P. Hall, 1933). But if $B(q, r)$ is infinite, it is conceivable that its lower central series might terminate after a finite number of steps with a group $\tilde{B}(q, r)$ which in this case would be infinite, whereas its quotient group

$$B^*(q, r) = B(q, r)/\tilde{B}(q, r)$$

would be finite. B^* would then have the property of being maximal in the sense that every finite group which can be generated by r of its elements and in which the order of all elements is a divisor of q would be a quotient group of B^*. This leads to be following question:

Restricted Burnside Problem. For which values of q and r does $B(q, r)$ have a maximal quotient group of finite order?

If there exists an infinite group $B(q, r)$ (with $q = p^m$, p a prime number) for which $B^*(q, r)$ is finite, then the group $\tilde{B}(q, r)$ would have the property of being infinite, having only elements whose order is a power of p, and being identical with its commutator subgroup $\tilde{B}'$. To show this, we observe that $\tilde{B}'$ is invariant in B, and since $\tilde{B}$ is a subgroup of finite index in B, it is finitely generated. Therefore, if $\tilde{B}' \neq \tilde{B}$, $\tilde{B}/\tilde{B}'$ would be finite and non-trivial, so that the order of $B/\tilde{B}'$ would be larger than the order of $B/\tilde{B}$.

For $r = 1$, Burnside's problem is trivial for all e. We shall therefore assume $r \geqq 2$. Then the case $e = 2$ is still trivial since $B(2, r)$ is always Abelian. For $e = 3$ we shall prove:

THEOREM 5.24. *$B(3, r)$ is finite and of order 3^s, where*

$$s = \binom{r}{1} + \binom{r}{2} + \binom{r}{3} = r(r^2 + 5)/6.$$

The fourth group of the lower central series of $B(3, r)$ is the identity. The quotient groups

$$B_k(3, r)/B_{k+1}(3, r), \qquad k = 1, 2, 3$$

of the lower central series have the order 3^{s_k} where

$$s_k = \binom{r}{k} = r(r - 1) \ldots (r - k + 1)/k!.$$

PROOF. Before proving Theorem 5.24 in full, we observe that a weaker statement can be derived rather easily. By applying Theorem 5.23 (Section 5.11), we can show that the restricted Burnside problem has an affirmative answer for the exponent 3 and any finite r. For this purpose, we introduce first

Definition 5.10. For every prime number p and any algebra $A_0(Z, r)$, the module $E_p(r)$ is defined as the module of Lie elements generated by the Engel module of class $p - 1$ and the elements of the form pv, where v is any homogeneous Lie element.

If we can show that, for $p = 3$, $E_3(r)$ contains all homogeneous Lie elements of degree 4 in $A_0(Z, r)$, then the proof of Theorem 5.23 shows that the relation module of $B(3, r)$ contains all Lie elements of degree 4. Therefore, the fourth group of the lower central series of $B(3, r)$ coincides with all the later ones, which is what we want to prove.

Let x, y, z, u, v be any free generators of A_0. We have from (5.11.12) and (5.11.13) for $p = 3$:

$$z \frac{\partial}{\partial y} [[x, y]y] = [[x, z]y] + [[x, y]z] \equiv 0 \bmod E_3. \tag{1}$$

Together with

$$[[x, y]z] + [[y, x]z] = 0,$$

(1) shows that the bracket product of three factors alternates mod E_3 under all permutations of the factors since it does so under two permutations which generate all permutations of three symbols.

Substituting $[u, v]$ for x in (5.11.17) we find

$$[[[u, v]z]y] + [[[u, v]y]z] \equiv 0 \bmod E_3. \tag{2}$$

Therefore, a simple bracket product of four factors alternates if the last two factors are exchanged and also, according to the preceding remark, if the first three factors are permuted in any manner. Therefore it alternates under all permutations of the four factors.

Now we substitute $[u, v]$ for z in (1). Since we have from Theorem 5.16 that

$$[[x, y], [u, v]] = [[[x, y]u]v] - [[[x, y], v]u],$$

we find

$$\begin{aligned} (3) \quad & [x, [u, v], y] + [[x, y], [u, v]] \\ & = -[[[u, v]x]y] + [[[x, y]u]v] - [[[x, y]v]u] \equiv 0 \bmod E_3. \end{aligned}$$

If we permute x, y, u, v in the first and in the third term of the last line of (3) so that they appear in the same order as in the middle term, and observe that this requires an odd permutation for the third term and an even one for the first, we obtain

$$[[[x, y]u]v] \equiv 0 \bmod E_3.$$

This proves that all Lie elements of degree 4 are in E_3.

We now turn to the proof of the stronger statement of Theorem 5.24. We shall follow the pattern of our argument above, but instead of deriving congruences mod E_3, we shall establish relations for the elements of $B(3, r)$. We shall prove:

LEMMA 5.11. *Let a, b, c, d be any four elements of $B(3, r)$. Then*
(i) *Conjugate elements commute, that is:*

$$(a, b^{-1}ab) = 1. \tag{4}$$

(ii) *The simple commutator (a, b, c) does not change if a, b, c are permuted cyclically.*

(iii) *All simple commutators (a, b, c) belong to the center. Equivalently, it is true that*

$$(a, b, c, d) = 1, ((a, b), (c, d)) = 1.$$

Proof of assertion (i). $(ab)^3 = 1$ implies $a^{-1}b^{-1}a^{-1} = bab$. Therefore,

$$(a, b^{-1}ab) = a^{-1}b^{-1}a^{-1}bab^{-1}ab = bab^2ab^2ab = b^{-1}(b^2a)^3b = 1.$$

Proof of assertion (ii). We have

$$\begin{aligned} d^{-1}b^{-1}a^{-1}b^{-1}ada^{-1}d &= d^{-1}(b^{-1}a^{-1})^2(a^{-1}d)^2 \cdot \\ &= d^{-1}abd^{-1}a = (d^{-1}ab)^3b^{-1}a^{-1}db^{-1} = b^{-1}a^{-1}db^{-1} \end{aligned}$$

and therefore, by equating the first and the last term,

$$a^{-1}b^{-1}ada^{-1} = bdb^{-1}a^{-1}db^{-1}d^{-1}. \tag{5}$$

Now we substitute bcb^{-1} for d in (5). We derive easily that

$$c^{-1}a^{-1}b^{-1}abcb^{-1}a^{-1}ba = c^{-1}b^{-1}cba^{-1}bcb^{-1}c^{-1}a. \tag{6}$$

Since it follows from $(b^2c)^3 = 1$ that

$$bcb^{-1}c^{-1} = b^{-1}c^{-1}bc,$$

we can write (6) in the form

$$(c, (b, a)) = ((b, c), a). \tag{7}$$

Since (i) implies that $(c, d) = (d^{-1}, c)$, it follows from (7), with $d = (b, a)$, that

$$(c, (b, a)) = ((b, a)^{-1}, c) = ((a, b), c) = ((b, c), a),$$

which proves assertion (ii).

Proof of assertion (iii). Let $u = ((a, b), (c, d))$, and let $g = (c, d)$. Then we have from (ii):

$$u = ((a, b), g) = ((b, g), a) = ((g^{-1}, b), a) = (((d, c), b), a) = (d, c, b, a).$$

According to (ii), u will not change if we permute d, c, b cyclically. We shall now show that it also remains unchanged under the permutation

$$a \to b, \quad b \to a, \quad c \to d, \quad d \to c.$$

In fact, if $(a, b) = v$, $(c, d) = w$, then

$$(b, a) = v^{-1}, \qquad (d, c) = w^{-1},$$

and it follows from (i) that

$$u = (v, w) = (v^{-1}, w^{-1}).$$

Now we have two even permutations of a, b, c, d which leave u unchanged, and since these generate the alternating group on four symbols, u remains unchanged under all even permutations of a, b, c, d. In particular, u does not change if we exchange c with a, and b with d. But then we exchange w with v which means that we map u onto its inverse. Therefore, we have $u = u^{-1}$, which together with $u^3 = 1$ implies $u = 1$; this proves (iii).

Now we shall prove Theorem 5.24. Lemma 5.11 shows that the orders of the quotient groups of the lower central series of $B(3, r)$ are at most equal to s_k for $k = 1, 2, 3$. For the first quotient group, this is trivial, since we have r generators. For the second quotient group, it follows because the number of basic commutators of weight 2 in r generators is simply s_2 in a free group, since these commutators are given by

$$(a_l, a_m),\ l > m$$

in a free group on free generators a_ρ, $\rho = 1, \ldots, r$.

Finally, the number of different basic commutators of weight 3 in the generators a_ρ is given by s_3, since, even in a free group, we can use a subset of the elements

$$((a_l, a_m), a_n) \tag{8}$$

as basic commutators, and in $B(3, r)$ these are invariant under cyclic permutations according to Lemma 5.11. Therefore, all those in which two subscripts coincides are equal to 1, and of the remaining elements (8) with different values of l, m, n any six with the same values of l, m, n are either equal to each other or inverses of each other. All that remains to be done is therefore the construction of a group of order 3^s on r generators in which every element $\neq 1$ is of order 3. For this purpose, we shall use a quotient ring A_0/K of the algebra

$$A_0(G(3), r)$$

on r free generators x_ρ with coefficients in the Galois field $G(3)$ of order 3, modulo an ideal K which is defined as follows:

K shall be the smallest ideal in A_0 containing firstly the elements

$$x_\rho{}^3,\ x_\rho{}^2x_\lambda,\ x_\rho x_\lambda x_\rho,\ x_\lambda x_\rho{}^2$$

and furthermore all homogeneous elements of a degree $\geqq 4$ and finally all of the elements

$$P_{\rho.\delta.\lambda} = x_\rho x_\delta x_\lambda + x_\rho x_\lambda x_\delta + x_\delta x_\rho x_\lambda + x_\delta x_\lambda x_\rho + x_\lambda x_\delta x_\rho + x_\lambda x_\rho x_\delta.$$

Within A_0/K, the elements b_ρ defined by

$$b_\rho = 1 + x_\rho$$

generate a group D. We have

$$b_\rho^{-1} = 1 - x_\rho,\ (b_\rho, b_\delta) = 1 + [x_\rho, x_\delta],\ ((b_\rho, b_\delta), b_\lambda) = 1 + [[x_\rho, x_\delta], x_\lambda].$$

It is easy to see that the x_ρ and the $[x_\rho, x_\delta]$, $\rho > \delta$, are linearly independent in A_0/K. The same is true for the set of the $[[x_\rho, x_\delta], x_\lambda]$ with $\rho > \delta > \lambda$, since a linear relation between several terms $[[x_\rho, x_\delta]x_\lambda]$ can exist only if the set of subscripts ρ, δ, λ is the same for all of them, and if $\rho > \delta > \lambda$ there exists only one such term which also is easily shown not to be in K. Therefore, the order of D is at least equal to 3^s, where s has been defined in Theorem 5.24. It remains to prove that in D the cube of every element equals 1. Let

$$g = 1 + u_1 + u_2 + u_3$$

be an element of D, where u_1, u_2, u_3 are, respectively, homogeneous elements of degree 1, 2, 3 in A_0/K. Since all terms of a degree $\geqq 4$ in the x_ρ vanish, we find (in A_0/K):

$$g^3 = 1 + u_1{}^3.$$

Let

$$u_1 = \sum_{\rho=1}^{r} \alpha_\rho x_\rho, \qquad \alpha \in G(3)$$

Then we see that, in A_0/K.

$$u_1{}^3 = \sum_{\rho > \delta > \lambda} \alpha_\rho \alpha_\sigma \alpha_\lambda P_{\rho,\delta,\lambda}$$

and therefore $u_1{}^3 = 0$. This shows that our group D has all the properties assigned to $B(3, r)$ in Theorem 5.24. ◀

The finiteness of $B(3, r)$ was proved first by Burnside, 1902, who also showed that $s \leqq 2^r - 1$. The present proof is due to Levi and van der Waerden, 1933.

Now we shall investigate the case $e = 4$. We have

THEOREM 5.25. (*Sanov's Theorem*) *Any finitely generated group B in which the order of every element is $\leqq 4$ is a finite group.*

PROOF. Sanov's theorem includes, of course, the statement that all groups $B(3, r)$ and $B(4, r)$ are of finite order. The proof is simple and of a purely combinatorial nature. We shall confine ourselves to the case where the fourth power of every element of the group B equals 1. The proof is based on the following

LEMMA 5.12. *Let B be a group in which $x^4 = 1$ for every element $x \in B$. Let D be a finite subgroup of order d in B. Assume that there exists an element $c \in B$ such that $c^2 \in D$, and that c and the elements of D generate B. Then B is finite.*

To prove Lemma 5.12, we observe that every element b of B can be written in the form

$$b = P_0 c Q_1 c P_1 c Q_2 c P_2 \ldots c Q_s c P_s c Q_{s+1},$$

where $P_0, \ldots, P_s, Q_1, \ldots, Q_{s+1}$ belong to D and where none of these elements except, possibly, for P_0 or Q_{s+1} equals 1. All we have to show is that every b admits a representation in which the number $s + 1$ of factors equal to c is bounded by a fixed number. We shall show $s < d + 1$ by exhibiting a transformation which reduces the number of c-factors if $s > d$. For this purpose, we note that for any $R \in D$, we have $c^3 - c^{-1}$, $(R^{-1}c^{-1})^4 = 1$, and so

$$cRc = R^{-1}cR^*cR^*,$$

where $R^* = c^2R^{-1} \in D$ since $c^2 \in D$. We shall use this relation for the purpose of transforming the expression for b in the manner indicated below:

$$\begin{aligned} &\ldots cQ_{l+k-1}cP_{l+k-1}cQ_{l+k}cP_{l+k}c \ldots \\ &= \ldots cQ_{l+k-1}cP_{l+k-1}cQ_{l+k}P_{l+k}^{-1}cUc \ldots \\ &= \ldots cQ_{l+k-1}cP_{l+k-1}P_{l+k}Q_{l+k}^{-1}cVcW \ldots \\ &= \ldots cQ_{l+k-1}Q_{l+k}P_{l+k}^{-1}P_{l+k-1}^{-1}cXcYcZ \ldots, \end{aligned}$$

where $U, \ldots, Z$ are certain elements of D which we need specify no further. The essential properties of our transformations are the following ones: They do not change the number of c-factors in the representation of b, and they show that we may find a representation of b in which two consecutive c-factors are separated by an element $S_{l,k}$ or $S_{l,k}^{-1}$, where, for any $l > 0$, $k \geqq 0$, $l + k \leqq s$, the element $S_{l,k}$ of D is defined by

$$S_{l,k} = P_l P_{l+1} \ldots P_{l+k} Q_{l+k}^{-1} \ldots Q_{l+1}^{-1} Q_l^{-1}.$$

Lemma 5.12 will be true if we can show that at least one $S_{l,k} = 1$ provided that $s > d$. For then our transformations allow us to find a representation of b with fewer than s factors c since

$$cS_{l,k}^{\pm 1}c = c^2 \in D.$$

To show that at least one $S_{l,k}$ equals 1, we consider the elements $S_{1,k}$ of D for $k = 1, \ldots, s$. If $s > d$, the order of D, at least two of them must be equal. Suppose $S_{1,k} = S_{1,l}$, where $k < l$. Then

$$P_1 \ldots P_{k+1}Q_{k+1}^{-1} \ldots Q_1^{-1} = P_1 \ldots P_{l+1}Q_{l+1}^{-1} \ldots Q_1^{-1}$$

and therefore,

$$S_{k+2,l-k-1} = P_{k+2} \ldots P_{l+1}Q_{l+1}^{-1} \ldots Q_{k+2}^{-1} = 1$$

which proves Lemma 5.12.

Now the finiteness of $B(4, r)$ follows by using induction with respect to r. For $r = 1$, the result is trivial. Let $r > 1$, and assume that $B(4, r - 1) = D_0$ is finite. Let c be a generator of $B(4, r)$ not appearing in $B(4, r - 1)$, and let $c_0 = c^2$. Then the subgroup D of $B(4, r)$ generated by D_0 and c_0 is finite according to Lemma 5.12 since $c_0{}^2 = 1$ and therefore $c_0{}^2 \in D_0$. But if D is finite, so is $B(4, r)$ since it is generated by c and D, and $c^2 = c_0 \in D$, from which fact the finiteness of $B(4, r)$ follows by a second application of Lemma 5.12. ◀

The proof of Theorem 5.25 also gives an estimate for the order of $B(4, r)$ which, however, is a very crude one. The finiteness of $B(4, 2)$ was proved earlier by J. A. de Séguier, 1904 (p. 72).

5.13. Burnside's Problem, Report on $e > 4$

For exponents $e > 5$, the finiteness for $B(e, r)$ has been proved only for $e = 6$. We state the result as

MARSHALL HALL'S THEOREM. *The order of $B(6, r)$ is*

$$2^a 3^{b+b'+b''} \tag{1}$$

where

$$a = 1 + (r - 1)\, 3^{r+r'+r''}, \qquad b = 1 + (r - 1)2^r$$

and

$$r' = r(r - 1)/2, \qquad r'' = r(r - 1)(r - 2)/6$$
$$b' = b(b - 1)/2, \qquad b'' = b(b - 1)(b - 2)/6.$$

For a (rather condensed) proof see M. Hall, 1957 or M. Hall, 1959 (Chapter 18.4). M. Hall's proof is based on the work of P. Hall and G. Higman, 1956, who has also proved that $B(6, r)$ has a maximal quotient group of finite order and that the order of this quotient group is given by (1). In the same paper, P. Hall and G. Higman, 1956, also proved the following

THEOREM OF P. HALL AND G. HIGMAN. *Let* $q = q_1q_2 \ldots q_k$, *where* $q_1, \ldots, q_k$ *are powers of different primes. Assume that, for a particular finite* r, *the restricted Burnside problem* (Section 5.12) *has an affirmative answer for the groups*

$$B(q_1, r), B(q_2, r), \ldots, B(q_k, r).$$

Then there exists a finite solvable quotient group $\tilde{B}(q, r)$ *of* $B(q, r)$ *such that all other finite solvable quotient groups of* $B(q, r)$ *are quotient groups of* $\tilde{B}(q, r)$. *For* $k = 2$, *the word "solvable" can be omitted in the statement of the theorem, since every finite group of order* q_1q_2 *is solvable.*

The paper of P. Hall and G. Higman, 1956, is fundamental for our present knowledge of finite solvable groups. We cannot go into the many important details of the paper, but we may mention the following result which is related to Burnside's problem.

There exists a group of exponent 15, i.e., $x^{15} = 1$ for all elements x, which is generated by two of its elements, the Sylow subgroups of which have Abelian commutator subgroups, and which is maximal in the sense that every other group with the same properties is a quotient group of this particular group. The order of this maximal group is $3^\alpha 5^\beta$, where

$$\alpha = 9,934,183,757,031,251, \qquad \beta = 568,225.$$

For the restricted Burnside problem (Section 5.12), the most comprehensive result can be formulated as

KOSTRIKIN'S THEOREM. *For every prime number* p *and any finite integer* $r > 0$, *the Burnside group* $B(p, r)$ *has a maximal quotient group of finite order.*

This theorem was first proved by Kostrikin, 1955, for $p = 5$, $r = 2$, and for $p = 5$ and arbitrary r by G. Higman, 1956. Higman's method, which uses associative rings, is different from the method used by Kostrikin which is based on the theory of Lie algebras. The starting point of Kostrikin's investigation is Theorem 5.23 (Section 5.11) and the question whether the quotient ring of $\Lambda_0(Z, r)$ with respect to $E_p(r)$ (Definition 5.12, Section 5.12), which is an ideal in Λ_0, will be of finite order. In three papers, Kostrikin, 1957a, 1957b, 1959 demonstrates that certain theorems about Lie algebras will imply the existence of a maximal finite quotient group of $B(p, r)$; the necessary theorems about Lie algebras are proved in the last one of these papers. An outline of Kostrikin's method may be obtained from the reviews of his papers in the *Mathematical Reviews*; these are quoted in the list of references at the end of this book. About the role of the "Engel identities" in the theory of Lie algebras see also Gruenberg, 1953, Baer, 1940, and P. M. Cohn, 1955.

Novikoff, 1959, published an outline of a proof of the following:

For $e \geqq 72$ *and* $r \geqq 2$, $B(e, r)$ *is of infinite order.*

For some implications of this theorem, in conjunction with Kostrikin's Theorem, see the remarks made after the formulation of the Restricted Burnside Theorem, in Section 5.12.

5.14. Topological Aspects

A large number of topological concepts have been associated with group theory. Of these, we shall mention only a few here; we shall review even fewer results, confining ourselves to some theorems which are closely related to free groups and to the methods developed in this chapter.

A topological group G can be defined by defining open sets (having the ordinary topological properties) in the group such that, for any element g in the group and for any open set S (which is a collection of group elements), both S^{-1} and gS are open sets again. The condition that hS be an open set if S is open means this: G can be considered as a group of transformations acting on G if we assign to $g \in G$ the mapping $x \rightarrow gx$ of the elements x of G onto the elements gx. The topology in G is then invariant under transformations of the group. [Of course, we could have postulated that Sg rather than gS is an open set if S is an open set. But since $(S^{-1}g^{-1})^{-1} = gS$, the two definitions are equivalent.]

A topology for a group G is called a *subgroup topology* if there exists a family Φ of subgroups V_i (where i runs through a set of subscripts which is not necessarily countable) such that the subgroups V_i and their cosets gV_i form a basis for the open sets in G. For this purpose, it suffices to impose the following conditions on the V_i:

(i) The intersection of all V_i is the identity of G.

(ii) If $V, V' \in \Phi$, then $V \cap V' \in \Phi$.

(iii) Given any $V \in \Phi$, where V is not of order 1, there exists a $V' \subset V$ which is not of order 1, is normal in G, and belongs to Φ.

M. Hall, 1950a, proved the following theorems about subgroup topologies:

T1. *Let G be a group which contains at most a countable infinity of elements and a sequence of normal divisors N_i, $i = 1, 2, 3, \ldots$ such that*

$$G = N_1 \supset N_2 \supset N_3 \supset \ldots,$$

where the N_i *are of finite index in* G *and where the intersection of all the* N_i *is the identity. Then the family* Φ *of all* N_i *defines a subgroup topology in* G. *The completion* $\tilde{G}$ *of* G *under this topology is compact. The topology is metrizable in such a manner that* $\tilde{G}$ *admits a topological mapping onto a Cantor set which lies in a finite interval of the line of real numbers.*

For free groups, subgroup topologies can be defined by using the following result:

T2. *Let* F *be a free group on countably many generators and let* Φ *be a family of subgroups* U_i, $i = 1, 2, 3, \ldots$, *such that*

$$F = U_1 \supset U_2 \supset U_3 \supset \ldots .$$

If, for all i, *and for any set of free generators of* U_i, *the length of every element* $\neq 1$ *of* U_{i+1} *is, in terms of these generators, not less than 2, then the intersection of all the* U_i *is the identity. If, in addition, the* U_i *are normal in* F, *then* Φ *defines a subgroup topology in* F.

Examples of families U_i can be obtained by taking for U_{i+1}, the ith derived group of F, or if F is finitely generated, by taking for U_{i+1} the intersection of all normal divisors of index p in U_i, where p is a prime number ≥ 2.

For more results and for references see M. Hall, 1950. For the topological concepts see Pontrjagin, 1946 and A. Weil, 1937, 1938.

A subgroup topology can always be constructed from the family of the groups of the lower central series, provided that these groups have only the identity in common. In addition, we can associate a Lie ring with the group. To do this, we start with a presentation of the group G as a quotient group F/N of a free group F. Using the notations of Section 5.8, in particular the definition of the nth relation module M_n introduced before the statement of Theorem 5.15, we define the modules Γ_n by

$$\Gamma_n = \Lambda_n/M_n. \tag{1}$$

Since the direct sum of the M_n forms an ideal in Λ_0 (Problem 5.8.1), the direct sum Γ of the Γ_n forms a Lie ring (which is a quotient ring of Λ_0). It can be shown (Mangus, 1939) that Γ is independent of the particular presentation of G as a quotient group of a free group. Also Γ is a graded Lie ring which is generated by the homogeneous elements of degree 1, that is, by the elements of Γ_1.

The construction which leads from $G = F/N$ to the Lie Algebra Γ has been extended in a remarkable manner by Lazard, 1954. We shall now give a brief account of some of the concepts used in his work and of a few (in fact, very few) of his results. For the full theory, for a bibliography, and for applications see Lazard, 1954. For the theory of Lie rings associated with a group see also Zassenhaus, 1939. For the topological concepts see Leray, 1950.

Let A be an algebra. For all later purpose, A will be either a Lie algebra or an associative algebra. We consider first the additive group (module) of A, and we shall assume that, on A, a positive *filtration* is given by a function $\nu(x)$ with $x \in A$ and with values $\nu = 0, 1, 2, 3, \ldots + \infty$ such that the following conditions are satisfied: The set of elements x of A for which $\nu(x) \geqq i$, $i = 0, 1, 2, \ldots$, is a submodule A_i of A such that

$$A = A_0 \supset A_1 \supset A_2 \ldots$$

and A_∞ is the intersection of all A_i. The function $\nu(x)$ satisfies the condition.

$$(2) \qquad \nu(x - y) \geqq \sup (\nu(x), \nu(y)), \qquad \nu(\alpha x) \geqq \nu(x),$$

where α is an element in the ring of operators (e.g., the integers) of the module A.

For the (associative or Lie) multiplication in A we postulate that

$$(3) \qquad \nu(xy) \geqq \nu(x) + \nu(y),$$

where xy denotes the (associative or Lie) product of x and y in A.

It will be convenient to assume that, for an associative algebra A with an identity 1, we have $\nu(1) = 0$, whereas we shall assume that for a Lie algebra A the set of elements x with $\nu(x) = 0$ is empty and therefore $A = A_1$.

Once a $\nu(x)$ satisfying (2) and (3) is given we can construct a *graded algebra* A_γ associated with A in the following manner.

We form the quotient groups B_i defined by

$$B_i = A_i/A_{i+1}$$

of the additive groups A_i. Then the general element of A_γ will be the vector b with infinitely many components $b_i \in B_i$, where addition and multiplication by operators α is defined in the usual manner. In other words, the additive group of A_γ will be the unrestricted direct sum of the modules B_i. If we want the distributive laws to hold for A_γ, multiplication can be defined by defining the products b_ib_j, where $b_i \in B_i$, $b_j \in B_j$. For this purpose, let respectively $a_i \in A_i$, $a_j \in A_j$ be arbitrary but fixed preimages of b_i, b_j in the homomorphic mappings

$$A_i \rightarrow B_i = A_i/A_{i+1}, \qquad A_j \rightarrow B_j = A_j/A_{j+1}.$$

Then $a_ia_j \in A_{i+j}$, and we define $b_ib_j \in B_{i+j}$ as the map of a_ia_j onto a coset of

$$B_{i+j} = A_{i+j}/A_{i+j+1}.$$

Clearly b_ib_j may be zero although a_ia_j is not zero (but in this case $a_ia_j \in A_{i+j+1}$). It is shown easily that b_ib_j depends only on the residue classes of $a_i \bmod A_{i+1}$ and $a_j \bmod A_{j+1}$, i.e., on b_i and b_j and not on the choice of a_i and a_j.

We shall call the b_i the *homogeneous components* of degree i of the element $b \in A_\gamma$. An element b in which only the ith component is $\neq 0$ is called homogeneous of degree i. It is a characteristic property of a graded algebra that the product of homogeneous elements of degrees i and j is either homogeneous of degree $i + j$ or zero. The algebras $A(Z, r)$ and $\Lambda(Z, r)$ of Section 5.4 provide examples of graded algebras. The proof that products always exist in graded algebras is the same as the proof in the special case of $A(Z, r)$.

Lazard's method of associating graded algebras with groups is based on the following definitions and constructions: Let G by a group and let H_i, $i = 1, 2, 3, \ldots$ be a sequence of subgroups such that

$$G = H_1 \supset H_2 \supset \ldots \supset \ldots \supset H_i \supset H_{i+1} \ldots \tag{4}$$

where H_{i+1} is normal in H_i and where

$$(H_i, H_j) \subset H_{i+j}, \qquad (i, j = 1, 2, 3, \ldots). \tag{5}$$

Such a sequence will be called an *N-sequence*. Obviously H_i/H_{i+1} is Abelian. We form the unrestricted direct sum B of the modules

$$B_i = H_i/H_{i+1}$$

and show that we can define a multiplication for the elements of B which makes a Lie algebra out of it. For this purpose let $b_i \in B_i$, $b_j \in B_j$ and let $x_i \in H_i$, $x_j \in H_j$ be elements of G with the property that the mappings

$$H_i \to H_i/H_{i+1}, \quad H_j \to H_j/H_{j+1},$$

map x_i onto b_i and x_j onto b_j, respectively. Then we define b_ib_j as the residue class of

$$H_{i+j}/H_{i+j+1}$$

which contains the commutator (x_i, x_j). By using the Witt-Hall identities (Theorem 5.1, Section 5.2) we can show that the product b_ib_j thus defined depends only on b_i and b_j (and not on x_i, x_j) and that the multiplication thus defined satisfies the postulates for a Lie algebra. This Lie algebra arising from B is graded and will be denoted by $\Lambda(H_i)$. Lazard, 1954, then shows that this Lie algebra has the following property:

T3. *Let G_i be the groups of the lower central series of G. Then the Lie algebra $\Lambda(H_i)$ belonging to the N-sequence H_i of G is generated by its homogeneous elements of degree* 1 *if and only if G_j* (for $j = 1, 2, 3, \ldots$) *is dense in H_j with respect to the subgroup topology induced by the sequence H_i.*

The relation between the Lie algebra $\Lambda(H_i)$ of G and the Lie algebras associated with the quotient groups of G is given by the following theorem:

T4. *Let G' be a homomorphic image of G. Let, respectively, H_i, H_i' be N-sequences in G and G' such that, under the homomorphism f mapping G onto G'*

$$f(H_i) \subset H_i', \qquad (i = 1, 2, 3, \ldots).$$

Then there exists a homomorphic mapping $\tilde{f}$ of $\Lambda(H_i)$ into $\Lambda(H_i')$ which preserves the degree of homogeneous elements in the two algebras. $\tilde{f}$ is an isomorphic mapping of $\Lambda(H_i)$ into $\Lambda(H_i')$ if and only if H_i is the preimage of H_i' under f. Furthermore, $\tilde{f}$ is a homomorphic mapping of $\Lambda(H_i)$ onto $\Lambda(H_i')$ if and only if $f(H_j)$ is dense in H_j' under the subgroup topology defined by the H_j'.

Lazard also shows that certain mappings of a group G into a filtered Lie algebra A can be used to define N-sequences H_i in G and that the Lie algebra $\Lambda(H_i)$ is a subalgebra of the graded Lie algebra A_γ arising from A and its filtration ν. We omit the details. However, we shall state Lazard's generalization of Theorem 5.7 (Section 5.5). We have:

T5. *Let A be an associative algebra with an identity 1 and a filtration $\nu(a)$, $a \in A$, with $\nu(a) = 0, 1, 2, 3, \ldots, +\infty$ and $\nu(1) = 0$.*

Let f be a homomorphic mapping of a group G into the multiplicative group of invertible elements of A, such that, for any elements x of G

$$\nu(f(x) - 1) \geqq 1.$$

Then the elements x of G for which

$$\nu(f(z) - 1) \geqq i$$

form a subgroup H_i of G, and the sequence H_i is an N-sequence. The Lie algebra $\Lambda(H_i)$ is a subalgebra of the graded Lie algebra L_γ associated with the filtered Lie algebra L which we obtain if we define product $a \circ b$ of elements $a, b \in A$ by $a \circ b = ab - ba$.

Lazard observes that Theorem T5 could have been derived from the following result:

Kaloujnine's Theorem. *Let G be a group, let H_i, $i = 1, 2, 3, \ldots$ be a sequence of normal divisors of G for which $H_i \supset H_{i+1}$. Let K be the group of automorphisms of G and let L_i be the subgroup of K which leaves each element of H_j in its residue class* mod. *H_{j+1}. Then the L_j form an N-sequence in the subgroup L_1 of K.*

For a proof see Kaloujnine, 1949, 1950.

For results about the various topologies defined in a group G by different N-sequences see Lazard, 1954.

Apart from subgroup topologies and filters, important topological concepts of a quite different character have been introduced into group theory. H. Hopf, 1941, 1944, defined for an arbitrary group G and an arbitrary (associative and commutative) ring J with an identity the *Betti groups of G with respect to J*. We shall not reproduce Hopf's definition here, but we may mention that if J is the ring of integers, the zero-th Betti group of G is its quotient group with respect to its commutator subgroup. Of the many results due to Hopf, we state here merely the following remarkable and simple result:

THEOREM OF H. HOPF. *Let G be a group presented in two different ways as a quotient group F/N and as a quotient group F^*/N^* of free groups F, F^*. Then*

$$(N \cap (F, F))/(F, N) \simeq (N^* \cap (F^*, F^*))/(F^*, N^*).$$

The "dual" theory to Hopf's theory of Betti groups of a group is the *cohomology theory* of Eilenberg and MacLane, 1947a, b, which has developed into a new discipline of algebra. See, for instance, Eilenberg and Cartan, 1956.

5.15. Free Differential Calculus

R. Fox, 1953, established a theory of derivations in the group ring of a free group which clarifies and extends many of the results of Section 5.5, and which admits a large number of applications, in particular to the isomorphism problem of groups. For these see R. Fox, 1954, and the references in R. Fox, 1953. In this section we shall review briefly the basic concepts of Fox's Theory.

Let G be a group and let JG be the group ring (group algebra) of G over the ring J of integers as ring of coefficients. As usual, we shall identify the unit element of G and the element 1 of J. Let α be a homomorphism of G onto a quotient group $G/N = Q$ with respect to a normal divisor N of G. Then α can be extended to a homomorphic mapping of the ring JG onto the group ring JQ of Q, and the kernel of this mapping is the two sided ideal N^* which is generated by the elements $b - 1 \in JG$ where b runs through a set of generators of N. Conversely, N^* is the smallest two sided ideal in JG such that the mapping $JG \to JG/N^*$ has the property of mapping an element $g \in G$ upon the identity if and only if $g \in N$. We shall refer to this situation by saying that *N determines N^** and *N^* determines N*.

The fundamental ideal G^* of JG is defined by the homomorphism α_0 which maps G onto the group of order 1. Obviously, α_0 maps JG onto the ring J of integers; the map $\alpha_0 u$ of an element $u \in JG$ shall be denoted by u^0.

The method developed by Fox, 1953 yields simple proofs of the following theorems:

(i) *Let G be a group and let G^* be the fundamental ideal of JG. Then the* nth *power G^{*n} of G^* determines the nth group G_n of the lower central series of G.*

(ii) *Let F be a free group. Then JF has no divisors of zero.* The first proof of this result is due to Higman, 1940.

(iii) *Let F be a free group, let N be a normal divisor of F and let F^*, N^* be the ideals determined by F, N, respectively. Then the product F^*N^* determines the commutator subgroup of N.* Earlier proofs of this result are due to Schumann, 1935 and to Blanchfield, 1949.

(iv) *An element $v - 1$ of JF belongs to N^* if and only if there exists an element $u \in N$ such that $v - u \in N^*F^*$.*

To prove these theorems, Fox, 1953, introduces the following concepts:

A *derivation* D in the group ring JG is defined as a mapping of JG into itself which has the following properties: For any elements u, v of JG,

$$D(u + v) = D(u) + D(v)$$

$$D(u \cdot v) - D(u) \cdot v^0 + u \cdot D(v).$$

where v^0 is the sum of the coefficients of v.

These definitions imply that $Dj = 0$ for $j \in J$ and that

$$D(g^{-1}) = -g^{-1}Dg$$

for any $g \in G$. The derivations in JG form a right JG-module if addition of two derivations D_1 and D_2 is defined by

$$(D_1 + D_2)u = D_1u + D_2u$$

and if the derivation $D' = Dv$ is defined by

$$D'(u) = D(u) \cdot v.$$

If F is a free group on free generators a_ν, $\nu = 1, 2, \ldots, n$, all derivations in JF can be determined explicitly. Fox, 1953, proved the following theorem:

(v) *To each generator a_ν of F there corresponds a derivation D_ν which is uniquely determined by the property*

$$D_\nu(x_\nu) = 1,\ D_\nu(x_\mu) = 0,\ \mu \neq \nu,\ \mu = 1, \ldots, n.$$

Every derivation D in JF can be defined by the formula

$$D(u) = \sum_{\nu=1}^{n} D_\nu(u)h_\nu$$

where the h_ν are arbitrary fixed elements of JF, i.e., the D_ν generate the derivations as a JF-module.

There exists a Taylor theorem for the elements of JF. In order to formulate it, we define recursively derivatives of higher order by the formula

$$D_{\nu_k,\nu_{k-1},\ldots,\nu_1}(u) = D_{\nu_k}(D_{\nu_{k-1},\ldots,\nu_1}(u)).$$

Then the "Taylor formula with remainder term" holds, i.e.,

$$\begin{aligned} u = (u)^0 &+ \Sigma_{\nu_1}(D_{\nu_1}(u))^0(a_{\nu_1} - 1) \\ &+ \Sigma_{\nu_2,\nu_1}(D_{\nu_2,\nu_1}(u))^0(a_{\nu_2} - 1)(a_{\nu_1} - 1) \\ &\cdots\cdots\cdots\cdots\cdots\cdots \\ &+ \Sigma_{\nu_{n-1},\ldots,\nu_1}(D_{\nu_{n-1},\ldots,\nu_1}(u))^0(a_{\nu_{n-1}} - 1)\ldots(a_{\nu_1} - 1) \\ &+ \Sigma_{\nu_n,\ldots,\nu_1}D_{\nu_n,\ldots,\nu_1}(u)(a_{\nu_n} - 1)\ldots(a_{\nu_1} - 1), \end{aligned}$$

where $\nu_1, \ldots, \nu_n$ run independently through the numbers $1, \ldots, n$. This terminating series can be extended to a formal infinite Taylor series. If u is an element of F and if $a_\nu - 1$ is replaced by the symbol x_ν, the resulting series is identical with the expansion of u in the algebra $A(Z, r)$ which was introduced in Section 5.5. The expressions obtained for the coefficients of this series in terms of higher order derivatives reveal various relations between these coefficients and can be used advantageously for many calculations. See Lyndon, 1954 and Fox, 1953.

The group ring JG is an Abelian group under addition. The elements of G operate on the left of JG in a natural manner; if $g \in G$ and $u \in JG$, the operator g maps u onto $gu \in JG$. The derivations D can be looked upon as functions on G to JG, and it is obvious (from the appropriate definitions in the cohomology theory of groups) that they are the one-dimensional cocycles of G in JG. In the case where G is a free group, this remark leads to an easy computation of the first cohomology group of G over JG. For cohomology theory in groups see Eilenberg and MacLane, 1947*a*, *b*, MacLane, 1949, Eckmann, 1945. For connections with the free differential calculus see Fox, 1953.

Chapter 6

Introduction to Some Recent Developments

Preliminary Remarks. In this chapter, we report briefly on a few topics which are closely connected with the theorems proved in the previous chapters but which, for one reason or another, have not been included in the main body of the text. We do not try to achieve completeness in any sense; the list of topics as well as the list of references and, even more so, the list of theorems taken from the papers quoted omits much. All we have tried to do is to describe in a language as non-technical as possible a few results which indicate the character of recent research in some sectors of group theory, and to refer the reader to a set of papers which may serve as a basis for fuller information. To make this chapter more readable, we have tried to keep the number of references to the earlier chapters at a minimum; this meant that we had to be repetitious in several places.

6.1. Word, Conjugacy, and Related Problems

Dehn's word problem, as formulated in Section 1.3, acquired fame when Novikov, 1952, 1955, and, independently, Boone, 1958 and Britton, 1958 proved that it is unsolvable in general. Specifically, these authors exhibited finitely presented groups G in which there is no general and effective procedure for determining whether a word in the given generators represents the unit element as a consequence of the given defining relations. Britton's example is particularly noteworthy since it is the free product with amalgamations of groups with solvable word problems. We shall not discuss the precise meaning of "general effective procedure," and we merely mention that an exact definition of this term is due to Turing, 1950. Earlier, V. Post, 1947 and Turing 1950, 1958 had shown that the word problem in semi-groups with cancellation is unsolvable in general.

First Rabin, 1958, and later Baumslag, Boone, and B. H. Neumann, 1958, used the discovery of Novikov to prove that, practically all problems concerning a finitely presented group are unsolvable in general, including the problems of deciding whether the group is trivial, finite, free, nilpotent, or simple. In fact, it seems that all "natural" questions about finitely presented groups have been shown to be unsolvable, with the exception of the question deciding whether a group is Hopfian (non-isomorphic to any proper quotient group). (This remark is not meant to suggest that the decision about the Hopfian property is a solvable problem.)

It is the purpose of this section to indicate for which groups, or classes of groups, certain of these problems have been solved. Before starting an account of these particular classes of groups in which the problems are solvable, we wish to mention an important result of a very general nature which is connected with the unsolvability proofs mentioned above. G. Higman, 1961, proved that a finitely generated group can be embedded in a finitely presented group, if and only if it has a "recursively enumerable" set of defining relations. This theorem has many remarkable consequences, one of them being the Novikov-Boone-Britton Theorem on the unsolvability of the word problem in certain groups. We shall not try to explain what "recursively enumerable" means; instead, we refer the reader to M. Davis, 1958, or to Post, 1944.

The word problem refers to a particular presentation of a group in terms of generators and defining relators rather than to the group itself. Nevertheless, in order to avoid clumsiness of expression, we shall talk about the word problem for free groups or groups with a single defining relation, meaning that the group is presented in terms of free generators or in terms of generators and a single defining relator, respectively.

In free groups, the solution of the word problem (implicit in Dehn, 1911) can be obtained by a particularly simple procedure which may be described as a *monotonic reduction process*. This means: given any non-empty word W in the generators which defines the unit element, one can replace a subword of W by a shorter word and thereby arrive at a new word W' of shorter length than W which also defines the unit element; moreover, the allowable replacements are given in a finite list. (In the case of a free group, the words $a_ia_i^{-1}$ and $a_i^{-1}a_i$, where the a_i are the free generators appearing in W may be replaced by the empty word.)

Dehn, 1912, discovered that the word problem for the fundamental groups of closed, two dimensional, orientable surfaces of a genus $g \geqq 2$ can also be solved by a monotonic reduction process. The groups in question are groups with a single defining relator. (See Section 5.8, Corollary 5.15 for a definition.) Dehn's proof was based on the fact that the graph of such a group can be pictured as a set of $4g$-gons in the non-Euclidean plane with $5g$ of these polygons meeting at every vertex.

Greendlinger, 1960a, succeeded in discovering an algebraic (combinatorial) argument which can be used to replace Dehn's geometric approach. In doing this, Greendlinger was able to prove a theorem much more general than Dehn's and yet of the same simple nature. In order to formulate some of his results, we introduce the following concepts and notations:

Let G be a group on generators a_ν ($\nu = 1, 2, 3, \ldots$) and let R_μ ($\mu = 1, \ldots, m$) be a finite set of defining relators for G in the a_ν. We may assume that the R_μ are words in the a_ν with the following properties:

(i) The R_μ are cyclically reduced and non-empty.

(ii) With every R_l, $1 \leqq l \leqq m$, the set of the R_μ also contains R_l^{-1} and all cyclic permutations of the word R_l.

Such a set of defining relators will be called a *symmetric* set. Every finite set of defining relators can be replaced by a symmetric set (involving the same generators) as follows: cyclically reduce the defining relators and then adjoin the missing inverses and cyclic permutations.

Let k be a fixed positive integer, and let $\{R_\mu\}$ be a symmetric set of defining relators for G. We shall say that G is (or more precisely, is presented as) a *less than* $1/k$ *group* if the following condition is satisfied:

(iii) If, in any product $R_\mu R_\lambda$ ($\mu, \lambda = 1, \ldots, m$), at least $1/k$ of the symbols in R_μ or in R_λ can be deleted by free reductions, then $R_\mu R_\lambda$ is freely equal to the empty word.

Now Greendlinger's solution for the word problem in less than $1/6$ groups is achieved by the monotonic reduction process based on the following theorem:

Let W be a word in the generators of a less than $1/6$ group G, which is cyclically reduced. If $W = 1$, then some cyclic permutation of W contains a subword V which is identical with a subword of a relator R_μ, such that the length of V is greater than one-half the length of R_μ.

Clearly this theorem solves the word problem in G, since we can replace V by the inverse of the remaining part of the R_μ in which V appears and thereby shorten the length of W.

Greendlinger's method of proof is of a combinatorial nature. His results are actually much stronger than the theorem stated above, which is the exact generalization of Dehn's theorem for the fundamental groups of closed orientable surfaces of genus $g > 2$. (These groups are less than $1/(4g - 1)$ groups.)

Greendlinger's combinatorial approach was preceded by a profound study made by Tartakovskiĭ, 1949, who solved the word problem for large classes of groups in which the relators do not "react" very strongly with each other. A special feature of Tartakovskiĭ's work is the assumption that the order of the generators are given beforehand. These orders may be finite and, in contradistinction to Greendlinger's approach, less than 7. However, Tartakovskiĭ's combinatorial arguments and results are of such an elaborate nature that it is impossible to describe them without lengthy preparations. For generalizations see Stender, 1953.

The papers by Britton, 1956, 1957, and Schiek, 1956, are related to Tartakovskiĭ's work with respect to results but not with respect to methods. Britton, 1957 proves a theorem on the normal closure Ω^* of a subset Ω of a free product Π of finitely or infinitely many arbitrary groups. If the products of pairs of elements in Ω satisfy certain cancellation (or rather, non-cancellation) conditions, then definite statements can be made about the normal form of elements in the subgroup Ω^*. This result is used by Britton, 1955, to solve the word problem for certain quotient groups Π/Ω^* which may be infinitely generated and related. He also shows that in some (infinitely presented) cases of such quotient groups the word problem is unsolvable. His results overlap with those of Tartakovskiĭ; in part, Britton's are strong generalizations.

Schiek, 1956, starts with a free product of free Abelian groups and adds defining relators for which certain non-cancellation postulates are made. Whereas Britton, 1956, had merely announced that an investigation of the cancellations occurring in a product of three relators would improve his results, Schiek actually carries out this proposal and arrives at a highly involved theory. He compares his results with those obtained by Tartakovskiĭ and shows that, in certain instances, his are definitely stronger. Schiek's method is independent of Britton's method.

To complete our report on the word problem, we first have to go back to some older results.

Given two groups G_1 and G_2 for which we can solve the word problem, we can also solve the word problem for the free product $G = G_1 * G_2$ (provided that G is presented in this form). If G is the generalized free product of G_1 and G_2, with isomorphic subgroups $H_1 \subset G_1$ and $H_2 \subset G_2$ amalgamated, we can again solve the word problem in G if in G_1 and G_2 we can decide whether a given element belongs to H_1 or H_2, respectively, and if we can solve the word problem in H_1 and H_2, and also if the isomorphism between H_1 and H_2 is given constructively. This is an immediate consequence of the standard form for elements in free products with amalgamations (see Section 4.2). The word problem for groups with a single defining relation can be solved by a repeated application of

(i) The solution of the word problem for generalized free products of groups with a single defining relation and amalgamated free subgroups and

(ii) Embedding of a group with a single defining relation in a larger one with the same property, such that a generator of the smaller group is a preassigned power of a generator of the embedding group. (See Section 6.2 or Magnus, 1932.) This embedding process can, in turn, be obtained by using generalized free products.

However, the general method for solving the word problem in groups with a single defining relator is already a rather complicated process, and it does not show at all that in the special case of less than 1/6 groups a much simpler solution is possible. Even more complicated are the results and proofs found by Grushko, 1938, which generalize those of Magnus, 1932. The method developed by H. Haken, 1952, for solving the word problem in certain classes of groups is closely related to the method based on the theory of generalized free products. The results are again too complicated to be presented in a brief review. But it may be mentioned that Haken solves the word problem for certain groups appearing in a paper by Brauner, 1928. These groups deserve special interest for the following reason: Brauner had shown that a knot or a linkage is associated with each singularity of algebraic type of an analytic function of two complex variables, and he computed presentations for some of these groups in terms of generators and defining relators.

Both Haken, 1952, and Schiek, 1956, emphasize the fact that they can solve the word problem for groups in which some relators are commutators, in contradistinction to the results of Tartakovskiĭ.

Whereas there now exists a large number of substantial contributions to the word problem, Dehn's second problem, the conjugacy (or transformation) problem has been treated with much less success. At first, it may seem that it should not be much more difficult to decide whether two given words in the generators of a group define conjugate elements than to decide whether they are equal. And the rather simple solution of the conjugacy problem for free groups (see Section 1.4) seemingly supports this conjecture. However, nothing is known in general about the solution of the conjugacy problem for groups with a single defining relator. The only general contribution to the problem has been made by Greendlinger, 1960b, who solved the conjugacy problem for less than 1/8 groups and indicated that it can be solved for less than 1/6 groups. Greendlinger's result is again a direct generalization of Dehn's solution for the conjugacy problem in fundamental groups of closed, orientable, two dimensional surfaces. Again, Greendlinger uses algebraic methods which are completely independent of the geometric construction used by Dehn, 1912.

Apart from Greendlinger's theorem it seems that the solution of the conjugacy problem is known only in isolated cases as, for instance, the group of braids on four strings (Froehlich, 1936).

Dehn's isomorphism problem has hardly been touched at all. We can, of course, decide whether two finitely presented groups which are known to be Abelian are isomorphic or not. We also know when two free groups (presented on free generators) are isomorphic. But it is not even known when two groups with a single defining relator are isomorphic. It may be conjectured that this is the case if and only if the following is true: Let a_ν, $a_\nu{}^*$ ($\nu = 1, \ldots, n$) be the generators of the first and of the second group, respectively, and let $R(a_\nu)$, $R^*(a_\nu{}^*)$ be the defining relators of the first and second group, respectively. Then there exists a Nielsen transformation on the a_ν, mapping them into words $b_\nu{}^*$ in the a_ν, such that $R^*(b_\nu{}^*)$ is freely equal (in the a_ν) to $R^{\pm 1}$. The conjecture is true if $\langle a_\nu; R\rangle$ and $\langle a_\nu{}^*; R^*\rangle$ are both isomorphic to free groups. Vaguely related to this conjecture, is the result referring to automorphisms (Magnus, 1930) for groups G with a single defining relator R on a set of generators a_ν, $\nu = 1, 2, 3, \ldots$. If G can also be presented as a group with a single defining relator R^* and on the same generating elements a_ν, then R^* is freely equal (in the a_ν) to a conjugate of $R^{\pm 1}$. (This implies that a Nielsen Transformation on the a_ν which defines an automorphism of G must transform R into a word freely equal to an R^*.) Greendlinger, 1961, generalized this result to less than 1/6 groups. It should be noted that a group on generators a_ν with a single defining relator may have automorphisms which cannot be expressed as Nielsen Transformations on the a_ν, even if we exclude the obvious case where the relator is a power of a generator (Rapaport, 1959). Whether the same can happen in less than 1/6 groups seems to be unknown.

S. Lipschutz, 1960, 1962, and Greendlinger, 1962, applied Greendlinger's results to show that less than 1/6 groups and less than 1/8 groups have certain important properties in common with free groups.

The question of when a group with a single defining relator can be decomposed in a nontrivial manner into a free product has been investigated by Shenitzer, 1955. His tests are based on a combination of Grushko's Theorem (Chapter 4) and the Whitehead-Rapaport Theorem (Chapter 3).

6.2. Adjunction and Embedding Problems

The multiplicative notation employed in the study of groups leads in a natural way to the notion of roots in groups. Thus if n is a positive integer and g is an element of a group G, then a solution x of the equation

$$x^n = g$$

is called an nth root of g. There may or may not exist an nth root of g in G, and if there exists one, it need not be unique. If g does not have an nth root in G, the question arises whether we can adjoin one. More precisely, we pose the following problem:

(I) *Given an element g in a group G. Does there exist a group G^* containing an isomorphic replica of G (denoted again by G) as a subgroup such that in G^*, g has an nth root x?*

The affirmative answer to this question plays a role in the solution for the word problem for groups with a single defining relation (see Sections 4.4 and 6.1). The first systematic investigation of Problem (I) was made by B. H. Neumann, 1943b. Using the theory of the generalized free product with amalgamations) he proved:

Given a group G and an element $g \in G$, there exists exactly one group G with the following properties:

(i) *G^* contains G and is generated by the elements of G and an additional element x such that $x^n = g$ where n is a fixed positive integer.*

(ii) *All groups sharing property (i) with G^* are quotient groups of G^*.*

As an immediate consequence of this theorem, Neumann, 1943b, states:

Every group G is isomorphic to a subgroup of a group D in which every element has an nth root for every $n = 1, 2, 3, \ldots$.

Such a group D is called a *divisible* group.

The general adjunction problem was formulated by B. H. Neumann, 1943b, as follows:

(II) *Let G be a group and let $\tilde{G}$ be the free product of G and a free group freely generated by elements x_i. Let $W_\nu(x, g)$, $(\nu = 1, 2, \ldots)$ denote a set of elements in $\tilde{G}$ none of which is an element of G. Consider the quotient group G^* of $\tilde{G}$ obtained by adding the relations*

$$W_\nu(x, g) = 1 \tag{1}$$

to those of $\tilde{G}$. Will it be true that no element of G is mapped on the identity under the mapping $\tilde{G} \to G^*$? If so, we shall say that (1) can be *solved* over G.

In this general form, the adjunction problem is, of course, much too difficult to be investigated successfully. However, Neumann, 1943b, contributed the following theorem to it:

Let A be a group of automorphisms of G. If (1) can be solved over G by constructing a group G^*, then (1) also can be solved in such a manner that

A acts as a group of automorphisms on G^* *(preserving the action of A on* $G \subset G^*$*).*

Problem (II) has been investigated by Schiek, 1962a, b in the special case where the set of the x_i consists of one symbol x only and the set of relations (1) consists only of a single relation $R(x, g) = 1$. Schiek, 1962a, deals with the case where the sum of exponents x in R is zero. This is clearly a case where $R = 1$ may not be solvable over G since the relation $R = 1$ may be of the form

$$xax^{-1}b^{-1} = 1$$

where a, b are elements of G of finite but different orders. Schiek, 1962b, investigated the case where the sum of the exponents of x in R is not zero. In both cases, a solution of (II) is given only after introducing additional assumptions concerning R. A rather general result (based on a very short proof) was found by F. Levin, who showed that $R = 1$ is always solvable over G if all exponents of x in R are positive. However, even the case of a single "unknown" x and a single relator R is far from being settled completely. In particular, the following question raised by M. Kervaire in connection with a topological problem is still unanswered:

Let G be a group, let $\tilde{G}$ be the free product of G and an infinite cyclic group generated by x, and let R be an element of $\tilde{G}$ such that the quotient group G^* of $\tilde{G}$ obtained by introducing the relation $R = 1$ is of order 1. Does this imply that G itself must be of order 1?

The adjunction problem (II) may be considered as one of many embedding problems. We shall not attempt to give a general definition of embedding problems; it would necessarily be a rather inane one. Instead, we shall state some important results which will also clarify the general nature of the problems involved.

G. Higman, B. H. Neumann, and H. Neumann, 1949, proved:

Let G be a group containing two isomorphic (not necessarily different) subgroups A and B. Let μ *be an isomorphic mapping of A onto B;* $\mu(a) = b$ *shall denote the map* $b \in B$ *of the element* $a \in A$*. Then there exists a group H containing G and an element t such that, for all* $a \in A$,

$$t^{-1}at = \mu(a).$$

If G is locally infinite (i.e., has no elements of finite order except 1), *the group H can be constructed in such a manner that it is also locally infinite.*

Every group G can be embedded in a group G′ such that, in G′, all elements of the same order in G are conjugate. If G is locally infinite or countable, the same is true for G′.

Any locally infinite group can be embedded in a group G^* in which all elements except the unit element are conjugate to each other. If G is countable, G^* can be constructed in such a manner that it is also countable.

Clearly, G^* is an infinite simple group (the only finite group having only two classes of conjugate elements is of order 2).

Another type of embedding theorem proved in the same paper is the following one:

Any countable group G can be embedded in a group H generated by only two elements. H can be constructed in such a way that the number of its defining relations does not exceed the number of defining relations for G.

The result is based on a remarkable lemma which states:

Let F be the free group on two free generators a, b. There exist infinitely many elements e_i, $(i = 1, 2, 3, \ldots)$ in F which are free generators of a free subgroup E of F such that any normal divisor N of E is the complete intersection of the normal closure $\bar{N}$ of N in F with E; (i.e., $N = \bar{N} \cap E$). Higman, Neumann and Neumann, 1949, show that the elements

$$e_i = a^{-1}b^{-1}ab^{-i}ab^{-1}a^{-1}b^{i}a^{-1}bab^{-i}aba^{-1}b^{i}, \qquad (i = 1, 2, 3, \ldots) \tag{2}$$

have the desired property. According to B. H. Neumann and H. Neumann, 1959, the commutators

$$(b^{-2i+1}ab^{2i-1}, a), \qquad (i = 1, 2, 3, \ldots) \tag{3}$$

could be used for the same purpose as the e_i in (2). To describe the main results of this paper, we need the concepts of a variety and of the product of varieties, which are defined in Section 6.3. The authors prove:

Every group of a variety V can be embedded in the derived group P' of a group P which belongs to the variety $V \circ A$, where A denotes the variety of Abelian groups.

Every countable group G in V can be embedded in a two generator group H belonging to $V \circ A^2$. If G is finite, H can be constructed so as to be finite too.

These theorems have many applications, one of which will be described in Section 6.5.

B. H. Neumann, 1960, also proved that every countable fully ordered group G can be embedded in a fully ordered two generator group H such that the order relations in H and G agree.

The fact that, for each $n = 1, 2, 3, \ldots$, and each element g of any group G, an nth root of g can be adjoined to G has led to a theory of groups with roots. This theory has been developed systematically by Baumslag,

1960. We shall outline briefly a few of the basic definitions and results given by him. For a wealth of details, and also for a bibliography, including references to the work of Baer, 1937, Černikov, 1946, Kantorovič, 1946, 1948, and Mal'cev, 1949, the reader is referred to his paper.

Let ω denote an arbitrary non-empty set of prime numbers. Let E_ω denote the set of groups in which every element has a pth root for all prime numbers $p \in \omega$. Let U_ω denote the class of groups in which (for all $p \in \omega$), a pth root, if it exists, is unique. The intersection of E_ω and U_ω will be denoted by D_ω. The groups in D_ω will be called D_ω-groups; they form a new type of algebraic structure. Apart from the customary composition of elements in a group, we have a set of "operators," such that to each prime number $p \in \omega$ there belongs an operator π (which will be written to the right of the group element x it acts on) such that

$$(x\pi)^p = x, \; x^p\pi = x. \tag{4}$$

The relations (4) characterize D_ω-groups. We shall say that a set X ω-generates a D_ω-group if its elements together with their images under the operators π generate the group. In any group G, a subgroup H is called an ω-subgroup of G if, for $p \in \omega$, $g \in G$, the relation $g^p \in H$ implies $g \in H$. If H is a normal subgroup of the D_ω-group G such that G/H is a D_ω-group, then H is called an *ω-ideal* of G.

According to G. Birkhoff, 1935, there must exist free algebras in the variety of D_ω-groups. A set X is called a free ω-generating set of a D_ω group F if X ω-generates F and if, for every D_ω-group H and every mapping θ of X into H there exists a homomorphism φ of F into H that coincides with θ on X. A D_ω-group F is called a D_ω-free group if it is freely ω-generated by some set X. As an analog to a well known theorem about ordinary free groups, Baumslag, 1960, proves:

Let F be a D_ω-free group which is freely ω-generated by the set X. Then the factor group of F by its commutator ideal is a direct product of $|X|$ groups isomorphic to Γ_ω, where $|X|$ denotes the cardinality of X and where Γ_ω is the additive group of those rational numbers whose denominators contain only prime numbers belonging to ω.

The commutator ideal of F is generated by the images under the operators π of the ordinary commutator subgroup F' of F. The quotient group F/F' has a more complicated structure which has also been determined by Baumslag.

A connection between D_ω-groups and E_ω-groups is established by the following theorem proved by Baumslag.

The homomorphic images of the D_ω-free groups are precisely all the E_ω-groups.

In his papers, Baumslag gives a constructive definition of a D_ω-free product of D_ω-groups which is the analog of the ordinary free product of ordinary groups. For theorems on the embedding of U_ω-groups in D_ω-groups, on the construction of U_ω, E_ω, and D_ω-groups and many other problems the reader is referred to Baumslag, 1960. For an application to nilpotent groups with roots see also Baumslag, 1961a.

6.3. Varieties of Groups

We may describe an Abelian group by saying that the commutative law is satisfied by the composition of group elements, or, alternatively, that the identity

$$x^{-1}y^{-1}xy = 1 \tag{1}$$

holds in the group, meaning that (1) is satisfied for an arbitrary choice of x, y as elements of the group. We may say that the identical relation (1) defines the "variety" of Abelian groups. Clearly, we can define other "varieties" in the same manner; for instance, groups in which commutators commute (metabelian variety) or in which the nth power of every element is equal to 1. In order to obtain a precise definition, we proceed as follows:

Let F be a free group on (at most countably many) free generators x_ν, $\nu = 1, 2, 3, \ldots$. Let $\{W_\mu\}$ be a set of words W_μ in the x_ν. Let G be a group such that, under any homomorphic mapping of F into G the W_μ are mapped onto the unit element of G. Then we shall say that G belongs to the variety $V(W_\mu)$ defined by the identical relations $W_\mu = 1$.

According to Birkhoff, 1935, a variety may (equivalently) also be defined as a set of groups such that with every group G, all of its subgroups and quotient groups are in the variety, and with every set $\{G_\rho\}$ of groups G_ρ of the variety the unrestricted direct product of the G_ρ also belongs to the variety.

B. H. Neumann, 1939b, showed:

(i) *The set of words W_μ can always be chosen in such a manner that $W_1 = x_1^n$, where $n = 0, 1, 2, \ldots$ is a fixed integer and all other W_μ are words representing elements of the commutator subgroup of F.*

(ii) *Every group G defines a variety $V(G)$ which is the intersection of all varieties containing G.*

(iii) *Every variety V is completely defined by a single group $\Phi_\infty(V)$ which is called the free group of the variety on countably (infinitely) many generators.*

$\Phi_\infty(V)$ is defined as follows: Take the free group F_∞ on countably many generators a_ν. Then the group $\Phi_\infty(V)$ can be defined (on the a_ν as generators) by the relations $W_\mu(X_\nu) = 1$, where the words W_μ define the variety V and the X_ν range over all possible words in the a_ν.

In the same manner we define, for $n = 1, 2, 3, \ldots$, the group $\Phi_n(V)$, called the *free group of rank n in the variety* V, by starting with a free group F_n of rank n.

Obviously, all countable groups of a variety are quotient groups of its free group of countably infinite rank.

If the set of words W_μ in the generators x_ν of the free group F is finite, we can define the variety $V(W_\mu)$ by a single identical relation $W^* = 1$. To do this, we simply use different sets of symbols for every word W_μ and multiply all the resulting W_μ together to obtain W^*. We shall call a variety which can be defined by using a finite set $\{W_\mu\}$ a *finitely based variety.* However, such a variety may not be definable by one of its finitely generated free groups. It was shown by B. H. Neumann that the finitely generated groups with the identity $(X^2Y^2)^2 = 1$ are nilpotent, but that the class of nilpotency increases with the number of generators.

If two varieties V and V' are different from each other, V must contain a finitely generated group G_0 which is not contained in V'. This is obvious, since there must exist an identical relation in V' which does not hold for all groups in V, and such an identical relation can involve only finitely many symbols X_ν.

Hanna Neumann, 1956a, assigned to every variety V an ascending chain of varieties V_n, $(n = 1, 2, 3, \ldots)$ and a descending chain of varieties V^n which satisfy the inclusion relations

$$V_1 \subset V_2 \subset V_3 \ldots \subset V \subset \ldots \subset V^n \subset \ldots \subset V^1, \tag{2}$$

and are defined as follows: V_n is the variety generated by the n-generator groups of V under the processes of taking quotient groups, subgroups, and unrestricted direct products; V^n is the variety of all groups whose n-generator subgroups belong to V.

Hanna Neumann showed that the ascending chain is finite (i.e., coincides with V after a finite number of steps) if and only if we can find a set $\{W_\mu\}$ defining V which involves only a finite number of symbols x_ν. If $V = V(G)$, where G is a finite group, the ascending chain is finite if and only if V is finitely based. The ascending chain will be finite if and only if V is defined by an n-generator group for some finite n. In this case, the free group of rank n of the variety V defines V completely.

Higman, 1959, and P. M. Neumann, 1963, gave examples where the ascending chain in (2) is properly infinite.

It is an unsolved and, apparently, very difficult problem to find out whether the descending series will always be finite or not.

For a finitely based variety, both the ascending and the descending chain in (2) will be finite although they may be of different length. An example is provided by the "Burnside variety" defined by the identical relation $X_1^3 = 1$. (H. Neumann, 1956a remarks that this follows from the results obtained by Levi and van der Waerden, 1933.)

Varieties $V(G)$ where G is a finite group were investigated by G. Higman and D. C. Cross, 1960. The authors found large classes of finite groups for which $V(G)$ is finitely based. A further result of their paper is:

Let a finite group G be called *critical* if it is not contained in the variety generated by its proper quotient groups. *Let V be a variety which is finitely based, in which all finitely generated groups are finite and which contains only a finite number of critical groups. Then every subvariety of V has the same properties.*

The variety V^* of metabelian groups can be defined by the single identity

$$((x_1, x_2), (x_3, x_4)) = 1, \tag{3}$$

where, as usual,

$$(x, y) = x^{-1}y^{-1}xy$$

denotes the commutator of x and y. H. Neumann, 1956a, showed that the free metabelian group of rank 3 belongs also to a variety V' which contains V^* properly and which is in turn, contained properly in the variety V'' defined by the free metabelian group M_2 of rank 2. Surprisingly, M_2 can be defined by the single identity

$$((x_1, x_2), (x_1^{-1}, x_2)) = 1, \tag{4}$$

(Higman, 1959). N_2 is therefore the free group of rank 2 in at least three different varieties, two of which are given respectively by the identical relations (3) and (4).

Lyndon, 1952, proved that a nilpotent variety N (i.e., a variety all groups of which are nilpotent) is always finitely based. In addition, it can be shown that the class of the groups in N is bounded. If C is the largest class of any group G in the variety (i.e., if the $(C + 1)$st term of the lower central series of all $G \in N$ is the identity but not always the Cth term) we shall say that the variety N is nilpotent of class C. Higman, 1956, showed that N is defined in this case by its free group of rank $C + 1$ and is generated by its C-generator groups. These results are not always optimal. However, if n is the smallest integer such that all varieties N of class C can be defined by its free group of rank n, then $n \to \infty$ as $C \to \infty$. This follows from a result due to Burrow, 1958, which states:

In the free group F on n free generators there exist words belonging to the Kn^2th group of the lower central series of F but not to the $(Kn^2 + 1)$st group and yet which are mapped onto an element of the $(Kn^2 + 1)$st group of the lower central series of any group F^* of fewer than n generators under any homomorphic mapping of F into F^*. Here $K = 1, 2, 3, \ldots$. For $n > 3$, the number Kn^2 may be replaced by Kn.

Let U and V be any two varieties (not necessarily distinct). The *product* $U \circ V$ is defined as the set of all groups which have a normal subgroup in U with a quotient group in V. The product $U \circ V$ is again a variety. Higman, 1959, showed:

$A \circ V$ is finitely based if A is nilpotent and V is finitely based.

Related results may be found in Higman and Cross, 1960.

B. H. Neumann, H. Neumann, and P. M. Neumann, 1962, proved the following theorem:

Uniqueness of variety factorization: Let U, V, U', V' be varieties of groups such that

$$U \circ V \subseteq U' \circ V'.$$

Then either

$$V \subseteq V'$$

or

$$U \subset U'$$

or

$$U = U' = \mathcal{O},$$

where $\mathcal{O}$ is the variety of all groups. Note that $U \subset U'$ means here that U is properly contained in U'.

In the same paper, the authors prove a large number of other results concerning the algebra of varieties. In this algebra, composition of varieties can be defined, apart from multiplication, by using the fact that the varieties form a lattice.

A variety U is called *indecomposable* if $U \neq E$ (the variety consisting of the group of order 1) and if $U = V \circ V'$ implies $V = E$ or $V' = E$. Non-trivial examples of indecomposable varieties are given by P. M. Neumann, 1963.

Our enumeration of results is far from being complete. For examples not fully included in the papers quoted see also Wever, 1950, and MacDonald, 1960, 1962.

6.4. Products of Groups

Golovin, 1950, investigated the question of whether the concepts of the direct product and of the free product of groups are special cases of a wider class of products. In order to present his results, we shall use the following notations. Given any two groups, A and B, we denote by $A \times B$ their direct product and by $A * B$ their free product. By $A \circ B$, we denote either the direct or the free or any other product still to be defined. Also, we shall use the notation A^G to denote the normal closure of A (the smallest normal subgroup containing A) in the group G, where A will usually be a subgroup of G, but may be any set of elements of G.

Now we shall list six properties of products of groups, all of which are satisfied by the direct and the free product:

I. Given any two groups $\tilde{A}$ and $\tilde{B}$, there exists a group G, denoted by $A \circ B$ and called the product of $\tilde{A}$ and $\tilde{B}$, such that G contains an isomorphic replica A of $\tilde{A}$ and an isomorphic replica B of $\tilde{B}$ and is generated by A and B.

II. The normal closure A^G intersects B in the identity, and B^G intersects A in the identity.

III. $A \circ B \simeq B \circ A$, under the isomorphism which maps the subgroups A, B of the first product into the subgroups A, B, respectively, of the second one. (We call this isomorphism the *natural* isomorphism.)

IV. If $\tilde{A}$, $\tilde{B}$, $\tilde{C}$ are any three groups (with $C \simeq \tilde{C}$), then

$$(A \circ B) \circ C \simeq A \circ (B \circ C),$$

again under the natural isomorphism.

V. Let M be any normal divisor of A and let N be any normal divisor of B. Then, if $G = A \circ B$,

$$(A/M) \circ (B/N) \simeq G/(M^G \cdot N^G)$$

under the natural isomorphism, mapping A/M and B/N in $(A/M) \circ (B/N)$ onto the subgroups

$$(A \cdot M^G \cdot N^G)/(M^G \cdot N^G),$$
$$(B \cdot M^G \cdot N^G)/(M^G \cdot N^G)$$

of $G/(M^G \cdot N^G)$, respectively.

VI. Let $H \subset A$ and $K \subset B$ be any subgroups of A and B, respectively. Then the subgroup S of $G = A \circ B$ generated by H and K is isomorphic to $T = H \circ K$, under the isomorphism which maps $H \subset S$ onto $H \subset T$ and $K \subset S$ onto $K \subset T$. (We call this isomorphism the *natural* isomorphism.)

Golovin, 1950, called a product satisfying I and II a *regular product*, and a product that also satisfies III and IV a *fully regular product*. If we use the notation (A, B) for the subgroup generated by the commutators (a, b) with $a \in A$, $b \in B$, we find that (A, B) is a normal divisor of $A * B$ and, under the natural isomorphism,

$$A \times B \simeq (A * B)/(A, B).$$

Golovin showed that any regular product $A \circ B$ has the property that, under the natural isomorphism,

$$A \circ B \simeq (A * B)/N,$$

where $N \subset (A, B)$ is a normal divisor of $A * B$. Ruth Struik, 1956, showed that there exists products for which I and II but not III, or I, II, and III but not IV, or I, II, III, and IV but not V is satisfied. In addition, she gave an example of a product satisfying I, II, III, and V but not IV. Regular products satisfying III but not IV were also constructed by Moran, 1956, and by Benado, 1956, 1957.

Golovin, 1950, constructed an infinite sequence of fully regular products all of which also satisfy condition V. His construction was put into different shapes by Moran, 1956, and Struik, 1956. Following Golovin, we shall define, for $k = 1, 2, 3, \ldots$, a product $A \underset{k}{\circ} B$, which will be called kth *nilpotent product* of A and B. To do so, we introduce the following notations:

Let G by any group and let H be a subgroup of G. We define, for $k = 0, 1, 2, \ldots$:

$${}_0H_G = H^G, \qquad {}_kH_G = ({}_{k-1}H_G, G).$$

Then the kth nilpotent product is defined by

$$A \underset{k}{\circ} B = (A * B)/{}_k(A, B)_G, \qquad \text{(Golovin)}$$

$$A \underset{k}{\circ} B = (A * B)/({}_0A_G, {}_kB_G) \cdot ({}_kA_G, {}_0B_G), \qquad \text{(Struik)}$$

$$A \underset{k}{\circ} B = (A * B)/(A, B) \cap {}_k(A * B)_G, \qquad \text{(Moran)}$$

where throughout G is used as an abbreviation for $A * B$. The equivalence of the three definitions can be derived from an identity proved independently by Struik, 1956, and Moran, 1956:

In $G = A * B$,

$${}_k(A, B)_G = \prod_{n+m=k} ({}_mA_G, {}_nB_G) = ({}_0A_G, {}_kB_G) \cdot ({}_0B_G, {}_kA_G).$$

Golovin, Moran, and Struik proved that, in general, $A \underset{k}{\circ} B$ and $A \underset{l}{\circ} B$ are not isomorphic under the natural isomorphism if $k \neq l$.

Moran, 1956, showed that Golovin's nilpotent products are special cases of a much more general class of products which he called *verbal products*. To construct them, we define first a fixed (but arbitrary) verbal subgroup $V(G)$ for every group G (see the definition in Section 2.2). Then we define the V-product $A \underset{V}{\circ} B$ by

$$A \underset{V}{\circ} B = (A * B)/(A, B) \cap V(A * B)).$$

For all possible types of verbal subgroups, $V(G)$, the V-product of groups is fully regular and satisfies postulate V. In addition, it is possible to write down explicitly the V-product of any set of groups $\tilde{A}_\alpha$, where α runs through an arbitrary set L of labels. (Since we shall use the free product of the A_α we may identify them with their replicas in their free product, omitting the use of the $\tilde{A}_\alpha$.) For this purpose, let G be the free product of all the A_α, and let $C(G)$ be the product of the normal closures of all (A_α, A_β) in G, where $\alpha \neq \beta$, $\alpha, \beta \in L$. We call $C(G)$ the *Cartesian subgroup* of the free product G; obviously, $G/C(G)$ is the (restricted) direct product of the A_α. Then Moran showed that the V-product of the A_α is given by

$$G/(C(G) \cap V(G))$$

Incidentally, Golovin, 1950, proved that $C(G)$ is always a free group.

Except for verbal products, no other examples of products satisfying postulates I to V are known. On the other hand, the free and the direct product are the only verbal products known to satisfy postulate VI. In fact, it is fairly easy to show that many verbal products do not satisfy postulate VI, not even if the factors in the V-product are restricted to groups whose V-subgroups are the identity. However, Moran 1959a, b, constructed large classes of regular products of groups satisfying postulates III and IV which are not verbal products in general.

Golovin, 1950, had shown that each decomposition of a given group into a regular product corresponds to a set of orthogonal idempotent endomorphisms. Benado, 1956, 1957, used this result as a starting point for an investigation of associative products, and for constructing examples of non-associative products.

In general, it is a difficult task to prove that two verbal products are different if the verbal subgroups used for their definition are different in a free group on sufficiently many generators. If we define the lth *soluble product* as the verbal product arising from the case where $V(G)$ is the lth derived group of G, then it can be shown (Moran, 1958c) that for $l \geqq 2$ the lth soluble product of Abelian groups contains a locally infinite subgroup. Since Golovin, 1950, 1951, had proved that the nilpotent products of a finite number of finite groups are themselves finite, it follows that the soluble products are, for $l \geqq 2$, not nilpotent products. R. Struik, 1959, proved that a large class of verbal products (defined by using "complex" commutators for the definition of the underlying verbal subgroups) are different from each other and from Golovin's nilpotent products. As a tool, she used expansions in associative rings, similar to the one used in Section 5.5.

All of the problems that have been studied in connection with the definition of the direct and of the free product also can be investigated for verbal and other fully regular products of groups. These investigations have been carried out in part for nilpotent, especially for 2-nilpotent, products by Golovin, 1951a, b. Maximality properties have been found by Moran, 1958c, who also showed that a group isomorphic to a verbal product $A \underset{V}{\circ} B$, where $\underset{V}{\circ}$ does not denote the free product, cannot be decomposed into a free product (of non trivial groups) at all unless A and B are of order 2.

The papers by Golovin, Bendo, Moran, and Struik quoted above contain a very large number of results, many of which are still of a somewhat isolated nature. We cannot go into the details here; we may mention, however, a result due to MacHenry, 1960, which points in a different direction. In its simplest version, it states that the tensor product of two Abelian groups is isomorphic to the commutator subgroup of the 2-nilpotent product of these Abelian groups.

6.5. Residual and Hopfian Properties

The concept of a variety of groups, which we discussed in Section 6.3, provides a useful classification for the bewildering manifold of groups. However, this usefulness has narrow limitations since there are so many simply defined groups (as, for instance, groups on more than two generators with a single defining relator) which, like the free groups, belong to no variety except the trivial one containing all groups. Also, the set of all finite groups (which constitutes a well investigated class of important groups with interesting properties) cannot be accommodated in any variety smaller than that of all groups. Two of the concepts which we are going to define now (residually finite and Hopfian) will describe classes of groups which include the finite groups and, although rather large, contain only groups which have something in common with finite groups.

We start with a rather general definition. Let Π be a *property* of groups. By that we mean something such as being finite, free, Abelian, or belonging to a specified variety of groups. (We shall restrict the concept of "property" later on, but at the moment this is not necessary.) Following P. Hall, we shall say that a group G is *residually* Π if G satisfies the following condition:

(I) *For any element* $g \neq 1$ *in* G *there exists a normal divisor* K *of* G *which does not contain* g *and is such that* G/K *has the property* Π.

Thus a group G is called *residually finite* if, for every $g \neq 1$ in G, there exists a normal divisor K of finite index in G which does not contain g. For finite groups this is trivially true. For infinite residually finite groups G, the set of all normal divisors K of finite index in G provides us with a subgroup topology of the type described in Section 5.14.

Finitely generated residually finite groups have two remarkable properties. One was discovered by Baumslag, 1963a. It may be stated as follows:

The automorphism group of a finitely generated residually finite group G *is residually finite.*

The proof is so simple that we shall reproduce it here. Let A be the group of automorphisms of G, and let α be an element of A which is not the identical automorphism. Then there exists at least one $g \in G$ such that

$$h = \alpha(g)g^{-1} \neq 1.$$

Let K be a normal subgroup of finite index n in G such that K does not contain h. Let K^* be the intersection of all normal subgroups with index n in G. According to M. Hall, 1949, K^* is again of finite index in G and is also characteristic in G. Now A induces a finite group $\bar{A}$ of automorphisms in G/K^*, and $\bar{A}$ is a quotient group of A. Since h is not in K it is also not in K^* and therefore the map $\bar{\alpha}$ of α in $\bar{A}$ is $\neq 1$. Therefore, A is residually finite.

Baumslag's result permits an easy and, in some cases, the only known proof for the residual finiteness of certain groups. For instance, the fact that free groups are residually finite (Levi, 1933) can be derived as follows: From the theory of the modular group it is known that the two matrices

$$b = \begin{pmatrix} 1 & 2 \\ 0 & 1 \end{pmatrix}, \qquad c = \begin{pmatrix} 1 & 0 \\ 2 & 1 \end{pmatrix}$$

generate a free group. These matrices define automorphisms of a free Abelian group of rank two, which obviously is residually finite. Therefore the free group of rank two is residually finite. Since subgroups of residually finite groups have the same property, all free groups of countable rank must be residually finite.

In the case of groups generated by matrices, we could prove residual finiteness simply by taking the quotient groups of congruence subgroups (an idea used extensively by Mal'cev, 1940). However, there are cases like the group of automorphisms of a free group on $n > 2$ free generators (about which, in spite of the results due to Nielsen, 1924 we know little) or of the group of automorphisms of a fundamental group of two dimensional orientable closed surfaces (about which we know even less) where Baumslag's argument is the only one available to prove residual finiteness.

The other important property of residually finite groups may be stated as follows (according to Mal'cev, 1940):

A finitely generated residually finite group G cannot be isomorphic with one of its proper quotient groups.

We shall say that such a group G is *Hopfian* or has the *Hopfian property*. The proof of our statement is again very simple. Let $N \neq 1$ be a normal divisor of G such that $\tilde{G} = G/N$ is isomorphic with G. We wish to show that this is impossible by proving that no element $g \neq 1$ of G can be mapped onto the unit element 1 of $\tilde{G}$. Let K be a normal divisor of finite index n in G such that K does not contain g. Let K^* be the intersection of all normal subgroups of an index $\leqq n$ in G. K^* is again of finite index n^* in G, and it does not contain g. Under the homomorphic mapping of G onto $\tilde{G}$, K^* is mapped onto a subgroup K^{**} of $\tilde{G}$. Since the index of a finite subgroup cannot be increased by a homomorphic mapping, K^{**} will contain the subgroup $\tilde{K}^*$ of $\tilde{G}$ which corresponds to $K^* \subset G$ under the *isomorphic* mapping of G onto $\tilde{G}$. The preimage P of $\tilde{K}^*$ in G is therefore contained in K^*. On the other hand, P contains N, and therefore g is not mapped onto 1 (under the homomorphism of G onto $\tilde{G}$).

The question whether the fundamental groups of closed, two dimensional, orientable surfaces are Hopfian is topologically significant. It has been raised and answered in the affirmative by H. Hopf, 1930, who used topological methods. Below, we shall quote algebraic results from which the same fact follows easily.

For a Hopfian group, it is easier to find automorphisms than for non-Hopfian groups. Let a_ν, $\nu = 1, \ldots, n$, be generators of a Hopfian group G, and $R_\mu(a_\nu)$ ($\mu = 1, 2, 3, \ldots$) be a set of defining relators. Then the mapping $a_\nu \to b_\nu$ (where the b_ν are words in the a_ν) will be an automorphism if an only if the b_ν generate G and, for all μ,

$$(1) \qquad R_\mu(b_\nu) = 1.$$

For a non-Hopfian group, this test is obviously insufficient, since the b_ν may satisfy relations not derivable from (1). This may happen even if the b_ν arise from the a_ν by a Nielsen transformation (Chapter 3). An example of this type can be derived from Higman, 1951.

It is easily seen that an infinitely generated residually finite group may be non-Hopfian. For instance, the free group on countably many generators a_ν ($\nu = 1, 2, 3, \ldots$) is mapped onto an isomorphic replica of itself by the homomorphism

$$a_1 \to 1, \quad a_2 \to a_1, \quad a_3 \to a_2, \ldots .$$

Finitely generated but infinitely related non-Hopfian groups were found by B. H. Neumann, 1950. A 3-generator, 2-relator non-Hopfian group was discovered by G. Higman, 1951. The simplest example of a non-Hopfian finitely presented group was described by Baumslag and Solitar, 1962 who showed:

The group on two generators a, b, with the single defining relation

$$a^{-1}b^2a = b^3 \tag{2}$$

is non-Hopfian.

In the same paper, the authors indicated that the group G_0 on two generators a, b with the defining relation

$$a^{-1}b^{12}a = b^{18} \tag{3}$$

is Hopfian but contains a normal subgroup N of finite index which is non-Hopfian. Incidentally, this shows that Hopfian groups need not be residually finite even if they are finitely generated. For, if G_0 were residually finite, N would be both residually finite and, as a subgroup of finite index, finitely generated.

As a by-product, Baumslag and Solitar, 1962, exhibit two finitely presented non-isomorphic groups which are homomorphic images of each other. (The first example of such an occurrence was found by B. H. Neumann.)

Residual properties of groups were systematically investigated first by Gruenberg, 1957. To formulate a few of his results we begin with the following definition:

(II) *A property* Π *is called a root property:*

(i) *If every subgroup of a group having property* Π *also has property* Π.

(ii) *Whenever* G *and* H *have property* Π, *their direct product* $G \times H$ *also has the property.*

(iii) *If* $G \supseteq H \supseteq K \supseteq 1$ *is a sequence of subgroups, each normal in its predecessor, and if* G/H *and* H/K *have property* Π, *then* K *contains a subgroup* L, *normal in* G, *such that* G/L *also has the property.*

Examples of root properties are finiteness, solvability (i.e., the sequence of derived groups terminates with 1 after a finite number of terms), and the property of "having order a power of a given prime p." This third property will be denoted by Π_p. Gruenberg, 1957, proved:

If Π is a root property, then every free product of residually Π-groups is itself residually Π if and only if every free group is residually Π.

This theorem applies if Π denotes the property of finiteness, or of being solvable, or the property Π_p for any prime number p. (For all of these Π, free groups are residually Π.)

Of the many other results obtained by Gruenberg, 1957, we mention the following two:

Every finitely generated nilpotent group, in which no element $\neq 1$ is of finite order, is residually Π_p for any p. If the group has elements of finite order, and if $p_1, \ldots, p_r$ are the different prime numbers dividing the orders of its elements, then the group is residually $\Pi_{p_1}, \Pi_{p_2}, \ldots, \Pi_{p_r}$.

Nilpotency is not a root property since (iii) in Definition (II) is not always satisfied for it. However, it is possible to prove a simple theorem about certain *polynilpotent* groups G which are defined by the fact that $G = G_0$ contains a descending series of groups

$$G_0 \supset G_1 \supset G_2 \ldots \supset G_m = 1$$

each of which is normal in the previous one and such that $G_{\mu-1}/G_\mu$ is nilpotent for $\mu = 1, \ldots, m$. We need a particular set of polynilpotent groups, namely, the *free polynilpotent groups of class-row*

$$(c_1, \ldots, c_m) \tag{4}$$

which are defined as follows. Let X be any free group. Let $\gamma_i X$ denote the ith group of the lower central series of X, and define recursively

$$\gamma_{i_k}\gamma_{i_{k-1}} \cdots \gamma_{i_1}X$$

as the i_kth group of the lower central series of

$$\gamma_{i_{k-1}} \cdots \gamma_{i_1}X.$$

Then the free polynilpotent group of class-row

$$(i_1 - 1, i_2 - 1, \ldots, i_m - 1)$$

is defined as the quotient group

$$X/\gamma_{i_m}\gamma_{i_{m-1}} \cdots \gamma_{i_1}X.$$

We shall call m the *length* of the polynilpotent group.

Now Gruenberg, 1957, proved:

A free polynilpotent group of class-row (4) *is residually Π_p for any prime p exceeding all of the integers $c_1, \ldots, c_{m-1}$.*

Gruenberg remarks that by using an unpublished result obtained by P. Hall the restriction on p can be shown to be unnecessary.

Gruenberg's theorem is a strong generalization of a theorem due to K. Hirsch, 1946, which states that all polycyclic groups are residually finite.

In turn, Gruenberg's theorem has been generalized by Baumslag, 1963c, who within a more general framework, proved:

Let Y be a free group, R a normal divisor of Y and S a fully invariant subgroup of R. If Y/R and R/S are residually finite, so is Y/S. If Y/R and R/S have the property Π_p (for the same prime number p), then Y/S also has this property.

The question of which residual properties are retained by generalized free products of nilpotent groups with amalgamated subgroups has been investigated by Baumslag, 1963c. In order to formulate at least some of his results, we introduce the following notations.

If Π, Π' denote properties of groups, we denote by $\Pi \circ \Pi'$ the class of all groups G which possess a normal subgroup N with property Π and a quotient group G/N with property Π'. In particular, we shall denote by Π_F the property of being residually finite, by Π_N the property of being nilpotent, and by Φ the property of being free.

If A and B denote any groups which have properties Π and Π', respectively, then

$$\sigma(A, B; \Gamma)$$

shall denote the class of groups which are free products of a group A and a group B with an amalgamated subgroup having property Γ.

By $\sigma(A, B)$ we mean the class of generalized free products of A and B with no restriction on the amalgamated subgroups. Now Baumslag, 1963c, proved:

If A and B are finite, then the groups $\sigma(A, B)$ are Π_F. If A and B are residually finite, then $\sigma(A, B$; finite) is residually finite. If A, B, are Π_N, then $\sigma(A, B)$ is in $\Pi_F \circ \Pi_F$ and even in $\Phi \circ \Pi_F$. But there exist examples where A, B are Π_N but $\sigma(A, B)$ is not residually finite. (In this case, A and B must be non-Abelian.)

A subgroup H of a group G will be called *closed in G* if, for any element g of G, $g^n \in H$, $n \neq 0$ implies $g \in H$. Baumslag, 1963c, also showed:

If A and B are nilpotent and finitely generated, then the groups

$$\sigma(A, B; \text{closed in } A \text{ and } B)$$

and

$$\sigma(A, B; \text{cyclic})$$

are residually finite.

This theorem is connected with earlier results found by Baumslag, 1962, which may be stated as follows:

Let F be free and let G be free Abelian of countably infinite rank. Then

$$\sigma(F, G; \text{cyclic and closed in } F \text{ and } G)$$

is residually free.

Since free groups are residually finite, this implies that the groups in question are also residually finite.

Finally, we mention the following result derived by Baumslag, 1962, because of its applications:

Let F and $\bar{F}$ be isomorphic free groups. Let the element $u \in F$ generate its own centralizer in F, and let $\bar{u}$ be the image of u in $\bar{F}$ under the isomorphism connecting F and $\bar{F}$. Then the free product K of F and $\bar{F}$ with the amalgamation defined by $u = \bar{u}$ is residually free and every two-generator subgroup of K is free.

This theorem implies that the fundamental groups of orientable, two dimensional, closed surfaces are residually free (of finite rank) and therefore Hopfian. It also implies that certain Fuchsian groups, on two generators c, d, with defining relations

$$c^l = d^m = (cd)^n = 1$$

are Hopfian.

The fact that the fundamental groups of closed, two dimensional, orientable surfaces are residually free and Hopfian was proved using a different method by K. Frederick, 1962.

P. Hall, 1961, proved a large number of theorems which are closely related to problems involving residual properties. Some of these arise from an investigation of the Frattini group (i.e., the intersection of maximal subgroups of a given group) which is nilpotent for finite groups but may have a much more complicated, although still well determined structure, for other groups. (For instance, it will be polynilpotent of length $\leq m - 1$ for finitely generated polynilpotent groups of length m.) However, of all of Hall's results, we mention here only one which refers to the Hopfian property:

There exists a 3-generator group G for which $[G'', G] = 1$ and which is non-Hopfian.

Here G'' denotes the second derived group of G, and $[G'', G]$ is the group generated by commutators of elements of G'' and G.

Baumslag and Solitar, 1962, announced the construction of a non-Hopfian group G on two generators for which $G'''' = 1$.

References

Note: In some cases, the reference to a paper is followed by a reference to its review in the *Zentralblatt* or in the *Mathematical Reviews* (M. R.).

Adelsberger, Hertha, 1930, Über unendliche diskrete Gruppen, *J. reine u. angew. Math.*, **163,** 103–124.

Albert, A. Adrian, 1937, *Modern Higher Algebra.* The University of Chicago Press.

Alexander, J. W., 1928, Topological Invariants of Knots and Links. *Trans. Amer. Math. Soc.*, **30,** 275–306.

Artin, E., 1925, Theorie der Zöpfe *Abh. Math. Sem., Hamburg Univ.*, **4,** 47–72.

Artin, E., 1947a, Theory of Braids. *Annals of Math.*, **48,** 101–126.

Artin, E., 1947b, The Free Product of Groups. *Amer. J. Math.*, **69,** 1–4.

Artin, E., 1947c, Braids and Permutations. *Annals of Math.*, **48,** 643–649.

Auslander, M., and R. C. Lyndon, 1955, Commutator Subgroups of Free Groups. *Amer. J. Math.*, **77,** 929–931.

Baer, R., 1927, Über Kurventypen auf Flächen. *J. reine u. angew. Math.*, **156,** 231–246.

Baer, R., 1929, Die Abbildungstypen-Gruppe der orientierbaren geschlossenen Flaeche vom Geschlechte 2. *J. reine u. angew. Math.*, **160,** 1–25.

Baer, R., 1927, Kurventypen auf Flaechen. *J. reine u. angew. Math.*, **156,** 231–246.

Baer, R., 1937, Abelian Groups without Elements of Finite Order. *Duke Math. J.*, **3,** 68–122.

Baer, R., 1938, Groups with Abelian Central Quotient Group. *Trans. Amer. Math. Soc.*, **44,** 357–412.

Baer, R., 1940, Nilpotent Groups and their Generalizations. *Trans. Amer. Math. Soc.*, **47,** 393–434.

Baer, R., 1944, The Higher Commutator Subgroups of a Group. *Bull. Amer. Math. Soc.*, **50,** 143–160.

Baer, R., 1945, Representation of Groups as Quotient Groups. *Trans. Amer. Math. Soc.*, **58,** 295–419.

Baer, R., 1949, Free Sums of Groups and their Generalization. *Amer. J. Math.*, **71,** 706–742.

Baer, R., 1950, Free Sums of Groups and their Generalization II, III *Amer. J. Math.*, **72,** 625–670.

Baer, R., and F. Levi, 1936, Freie Producte und ihre Untergruppen. *Compositio Mathematica*, **3,** 391–398.

Baker, H. F., 1904, Alternants and Continuous Groups. *Proc. London Math. Soc.* (2), **3,** 24–47.

Baker, R. P., 1931, Cayley Diagrams on the Anchor Ring. *Amer. J. Math.*, **53,** 645–669.

Baumslag, G., 1960, Some Aspects of Groups with Roots. *Acta Math.*, **104,** 217–303.

Baumslag, G., 1961, Some Remarks on Nilpotent Groups with Roots. *Proc. American Math. Soc.*, **12**, 262–267.

Baumslag, G., 1962, On Generalized Free Products. *Math. Z.*, **78**, 423–438.

Baumslag, G., 1963a, Automorphism Groups of Residually Finite Groups. *J. London Math. Soc.*, **38**, 117–118.

Baumslag, G., 1963b, On the Residual Finiteness of Generalized Free Products of Nilpotent Groups. *Trans. Amer. Math. Soc.*, **106**, 193–209.

Baumslag, G., 1963c, Wreath Products and Extensions. *Math. Z.*, **81**, 286–299.

Baumslag, G., and D. Solitar, 1962, Some Two-generator One-relator Non-Hopfian Groups. *Bull. Amer. Math. Soc.*, **68**, 199–201.

Benado, M., 1956, Über die allgemeine Theorie der regulären Produkte von Herrn O. N. Golovin. I. *Math. Nachr.*, **14**, 213–234.

Benado, M., 1957, Über die allgemeine Theorie der regulären Produkte von Herrn O. N. Golovin. II. *Math. Nachr.*, **16**, 137–194.

Bergau, P., and J. Mennicke, 1960, Über topologische Abbildungen der Brezelfläche vom Geschlecht 2. *Math. Z.*, **74**, 414–435.

Birkhoff, G., 1935, On the Structure of Abstract Algebras. *Proc. Cambridge Philos. Soc.*, **31**, 433–454.

Birkhoff, G., 1937, Representability of Lie Algebras and Lie Groups by Matrices. Annals of Math., **38**, 526–529.

Blanchfield, R. C., 1949, Applications of Free Differential Calculus to the Theory of Groups. Senior Thesis, Princeton University.

Blanchfield, R. C., 1951, Invariants of Self-linking. *Annals. of Math.*, **53**, 556–564.

Blankinship, W. A., and R. H. Fox, 1950, Remarks on Certain Pathological Open Subsets of Three-space and their Fundamental Groups. *Proc. Amer. Math. Soc.*, **1**, 618–624.

Bohnenblust, F., 1947, The Algebraic Braid Groups. *Annals of Math.*, **48**, 127–136.

Boone, W. W., 1955, Certain simple unsolvable problems of group theory. *Indig. Math.*, **16**, 231–237, 492–497; **17**, 252–256; **19**, 22–27, 227–232.

Brahana, H. R., 1926, Regular Maps on an Anchor Ring. *Amer. J. Math.*, **48**, 225–240.

Brandt, Angeline, 1944, The Free Lie Ring and Lie Representations of the Full Linear Group. *Trans. Amer. Math. Soc.*, **56**, 528–536.

Brauner, K., 1928, Zur Geometrie der Funktionen zweier komplexer Veraenderlicher. *Abhandlungen Math. Sem. d. Hamburg. Univ.*, **6**, 1–55.

Britton, J. L., 1956, Solution of the Word Problem for Certain Types of Groups. I. *Proc. Glasgow Math. Assoc.*, **3**, 45–54.

Britton, J. L., 1957, Solution of the Word Problem for Certain Types of Groups. II. *Proc. Glasgow Math. Assoc.*, **3**, 68–90.

Britton, J. L., 1958, The Word Problem for Groups. *Proc. London Math. Soc.*, **8**, No. 32 (3rd Series), 493–506.

Burau, W., 1933, Über Zopfinvarianten. *Abh. Math. Sem., Hamburg Univ.*, **9**, 117–124.

Burau, W., 1936, Über Zopfgruppen und gleichsinnig verdrillte Verkettungen. *Abh. Math. Sem., Hansischen Univ.*, **11**, 171–178.

Burkhardt, H., 1890, Grundzüge einer allgemeinen Systematik der hyperelliptischen Funktionen erster Ordnung. *Math. Ann.*, **35**, 198–296 (in particular, pp. 209–212.)

Burnside, W., 1902, On an Unsettled Question in the Theory of Discontinuous Groups. *Quart. J. Math.*, **33**, 230–238.

Burnside, W., 1911, *The Theory of Groups of Finite Order.* Second Edition, Cambridge, England. (Reprinted 1955, Dover Publications, New York.)

Burrow, Martin D., 1958, Invariants of Free Lie Rings. *Comm. Pure and Appl. Math.*, **11,** 419–431.

Campbell, J. E., 1898, On a Law of Combination of Operators. *Proc. London Math. Soc.*, **29,** 14–32.

Černikov, V. M., 1946, Complete Groups with an Ascending Central Series. *Mat. Sbornik,* **18,** 397–422.

Chen, Kuo-Tsai, 1951, Integration in Free Groups. *Annals of Math.*, **54,** 147–162.

Chen, K. T., 1952, Commutator Calculus and Link Invariants. *Proc. Amer. Math. Soc.*, **3,** 44–55.

Chen, K. T., 1954a, A Group Ring Method for Finitely Generated Groups. *Trans. Amer. Math. Soc.*, **76,** 275–287.

Chen, K. T., 1954b, Iterated Integrals and Exponential Homomorphisms. *Proc. London Math. Soc.* (3rd Ser.), **4,** 502–512.

Chevalley, C., 1956, *Fundamental Concepts of Algebra.* Academic Press, New York. (241 pp.)

Chow, Wei Liang, 1948, On the Algebraical Braid Group. *Annals of Math.*, (2), **49,** 654–658.

Clebsch, A., and P. Gordan, 1866, *Theorie der Abelschen Funktionen.* (p. 304). Teubner, Leipzig.

Cohn, P. M., 1952, A Theorem on the Structure of Tensor Spaces. *Annals of Math.* (2), **56,** 254–268.

Cohn, P. M., 1954a, Sur le critère de Friedrichs pour les commutateurs dans un algèbre associative libre. *Comtes Rendus des Séances de l'Académie des Sciences,* **239,** 743–745.

Cohn, P. M., 1954b, A Countably Generated Group which cannot be Covered by Finite Permutable Subsets. *J. London Math. Soc.*, **29,** 248–249.

Cohn, P. M., 1954c, On Homomorphic Images of Special Jordan Algebras. *Canadian J. of Math.*, **6,** 253–264.

Cohn, P. M., 1955, A Non-nilpotent Lie Ring Satisfying the Engel Condition and a Non-nilpotent Engel Group. *Proc. Cambridge Philos. Soc.*, **51,** 401–405.

Coxeter, H. S. M., 1936, The Abstract Groups $R^m = S^m = (R^jS^j)^{p_j} = 1, S^m = T^m = (S^jT)^{2p_j} = 1$, and $S^m = T^2 = (S^{-j}TS^jT)^{p_j} = 1$. *Proc. London Math. Soc.* (2), **41,** 278–301.

Coxeter, H. S. M., 1961, *Introduction to Geometry.* John Wiley, New York, 443 pp.

Coxeter, H. S. M., and W. O. J. Moser, 1965, *Generators and Relations for Discrete Groups.* Ergebnisse der Mathematik und ihrer Grenzgebiete. New Series, No. 14 (Springer).

Crowell, R. H., and R. H. Fox 1963, *Introduction to Knot Theory.* Ginn and Co., Boston, New York. (182 pages)

Davis, Martin, 1958, *Computability and Unsolvability.* McGraw-Hill, New York. (210 pages) pp. 73–77.

Dehn, M., 1911, Über unendliche diskontinuierliche Gruppen. *Math. Ann.*, **71,** 116–144.

Dehn, M., 1912, Transformation der Kurven auf zweiseitigen Flaechen. *Math. Ann.*, **72,** 413–421.

Dehn, M., 1914, Die beiden Kleeblattschlingen. *Math. Ann.*, **75,** 402–413.

Dehn, M., 1931, Über einige neuere Forschungen in den Grundlagen der Geometrie. *Matematisk Tideskrift B.* No. 3–4, 1931.

Dehn, M., 1939, Die Gruppe der Abbildungsklassen. *Acta Math.*, **69,** 135–206.

De Séguier, J. A., 1904, *Théorie des groupes finis.* Vol. I. Eléments de la théorie des groupes abstraits, Paris.

Dieudonné, J., 1953, On Semi-simple Lie Algebras. *Proc. Amer. Math. Soc.*, **4,** 931–932.

Dieudonné, J., 1956, Groupes de Lie et hyperalgèbras de Lie sur un corps de caractéristique $p > 0$. (V) *Bull. Soc. Math. France*, **84,** 207–239.

Doniakhi, K. A., 1940, Linear Representation of the Free Product of Cyclic Groups. Leningrad State Univ. Annals (Uchenye Zapiski) *Math. Ser.*, **10,** 158–165. (Russian).

Douglas, Jesse, 1951, On Finite Groups with Two Independent Generators. *Proc. Nat. Acad. Sci. U.S.A.*, **37,** 604–610, 677–691, 749–760, 808–813.

Dynkin, E. B., 1947, Calculation of the Coefficients in the Campbell-Hausdorff Formula. *Doklady Akad. Nauk. USSR* (N.S.), **57,** 323–326. (Russian)

Easterfield, T. E., 1938, The Orders of Products and Commutators in Prime Power Groups. *Proc. Cambridge Philos. Soc.*, **36,** 14–26.

Eckmann, B., 1945, Der Cohomologiering einer beliebigen Gruppe. *Comment. Math. Helv.*, **18,** 232–282.

Eilenberg, S., and H. P. Cartan, 1956, *Homological Algebra.* Princeton Univ. Press.

Eilenberg. S., and S. Maclane, 1949a, b, Cohomology Theory in Abstract Groups. I, II. *Annals of Math.*, **48,** 51–78, 326–341.

Falk, Gottfried, 1951, Über Ringe mit Poisson-Klammern. *Math. Ann.*, **123,** 379–391.

Federer, H., and B. Jonsson, 1950, Some Properties of Free Groups. *Trans. Amer. Math. Soc.*, **68,** 1–27.

Fer, F., 1958, Résolution de l'équation matricielle $dU/dt = pU$ par produit infini d'exponentielles matricielles. *Acad. Roy. Belg. Bull. Cl. Sci.* (5), **44,** 818–829.

Finkelstein, D., 1955, On Relations between Commutators. *Comm. Pure and Appl. Math.*, **8,** 245–250.

Ford, L. R., 1929, *Automorphic Functions.* Chelsea, New York (2nd ed. 1951).

Fouxe-Rabinovitch, *see* Fuchs-Rabinovitch.

Fox, R. H., 1953, Free Differential Calculus I. *Annals of Math.*, **57,** 547–560.

Fox, R. H., 1954, Free Differential Calculus II. *Annals of Math.*, **59,** 196–210.

Fox, R. H., *see* Blanchefield, R. C.

Fox, R. H., *see* Blankinship.

Fox, R. H., *see* Crowell, R. H.

Fox, R. H., *see* Torres, G.

Frasch, H., 1933, Die Erzeugenden der Hauptkongruenz Gruppen für Primzahlstufen. *Math. Ann.*, **108,** 229–252.

Frederick, Karen, 1963, Hopfian Property of a Class of Fundamental Groups. *Comm. Pure* and *Appl. Math.*, **16,** 1–8.

Fricke, R., and F. Klein, 1897, *Vorlesungen über die Theorie der automorphen Funktionen.* Vol. 1. Leipzig. Quoted in text as Fricke, 1897.

Friedrichs, K. O., 1953, Mathematical Aspects of the Quantum Theory of Fields, Part V. *Comm. Pure and Appl. Math.*, **6,** 1–72.

Fröhlich, W., 1936, Über ein speziellles Transformations- problem bei einer besonderen Klasse von Zöpfen. *Monatshefte f. Math. u. Phys.*, **44,** 225–237.

Fuchs-Rabinovitch, D. J., 1940a, On the Determinators of an Operator of the Free Groups. *Rec. Math.* [*Mat. Sbornik*] N.S., **7** (49), 197–208. (English. Russian Summary.)

Fuchs-Rabinovitch, D. J., 1940b, On a Certain Representation of a Free Group. *Leningrad State Univ. Annals [Uchenye Zapiski] Math. Ser.*, **10,** 154–157. (Russian).

Fuchs-Rabinovitch, 1940c, Über die Nichteinfachheit einer lokal freien Gruppe. *Rec. Math. [Mat. Sbornik]* N.S., **7** (49), 327–328. (Russian. German Summary.)

Fujiwara, I., 1953, On the Evaluation of the Operator Function log $(e^y e^z)$. *Progress of Theoretical Physics*, **9,** No. 6, 976.

Gaschütz, W., 1956, Zu einem von B. H. und H. Neumann gestellten Problem. *Math. Nachr.*, **14,** 249–252.

Gassner, Betty Jane, 1961, On Braid Groups. *Abh. Math. Sem., Hamburg Univ.*, **25,** 10–22.

Gold, P., 1960, unpublished.

Goldberg, K., 1956, The Formal Power Series for log $e^x e^y$. *Duke Math. J.*, **23,** 13–22.

Golovin, O. N., 1950, Nilpotent Products of Groups. Mat. Sbornik N.S. 27(69) 427–454. (M.R. 12, 672) *Amer. Math. Soc. Translations*, Ser. 2, **2,** 89–116 (1956).

Golovin, O. N., 1951a, The Metabelian Products of Groups. Mat. Sbornik N.S. 28(70), 431–444. (M.R. 13, 105) *Amer. Math. Soc. Translations*, Ser. 2, **2,** 117–132 (1956).

Golovin, O. N., 1951b, On the Isomorphisms of Nilpotent Decompositions of Groups. Mat. Sbornik N.S. 28(70), 445–452. *Amer. Math. Soc. Translations*, Ser. 2, **2,** 133–140 (1956).

Golovin, O. N., and N. P. Goldina, 1955, Subgroups of Free Metabelian Groups. *Mat. Sbornik* N.S., **37** (79), 323–336 (Russian) MR 17, 234.

Green, J. A., 1952, On Groups with Odd Prime-power Exponent. *J. London Math. Soc.*, **27,** 476–485.

Greenberg, Leon, 1960, Discrete Groups of Motions. *Canadian J. Math.*, **12,** 414–425.

Greendlinger, M., 1960a, Dehn's Algorithm for the Word Problem. *Comm. Pure and Appl. Math.*, **13,** 67–83.

Greendlinger, 1960b, On Dehn's Algorithms for the Conjugacy and Word Problems. With Applications. *Comm. Pure and Appl. Math.*, **13,** 641–677.

Greendlinger, M., 1961, An Analogue of a Theorem of Magnus. *Archiv d. Mathematik*, **12,** 94–96.

Greendlinger, M., 1962, A Class of Groups all of whose Elements have Trivial Centralizers. *Math. Z.*, **78,** 91–96.

Griffiths, H. B., 1955, A Note on Commutators in Free Products. II. *Proc. Cambridge Philos. Soc.*, **51,** 245–251.

Griffiths, H. B., 1956, Infinite Products of Semi-groups and Local Connectivity. *Proc. London Math. Soc.* (3) **6,** 455–480.

Grossman, I. and W. Magnus, 1964, *Groups and Their Graphs*. Random House, New York.

Grosswold, Emil, 1950, On the Structure of some Subgroups of the Modular Group. *Amer. J. Math.*, **72,** 809–834.

Grün, O., 1940, Zusammenhang zwischen Potenbildung und Kommutatorbildung. *J. reine u. angew. Math.*, **182,** 158–177.

Gruen, O., 1953, Beitraege zur Gruppentheorie V: Ueber endliche p- Gruppen. *Osaka Math. J.*, **5,** 117–146.

Gruenberg, K. W., 1953, Two Theorems on Engel Groups. *Proc. Cambridge Philos. Soc.*, **49,** 377–380.

Gruenberg, K. W., 1957, Residual Properties of Infinite Soluble Groups. *Proc. London Math. Soc.* (3), **7,** 29–62.

Grushko, I., 1938, La résolution du problème d'identité dans les groupes à plusieurs relations d'un type spécial. *Rec. Math. Moscow N.S.*, **3** 543–550. (Russian, French Summary, *Zentralblatt*, **19,** 156.)

Haken, H., 1952, Zum Identitätsproblem bei Gruppen. *Math. Z.*, **56,** 335–362.

Hall, M., 1949, Subgroups of Finite Index in Free Groups. *Canadian J. Math.*, **1,** 187–190.

Hall, M., 1950a, A Topology for Free Groups and Related Groups. *Annals of Math.*, **52,** 127–139.

Hall, M., 1950b, A Basis for Free Lie Rings and Higher Commutators in Free Groups. *Proc. Amer. Math. Soc.*, **1,** 575–581.

Hall, M., 1952, A Combinatorial Problem on Abelian Groups. *Proc. Amer. Math. Soc.*, **3,** 584–587.

Hall, M., 1953, Subgroups of Free Products. *Pacific J. Math.*, **3,** 115–120.

Hall, M., 1957, Solution of the Burnside Problem for Exponent 6. *Proc. Nat. Acad. Sci.*, **43,** 751–753.

Hall, M., 1958, Solution of the Burnside Problem for Exponent six. *Illinois J. of Math.*, **2,** No. 4B. Miller Memorial Issue, pp. 764–786.

Hall, M., 1959, *The Theory of Groups.* New York, Macmillan Co.

Hall, M., and Tibor Rado, 1948, On Schreier Systems in Free Groups. *Trans. Amer. Math. Soc.*, **64,** 386–408.

Hall, P., 1933, A Contribution to the Theory of Groups of Prime Power Order. *Proc. London Math. Soc.*, Ser. 2, **36,** 29–95.

Hall, P., 1954, The Splitting Properties of Relatively Free Groups. *Proc. London Math. Soc.* (3), **4,** 343–356.

Hall, P., 1958a, Some Word Problems. *J. London Math. Soc.*, **33,** 482–496. (Presidential address.)

Hall, P., 1958b, Some Sufficient Conditions for a Group to be Nilpotent. *Illinois J. Math.*, **2,** 787–801.

Hall, P., 1961, The Frattini Subgroups of Finitely Generated Groups. *Proc. London Math. Soc.* (3), **11,** 327–352.

Hall, P., and Graham Higman, 1956, On the p-length of p-soluble Groups and Reduction Theorems for Burnside's Problem. *Proc. London Math. Soc.* (3) **6,** No. 21, 1–42.

Hardy, G. H., and E. M. Wright, 1954, *An Introduction to the Theory of Numbers.* Oxford. 3rd edition.

Hausdorff, F., 1906, Die symbolische Exponentialformel in der Gruppentheorie. *Berichte der Sächsischen Akademie der Wissenschaften* (*Math. Phys. Klasse*) Leipzig, Vol. 58, 19–48.

Higgins, P. J., 1954, Lie Rings Satisfying the Engel Condition. *Proc. Cambridge Philos. Soc.*, **50,** 8–15.

Higgins, P. J., and R. C. Lyndon, 1962, Equivalence of elements under automorphism of a free group. Mimeographed notes, Queen Mary College, London.

Higman, G., 1940, The Units of Group-Rings. *Proc. London Math. Soc.*, **46,** 231–248.

Higman, G., 1951a, A Finitely Generated Infinite Simple Group. *J. London Math. Soc.*, **26,** 61–64.

Higman, G., 1951b Almost Free Groups. *Proc. London Math. Soc.* **1,** 284–290.

Higman, G., 1951c, A Finitely Related Group with an Isomorphic Proper Factor Group. *J. London Math. Soc.*, **26,** 59–61.

Higman, G., 1952, Unrestricted Free Products, and Varieties of Topological Groups. *J. London Math. Soc.*, **27,** 73–81.

Higman, G., 1953, On a Problem of Takahasi. *J. London Math. Soc.*, **28,** 250–252.

Higman, G., 1956, On Finite Groups of Exponent 5. *Proc. Cambridge Philos. Soc.*, **52,** 381–390.

Higman, G., 1959, Some Remarks on Varieties of Groups. *Quart. J. Math.* (Oxford Ser.), **10,** 165–178.

Higman, G., and D. C. Cross, 1960, Identical Relations in Finite Groups. *Atti del Convegno sulla teoria dei gruppi finiti.* Firenze, pp. 92–100.

Higman, G., Neumann, B. H., and Hanna Neumann, 1949, Embedding Theorems for Groups. *J. London Math. Soc.*, **24,** 247–254.

Higman, G., and B. H. Neumann, 1954, On Two Questions of Itô. *J. London Math. Soc.*, **29,** 84–88.

Higman, G., and A. H. Stone, 1954, On Inverse Systems with Trivial Limits. *J. London Math. Soc.*, **29,** 233–236.

Higman, G., 1956, *see* P. Hall

Hirsch, K. A., 1938, 1946, On Infinite Soluble Groups. II, III. *Proc. London Math. Soc.* (2), **44,** 336–344; **49,** 184–194.

Hirsch, K., 1950, Eine kennzeichnende Eigenschaft nilpotenter Gruppen. *Math. Nachr.*, **4,** 47–49.

Hopf, H., 1941, Fundamentalgruppe und zweite Bettische Gruppe. *Comment. Math. Helv.*, **14,** 257–309.

Hopf, H., 1949, Über die Bettischen Gruppen, die zu einer beliebigen Gruppe gehören. *Comment. Math. Helv.*, **17,** 39–79.

Hopkins, C., 1937, Concerning Uniqueness-bases of Finite Groups with Applications to p-groups of Class 2. *Trans. Amer. Math. Soc.*, **41,** 287–313.

Howson, A. G., 1954, On the Intersection of Finitely Generated Free Groups. *J. London Math. Soc.*, **29,** 428–434.

Hu, Sze-Tsen, 1959, *Homotopy Theory.* (347 pp.) Academic Press, New York and London, 347 pp.

Hua, L. K., and I. Reiner, 1949, On the Generators of the Symplectic Modular Groups. *Trans. Amer. Math. Soc.*, **65,** 415–426.

Hua, L. K., and I. Reiner, 1951, Automorphisms of the Unimodular Group. *Trans. Amer. Math. Soc.*, **71,** 331–348.

Hua, L. K., and I. Reiner, 1952, Automorphisms of the Projective Unimodular Group. *Trans. Amer. Math. Soc.*, **72,** 467–473.

Huppert, B., 1953a, Über das Produkt von paarweise vertauschbaren zyklischen Gruppen. *Math. Z.*, **58,** 243–264.

Huppert, B., 1953b, Über die Aufloesbarkeit faktorisierbarer Gruppen. *Math. Z.*, **59,** 1–7.

Huppert, B., 1953c, Monomiale Darstellung endlicher Gruppen. *Nagoya Math. J.*, **6,** 93–94.

Hurewitz, W., 1931, Zu einer Arbeit von O. Schreier. *Abh. Math. Sem.*, Hamburg Univ., **8,** 307–314.

Hurwitz, A., 1893, Über algebraische Gebilde mit eindeutigen Transformationen in sich. *Math. Ann.*, **41,** 403–442.

Itô, N., 1955, Über das Product von zwei abelschen Gruppen. *Math. Z.*, **62,** 400–401.

Jacobson, N., 1937, Abstract Derivation and Lie Algebras. *Trans. Amer. Math. Soc.*, **42,** 206–229.

Jacobson, N., 1941, Restricted Lie Algebras of Characteristic p. *Trans. Amer. Math. Soc.*, **50,** 15–25.

Jacobson, N., 1949, Lie and Jordan Triple Systems. *Amer. J. of Math.*, **71,** 149–170.

Jacobson, N., 1962, *Lie Algebras*. Interscience (John Wiley), New York. 331 pp.

Kaloujnine, L., 1949, Sur quelques propriétés des groupes d'automorphisms d'un groupe abstrait. *C.R. Acad. Sci.*, Paris, **230**, 2067–2069.

Kaloujnine, L., 1950, Sur quelques propriétés des groupes d'automorphismes d'un groupe abstrait. (Généralisation d'un théorème de M. Ph. Hall) *C.R. Acad. Sci.*, Paris, **231**, 400–402.

Kaloujnine, L., 1953, Über gewisse Beziehungen zwichen einer Gruppe und ihren Automorphismen. *Ber. Math. Tagrung.*, Berlin, 164–172.

Karrass, A., and D. Solitar, 1956, Some Remarks on the Infinite Symmetric Groups. *Math. Z.*, **66,** 64–69.

Karrass, A., and D. Solitar, 1957, Note on a Theorem of Schreier. *Proc. Amer. Math. Soc.*, **8,** 696–697.

Karrass, A., and D. Solitar, 1958a, Subgroup theorems in the theory of groups given by defining relations. *Comm. Pure and Appl. Math.*, **11,** 547–571.

Karrass, A., and D. Solitar, 1958b, On free products. *Proc. Amer. Math. Soc.*, **9,** 217–221.

Karrass, A., Magnus, W., and D. Solitar, 1960, Elements of Finite Order in Groups with a Single Defining Relation. *Comm. Pure and Appl. Math.*, **13,** 57–66.

Keller, O. H., 1954, Eine Darstellung der Komposition endlicher Gruppen durch Streckenkomplexe. *Math. Ann.*, **128,** 177–199.

Klein, F., and R. Fricke, 1890–92, *Vorlesungen über die Theorie der Elliptischen Modulfunktionen* (2 volumes). Leipzig.

Koehler, 1947, unpublished.

Kontorovič, P. G., 1946, Groups with a Separation Basis. III. *Mat. Sbornik*, **19,** 287–308.

Kontorovič, P. G., 1948, On the Theory of Noncommutative Torsion Free Groups. *Doklady Akad. Nauk. SSSR*, **59,** 213–216.

Kostrikin, A. I., 1955, Solution of the Restricted Burnside Problem for the Exponent 5. *Izvestia Akad. Nauk. SSSR*, **19,** 233–244. M.R. 17, 126.

Kostrikin, A. I., 1957a, On the Connection between Periodic Groups and Lie Rings. *Izvestia Akad. Nauk. SSSR, Ser. Mat.* **21,** 289–310.

Kostrikin, A. I., 1957b, Lie Rings Satisfying the Engel Condition. *Izvestia Akad. Nauk. SSSR, Ser. Mat.* **21,** 515–540. (Russian; M.R. 20, p. 279.)

Kostrikin, A. I., 1959, The Burnside Problem. *Izvestia Akad. Nauk. SSSR, Ser. Mat.* **23,** 3–34. (Russian; M.R. 24, p. 358.)

Kuhn, H. W., 1952, Subgroup Theorems for Groups Presented by Generators and Relations. *Annals of Math.*, **56,** 22–46.

Kurosh, A., 1934, Die Untergruppen der freien Produkte von beliebigen Gruppen. *Math. Ann.* **109,** 647–660.

Kurosh, A., 1937, Zum Zerlegungsproblem der Theorie der freien Produkte. *Rec. Math. Moscow*, N.S. 2 (1937), 995–1001.

Kurosh, A., 1953, *Gruppentheorie*. German translation of the first edition of th. Russian book. Berlin.

Kurosh, A., 1955, *The Theory of Groups*, Vol. I. Translated and edited by K. Hirsch. Chelsea, N.Y.

Lazard, M., 1952a, Sur les algèbres enveloppantes universelles de certaines algèbres de Lie. *Computes rendus des Séances de l'Académie des Sciences*, **234,** 788–791.

Lazard, M., 1952b, Sur les groupes analytiques dans les modules filtrés. *Comptes rendus des Séances de l'Académie des Sciences*, **235,** 1465–1467.

Lazard, M., 1954, Sur les groupes nilpotents et les anneaux de Lie. *Annales Sci. L'Ecole Normale Superieure* (3), **71,** 101–190.

Lehner, J., 1964, *Discontinuous Groups and Automorphic Functions Amer. Math. Soc.*, Providence, R.I., 425 pp.

Leray, J., 1950, L'anneau spectral et l'anneau filtré d'homologie d'un espace localement compacte et d'une application continue. (Chapter I, §1) *J. Math. pures et appl.*, **29,** 10–14.

Levi, F., 1930, Über die Untergruppen der freien Gruppen. *Math. Z.*, **32,** 315–18.

Levi, F., 1933, Über die Untergruppen der freien Gruppen. II. *Math. Z.*, **37,** 90–97.

Levi, F., 1941, On the Number of Generators of a Free Product, and a Lemma of Alexander Kurosh. *J. Indian Math. Soc.* (N.S.), **5,** 149–155.

Levi, F., 1942, Groups in which the Commutator Operation Satisfies Certain Conditions. *J. Indian Math. Soc.* (N.S.), **6,** 87–92.

Levi, F., and B. van der Waerden, 1933, Über eine besondere Klasse von Gruppen. *Abh. Math. Sem., Hamburg. Univ.*, **9,** 154–158.

Levin, Frank, 1962, Solutions of Equations over Groups. *Bull. American Math. Soc.*, **68,** 603–604.

Lipschutz, S., 1960, Elements in S-groups with Trivial Centralizers. *Comm. Pure and Appl. Math.*, **13,** 679–683.

Lipschutz, S., 1961, On a finite matrix representation of the braid groups. *Archiv d. Math.* **12,** 7–12.

Lipschutz, S., 1962, On Square Roots in Eighth-Groups. *Comm. Pure and Appl. Math.*, **15,** 39–43.

Lyndon, R. C., 1950, Cohomology Theory of Groups with a Single Defining Relation. *Annals of Math.*, **52,** 650–665.

Lyndon, R. C., 1952, Two Notes on Nilpotent Groups. *Proc. Amer. Math. Soc.*, **3,** 579–583.

Lyndon, R. C., 1953, On the Fouxe-Rabinovitch Series for Free Groups. *Portugaliae Math.*, **12,** 115–118. (M.R. 15, No. 9, p. 776.)

Lyndon, R. C., 1954a, On Burnside's Problem. *Trans. of the Amer. Math. Soc.*, **77,** 202–215.

Lyndon, R. C., 1954b, Identities in Finite Algebras. *Proc. of the Amer. Math. Soc.*, **5,** 8–9.

Lyndon, R. C., 1955a, A Theorem of Friedrichs. *Michigan Math. J.*, **3,** 27–29.

Lyndon, R. C., 1955b, unpublished.

Lyndon, R. C., 1962, Dependence and Independence in Free Groups. *J. reine u. angew. Math.*, **210,** 148–174.

Lyndon, R. C., *see* Auslander, M., 1955, and Higgins, 1962.

MacDonald, I. O., 1960, 1962, On Certain Varieties of Groups. I, II. *Math. Z.*, **76,** 270–282; **78,** 175–188.

MacHenry, T., 1960, The Tensor Product and the 2nd Nilpotent Product of Groups. *Math. Z.*, **73,** 134–145.

MacLane, S., 1949, Cohomology Theory in Abstract Groups. III. *Annals of Math.*, **50,** 736–761.

MacLane, S., 1950, Duality for Groups. *Bull. Amer. Math. Soc.*, **56,** 485–516.

MacLane, S., 1958, A proof of the subgroup theorem for free products. *Mathematica*, **5,** 13–19.

MacLane, S., and S. Filenberg, 1947a, Cohomology Theory in Abstract Groups. I. *Annals of Math.*, **48,** 51–78.

MacLane, S., and S. Eilenberg, 1947b, Cohomology Theory in Abstract Groups. II. Group Extensions with a Non-Abelian Kernel. *Annals of Math.*, **48**, 326–341.

Magnus, W., 1930, Über diskontinuierliche Gruppen mit einer definierenden Relation (Der Freiheitssatz). *J. reine u. angew. Math.*, **163**, 141–165.

Magnus, W., 1931, Untersuchungen über einige unendliche diskontinuierliche Gruppen. *Math. Ann.*, **105**, 52–74.

Magnus, W., 1932, Das Identitäts problem für Gruppen mit einer definierenden Relation. *Math. Ann.*, **106**, 295–307.

Magnus, W., 1934a, Über Automorphismen von Fundamental-Gruppen berandeter Flächen. *Math. Ann.*, **109**, 617–646.

Magnus, W., 1934b, Über n-dimensionale Gittertransformationen. *Acta Math.*, **64**, 353–367.

Magnus, W., 1934c, Über den Beweis des Hauptidealsatzes. *J. reine u. angew. Math.*, **170**, 235–240.

Magnus, W., 1935, Beziehungen zwischen Gruppen und Idealen in einem speziellen Ring. *Math. Ann.*, **111**, 259–280.

Magnus, W., 1937, Über Beziehungen zwischen höheren Kommutatoren. *J. reine u. angew. Math.*, **177**, 105–115.

Magnus, W., 1939, Über freie Faktorgruppen und freie Untergruppen gegebener Gruppen. *Monatshafte für Math. u. Phys.*, **47**, 307–313.

Magnus, W., 1940, Über Gruppen und zugeordnete Liesche Ringe. *J. reine u. angew. Math.*, **182**, 142–149.

Magnus, W., 1950, A Connection between the Baker-Hausdorff Formula and a Problem of Burnside. *Annals of Math.*, **52**, 111–126; **57**, 606 (1953).

Magnus, W., 1954, On the Exponential Solution of Differential Equations for a Linear Operator. *Comm. Pure and Appl. Math.*, **7**, 649–673.

Mal'cev, A. I., 1940, On Isomorphic Matrix Representations of Infinite Groups. *Mat. Sb.* (N.S.), **8** (50), 405–421.

Mal'cev, A. I., 1949, Nilpotent Torsion Free Groups. *Izvestiya Akad. Nauk. SSSR Ser. Mat.*, **13**, 201–212.

Markoff, A., 1945, Foundations of the Algebraic Theory of Tresses. (Russian, reviewed in Math. Reviews, **8**, p. 131) *Trav. Inst. Math. Stekloff*, **16**, 53 pages.

Maschke, H., 1896, The Representation of Finite Groups, Especially of the Rotation Groups of the Regular Bodies of Three and Four Dimensional Space, by Cayley's Color Diagrams. *Amer. J. Math.*, **18**, 156–194.

Mauler, H., 1953, Eine Darstellung für Identitäten zwischen den Kommutatoren eines Ringes. *Math. Ann.*, **126**, 410–417.

Meier-Wunderli, H., 1949, Über die Gruppen mit der identischen Relation $(x_1, x_2, \ldots x_n) = (x_2, x_3, \ldots, x_n, x_1)$. *Vierteljahrschrift der Naturforschenden Gesellschaft in Zuerich*, **94**, 211–218.

Meier-Wunderli, H., 1950, Über endliche p-Gruppen, deren Elemente der Gleichung $x^p = 1$ genügen. *Comm. Math. Helv.*, **24**, 18–45.

Meier-Wunderli, H., 1952, Note on a Basis of P. Hall for the Higher Commutators in Free Groups. *Comm. Math. Helv.*, **26**, 1–5.

Moran, S., 1956, Associative Operators in Groups. I. *Proc. London Math. Soc.*, **6**, 581–596.

Moran, S., 1958a, Duals of a Verbal Subgroup. *J. London Math. Soc.*, **33**, 220–236.

Moran, S., 1958b, The Homomorphic Image of the Intersection of a Verbal Subgroup and the Cartesian Subgroup of a Free Product. *J. London Math. Soc.*, **33**, 237–245.

Moran, S., 1958c, Associative Operations in Groups. II. *Proc. London Math. Soc.* (3), **8,** 548–568.

Moran, S., 1959a, Associative Operations on Groups. III. *Proc. London Math. Soc.* (3), **9,** 287–317.

Moran, S., 1959b, Associative Regular Operations on Groups Corresponding to a Property P. *Proc. Amer. Math. Soc.*, **10,** 796–799.

Moser, W. O. J., 1957, *see* Coxeter, H. S. M.

Murasugi, K., 1964, The center of a group with a single defining relation. *Math. Ann.*, **155,** 246–251.

Neumann, B. H., 1932, Die Automorphismengruppe der freien Gruppen. *Math. Ann.*, **107,** 367–386; *see* also *Zentralblatt fuer Math.*, **5,** 244, 1933.

Neumann, B. H., 1937a, Groups whose Elements have Bounded Orders. *J. London Math. Soc.*, **12,** 195–198.

Neumann, B. H., 1937b, Identical Relations in Groups. I. *Math. Ann.*, **114,** 506–525.

Neumann, B. H., 1943a, On the Number of Generators of a Free Product. *J. London Math. Soc.*, **18,** 12–20.

Neumann, B. H., 1943b, Adjunction of Elements to Groups. *J. London Math. Soc.*, **18,** 4–11.

Neumann, B. H., 1949a, On Ordered Groups. *Amer. J. of Math.*, **71,** 1–18.

Neumann, B. H., 1949b, On Ordered Division Rings. *Trans. Amer. Math. Soc.*, **66,** 202–252.

Neumann, B. H., 1950, A two-generator Group Isomorphic to a Proper Factor Group. *J. London Math. Soc.*, **25,** 247–248.

Neumann, B. H., 1952, A Note on Algebraically Closed Groups. *J. London Math. Soc.*, **27,** 227–242.

Neumann, B. H., 1953a, On a Problem of Hopf. *J. London Math. Soc.*, **28,** 351–353.

Neumann, B. H., 1953b, A Note on Means in Groups. *J. London Math. Soc.*, **28,** 472–276.

Neumann, B. H., 1954a, An Essay on Free Products of Groups with Amalgamations. *Phil. Trans. Royal Soc. of London*, No. 919, **246,** 503–554, June 15, 1954.

Neumann, B. H., 1954b, An Embedding Theorem for Algebraic Systems. *Proc. London Math. Soc.*, **4,** 138–153.

Neumann, B. H., 1954c, Groups Covered by Finitely Many Cosets. *Publicationes Mathematicae*, **3,** 247–249.

Neumann, B. H., 1955, Groups with Finite Classes of Conjugate Subgroups. *Math. Z.*, **63,** 76–96.

Neumann, B. H., 1956, On a Conjecture of Hanna Neumann. *Proc. Glasgow Math. Assoc.*, **3,** 13–17.

Neumann, B. H., 1960, Embedding Theorems for Ordered Groups. *J. London Math. Soc.*, **35,** 503–512.

Neumann, B. H., 1962, On a Theorem of Auslander and Lyndon. *Archiv d. Math.*, **13,** 4–9.

Neumann, B. H., *see* Higman, G.

Neumann, B. H., and Hanna Neumann, 1950, A Remark on Generalized Free Products. *J. London Math. Soc.*, **25,** 202–204.

Neumann, B. H., and Hanna Neumann, 1954, Partial Endomorphisms of Finito Groups. *J. London Math. Soc.*, **29,** 434–440.

Neumann, B. H., and Hanna Neumann, 1951, Zwei Klassen Charakteristischer Untergruppen und ihrer Faktorgruppen. *Math. Nachr.*, **4,** 106–125.

Neumann, B. H., and Hanna Neumann, 1959, Embedding Theorems for Groups. *J. London Math. Soc.*, **34,** 465–479.

Neumann, B. H., H. Neumann, and P. M. Neumann, 1962, Wreath Products and Varieties of Groups. *Math. Z.*, **80,** 44–62.

Neumann, Hanna, 1948, Generalized Free Products with Amalgamated Subgroups. *Amer. J. Math.*, **70,** 590–625.

Neumann, Hanna, 1949, Generalized Free Products with Amalgamated Subgroups. *Amer. J. Math.*, **71,** 491–540.

Neumann, Hanna, 1950, Generalized Free Sums of Cyclical Groups. *Amer. J. Math.*, **72,** 671–685.

Neumann, Hanna, 1956a, On Varieties of Groups and their Associated Near Rings. *Math. Z.*, **65,** 36–39.

Neumann, Hanna, 1956b, On the Intersection of Finitely Generated Free Groups. *Publ. Math. Debrecen*, **4,** 186–189.

Neumann, Hanna, 1962, On a Theorem of Auslander and Lyndon. *Archiv d. Math.*, **13,** 1–3.

Neumann, Hanna, *see* Neumann, B. H.

Newman, M. H. A., and J. H. C. Whitehead, 1937, On the Group of a Certain Linkage. *Quart. J. Math.*, **8,** 14–21.

Neumann, P. M., 1963, Some Indecomposable Varieties of Groups. *Quart. J. Math.* (Oxford, 2nd Series), **14,** 46–50. *See also* B. H. Neumann and Hanna Neumann.

Nielsen, J., 1918, Die Isomorphismen der allgemeinen unendlichen Gruppe mit zwei Erzeugenden. *Math. Ann.*, **78,** 385–397.

Nielsen, J., 1921, Om Regning med ikke kommutative Faktoren og dens Anvendelse i Gruppeteorien. *Matematisk Tidsskrift*, **B,** 77–94.

Nielsen, J., 1924a, Die Isomorphismengruppe der freien Gruppen. *Math. Ann.*, **91,** 169–209.

Nielsen, J., 1924b, Die Gruppe der dreidimensionalen Gittertransformationen. *Kgl. Danske Videnskabernes Selskub., Math. Fys. Meddelelser V*, **12,** 1–29.

Nielsen, J., 1924c, Über topologische Abbildungen geschlossener Flächen. *Abh. Math. Sem., Hamburg. Univ.*, **3,** 246–260.

Nielsen, J., 1927, 1929, 1931, Untersuchungen zur Topologie der geschlossenen zweiseitigen Flächen, I, II, III. *Acta Math.* **50,** 189–358; **53,** 1–76; **58,** 87–167.

Nielsen, J., 1955, A Basis for Subgroups of Free Groups. *Math. Scand.*, **3,** 31–43.

Novikov, P. S., 1954, Unsolvability of the Conjugacy Problem in the Theory of Groups. *Izv. Akad. Nauk. SSSR, Ser. Mat.*, **18,** 485–524 (Russian). MR 17, 706.

Novikov, P. S., 1955, On the Algorithmic Unsolvability of the Word Problem in Group Theory. Trudy Mat. Inst. im Steklov No. 44., 143 pp. *Izdat. Akad. Nauk. SSSR, Moscow* (Russian), MR **17,** 706. English text in "Russian Translations," Series 2, Vol. 9, 1958. Edited by the American Math. Soc.

Novikov, P. S., 1959, On Periodic Groups. *Dokl. Akad. Nauk. SSSR*, **127,** 749–752. (M.R. 21, 5680.)

Petresco, J., 1955, Sur les commutateurs. *Math. Z.*, **61,** 348–356.

Pontrjagin, L., 1946, *Topological Groups.* Princeton University Press.

Post, E. L., 1944, Recursively Enumerable Sets of Positive Integers and their Decision Problems. *Bull. Amer. Math. Soc.*, **50,** 284–316.

Post, E. L., 1947, Recursive Unsolvability of a Problem of Thue. *J. Symbolic Logic*, **12,** 1–11.

Rabin, M. O., 1958, Recursive Unsolvability of Group Theoretic Problems. *Annals of Math.*, **67,** 172–194.

Rademacher, H., 1929, Über die Erzeugenden von Kongruenzuntergruppen der Modulgruppe. *Abh. Math. Sem., Hamburg Univ.*, **7,** 134–148.

Rado, T., *see* Hall, M.

Rapaport, Elvira S., 1958, On Free Groups and their Automorphisms. *Acta Math.*, **99,** 139–163.

Rapaport, Elvira S., 1959, Note on Nielsen Transformations. *Proc. Amer. Math. Soc.*, **10,** No. 2, 228–235.

Ree, R., 1957, Commutator Groups of Free Products of Torsion Free Abelian Groups. *Annals of Math.* (2), **66,** 380–394.

Reidemeister, K., 1932a, *Einfuehrung in die kombinatorische Topologie*, Braunschweig.

Reidemeister, K., 1932b, *Knotentheorie Ergebnisse der Mathematik und ihrer Grenzgebiete*, Vol. 1, No. 1. Berlin. Springer.

Reidemeister, K., 1934, Homotopiegruppen von Komplexen. *Abh. Math. Sem., Hamburg Univ.*, **10,** 211–215.

Reidmeister, K., 1950, Complexes and Homotopy Chains. *Bull. Amer. Math. Soc.*, **56,** 297–307.

Reiner, I., 1954, Symplectic Modular Complements. *Trans. Amer. Math. Soc.*, **77,** 498–505.

Reiner, I., 1955a, Maximal Sets of Involutions. *Trans. Amer. Math. Soc.*, **79,** 459–476.

Reiner, I., 1955b, Automorphisms of the Modular Group. *Trans. Amer. Math. Soc.*, **80,** 35–50.

Reiner, I., *see* Hua, L. K.

Sanov, I. N., 1940, Solution of Burnside's Problem for $n = 4$. *Učenyi Zapiski Leningrad Gos. Univ.*, No. 55, 166–170. (Russian.)

Sanov, I. N., 1951, On a Certain System of Relations in Periodic Groups with Period a Power of a Prime Number. *Izvestia Akad. Nauk. SSSR, Ser. Mat.* **15,** 477–502. (Russian.)

Sanov, I. N., 1952, Establishment of a Connection between Periodic Groups with Period a Prime Number and Lie Rings. *Izvestia Akad. Nauk. SSSR, Ser. Math.* **16,** 23–58 (Russian); M.R. **13,** 721.

Schiek, H., 1953, Bemerkung über eine Relation in freien Gruppen. *Math. Ann.*, **126,** 375–76.

Schiek, H., 1955a, Gruppen mit Relationen $x^3 = 1$, $(xy)^3 = 1$. *Archiv der Math.*, **6,** 341–347.

Schiek, H., 1955b, Gruppen mit Relationen $(abc)^2 = e$. *Math. Nachr.*, **13,** Heft 3/4, 247–256.

Schiek, H., 1956, Aehnlichkeitsanalyse von Gruppenrelationen. *Acta Math.*, **96,** 157–251.

Schiek, H., 1962a, Adjunktionsproblem und inkompressible Relationen. *Math. Ann.* **146,** 314–320.

Schiek, H., 1962b, Das Adjunktionsproblem der Gruppentheorie. *Math. Ann.*, **147,** 158–165.

Schreier, O., 1927, Die Untergruppen der freien Gruppen. *Abh. Math. Sem., Hamburg Univ.*, **5,** 161–183.

Schumann, H. G., 1937, Über Modulen und Gruppenbilder. *Math. Ann.*, **114,** 385–413.

Scott, W. R., 1955, On Infinite Groups. *Pacific J. Math.*, **8,** 589–598.

Seifert, H., and W. Threlfall, 1947, *Lehrbuch der Topologie.* Chelsea, N.Y.

Séminaire 'Sophus Lie,' 1955, *Théorie des algèbres de Lie.* Secrétariat Mathématique, 11 rue Pierre Curie, Paris, 5e.

Shenitzer, A., 1955, Decomposition of a Group with a Single Defining Relation into a Free Product. *Proc. Amer. Math. Soc.*, **6,** 273–279.

Siegel, C. L., 1943, Symplectic Geometry. *Amer. J. Math.*, **65,** 1–86.

Siegel, C. L., 1950, Bemerkung zu einem Satz von Jakob Nielsen. *Matematisk Tidsskrift*, B, pp. 66–70.

Sinkov, A., 1936, The Groups Determined by $S^l = T^m = (S^{-1}T^{-1}ST)^p = 1$. *Duke Math. J.*, **2,** 74–83.

Siršov, A. I., 1953, Subalgebras of Free Lie Algebras. *Mat. Sbornik N.S.*, **33** (75), 441–452 (Russian); M.R. **15,** 596.

Širšov, A. I., 1954, Subalgebras of Free Commutative and Free Anti-commutative Algebras. *Mat. Sbornik N.S.*, **34** (76), 81–88 (Russian); M.R. **15,** 929.

Solitar, D., *see* Karrass, A.

Solitar, D., *see* Baumslag, G.

Specht, W., 1948, Die linearen Beziehungen zwischen hoeheren Kommutatoren. *Math. Zeitscrift*, **51,** 367–403.

Specht, W., 1956, *Gruppentheorie.* Springer.

Stender, 1953, On the Application of the Sieve Method to the Solvability of the Word Problem for Certain Groups with a Denumberable Set of Generating Elements and a Denumerable Set of Defining Relations. *Mat. Sbornik N.S.*, **32** (74), 97–108 (Russian); M.R. **14,** 723.

Stone, A. H., *see* Higman, G.

Struik, R. R., 1956, On Associative Products of Groups. *Trans. Amer. Math. Soc.*, **81,** 425–452.

Struik, R. R., 1959, On Verbal Products of Groups. *J. London Math. Soc.*, **34,** 397–400.

Takahasi, M., 1944, Bemerkungen über den Untergruppensatz in freien Produkten. *Proc. Acad. Tokyo*, **20,** 589–594.

Takahasi, M., 1951a, Note on word subgroups in free product subgroups. *J. Inst. Polytech.* Osaka City Univ. Ser. A, Math., **2,** 13–18.

Takahasi, M., 1951b, Note on chain conditions in free groups. *Osaka Math. J.*, **3,** 221–225.

Tartakovskiǐ, V. A., 1949, The Sieve Method in Group Theory, Amer. Math. Soc., Translation No. 60, 1952. Translation of three articles in *Mat. Sbornik N.S.* **25** (67) 3–50, 251–274 and *Izvestiya Akad. Nauk. SSSR, Ser. Mat.*, **13,** 483–494.

Thrall, R. M., 1942, On Symmetrized Kronecker Powers and the Structure of the Free Lie Ring. *Amer. J. Math.*, **64,** 371–388.

Threlfall, W., 1932, Gruppenbilder. *Abh. Math. Phys. Kl. Saechs. Akad. Wiss.*, **41,** No. 6, 1–54.

Tietze, H., 1908, Über die topologischen Invarianten mehrdimensionalen Mannigfaltigkeiten. *Monatsh. f. Math. u. Physik*, **19,** 1–118.

Torres, G., and R. H. Fox, 1954, Dual Presentations of the Group of a Knot. *Annals of Math.*, **59,** 211–218.

Wagner, D. H., 1957, On Free Products of Groups. *Trans. Amer. Math. Soc.*, **84,** 352–378.

van der Waerden, 1948, Free Products of Groups. *Amer. J. Math.*, **79,** 527–528.

Weil, A., 1937, *Sur les espaces à structure uniforme et sur la topologie générale,* Actualitées scientifiques et industrielles, No. 551, Paris.

Weil, A., 1938, *L'intégration dans les groupes topologiques et ses applications,* Actualités scientifiques et industrielles, No. 869, Paris.

Weir, A. J., 1956, The Reidemeister and Kurosh subgroup theorems. *Mathematika*, **3**, 47–55.

Wever, F., 1947, Operatoren in Lieschen Ringen. *J. reine u. angew. Math.*, **187**, 44–55.

Wever, F., 1949, Über Invarianten in Lieschen Ringen. *Math. Ann.*, **120**, 563–580.

Wever, F., 1950, Über Regeln in Gruppen. *Math. Ann.*, **122**, 334–339.

Whitehead, J. H. C., 1936a, On Certain Sets of Elements in a Free Group. *Proc. London Math. Soc.*, **41**, 48–56, Ser. 2.

Whitehead, J. H. C., 1936b, On Equivalent Sets of Elements in a Free Group. *Annals of Math.*, **37**, 782–800.

Whitehead, J. H. C., *see* Newman, M. H. A.

Wiegold, J., 1961, Some Remarks on Generalized Products of Groups with Amalgamations. *Math. Z.*, **75**, 57–78.

Witt, E., 1937, Treue Darstellung Lieschen Ringe. *J. reine angew. Math.*, **177**, 152–160.

Witt, E., 1953, Treue Darstellungen beliebiger Lieschen Ringe. *Collectanea Math.*, **6**, 107–114.

Witt, E., 1956, Die Unterringe der freien Lieschen Ringe. *Math. Z.*, **64**, 195–216.

Zassenhaus, H., 1939, Über Liesche Ringe mit Primzahlcharakteristik. *Abh. Math. Sem. Hansischen Univ.*, **13**, 1–100.

Zassenhaus, H., 1957, A Theorem of Friedrichs. *Trans. Royal Soc. Canada*, **51**, 55–64.

List of Theorems, Corollaries, Lemmas, and Definitions

Abbreviations: Th = Theorem L = Lemma
C = Corollary D = Definition

The number of a theorem, etc., is followed by the number of the page on which it appears.

Theorems which bear the name of an author are listed in the index; e.g. Hopf's Theorem page 393.

List of Symbols and Abbreviations

Whenever the definition given for a symbol is not complete, a number refers to the page where a complete definition may be found.

$A_0(R, r)$	299.
$A_0(R, \infty)$, $A(R, r)$	301.
$A \times B$	direct product of groups A and B.
$A * B$	free product of groups A and B, 180.
a^b	conjugate of a by b, $b^{-1}ab$.
$B(e, r)$	Burnside group of exponent e and rank r, 379.
C	field of complex numbers.
gcd	greatest common divisor.
$E_p(r)$	definition 5.10, 381.
$F(g)$, $F(W)$	free factor containing g, W, 181.
F_n	free group on n generators, 33.
$G^{(n)}$	nth derived group of the group G, 293.
G_n	nth group of the lower central series of the group G, 293.
G_q	Galois field of order q.
JG	group ring .(algebra) of the group G with integer coefficients.
$L(W)$	length of the word W, 4.
Q	field of rational numbers.
R	an arbitrary integral domain.
T_1, T_2, T_3, T_4	Tietze transformations, 49, 50.
X	basic ideal, definition 5.5, 302.
Z	ring of integers.
α, β, γ	type of generator, type of representative function; 229.
β^n	bracket arrangement of weight n, definition 5.1, 288.
$\beta^n(A_1, \ldots, A_n)$	commutator of weight n in the components $A_1, \ldots, A_n$ (subgroups), definition 5.3, 289.
$\beta^n(a_1, \ldots, a_n)$	commutator of weight n in the components $a_1, \ldots, a_n$ (elements), definition 5.2, 289.
$\beta^n(\xi_1, \ldots, \xi_n)$	337.

δ	α, β, or γ-generator; α, β, or γ-representative function; 229.
$\delta(W)$	deviation of W, definition 5.6, 313.
$\delta_n(W)$	340.
$\Lambda_0(R, r)$	300.
$\Lambda_0(R, \infty)$, $\Lambda(R, r)$	302.
$\lambda(W)$	syllable length of W, 182.
μ	standard mapping of Λ_0 into A_0, lemma 5.5, 318.
$\mu(d)$	Moebius function, 329.
$\rho(W)$	reduced form of W; in a free group, 34; in a free product, 183; in an amalgamated product, 202.
Σ_n	symmetric group on n symbols.
$\sigma(W)$	cyclically reduced form of W; in a free group, 36; in a free product, 189.
τ	rewriting process, 86, 90, 227, 230.
$\Omega_{k,m}$	359.
1	unit element of a group; empty word, 4.
$*(A, B, H, K;\ \varphi)$	amalgamated product of groups A and B, with the subgroup H of A amalgamated with the subgroup K of B under the isomorphism φ of H onto K, 207.
$*W$	neutral representative of W, 229.
$\sim$	equivalent, 12.
$\approx$	freely equal, 34.
$\simeq$	isomorphic.
$\rightleftarrows$	commutes with, 163.
$\{\ \}$	curly bracket operator, definition 5.8, 361.
(a, b)	commutator of a and b, $a^{-1}b^{-1}ab$.
$(g_1, \ldots, g_n)$	simple commutator of $g_1, \ldots, g_n$, 293.
$[u, v]$	bracket product of u and v, $uv - vu$, where u, v are in an associative algebra.
$\{W\}$	equivalence class of a word W, 13.
$(\eta_1, \ldots, \eta_n)$	simple product of elements $\eta_1, \ldots, \eta_n$ in a Lie algebra, 359.

Index